WHO WROTE THE BOOK OF LIFE?

A HISTORY OF THE GENETIC CODE

WRITING SCIENCE

EDITORS Timothy Lenoir and Hans Ulrich Gumbrecht

WHO WROTE THE BOOK OF LIFE?

A HISTORY OF THE GENETIC CODE

Lily E. Kay

STANFORD UNIVERSITY PRESS
STANFORD, CALIFORNIA

Stanford University Press
Stanford, California

Printed in the United States of America
CIP data are at the end of the book

For Kurt and Paulette Olden

ACKNOWLEDGMENTS

I could not have researched and written this project during a six-year period were it not for the enormous range of collegial, institutional, and financial resources I enjoyed. Such support represented a much appreciated vote of confidence, which countered the occasional skeptical voices. The book has many contributors, including colleagues and friends, students, archivists, and, of course, several of the scientists whose work directly or tangentially figures in the text.

I owe a great debt to Manfred Eigen for his generous hospitality during my spring 1992 visit at the Max-Planck Institute für Biophysikalische-Chemie in Göttingen. Although, chronologically, his scientific work is recounted only at the end of the book, his contributions to this project—in the form of lent publications, access to his library collection, and provocative discussions—formed the project's conceptual starting point. I am also grateful to Marshall Nirenberg for providing me with his laboratory diaries and for enduring extended interviews conducted over a three-year period. Heinrich Matthaei, too, gave freely of his time and supplied me with copies of his laboratory notebooks. I also have benefited from lively discussions with Sydney Brenner and Joshua Lederberg and from their personal archival treasures, as well as from the historical nuggets of Martynas Yčas. Communications with Heinz Fraenkel-Conrat, Morris Halle, Henry Linschitz, Wayne O'Neil, Leslie Orgel, Robert Sinsheimer, Heinz von Foerster, and members of Brookhaven National Laboratory contributed to different aspects of this historical reconstruction; and spirited conversations with Ernst Mayr have kept an evolutionary perspective on molecular biology within close range. Each of these scientists has been remarkably open-minded to my queries and challenges despite some differences in outlook. Responsibility for the final scientific and historical interpretations is mine alone.

Numerous scholars read and commented on various parts of the manuscript; I have not always followed all of their suggestions, but I am very grateful to William Aspray, Mario Biagioli, James Bono, Yoonsuhn Chung, Angela Creager, Lorraine Daston, Soraya de Chadarevian, Paul Forman, Peter

Galison, Jean-Paul Gaudillière, Herbert Gottweis, Loren Graham, Morris Halle, Donna Haraway, Victoria Harden, Ruth Harris, Thomas Hughes, Henry Krips, Joshua Lederberg, Timothy Lenoir, Michael Mahoney, Helmut Müller-Sievers, Marshall Nirenberg, Robert Richards, Henning Schmidgen, Skuli Sigurdsson, Denis Thieffry, and Mary Winsor. I am indebted to Robert Olby, Silvan Schweber, Michael Fischer, and most of all to Hans-Jörg Rheinberger, for thoughtful readings of the manuscript, animated discussions, and the fine balance of criticism and support.

The financial and institutional support for this project was extensive in scope and duration. At its embryonic planning stage it benefited from MIT's Provost Fund and an Old Dominion Fellowship; soon after it received a generous grant from NIH-ELSI branch, cosponsored by NSF (1993–95). These funds and leave time facilitated travel to archives, interviews, and conferences as well as the long writing periods. I especially thank archivists Madeleine Brunerie, Helen Samuels, Tom Rosenbaum, and Clifford Mead for their resourcefulness and engagement with this project. Equally important, these funds provided for several research assistants. I am grateful to MIT's undergraduate students Ashwin Balogopal, Ahlam Hashem, and Smruti Vidwans; graduate students Steven Collier, Evan Ingersoll, and most of all Eric Kupferberg for their excellent research and patient photocopying from scientific and popular sources; and to Slava Gerovitch for always raising critical questions. The remarkable intellectual community of the Max-Planck Institute for the History of Science in Berlin has nurtured this project at its final stage. I also thank Judith Stein, Phyllis Klein, and Betsy Keats, the support staff of MIT's Program in Science, Technology, and Society, for their administrative and editorial services. Debbie Meinbresse deserves particular thanks for her smiling help and expert preparation of the manuscript. Helen Tartar and Nathan MacBrien of Stanford University Press have been an author's dream, joyfully shepherding the manuscript through the editorial and production phases with refined academic and aesthetic sensibilities.

Finally, I thank Charles Weiner, Alan Attie, John Eskridge, and especially Peter Kuznick for the constancy and scope of their friendship and collegiality, and my family, Kurt and Paulette Olden, for their spirited generosity. I can only hope that all this support is judged worthwhile in what one sympathetic grant reviewer called "the marketplace of ideas."

L.E.K.

CONTENTS

PREFACE xv
ABBREVIATIONS xxi

1. The Genetic Code: Imaginaries and Practices 1

2. Spaces of Specificity: The Discourse of Molecular Biology Before the Age of Information 38

3. Production of Discourse: Cybernetics, Information, Life 73

4. Scriptural Technologies: Genetic Codes in the 1950s 128

5. The Pasteur Connection: *Cybernétique Enzymatique*, *Gène Informateur*, and Messenger RNA 193

6. Matter of Information: Writing Genetic Codes in the 1960s 235

7. In the Beginning Was the Wor(l)d? 294

Conclusion 326

NOTES 335
WORKS CITED 381
INDEX 427

ILLUSTRATIONS

1. The genetic code 4
2. Erwin Schrödinger, 1930s 60
3. Stern's representation of "modulated" nucleic acid chains incorporating different "gene codes" 68
4. Dounce's scheme of protein synthesis via an RNA intermediate 71
5. Norbert Wiener, 1950s 74
6. Claude E. Shannon, 1950s 93
7. Schematic diagram of a general communication system 97
8. John von Neumann, ca. 1952 103
9. Henry Quastler with his wife, Gertrud, 1950s 117
10. George Gamow, ca. 1955 130
11. The twenty different types of diamonds; and diamonds arranged along a schematic representation of the double helix 137
12. Members of the RNA Tie Club, ca. 1955 142
13. Drawing of the RNA Tie Club from Gamow's notebooks 143
14. Twenty possible triads of triangular code; and schematic diagram of triangular code 145
15. Gamow's illustration of Teller's sequential (Russian bath) code 146
16. Alexander Rich, Francis Crick, Sydney Brenner, ca. 1955 155
17. Codes without commas 161
18. Schematic representation of fully overlapping, partially overlapping, and nonoverlapping triplet codes 164
19. Heinz Fraenkel-Conrat and Wendell M. Stanley, ca. 1957 181
20. Tobacco mosaic virus 183
21. Sequence of 158 amino acid residues 188
22. The bactogène apparatus for continuous bacterial culture; and the chemostat device for continuous bacterial culture 200

23. François Jacob, Jacques Monod, and Andre Lwoff, 1965 208
24. Genetic transfer in bacterial conjugation 210
25. Biochemical genotypes and phenotypes of the galactosidase-permease system 218
26. Structure of the Lac segment of the E. coli chromosome 219
27. Models of the regulation of protein synthesis 223
28. Monod's model for the transcription of messenger RNA 225
29. Leo Szilard, Jacques Monod, and François Jacob at the 1961 Cold Spring Harbor Symposium 232
30. Matthaei's experiment 27Q 252
31. Heinrich Matthaei and Marshall W. Nirenberg, ca. 1961 253
32. Severo Ochoa, ca. 1960 258
33. Amino acid incorporation in E. coli system 260
34. Triplet code letters for amino acids 261
35. Genetic code for fifteen amino acids 263
36. "Reading frame" of nonoverlapping triplet code 267
37. Model of frame-shift mutations 268
38. Multiple Millipore filtration apparatus ("multiplator") 284
39. Hypothetical models of alanine transfer RNA 288
40. Roman Jakobson, ca. 1960 298
41. The final two complementary states . . . in the I Ching code 316
42. The I Ching transcribed into the genetic code 317
43. Manfred Eigen, 1985 320

PREFACE

In a sense, this book is a genealogy of the future. The image of a genomic "Book of Life"—laden with biblical resonances—emerged in the 1960s and now animates human genome projects, which are so often viewed as a mammoth task of information and word processing. Driven by global capital, these biomedical projects are perceived as a mission of "reading" and "editing," while the information they amass affects our basic notions of humanness, illness, and health. Beyond the material control of life, there is now a quest for controlling information—the DNA sequence, the "word"—frequently perceived as life's logos. Though problematic, this view of the genome as an information system, a linguistic text written in DNA code, has been guiding theories and practices of molecular biologists since the 1950s. My study aims to explain the emergence and spread of these scriptural genomic visions, as well as their scope and limits (in fact, this work was generously funded by the NIH-ELSI branch as part of their commitment to critical scholarship and public discussion on the topic). This book is not meant to be *the history* of the genetic code; the sheer scope of the relevant research precludes its being comprehensive or definitive. Rather it is *a history* of one of the most important and dramatic episodes in modern science recounted from a novel vantage point: the dawn of the information age and its impact on representations of nature and society.

I situate this work on the genetic code (1953–67) not only within the history of life science but also along the rise of communication technosciences (cybernetics, information theory, and computers), at the intersection with cryptanalysis and linguistics, and within the social history of postwar United States and Europe. The gestalt switch to information thinking in biology, with all its paradoxes and aporias, was even more fundamental than the subsequent (1953) paradigm shift from protein to DNA. (The shift to information-thinking also pervaded other disciplines.) In fact, it is now hard to imagine that genes did not always transfer information or that there were other ways of knowing and doing. I view the process by which the central biological problem of protein synthesis came to be represented as an information code and writing technology—and consequently a Book of Life—as historically specific: it is part of the cultural production of the Nuclear Age and the legacy of the cold war; its power amplified by the "Book's" theistic overtones across the millennia. As such, my work is a cultural history of the genetic code and belongs to the growing literature on cold-war science and culture, to histories of life science and information technosciences. In this sense, too, it is a genealogy of the future, tracing some of the trails along the ascent of the "Information Society," perhaps toward postmodernity.

These novel informational and cultural perspectives on the genetic code entail borrowing judiciously from the conceptual frameworks and analytical tools of poststructuralism: Derrida's critique of "writing"; and Foucault's analysis (after Canguilhem) of epistemic-epochal ruptures in representation, his idea of discourse, and particularly his concept of biopower. Though they are not everyone's cup of tea, these theoretical approaches will enable the sympathetic reader to engage in additional layers of meaning interwoven in this documentary history; however, much of the narrative and scientific reconstruction is accessible without them. The theoretically disinclined may skip Chapter 1 (except for its introductory material) and get on with "the story," beginning in Chapter 2. Here I argue that the 1950s, a watershed period during which a rupture in representations of life shifted from purely material and energetic to the informational, resulted in a molecular vision of life supplemented by an informational gaze. I view the information discourse, simultaneously coursing through several contemporary fields in life and social science, as an emergent form of biopower: the power of genetic information, now routinely transacted by champions of human genome projects. Based on my study, and complemented by related scientific scholarship, this biopower proves to be a problematic form of power because it has promised a great deal more than it can reasonably deliver.

The question Who wrote the "Book of Life"? highlights the key conundrum of genomic writing. Of course, originally, the metaphorical meaning of the Book of Life and the Book of Nature referred to the material record of God's creation. As such, the Book became accessible to human reading only after the scientific revolution of the seventeenth century, after the language of science—mathematics and experiment—equipped natural philosophers with tools for unlocking its coded secrets. Views of Nature as ciphers and hieroglyphs persisted into the eighteenth and nineteenth centuries, when the practice of science was a form of the creator's glorification, his authorship of the Book implicit and uncontested. But in the post-Darwinian era and in the secular culture of molecular biology, the Book of Life metaphor has seldom retained such overt creationist and theistic interpretations. While the power of the metaphor also inheres in its affinities to the sublime, few molecular biologists would assign authorship of the genomic Book of Life to God; though they may regard its content—information—as an ontological entity, even a cosmological principle. Thus the Book of Life leads back to the age-old conundrum of creation versus revelation: In the beginning was the Word? If the genome was written, what is the source of this writing, what is its agency and its materiality?

This question may be answered from three vantage points: objectivist, constructivist, and deconstructivist. The objectivist (Platonist or logocentric) view would assign the authorship of the Book to Nature itself. Through chance conjunction of prebiotic events (some would say informationally driven) there arose the (random) DNA sequence—the word—that eventually instructed the primitive proteins. Millions of years of evolution, of mutations and natural selection, culminated in increasingly complex and informationally rich organic aggregates, multicellular forms, even linguistically endowed life. From this standpoint, Nature and the Book of Nature are one and the same; Nature is source, agent, *and* outcome of these temporal molecular scriptures: the author and the writing itself. It has existed independently of human agency, preceding human intervention and awaiting decoding. The material and theoretical tools of molecular biology finally enabled its unambiguous reading. This view, however, does not probe the process of representation and construction of the tools themselves, their physical and discursive fashioning; nor does it question the cognitive assumptions, technological imperatives, and disciplinary and social commitments behind these informational and scriptural representations of heredity.

The constructivist view, on the other hand, would assign the agency of this molecular writing to scientists themselves. This position would not necessarily deny that objects exist external to thought; it would not negate the existence of genes or the correlation between codons and amino acids. But it would deny the objectivist claim that these entities and phenomena present themselves to practitioners as transparent reading, unmediated by scientists' own modes of representation: theoretical, material, discursive, and social. According to the constructivist view, rather than simply deciphering the DNA language, or reading a preexisting genomic text, researchers were actively producing the representation of genetic phenomena as *writing*: they were *constructing* imageries of the text, its messages, letters, and words. According to the constructivist view, then, it is molecular biologists who wrote the Book of Life.

The deconstructivist, or poststructuralist, perspective regards representation as a dialectic of *episteme* and *techne*, thus further problematizing representation, systems of meaning making, and agency. While this view, too, would not deny that objects such as genes or proteins exist external to thought, it would question the objectivist premise that they could constitute themselves as objects—such as a code—outside any discursive conditions of emergence. It would also challenge the constructivist view of knowledge production as human determinism. For such discursive conditions are not purely linguistic, according to this standpoint; instead they pierce the entire material density of the multifarious institutions, rituals, and practices through which discursive formations are structured. Thus, a poststructural approach to the genomic Book of Life is grounded in the conviction that once a commitment to a particular representation of life is made—material, discursive, and social—it assumes a kind of agency that both enables and constrains the thoughts and actions of biologists. In a sense, it is the representation itself that guides the imagination and reasoning, as was the case with the idioms of "information" and "language." In this view, scientists form only one of many elements, among them organisms, tools, theories, language, disciplines, institutions, and politics, defining the representational space within which *technoepistemic* events take place. The actors' freedom of movement, from experimental design to data interpretation and presentation, is always already mediated through that space. Thus from the poststructuralist perspective, which informs this book, it is the writing itself that writes. Namely, once molecular biologists adopted the scriptural representations of the genetic code, once they committed themselves, consciously or not, to the information discourse and to the attendant analogies of genomic writing and reading, these representations became constitutive of the decoders' reasoning; their work was shaped by the new biosemiotics of communication.

In the end, from all three standpoints—objectivist, constructivist, and poststructuralist—the Book of Life has always functioned as a potent metaphor in the mind and the laboratory, a way of knowing and doing. Yet it is an age-old aporia about a mute language and authorless book, a difficulty with concrete epistemic, cultural, and economic consequences. For even if the genome were to be a text and DNA a language, reading the "Book of Life" would be hardly unambiguous, for language is context-dependent and words are polysemic. Once the genetic, cellular, organismic, and environmental complexities of DNA's context-dependence are taken into account, biological meaning is hard to extract from the molecular syntax. (Pure upward causation is an insufficient explanation.) Even on the most basic biochemical level, protein folding—the key to physiology and drug design—is not deducible from the DNA sequence, or "word," let alone higher-level genomic functions and multicellular organization. When "context" itself is problematized (e.g., What is the system's "outside" and "inside"?) and when epigenetic networks are included in the analysis, the dynamic processes linking genotype to phenotype become enormously complex. Genetic messages might read less like an instruction manual and more like poetry, in all their exquisite polysemy, ambiguity, and biological nuances. My historical critique might therefore serve to further question the role of DNA as the prime mover of all the subtle diversities of life, from archeopteryx to thermophiles, and thereby help loosen the grip of genetic determinism in the "marketplace of ideas."

Finally, a brief meditation on language: while engaging in a critique about the language of DNA, I have tried, whenever possible, to avoid using the very terminology scrutinized in this study—"information," "language," "code"—or at least to signal its metaphorical status and historical usages. Ultimately, these overlapping resonances set a limit on linguistic resolution. So my book, too, as with the genomic "Book of Life," is subject to the same complications of discursive origins and their context-dependence. It has not always been possible to escape from this linguistic hall of mirrors.

ABBREVIATIONS

AEC	Atomic Energy Commission
AIP	American Institute of Physics
AMA	American Medical Association
AMP	Applied Mathematics Panel
APS	American Philosophical Society
BLA	Bell Laboratory Archive
CIT	California Institute of Technology
CIW	Carnegie Institution of Washington
CNRS	Centre National de la Recherche Scientifique
DGRST	Délégation Général à la Recherche Scientifique et Technique
DOD	Department of Defense
EDVAC	Electronic Discrete Variable Arithmetic Computer
EMBO	European Molecular Biology Organization
ENIAC	Electronic Numerical Integrator and Calculator
FCP	French Communist Party
ICBM	Intercontinental Ballistic Missile
JPL	Jet Propulsion Laboratory
LC	Library of Congress
MIT	Massachusetts Institute of Technology
NAS	National Academy of Sciences
NASA	National Aeronautics and Space Administration
NATO	North Atlantic Treaty Organization
NDRC	National Defense Research Council
NHI	National Health Insurance
NIH	National Institutes of Health

NSA	National Security Administration
NSC	National Security Council
OSRD	Office of Scientific Research and Development
OSU	Oregon State University
PHS	Public Health Service
PNAS	*Proceedings of the National Academy of Sciences*
RAC	Rockefeller Archive Center
SAIP	Service des Archives de L'Institut Pasteur
SIS	Signal Intelligence Service
TMV	Tobacco Mosaic Virus
TYV	Turnip Yellow Virus
UCB	University of California at Berkeley
UCSD	University of California at San Diego
UIA	University of Illinois Archive
USAF	United States Air Force
USDA	United States Department of Agriculture

WHO WROTE THE BOOK OF LIFE?

A HISTORY OF THE GENETIC CODE

CHAPTER ONE

The Genetic Code: Imaginaries and Practices

The human genome is now generally viewed as an information system and, more specifically, as a "Book of Life" written in the language of DNA, or DNA code, to be read and edited. A 1989 episode of the PBS television program *Nova*, entitled "Decoding the Book of Life," promotes the Human Genome Project as a scriptural mission. In a key scene, a personable white-coated molecular biologist slides one of the volumes of the Book of Life from the shelf and explains how to identify genetic misprints in the sequences of DNA text.[1] On other occasions, Harvard molecular biologist and Nobel laureate Walter Gilbert—probably the first to argue for copyrighting genetic sequences—predicts that it will soon be possible to put three billion bases on a single CD. He posits, "One will be able to pull a CD out of one's pocket and say, 'Here is a human being; it's me!' . . . To recognize that we are determined . . . by a finite collection of information that is knowable will change our view of ourselves. It is a closing of an intellectual frontier, with which we will have to come to terms."[2] These are visions of "informational man," an image elicited not just in the service of public understanding of science. Addressing a large professional gathering, biotechnology champion David Jackson argues, "To be fluent in a language, one needs to be able to *read*, to *write*, to *copy*, and to *edit* in that language. The functional equivalents of each of those aspects of fluency have now been embodied in technologies to deal with the language of DNA."[3] Beyond exegesis the "Book of Life" awaits revision.

The emergence in the 1980s of the subspecialty of DNA linguistics underscores that the "language of DNA" is not merely a popularization or rhetoric of persuasion, but rather a representation qua intervention with operational force. Admittedly not mainstream molecular biology, this field of inquiry illustrates the point: a metaphor literalized. First championed within linguistic structuralism in the 1960s by the acclaimed linguist Roman Jakobson, DNA linguistics was soon recast within the rigors of the Chomskian paradigm. Theoretical biologists have turned to generative grammar as a framework for understanding genomic organization and expression regulation in prokaryotic and eukaryotic systems, in search of biological meaning. This quest has become even more urgent as the sequences cascade from various human genome projects and their status—coding, regulatory, normal, abnormal, or so-called junk DNA (95–97% of the genome)—needs to be established. These examples spotlight the growing presence of bioinformatics as well as the pervasive notion of DNA language. They further reveal, as Robert Pollack observes, the often naive faith in the unambiguous reading and word processing of the genomic text in the form of spell-checking, deleting, adding, and splicing DNA sequences.[4]

These informational and scriptural representations of heredity are neither new nor unproblematic. The metaphor of the "Book of Life" and its subsequent variant, the "Book of Nature," reaches back to antiquity; its aporias have been examined by ancient and modern scholars.[5] But this metaphor of transcendent writing acquired new, seemingly scientifically legitimate meanings through the discourse of information. In other contemporary fields of life and social sciences and in the culture at large, entities and processes were similarly being recast as information systems. Information theorists, cryptologists, linguists, and life scientists criticized the difficulties (some would say inappropriateness) of these borrowings in molecular biology, arguing that the genome's information content cannot be assessed since the key parameters (e.g., signal, noise, message channel) cannot be properly quantified. DNA is not a natural language: it lacks phonemic features, semantics, punctuation marks, and intersymbol restrictions. So unlike any language, "letter" frequency analyses of amino acids yield only random statistical distributions. Furthermore, no natural language consists solely of three-letter words. Finally, if it were purely a formal language, then it would possess syntax only but no semantics. Thus the informational representations of the genome do not stand up under rigorous scrutiny. From linguistic and cryptanalytic standpoints, the genetic code is not a code: it is simply a table of correlations, though not nearly as systematic or predictive as the periodic table, for example, because of contingencies, degeneracies, and ambiguities in the structure of the so-called genetic code. These culturally animated imaginaries, nevertheless, have persisted, making it now seem inconceivable that genes did not always transfer information, or that the relation between DNA and protein could be something other than a code. Yet, there were (and probably could be) other ways of knowing. These particular representations were historically specific and culturally contingent. The genetic code is a "period piece," a manifestation of the emergence of the information age.

My thesis is that molecular biologists used "information" as a metaphor for biological specificity. However, "information" is a metaphor of a metaphor and thus a signifier without a referent, a *catachresis*. As such, it became a rich repository for the scientific imaginaries of the genetic code as an information system and a Book of Life. The information discourse and the scriptural representations of life were inextricably linked. Metaphors, as I will examine, are ubiquitous in science, but not all metaphors are created equal. Some, like the information and code metaphors, are exceptionally potent due to the richness of their symbolisms, their synchronic and diachronic linkages, and their scientific and cultural valences. They are the elements of what James Bono termed the "cultural poetics of science."[6] Though remarkably compelling and productive as analogies, "information," "language," "code," "message," and "text" have been taken as ontologies. The consequences are far-reaching, for the limits of these analogies also challenge the mastery of the genomic "Book of Life," the technological and commercial goals of its "reading" and "editing."

The conceptualization, breaking, and completion of the genetic code, 1953–67, was one of the most important and dramatic episodes in twentieth-century science and a manifestation of the stupendous reaches of molecular biology. The so-called code—actually a table of correlations—outlined the logic of gene-based protein synthesis, providing the key to what was widely perceived to be the "secret of life." It showed how the four bases of RNA, A, U, C, G, permuted three at a time, yielded sixty-four triplet codons, specifying the assembly of twenty amino acids into myriads of exquisitely specific proteins. Based primarily on studies of bacteria and viruses, this synoptic scheme has nevertheless been regarded as an (almost) universal code, applying to nearly all plant and animal life. From transcription to translation (themselves scriptural operational representations), the code tied mechanisms of genetic replication, mutation, and regulation to nucleic acid and protein syntheses, thus bridging molecular genetics and biochemistry; genetic "information"—or what was previously perceived as biological and chemical specificity—served as a discursive link between these two previously distant fields. Beyond the academy, as captured in the pages of the *New York Times* and *Time* magazine, the genetic code also signaled the potential for genetic engineering, even before the advent of recombinant-DNA technologies in the 1970s. Genetic information signified an emergent form of biopower: the material control of life would be now supplemented by the promise of controlling its form and logos, its information (the DNA sequence, or the "word") (see Fig. 1).

To understand and critically assess the formation of the genetic Book of Life and its so-called informational and linguistic attributes, I trace their lineages to the 1950s and 1960s, to the theoretical and experimental phases, respectively, of research on the genetic code. Undoubtedly, Erwin Schrödinger's celebrated proposal of a Morse-like "code-script" for heredity already in 1943—when DNA was of little interest, when genes were thought to be proteins, and information theory had not yet come into being—was a sig-

1st position (5′ end)	2nd position				3rd position (3′ end)
	U	C	A	G	
U	Phe	Ser	Tyr	Cys	U
	Phe	Ser	Tyr	Cys	C
	Leu	Ser	STOP	STOP	A
	Leu	Ser	STOP	Trp	G
C	Leu	Pro	His	Arg	U
	Leu	Pro	His	Arg	C
	Leu	Pro	Gin	Arg	A
	Leu	Pro	Gin	Arg	G
A	Ile	Thr	Asn	Ser	U
	Ile	Thr	Asn	Ser	C
	Ile	Thr	Lys	Arg	A
	Met	Thr	Lys	Arg	G
G	Val	Ala	Asp	Gly	U
	Val	Ala	Asp	Gly	C
	Val	Ala	Glu	Gly	A
	Val	Ala	Glu	Gly	G

Abbreviations

U Uracil (for DNA, read T [= Thymine] instead of U)
C Cytosine
A Adenine
G Guanine

Ala	Alanine	Gly	Glycine	Pro	Proline
Arg	Arginine	His	Histidine	Ser	Serine
Asn	Asparagine	Ile	Isoleucine	Thr	Threonine
Asp	Aspartic acid	Leu	Leucine	Trp	Tryptophan
Cys	Cysteine	Lys	Lysine	Tyr	Tyrosine
Gln	Glutamine	Met	Methionine	Val	Valine
Glu	Glutamic acid	Phe	Phenylalanine	STOP	means "end chain"

FIGURE 1. Adapted from Francis H. C. Crick, *What Mad Pursuit* (New York: Basic, 1988), p. 170. Courtesy of the author.

nificant moment in the history of biology. But the impact of this nineteenth-century engramic imagery on studies of the genetic code in the 1950s is unclear. His constructs belonged to an older epistemic and cultural epoch, far removed from the DNA paradigm and from its articulations by a scientific generation gripped by the information discourse. Retrospectively, Whig mythologies spun around Schrödinger's *What Is Life?* as *the* precursor to the genetic code have obscured the historical nature of his own preoccupations and the scientific and social context of the 1950s: the new world picture within which "genetic decoding" took place.[7]

In that postwar world order, the material, discursive, and social practices of molecular biology were transformed. Information theory, cybernetics, systems analyses, electronic computers, and simulation technologies fundamentally altered the representations of animate and inanimate phenomena. These new communication sciences began to reorient molecular biology (as they did, to various degrees, other life and social sciences) even before it underwent a paradigm shift (1953) from protein- to DNA-based explanations of heredity. It is within this information discourse that the genetic code was constituted as an object of study and a scriptural technology, and the genome textualized as a latter-day Book of Life. The disciplinary terrain and representational space of molecular biology changed, as well, partly through the growing participation of physical scientists. Worldwide, its institutional structures were reconfigured within cold-war organizations, military patronage, and the unprecedented commitment of government resources for scientific research. In short, from the 1950s on, the diachronic resonances of the Book of Life as transcendent writing were amplified by the synchronic articulations of DNA as a programmed text, and information became the animating *Primum Mobile*. The genetic code became the site of life's command and control.

This book is organized chronologically and consists of seven chapters. In order to understand the transition to the information discourse—what it displaced, absorbed, created, and resignified—Chapter 2, "Spaces of Specificity: The Discourse of Molecular Biology Before the Age of Information," reviews the previous era. It dissects the discourse of molecular biology and the genetic code in the 1940s, when the biosemiotic repertoire of communication sciences was absent. Situating the representations and practices of molecular biology within the institutional structures, disciplinary commitments, and social experiences of the interwar period, Chapter 2 places an emphasis on the cultural project of the Rockefeller Foundation (and thus draws upon my earlier studies). Answering the question, "What did genes do before they transferred information?" this chapter traces the workings of the gene to the key concept of biological and chemical specificity. An overarching theme guiding life-science research since the turn of the century, *specificity* was configured mainly within the older discourse of organization. The concept of specificity, I argue, was supposedly later interchanged in a complicated manner with *information*.

Chapter 3, "Production of Discourse: Cybernetics, Information, Life," examines the rise of cybernetics and information theory in the late 1940s and the emergence of the discourse of information in biology in the early

1950s within new regimes of signification, namely the postwar industrial-military-academic complex. It focuses on the works of Norbert Wiener, Claude Shannon, and John von Neumann and the application of their ideas to molecular biology by Henry Quastler and his colleagues. The chapter provides a solid point of both epistemic and social departure for understanding some of the overlaps and the many divergences between the mathematical theory of communication and its applications to biology, as well as the distinction between information theory and the information discourse.

Chapter 4, "Scriptual Technologies: Genetic Codes in the 1950s," recounts how work on the genetic code in the theoretical phase of the 1950s drew on information theory and its various tropes, and situates the decoding efforts within cold-war military culture. By analyzing the works of George Gamow, the RNA Tie Club, and their collaborators (physicists, mathematicians, cryptologists, computer analysts, and communication engineers), and by examining the decoding research in the tobacco mosaic virus (TMV), I show how the older central problem of biological specificity was (re)represented, or recast through the information discourse. Sometimes used technically, but mostly metaphorically, genetic specificity was revisioned as a scriptural technology poised at the interface of several interlocking postwar discourses: physics, mathematics, information theory, cryptanalysis, electronic computing, and linguistics. However, the most powerful computerized cryptanalytic technologies could not break the so-called genetic code, because technically speaking it is not a code.

Chapter 5, "The Pasteur Connection: *Cybernétique Enzymatique*, *Gène Informateur*, and Messenger RNA," serves two principal functions. First, it reconstructs the studies of Jacob and Monod in the 1950s, which demonstrated the genetic regulation of enzymatic synthesis (the operon model) in E. coli and led to the conceptualization and identification of messenger RNA through multiple international collaborations. The concept and preparations of artificial RNA messengers reoriented the approach to the coding problem and resulted in breaking the code in 1961. Second, this chapter shows how cybernetic models and informational representations provided the interpretive frameworks for experimental results and for designing subsequent experiments. Within the discourse of information, protein synthesis began to be recast in the late 1950s as a programmed communication system.

Chapter 6, "Matter of Information: Writing Genetic Codes in the 1960s," recounts the work on the genetic code during the second experimental phase, situating that work within the heavy scientific investments of the post-Sputnik era and alongside emergent public debates over biological engineering. This chapter traces Marshall Nirenberg's approaches to the problem of protein synthesis since his arrival at NIH in 1957, the dramatic events surrounding Nirenberg and Matthaei's breaking of the code in the spring of 1961, and the fierce race among numerous researchers at leading laboratories to complete the code. Completing the code took six years (instead of the anticipated one year), additional technical feats, and the participation of hundreds of researchers, whose efforts attest that, though this phase of genetic code work was primarily biochemical (supplemented with genetic analyses), the discursive structures erected during the theoretical phase guided laboratory practices during this second period. Hereditary material became informational, and the informational representations of the code were literally materialized. By the late 1960s, owing much to the enthusiasm of the acclaimed linguist Roman Jakobson, the genome had been authorized as a text written in a natural language, amenable (as some claim) to linguistic analysis in quest of biological "semantics." Once linguistics, like biology, came under the spell of the information discourse in the 1950s, the two entities—language and DNA—emerged, not surprisingly, with similar features. Thus the history of the genetic code does not come to an end with the completion of the last of its codons ("words"). Instead, its history had become intertwined with the history of linguistics.

Chapter 7, "In the Beginning Was the Wor(l)d?" spotlights and critiques key episodes from the 1960s to the present in the textualization of the genome, with a focus on Roman Jakobson. First, I examine his attempts to recast linguistics (structuralism) as a science based on information theory, then I look at how he applied this reconfigured linguistics to the genetic code. He was not merely analogizing here but rather viewing the code as a natural language. The chapter follows the information discourse and genetic textualization toward their ultimate reaches: information as the origin of life, as the unit of evolution and natural selection. I examine the chimera of the Book of Life as a text snagged in paradoxes of speechless communication, authorless writing, and the act of (re)creation as revelation.

THE GENETIC CODE AND THE COLD WAR

> When you see almost every one of your magazines, no matter what they are advertizing, has a picture of the Titan missile or the Atlas [missile] or solid fuel or other things, there is becoming a great influence, almost an insidious penetration of our minds that the only thing this country is engaged in is weaponry and missiles. —*Dwight Eisenhower*[8]

Scholars are now rethinking the cold war's history, meanings, and impacts. Beyond diplomacy, security, and military commitments, the cold war affected every facet of American national life and international relations, including nearly all academic disciplines and most particularly the natural and social sciences. After Paul Forman first called attention in 1987 to the military-driven transformation of postwar physics, science studies scholars, until very recently, examined mainly the central role of the physical sciences in the cold war. However even the fields of life and social sciences, as well as history and the history of science (as Peter Novick pointed out)—disciplines quite removed from strategic defense and weapons design—were affected to some degree. Their institutional frameworks, organizational structures, modes of patronage, magnitude and style of funding, scientific imaginaries and discourses, for example, all were altered during the cold war. Molecular biology and work on genetics share this legacy. I do not mean to suggest that all those who contributed to the genetic code supported the military consensus; indeed, Jacques Monod and Leo Szilard were political activists against the cold war. Nevertheless, the scientists all shared some of the imaginaries and discourses of that cultural hegemony.[9]

Two consecutive phases—the formalistic and the material—marked the study of the genetic code, its textualization, and its recasting as an information system; yet these phases differed in their disciplinary, institutional, and social linkages. The mathematical/genetic phase, 1953–61, spanned the period from the determination of DNA's structure by James Watson and Francis Crick and its signification as the code that carried genetic information, up to the dramatic breaking of the code by Marshall Nirenberg and Heinrich Matthaei in May 1961. The work in this phase was inspired by George Gamow and shaped by the efforts of about two-dozen researchers: physicists, mathematicians, cryptologists, systems analysts, and physical chemists, often working at the hub of defense projects. Some of the most eminent theoreticians sharpened their wits on the recalcitrant code, though crucial contributions came from the ingenuities of the handful of biologists and biochemists.

During this phase researchers attempted to deduce the code inferentially, by strictly comparing DNA input with protein output, to solve it cryptographically by never opening the black box of protein synthesis. The amino acid sequence of insulin, the first protein to be sequenced (by Frederick Sanger in Cambridge, England, in the early 1950s), became the initial source of "output" information in that black-box approach. Soon after, the amino acid sequence of the coat protein of TMV (partially known in the mid-1950s and completed in 1960) served as a so-called Rosetta stone for establishing correlations between amino acids of the viral coat and nucleotide triplets in its RNA core (related viral strains provided comparative data). In the late 1950s, chemical mutagens, notably nitrous acid, served as powerful tools for altering the viral RNA bases and correlating those with amino acid changes; laboriously obtained amino acid replacement data became grist to the theoreticians' mill. The mechanisms of replication and mutation in bacteriophage (minute viruses that infect susceptible E. coli bacteria)—the remarkably productive experimental system of molecular geneticists—offered insights into the relations between proteins and nucleic acids.

But, by and large, the MANIAC computer at Los Alamos was the most impressive hardware during the first phase. Technical manipulations—namely, cryptanalysis, information theory and mathematical theories of coding, Monte Carlo simulations, statistical analyses, and symbolic logic—composed this "paper and pencil" approach to genetic decoding. Through these studies, which proceeded at a relatively leisurely pace, the genome was textualized, as researchers transported the information discourse, its tropes and semiotics, into molecular biology, reconfiguring its representational space. As with other contemporary forms of knowledge production, the genetic code, as an icon of biological command and control, can be also viewed as part of the cultural experience of the cold war.

Originating in 1946, the term *cold war* materialized out of the extensive militarization of the West. The 1950s marked the rise of the American-centered political hegemony within a new world system: Pax Americana. Thomas J. McCormick has outlined its salient features, namely, economic supremacy, high-tech, high-profit industries, clear military superiority, and ideological and cultural power. Britain—heavily dependent on American dollars, goods, technology, and, above all, defense—was America's strongest ally in the Western hegemony; that country's enormous defense expenditures rendered it by the late 1950s Europe's leading military power. (As it so happened, British and American molecular biologists also formed the strongest scientific alliance.) The other principal foreign-policy goals of the 1950s consisted of overcoming French resistance to American domination of NATO and the rearming and reintegration of Germany.[10]

Military spending formed the cornerstone in both postwar foreign and domestic policies and the economic boom. In 1949 when the Department of Defense was established (merging the departments of Navy, Army, and Air Force), the United States was spending about $13 billion annually on defense. National Security Council document 68 (NSC-68), one of the most important documents in American history, which was approved in 1950 by President Truman, expressed a firm commitment to building what came to be called the national security state of the cold war, its price tag ranging from $37 to $50 billion per year. At the time, it remained unclear how to extract that kind of money from a fiscally conservative Congress, but as Secretary of State Dean Acheson would later remark, "Korea came along and saved us." Within President Eisenhower and Secretary of State John Foster Dulles's geopolitics of "brinksmanship" and the "domino theory" of totalitarianism, the Defense Department's budget soared to more than $50 billion by 1953 (thus doubling since 1951) and received an additional boost in 1958 with the passage of the National Defense Security Act, a response to the Soviets' launching of *Sputnik I* on October 4, 1957.[11]

The support of science in the 1950s came mainly from the Department of Defense, the military-controlled Atomic Energy Commission (AEC), and the National Aeronautics and Space Administration (NASA). Monies were spent on nuclear-weapons research, space research, ever-faster electronic computers, germ warfare, biological research on radiation, and techniques of social control, thus placing scientific research in the fields of physical, life, and social sciences at the center of cold-war knowledge production. As Paul Forman and Stuart Leslie have documented, this massive buildup of military technologies and bureaucracies enabled the expansion of American physical science. Science assumed a pivotal role both within the hegemony of the national security state and in the NATO alliance. Paul Edwards has argued that computer development was likewise shaped by and integral in shaping militarized consensus, that computer designs and materialities were grounded in discourses of the world as a closed bipolar system riveted by American and Soviet spheres of domination.[12]

A similar but attenuated argument can be made for the life (and social) sciences, as Donna Haraway observed fifteen years ago.[13] Both qualitatively and quantitatively, the structure of patronage for molecular biology changed markedly as the Rockefeller Foundation eased out of the life sciences in the United States. During the 1950s, biologists, particularly geneticists, became firmly tied to the military establishment. Between 1950 and 1955 the combined support of the AEC and Department of Defense accounted for 53 percent of total federal funds ($120 million) for the biological and medical sciences (excluding agriculture). From 1955 to 1960, military sponsorship stabilized at about 29 percent of the total (now $440 million), as other federal sources, principally the National Institutes of Health (NIH) but also the National Science Foundation (NSF), NASA, and several private foundations assumed greater roles in funding the biomedical fields. Besides block grants to universities and AEC fellowships, molecular life sciences thrived in the military-AEC sponsored national laboratories, notably Argonne, Brookhaven, and Oak Ridge.[14] The AEC was a major patron of genetics, as John Beatty has documented. In the 1950s the AEC funded half of federally funded genetic research in the United States (as well as many projects worldwide); about 20 percent of the active membership of the Genetic Society of America engaged in AEC-supported research. They were all subject to security clearance and the loyalty oath, as McCarthyism swept through federal institutions. Though not as deeply as the physical sciences, life sciences in the 1950s, too, were embedded within what Senator J. William Fulbright termed the *industrial-military-academic complex* and what Admiral Rickover critically labeled the *military-scientific complex.*[15]

Influence flowed both ways at institutional conjunctions of science and the military. The relationship altered the modus operandus, policies, and attitudes to research in the military and in science. What was the impact of military patronage on the organization and content of knowledge? How did this massive and systematic sponsorship by the military shape scientific research and even, to a considerable degree, the work on the genetic code in the 1950s? As Mario Biagioli observes, patronage is not simply a resource external to the scientists' projects. For example, Medicean courts and military courts share some general traits: in both the power of patronage manifested itself within the organization and practices of science. The web of military institutions sponsoring scientific research defined the conditions of possibility for the production of particular forms of knowledge. As Ian Hacking has suggested (echoing Eisenhower), beyond overt gross military manifestations such as federal, academic, and industrial budgets, and which specific fields and scientists received funding, military power extended into the world of the mind. It did so in biology through its various discourses and representations, notably the discourse of information and the technoscientific imaginary of communication and control systems.[16] That imaginary is illuminated by Baudrillard's observation:

> It is in effect in the genetic code that the "genesis of simulacra" today finds its most accomplished form. At the limits of an always more extensive abolition of references and finalities, of the loss of resemblances and designation, we find the digital program-sign, whose value is purely tactical, at the intersection of the other signals (corpuscles of information/test) and whose structure is that of macro-molecular code of command and control.[17]

Despite their mathematical prowess, access to the most recent findings about genetic mechanisms of viruses and bacteria, and command of cutting-edge computer analyses and simulation technologies, leading scientists failed to break the code. This is because from linguistic and cryptanalytic standpoints, the genetic code is not a code; it is, rather, a powerful metaphor for the correlations between nucleic and amino acids. However, despite the acknowledged pitfalls in applying information theory, linguistics, and cryptanalysis to molecular biology in the 1950s, these informational and scriptural representations of heredity set roots and proliferated. They did so mainly as a result of their transdisciplinary and cultural resonances and because of their efficacy as models and analogies in the process of biological meaning making. As such, they set the conceptual framework and discursive structures for the second phase of decoding and into present-day genomics.

In contrast with the first phase, the second phase, 1961–67, was primarily material. Here, analyses of molecular genetics informed biochemistry, though a few theoreticians continued to be lured by the coding problem. This phase was also differently situated on the scientific knowledge map. As researchers despaired over breaking the code by the elegant formalistic approach, Nirenberg and Matthaei quietly cracked open the black box of protein synthesis in all its organic complexity. The experimental space of molecular biology once again became reconfigured, now marked almost exclusively by researches on the E. coli cell-free system, utilizing some of the processes and products developed by biochemists and molecular geneticists in the 1950s. The amino acid replacement data from TMV and the partial but remarkably detailed maps of the E. coli and phage genomes (first established through laborious analyses of naturally occurring mutations) served as important guides to the correlations between amino acids and nucleotide triplets. Stocks of precisely defined E. coli mutants—like Drosophila stocks in the 1910s and 1920s—became molecular geneticists' standardized means of productions.[18] The composite experimental system of phage and bacteria was elegantly productive; "zygotic technologies," methods of transferring viral genes between bacteria through mating, yielded crucial clues into the mechanisms of gene regulation, expression, and enzymatic synthesis. Stocks of precisely defined phage mutants (naturally and chemically induced) circulated among members of the phage network; the "low-tech" but intricate recombination techniques with mutant phages in the 1960s established inferentially key features of the code. These pivotal genetic contributions notwithstanding, the second phase was predominantly biochemical, and well-equipped and well-staffed laboratories defined the vanguard of the decoding project.

Methods for separating closely related entities, such as analytical centrifugation, chromatography, and gel-electrophoresis, were refined during that time. Radioisotopes became commercially available (notably the set of twenty labeled amino acids), enabling the tracking of trajectories of nucleic acids and proteins; scintillation counters had become standard laboratory furniture. Powerful new technologies of nucleic acid synthesis reoriented biochemical practice in the late 1950s, assuming a central role in second-phase studies of the genetic code. Methods of chemical synthesis of nucleotide polymers were established by the early 1950s, and synthetic polynucleotides of known composition and structure—RNA messengers—were available in several laboratories. But with the isolation in the mid-1950s of the nucleotide-assembling enzymes DNA polymerase and polynucleotide phosphorylase, and at the end of the decade RNA polymerases, technologies of nucleotide synthesis reached new levels of efficiency and potency through a combination of chemical and enzymatic methods. Khorana was a leading expert in the field. Indeed, the significance of these enzymes transcended their epistemic value, for they quickly became "technical things" (to employ Hans-Jörg Rheinberger's analytics).[19] They formed implements for crafting custom-made polynucleotides that, in turn, served as messengers, or probes, into the genetic code, forming a continuous chain of signification. Conversely, DNAase and RNAase (enzymes that chew nucleotide chains) served as tools for manipulating genetic signals in cell-free protein synthesis and for determining the sequences of nucleotide chains (Holley's contributions to the genetic code were based on such sequencing techniques). Defined synthetic polynucleotides (messengers) in conjunction with the cell-free E. coli system (stabilized and fine-tuned in 1960) served as a Rosetta stone for breaking and completing the genetic code by 1967.

The second decoding period also differed from the first phase in its institutional and social settings. By 1957, as military support for life science streamlined at 19 percent, NIH became a major patron of biomedical research at home and abroad. NIH entered a period of unprecedented growth, driven by the general boost to science and technology from the space program—primarily a response to the Soviets' launching of *Sputnik*. Each year from 1957 to 1963 NIH's budget increased by an average of 40 percent annually; appropriations grew from $98 million in 1956 to $930 million in 1963, with twelvefold expansion in grants for extramural research. Congressional cuts in 1963 curtailed that exponential growth (stabilizing at 6 percent annual increase), but the cuts' immediate and lasting effect on Europe was the intensified support for molecular biology by European governments. By then American molecular biology had become the paradigm for biology around the world.[20]

Though life science's institutional maps had shifted, the information discourse persisted. Guided by the discursive framework and scriptural representations erected in the first phase, biochemistry too "went informational." Visualizing protein synthesis as a communication system, biochemists during the second phase were engaged in a fierce race to complete the code. Multiple authorships and literally hundreds of life scientists—principal investigators, international flow of postdoctoral fellows, and technicians—contributed to the completion of the "code of life." The 1966 Cold Spring Harbor symposium, with record attendance of three hundred, was devoted to the code. By 1967 most of the sixty-four codons were correlated with the twenty amino acids and initiation and termination codons elucidated. In 1968, with unprecedented swiftness, the Nobel Prize in Physiology or Medicine was awarded to Marshall Nirenberg, Har Gobind Khorana, and Robert Holley. By then the genome had become widely perceived as an information system, an authorless Book of Life written in a speechless language of DNA.

The notion of "code" carried multiple historical allusions and contemporary referents, eliciting imagery of transcendent knowledge, Mosaic tablets, positivists' ideals of nature's laws, secret writings, period intrigues with espionage and cryptology, ideas from linguistics, information theory, and cybernetics. At times it was a language, or a tape storing information; at other times it was viewed as a DNA code, RNA code, or the protein code, though it also referred to the correlation between nucleic acids and proteins. A code is, by definition, a relation, or a set of rules of transformation from plaintext to cryptogram; always operating on defined linguistic entities (e.g., words, sentences). It is neither a thing nor a language. This diversity of meanings could be confusing, but for the scientists involved the referents were clear by context and practice. Used loosely, tautologically, and inconsistently, the code was caught in a web of signifiers. This multiplicity of significations, definitional slippages, shifting meanings, and aporias ultimately served to destabilize the validity and predictive power of the genomic writings.

To understand these informational and scriptural representations of heredity, their emergence, scope and limits, appeal and dismissals, I now turn to some of the key concepts undergirding the historical narrative and analyses in subsequent chapters: discursive practices; discourse of information; biopower; and metaphors. Representations of the genetic code as language and writing technology and representation of the genomic Book of Life as an information system came into being simultaneously through the spread of the information discourse, but not so much from the direct, and often aborted, applications of the mathematical theory of communication to molecular biology in the 1950s and 1960s. Therefore, it is essential that the distinction between the two—information theory and information discourse—be clear and precise. To this end I examine next these concepts and their relations to information theory and molecular biology, followed by a brief historical sketch of nature's textuality and its relation to the genomic Book of Life.

DISCOURSE OF INFORMATION, METAPHORS, AND MOLECULAR BIOLOGY

> The word "information" in this theory [mathematical theory of communication] is used in a special sense that must not be confused with its ordinary usage. In particular, information must not be confused with meaning.
>
> —*Warren Weaver*[21]

> The ideas and terminology of information theory would not have caught on [in biology] as they have done unless they were serving some very useful purpose. It seems to me that they are highly appropriate in their proper context . . . and in speaking of genes and chromosomes the language of information theory is often extremely apt.
>
> —*Sir Peter B. Medawar*[22]

From information theorists to life scientists, nearly everyone agreed that the ideas and terminology of information theory became pervasive in the 1950s and 1960s. Representations of physical, biological, and social phenomena with the semiotics and models of cybernetics and information theory were widespread in molecular biology. Furthermore, nearly every discipline in the social sciences, (sociology, psychology, anthropology, linguistics, political science, and economics) as well as in the life sciences (immunology, endocrinology, embryology, physiology, neuroscience, evolutionary biology, ecology, and molecular genetics) flirted with the seductive ideas of cybernetics and information theory in the 1950s, each with different degrees of commitment. These trends were particularly strong in the United States and the USSR and were visible also in England and France.[23]

While scientists concurred on the ubiquity of informational and cybernetic terminologies in biology, their opinions diverged markedly on the appropriateness and utility of these applications; comments have ranged from dismissal, to tolerance, to celebration (see discussion below). Rather than adjudicate between experts' conflicting assessments, it is far more fruitful to examine up-close several of their distinct features. I will examine not only the operational utility of multiple informational tropes and scriptural representations in molecular biology but also their disciplinary efficacy and social value. When examined from a broader historical perspective—as the emergence of an information discourse—the power of these discursive practices transcends their significance in the laboratory (whether scientifically warranted or not). But what is meant by "discursive practices" and "discourse of information"? How is it related to metaphors? And how is it linked to the epistemic, disciplinary, and social terrains of molecular biology and its manifestation as new form of biopower?

Encompassing activities such as naming, describing, interpreting, analogizing, and signifying, discursive practices have formed the conceptual frameworks guiding molecular biologists in their theorizing, experimental design, and interpretations during both phases of the genetic code. Discursive practices were not merely rhetorical tools for scientific persuasion, disciplinary demarcations, or literary devices for popularization, though they have served such purposes as selection and founding of journals; personal and disciplinary expository styles; etiquettes for multiple authorships and rules for credit allocation within the hierarchy of principal investigators, associates, visitors, postdoctoral fellows, graduate students, and technicians; formal and informal seminars; and oratorical skills and lectures tailored for various audiences. Cold Spring Harbor symposia served as the principal stage for such discursive displays (some thought they brought out the worst in people). In fact, despite the growing technoepistemic overlap between biochemistry and molecular biology in the late 1950s, their discursive differences were still pronounced; the brash young Turks of molecular biology often sported unconventional presentational styles.

Their new modes of reasoning and argumentation, derived from genetics and the mathematical sciences, seemed suspiciously elegant (especially to biochemists); their bold theorizing and generalizations from limited samples set molecular biologists apart from their parent life sciences (e.g., microbiology, microbial genetics, and biochemistry). These discursive practices also served to establish disciplinary authority and to consolidate institutional power. Commenting on such differences, biochemist Robert G. Martin (Nirenberg's colleague who worked on the code) pointed out, "leading molecular biologists have rarely achieved their status through science alone. Equally important have been their facility with the written word, their eloquence of speech, their political savvy, their caustic wit—and plain luck."[24] Indeed, there is little doubt that the enormous impact and cultlike followings of scientists like Max Delbrück, George Gamow, Francis Crick, Leo Szilard, François Jacob, and Jacques Monod derived in part from their powers of articulation and persuasion (remembering too that biochemists and microbiologists had their heroes).

But discursive practices should not be crudely cast as interest-driven, or strategic manipulations, for they are not always conscious acts, but rather ones that reflect larger cultural forces at work. Beyond an individual's influence, a large-scale scientific and cultural shift in representation—the information discourse—enabled work on the genetic code. As Carl R. Woese, an important contributor to the code through information theoretical approaches, explained in his influential book, *The Genetic Code* (1967), there were two classes of "informational molecules": DNA and RNA, "tapes" and "tape readers," respectively, both governed by the rules of information processing. He wrote, "In cellular tape-reading process, an input tape [DNA or its RNA "transcript"] feeds linearly through the tape reader [ribosomes]; the reading in all cases consists of producing an output tape [protein] whose monomer units and mapping rules are characteristic of the tape reader but whose information content, of course, reflects exactly that of the input tape."[25] These goal-directed and self-regulating representations of a cellular computer program did not function merely as rhetoric but as discourse, as a way of thinking and doing.

"Heredity is described today in terms of information, message, and code," proclaimed François Jacob (recipient of the 1965 Nobel Prize with Andre Lwoff and Jacques Monod for their studies of genetic regulation) in his widely read book, *The Logic of Life* (1970). Comparing the genetic code to a computer program, Jacob argued that "heredity functions like the memory of a computer. . . . Organs, cells and molecules are thus united by a communication network."[26] Meanwhile, Monod, in his notorious book, *Chance and Necessity* (1970), characterized the organism as nothing but "a cybernetic system governing and controlling the chemical activity at numerous points." Within the technoscientific imaginaries of the missile age, gene-enzyme regulations became "systems [that are] comparable to those employed in electronic automation circuitry, where the very slight energy consumed by a relay can trigger a large-scale operation, such as, for example, the firing of a ballistic missile."[27] Both credited Norbert Wiener, Leo Szilard, and Leon Brillouin ("founding fathers" of information theory) for this profound reorientation of biology.

Recounting the "Breaking of the Code" in his book, *The Language of Life* (1966), geneticist and Nobel laureate George Beadle reflected:

> What has happened in genetics during the past decade has been the discovery of a Rosetta stone. The unknown language was the molecular one of DNA. Science can now translate at least a few messages written in DNAese into the chemical language of blood and bone and nerves and muscle. One might also say that the deciphering of the DNA code has revealed our possession of a language much older than hieroglyphics, a language as old as life itself, a language that is the most living language of all—even if its letters are invisible and its words are buried deep in the cells of our bodies.[28]

Robert Sinsheimer, a molecular biologist who attempted to solve the code using information theory and was also progenitor of the Human Genome Project, described the human chromosomes in his 1967 book, *The Book of Life*, as a storehouse of information:

> The book of life. In this book are instructions, in a curious and wonderful code, for making a human being. In one sense—on a sub-conscious level—every human being is born knowing how to read this book in every cell of his body. But on the level of conscious knowledge it is a major triumph of biology in the past two decades that we have begun to understand the content of these books and language in which they are written.[29]

These discursive practices were not ex post facto rhetorical veneers on sober scientific facts, nor were they constructed primarily to appeal to wider audiences: Monod, Jacob, and many other molecular biologists had deployed cybernetic models, informational tropes, linguistic and communication representations of nucleic acids and protein synthesis in their experimental and interpretive framework since the 1950s. The historicity of these discursive and material practices too was captured in Jacob's reflections: "Today the world is message, codes and information. Tomorrow what analysis will break down our objects to reconstitute them in a new space?"[30]

Thus the term *discourse* (and discursive practices) is not used here generically, merely as communication of ideas, speech, or rhetorics but in a Foucaultian sense of a *system* of linguistic *dispersion*, dispersion grounded in micropractices that vary with different cultural or historical eras. Foucault argued, "Whenever one can describe, between a number of statements, such a system of dispersion, whenever, between objects, types of statement, concepts, or thematic choices, one can define a regularity (an order, correlations, positions and functionings, transformations), we will say . . . that we are dealing with a *discursive formation* [emphasis added]."[31] Discourse refers to statements (and tropes) that in a particular historical period come to be consistently configured together, such as gene (hormone, antibody, or brain), information transfer, message, text, program, and so on. Always embedded in material, disciplinary, and social practices, discourse, in this sense, is also productive. It opens up conditions of possibility for the emergence of new objects and new representations of nature, "relating them to the body of rules that enable them to form as objects of a discourse and thus constitute the conditions of their historical appearance."[32] There were no genetic messages in the 1930s; genes did not transfer information before the 1950s; they only possessed biochemical specificities.

Discourses establish cultural efficacy through regimes of signification. These refer to the body of practices and representations a society at a particular historical period accepts and validates (e.g., gene-protein interactions as die-casting and lock-and-key models in the 1940s and as electronic communication systems in the 1950s); and the mechanisms enabling one to distinguish true and false statements and the means for their sanctioning (e.g., accepting linguistic representations of the genetic code despite contradictory empirical evidence). Regimes of signification also refer to the techniques and procedures accorded value in the acquisition of knowledge (e.g., computerized cryptanalysis or synthetic polynucleotides messengers); and the status of those charged with saying what counts as true (e.g., physicists and mathematicians versus biochemists or microbiologists). Knowledge is therefore generated through a system of ordered procedures for the production, regulation, circulation, and operation of statements. The products of science and technology are sociotechnical; they work because they are embedded not only in material practices but also in cultural practices that stabilize and naturalize the technologies for producing knowledge and power. In the case of the genetic code, it is biopower.[33]

In *The History of Sexuality*, Foucault explains the concept of biopower by characterizing the transition to the modern industrial era as a profound transformation of the mechanisms of power. The old privileges of sovereign power—the right to take life or to let live—were replaced by the modern power to foster life or disallow it to the point of death: this is what Foucault calls "bio-power." Such power over life, he argues, evolved in two basic forms constituting two poles linked together by a whole intermediary cluster of relations. One (and the earlier) pole has centered on the body as a machine: its disciplining; the optimization of its capacities; increasing its usefulness and docility; its integration into systems of efficient and economic control; and integration also into controls manifest in discursive and material practices and supported by the procedures of power that characterized the disciplines pertaining to what he calls the "anatomo-politics of the human body." The second (and later) pole has centered on biological processes, such as propagation, births and mortality, levels of health and life expectancy, the supervision of which has been effected through regulatory controls, that is, the biopolitics of populations.

The disciplines of the body—e.g., physiology, anatomy, biochemistry, genetics—and the regulations of the populations—e.g., evolutionary studies, statistics, actuarial practices, demographics, mortality and morbidity records—have constituted the two poles around which the organization of power over life has been deployed. This also explains the extreme importance that sex and reproduction have assumed as biological and social issues. Sex sits at the very pivot of the biopower axis, connecting the control of bodies, on the one hand, with control of populations, on the other. If a nation's institutions of power, among them government, industry, universities, and foundations, ensure the maintenance of production, then biopower has supplied the practices, discursive and material, at every level of the social body, including family, church, workplace, army, hospitals, schools, police, and the technologies of the self.[34]

The information discourse is used here as a historically and culturally situated *system of representations*, which in the 1950s became configured together and increasingly intuitive and commonsensical, and as an emergent form of *biopower*, where material control was supplemented by the control of genetic information. Many physical, biological, and social phenomena were redescribed within the system of metaphors, models, analogies, and semiotics, derived from information theory and cybernetics. Given the centrality of these metaphors within the information discourse and in the construction (and, ultimately, the deconstruction) of the genetic Book of Life, their workings and productivity calls for critical study. I will first examine their technical features and scientific utility and what they metaphorized, following with a brief look at their disciplinary and social dimensions. Despite the many epistemic difficulties inherent in the information discourse, I hope to convey an appreciation for its potency and limitations in both molecular biology and the wider scientific and popular culture.

To begin with, information theory has metaphorized the concept of information. Dating back to the late fourteenth century, the word *information* long signified the action of *in*forming: formation or molding of mind and character, training, instruction (including divine instruction and inspiration), communicated knowledge, news, and intelligence (in contrast to data); and, since the turn of the twentieth century, was deployed in this generic sense in physics, mathematical logic, electrical engineering, and biology. As such, it was a linguistic currency recognized through the tripartite hierarchy of human communications: the syntactic level (relations between signs); the semantic level (relations between signs and designata); and the pragmatic level (relations between signs and their receivers).[35] But in the late 1920s, as a result of studies at Bell Laboratories on telegraphic transmission, the concept of information began to be decoupled from its meaning as "intelligence" and to signify purely syntactic arrangements of symbols ("logical instruction to select") suited for electronic communications, a development culminating during World War II in Claude Shannon's mathematical theory of communication. As Bill Aspray has persuasively argued, only in the decade following the war did information become for the first time a physical parameter and a precisely defined concept amenable to scientific study.[36] As such, its study and manipulations were wedged at the intersection of several lines of military-sponsored research on both machines and living organisms: the mathematical theory of communication, modeling of the brain, linguistics, artificial intelligence, guidance and control systems, cybernetics, automata theory, and behaviorism.

But information theorists used "information" (and the communication idioms associated with it) metaphorically, subverting its sense of meaningful communication. Contrary to its generic use, "information" in the mathematical theory of communication implied that information had to be thought of in a manner entirely divorced from content and subject matter. As Warren Weaver, director of the Rockefeller Foundation's Natural Science Division, explained, "The word 'information' in this theory is used in a special sense that must not be confused with its ordinary usage. In particular, information must not be confused with meaning." [37] In fact, two messages, one heavily loaded with meaning and the other pure nonsense—a Shakespearean sonnet and a random collection of letters—can be exactly equivalent from an information theory standpoint. As Claude Shannon, cofounder with Norbert Wiener of information theory, made clear: "The semantic aspects of communication are irrelevant to engineering aspects" (though the converse is not true).[38] Even when applying information theory to cryptanalysis and human communications, he carefully skirted any problems of semantics. (Problems of pragmatics were irrelevant as well; from information theory's standpoint, it mattered not whether a message originated with or reached a human or a monkey.)

Moreover, information is not an entity. Wires do not carry information as freight trains transport coal. The Wiener-Shannon measure of information is a purely stochastic phenomenon concerning the statistical rarity of signals. What these signals signify or mean, or what their value or truth is, cannot be gleaned from the theory of communication. In the mid-1950s, at the height of the popularity of cybernetics and information theory, the prominent information theorist Colin Cherry warned against its widespread misuses: the "vagueness arising when human beings or biological organisms are regarded as 'communication systems'"—a critique still echoed by Heinz von Foerster, one of the leaders of cybernetics.[39] The noted logician Yehoshua Bar-Hillel, too, was a vociferous critic of colloquial misuses of information theory. He observed, "Unfortunately, however, it often turned out that impatient scientists in various fields applied the terminology and the theorems of the statistical (communication) theory to fields in which the term 'information' was used, pre-systematically, in a semantic sense . . . or even in a pragmatic sense."[40] Information theory, therefore, cannot serve to legitimate the DNA text or the Book of Life as a source of biological meaning. Even if it were possible to determine mathematically (in bits) the information content of a genomic message or a "sentence" in the Book of Life, this would not yield any semantics, not unless its context (genomic, cellular, organismic, environmental) could be properly specified.

But, of course, information theory, too, as most forms of knowledge (including scientific), is inescapably discursive and grounded in metaphors. Information theory metaphorizes the conventional notion of information by borrowing the semiotics of human language to describe highly technical, restrictive, and nonhuman processes. Yet the richness afforded by the multiple meanings and ambiguity of the information catachresis proved irresistible. The notions of information, its storage and transfer, conjured compelling and deceptively accessible imagery of communications that swiftly reshaped scientific and popular representations of nature and society.

But the metaphorical nature of information in information theory and its metaphorical applications to biological phenomena are not exceptional cognitive events. The observation that language and metaphors shape our temporal relations to the natural and social world is, by now, a truism. Some scholars go so far as to assert that because metaphor so pervades everyday life—not just in language but in thought and action—"our ordinary conceptual system, in terms of which we both think and act, is fundamentally metaphorical in nature." Based on analyses of numerous examples, George Lakoff and Mark Johnson have argued that most of our fundamental concepts are experiential. As such, they are systematically organized in terms of orientational metaphors (e.g., spatialization) and ontological metaphors (e.g., structural and container metaphors), all rooted in physical and cultural experience. "Concepts are not defined in terms of inherent properties," they conclude. "Instead, they are defined primarily in terms of interactional properties."[41]

In the history and philosophy of science, the discussions of models and metaphors have focused almost exclusively on problems of theory construction rather than on material practice.[42] Initiated by Max Black in 1962, these studies were developed by Mary Hesse, who has shown how "theoretical explanation in science is a metaphoric redescription of the domain of the explanandum" (the phenomenon to be explained). The metaphor connects theory with nature, as the language used to describe a primary system is transferred to words normally used in secondary systems. She argues, "Sound (primary system) is propagated by wave motion (taken from a secondary system)," "gases [primary] are collections of randomly moving massive particles [secondary]."[43] I will elaborate in Chapter 4 that in information theory a sequence of alternatives for making nonrandom selections is a "message"; the set of such alternatives is the "alphabet"; and the set D of k-letter words in genetic codes constructed based on mathematical coding theory is a "dictionary" ("set D" refers to the sixty-four triplet nucleotides, or "words").

In this transference process, metaphors may work both ways: they select and emphasize or suppress features of the primary; new slants on the primary are illuminated so that the primary is seen through the frame of the secondary. With time, if the transference results are successful and so firmly rooted as to seem nearly ontological in nature, then the secondary can also be reshaped by the primary. Thus biological specificity became informational, and information, message, and code eventually became biological concepts. The meaning of "message" has shifted from oral to written, to printed, to telegraphic, to cybernetic (partly animate, partly inanimate systems), as did scriptural representations of words and texts. These shifts in scriptural media have had a profound effect on the pragmatic level of communication, namely on the widening meanings of the message. (Interestingly, from computer metaphors of DNA in the 1960s, we have now moved to DNA computing in the 1990s; from viruses as information packets to computer viruses.) By 1986, Arbib and Hesse had asserted, "Scientific revolutions are, in fact, metaphoric revolutions, and theoretical explanation should be seen as metaphoric redescription of the domain of phenomena."[44]

By that time, the scholarship on metaphors had mushroomed into something of an industry. Focusing on the Wiener-Shannon theory, linguist Michael J. Reddy has spotlighted the problem of the two-way traffic between the primary and secondary linguistic domains. He has shown "that the stories English speakers tell about communication are largely determined by semantic structures of the language itself . . . that merely by opening our mouths and speaking English we can be drawn into a very real and serious conflict." Indeed, he argues, the mathematical theory of communication serves a convincing case for its own metaphoricity, a point made by Jacques Derrida a decade earlier. Jean-François Lyotard and Jean Baudrillard also spoke about the crisis of representation engendered by linguistic technologies in their respective work.[45]

Echoing and citing Cherry's work, completed nearly forty years earlier, Reddy shows how in information theory the "message" (sequence of alternatives) cannot be sent; how "signals" (which are mobile) do not *contain* the message, and how

> the destructive impact of ordinary language on any extensions of information theory begins with the very terms the originators (Shannon & Weaver, 1949) chose to name parts of the paradigm. They called the set of alternatives . . . the *alphabet*. It is true that in telegraphy [pre-information theory] the set of alternatives is in fact the alphabet; and telegraphy was their paradigm example. But they made it quite clear that the word "alphabet" was for them a technical coinage which was supposed to refer to *any* set of alternative states, behaviors, or what have you. But this piece of nomenclature is problematic when one turns to human communications [emphasis added].[46]

This critique applies also to the use of "code" in information theory (but not a Morse code) where, according to Weaver, it is used to change a "message" into a "signal." But a code is a relationship between two distinct linguistic systems; it does not "change" anything into anything else, neither do encoding or decoding. They simply amount to more metaphors.

Heinz von Foerster has traced this linguistic conundrum to the military origins of information theory.

> How such brilliant thinkers who created this novel science, which is deceptively called "information theory," could confuse two concepts of such profound semantic difference as "signal" and "information" is difficult to grasp unless we remember the historical context of the development of this theory: these concepts, together with those of the general purpose computer, evolved during World War II. And during wartime a particular mode of language—the imperative, or the command—tends to predominate over others (the descriptive, the interrogative, the exclamatory, etc.). . . . The command mode can ex-

> ist only in a "trivial" system, one in which it is assumed that output is uniquely determined by input [the behaviorist dream], in this case by command.

"Epistemology is a political issue," von Foerster concluded; or an ontology of the enemy, as Peter Galison aptly put it. It is also a lesson that David Noble, Langdon Winner, and Donald MacKenzie, among others, have expounded about the political nature of technoepistemologies.[47]

When applied metaphorically to biological phenomena, "information" becomes even more problematic: it seems actually to restore its first sense as intelligence and meaning, but as such it violates the precepts of information theory, which supposedly and initially legitimized the biological applications. It thus becomes a metaphor of a metaphor, a catachresis, and a signifier without a referent. Taking a broad view, philosopher Richard Boyd holds a much more positive stand on the application of cybernetic and computer metaphors in science. Thinking within the limits of theory-construction, he argues that such metaphors are *constitutive* of the theories they express. Rather than being merely exegetical, they constitute, at least for a time, a part of the linguistic machinery of scientific theory. As an example of such theory-constitutive metaphors, Boyd cites the incorporation of the information-cybernetic metaphor in cognitive psychology. Nearly the same analysis applies to molecular biology (if one substitutes the terms *genome*, *heredity*, *genetic processes*, and so on, for the terms *learning*, *thought*, *cognitive processes*, *consciousness*, and so forth).

1. The claim that thought [or heredity] is a kind of "information processing" and that the brain [or the genome] is a sort of "computer."
2. The suggestion that certain motoric or cognitive processes [or genetic processes] are "preprogrammed."
3. Disputes over the issue of the existence of an internal "brain-language" [genomic language] in which "computations" [combinatorial rearrangements] are carried out.
4. The suggestion that certain information is "encoded" or "indexed" in "memory store" [genome] by "labeling," whereas other information is "stored" in "images."
5. Disputes about the extent to which developmental "stages" [or gene expression] are produced by the maturation of new "preprogrammed" "subroutines," as opposed to the acquisition of learned "heuristic routines," or the development of greater "memory storage capacities" or better "information retrieval procedures."
6. The view that learning [biological development] is a "self-organizing machine."

7. The view that consciousness [genetic regulation] is a "feedback" phenomenon.[48]

Similar analogies may be found in other biological (and social) fields. Nobel laureate immunologist F. Macfarlane Burnet (based on earlier attempts of technical applications of information theory to biology) tried to develop a communication theory for immunology in the late 1950s by treating (metaphorically) the problem of immunological specificity in terms of "coded information transfer" of patterns. Similar attempts marked the applications of informational and cybernetic models in endocrinology. Operations research expert Herbert Simon, who also tried his hand at breaking the genetic code, remarked, "I continue to be intrigued by some of the analogies between the kinds of complex systems I am accustomed to consider in my research—large scale human organization—and the complex system you [Gamow and Yčas] are studying." Such computer metaphors, according to Boyd, display interactionist properties and thus shape salient features of science. By trafficking analogies between different areas of science, they have encouraged the discovery of new features in the primary, cognitive science or molecular biology, and in the secondary, information-cybernetics, subjects, he argues.[49]

These are helpful perspectives for assessing the epistemic roles of communication metaphors in representing objects and mechanisms in molecular biology, notably to the relations between genes and proteins ("the code"). And they stand in marked contrast to the few attempts to use information theory in molecular biology in its intended mathematical form. Of the numerous researchers who transported the information discourse into molecular biology in the 1950s and 1960s, only radiation biologist Henry Quastler and later those who applied mathematical and informational theories of coding to the genetic code did so by treating information in its technical sense. As the first to apply information theory proper to biology (in 1949), Quastler did so by reworking the central concept of biochemical specificity in information theoretical terms, applying it to genes and chromosomes still within the protein paradigm of genetics. Since information in the technical sense, like specificity, was a manifestation of the degree of differentiation and orderliness (or negative entropy) of biological molecules, it could be used as specificity's quantitative measure, he reasoned. (This conceptual equivalence of information and molecular specificity became an accepted view.)

Quastler's mission of transforming molecular biology into an authentic information science was reflected in his often-cited articles, books, and symposia.[50] While achieving a measure of acclaim in the 1950s, his project eventually fell into obscurity, mainly because some of its premises and data quickly became outdated. Moreover, its information-theoretical analyses offered no experimental agenda and thus seemed of little relevance to contemporary research in molecular biology. By 1956, even Quastler and other enthusiasts conceded that, given the enormous difficulties of quantitative applications of information theory to biology, it might be preferable to use it qualitatively (namely, metaphorically). Likewise, other information-theoretical applications by mathematicians, communication engineers, and molecular biologists to the genetic code were similarly historically eclipsed. But mathematical information theory in biology did not vanish. Instead, it has continued as a separate area within the field of theoretical and computational biology. (The works of Heinz von Foerster, Lila Gatlin, Henri Atlan, and Michael Arbib in the 1960s formed some early foundations enduring today.)

What remained in the wake of these aborted attempts to apply information theory to molecular biology was not a blank slate, but the information discourse: the system of representation—*information*, *messages*, *texts*, *codes*, *cybernetic systems*, *programs*, *instructions*, *alphabets*, *words*—that first emerged in the late 1940s. From information theorists' standpoint, it was merely a rhetorical shell; its technical content emptied out. But as such, information in molecular biology served as a potent metaphor for the century-old ideas of chemical and biological specificity and as a (re)validation of molecular nature as text (the Book of Life of the computer age).

Thus these discursive practices of information and language were neither external to researchers' analyses nor merely exegetical. Rather, they became constitutive of the reasoning and modes of signification of researchers by supplying productive models, analogies, and interpretive frameworks; though this borrowing was by no means a simple transfer but more of a two-way reworking. As Georges Canguilhem observed, "A model only becomes fertile by its own impoverishment. It must lose some of its own specific singularity to enter with the corresponding object into a new generalization."[51] The code and its metaphors instantiated themselves within the information discourse (in both theoretical and experimental phases of the code), but had to be continuously redefined to overcome what Derrida called the "aphoristic energy of its own writing" (see discussion below). The constant movement of definitional differences, the slippages in scriptural designations (letters, words, texts) attested to the resistances and plasticities in these analogical constructions. But the commitments to these textual representations persisted despite the unwarranted significations of molecules as linguistic entities.

Peter Medawar argued, "The ideas and terminology of information theory would not have caught on [in biology] as they have done unless they were serving some very useful purpose. It seems to me that they are highly appropriate in their proper context . . . and in speaking of genes and chromosomes the language of information theory is often extremely apt."[52] Similarly, and implicitly concurring with many biologists, Woese argued, "What has not been generally appreciated is that the subsequent spectacular advances in the field, occurring in the second period [the biochemical phase of the study of the genetic code, 1961–67], were interpreted and assimilated with ease, their value appreciated and new experiments readily designed, precisely because of the conceptual framework that had already been laid [in the 1950s]."[53]

Not everyone agreed. Generally speaking, biochemists until the 1960s (with few exceptions) found its communication tropes of little relevance to their analyses of the static aspects of biological macromolecules, such as the compositions and structures of proteins and nucleic acids. Even enzymologists, with their focus on dynamic properties and reaction mechanisms bypassed informational representations in the 1950s. As Canguilhem observed, cybernetic models were more fruitful in the study of function (e.g., genetics) than in investigations of structure or the relation of structure to function (e.g., biochemistry).[54]

The infiltration of informational representations into biochemistry was complex and uneven. Since the late 1940s, Sol Spiegelman, who was greatly influenced by von Neumann, incorporated informational and cybernetic models into his studies of protein synthesis. Likewise, Alexander Dounce began to do so around the mid-1950s. In working out the properties of the tobacco mosaic virus (TMV) in the mid-1950s, Heinz Fraenkel-Conrat had little use for informational models, despite his proximity to members of the RNA Tie Club who were then analyzing genetic codes from a scriptural vantage point. But after 1959 he did represent his project in terms of coding and information transfer. Marshall Nirenberg began to think informationally around 1958. Curiously, Erwin Chargaff, who in the mid-1950s signified biological specificity in terms of information transfer, confessed in 1963 to his "share of guilt" in promoting "the concept of 'biological information' [which] raised its head and began to sport a multicolored beard which has become ever more luxurious despite numerous applications of Occam's razor."[55] His acerbic caricatures of "information transfer" only intensified with time, targeted primarily at flashy molecular biologists. By then many biochemists had come to think of nucleic acids as "informational macromolecules"; information serving as a conceptual link to the genetics.

The distinguished biochemist and historian of biochemistry Marcel Florkin strongly criticized the "molecular biosemiotics" created by Wiener's claim "that the concepts of information and cybernetics are applicable to organisms." Biochemistry possessed its own semiotic repertoire, Florkin argued, pointing out, "Linguistics is dealing with linguistic signs, i.e. with psychological entities, relating in the receptor's mind a psychic acoustic image and a concept. . . . We shall therefore pray for the banishment of the abusive use of the term 'language' from the field of molecular biosemiotics."[56] Joseph Fruton concurred. Echoing several information theorists, he has been skeptical about the emphasis on information transfer in biology. Granting that such idealized models stimulated important empirical discovery, he nevertheless writes:

> The future historian of the origins of molecular biology will, I hope, examine critically the purported role of information (or communication) theory. . . . Although the mathematics of information theory . . . appears to have had little application in biological research, the language they introduced was eagerly adopted by those engaged in the study of the genetics and metabolism of bacteria and viruses. . . . Such terms as code, message, and noise acquired a biological meaning. Although biologists now use such terms *largely as metaphors*, the future historian may be well advised to examine more closely the ways in which the "intellectually satisfying schemes" offered during the 1950s and 1960s, and based on the ideas of information theory, influenced empirical research at that time [emphasis added].[57]

This is a tall order for historians of biology and the one taken up in this study.

To sum up, one can certainly agree that the concept of information and its many related tropes taken from the mathematical theory of communication were applied to molecular biology largely metaphorically. When technically applied (Quastler's project, information-based cryptanalysis, and mathematical theories of coding) the scientific results did not endure, though the discursive framework did. The idea of code and its attendant linguistic significations were similarly applied metaphorically and inconsistently; their meanings constantly destabilized by divergent designations. On the other hand, there is no doubt of their operational utility in molecular biology. Their potency was enhanced through their disciplinary linkages and social valences.

Indeed, analyses of life scientists, linguists, and philosophers only partially illuminate the scope and limits of the information metaphor. They do not address the powerful social and cultural functions of scientific and technological metaphors (studies which form a sizable subset of the metaphor industry). David Edge highlighted the centrality of the technological metaphor in social control, pointing out that where such metaphors have been successful they have been taken literally. Religious symbols such as "the book of life" are classic examples, as are society as machine, the body as machine, the universe as machine (clockwork), and the railway metaphor within the emerging class structure of modern America.[58] Charles Rosenberg's analysis of the popularity of the human nervous system as an electrical (telegraphic) network—with all its moralizing tropes directed against the excesses of modernity—has illuminated the power of metaphors in establishing and reinforcing moral and social order. More recently Nancy Leys Stepan has shown how social values are inscribed, validated, and circulated through scientific metaphors. Charting the two-way traffic between analogies of race and gender in the biological sciences in the nineteenth and twentieth centuries (the back-and-forth projections of inferiority of blacks and women) she shows how they buttressed each other. She also underscores that metaphor is constitutive of both scientific theory and the categories of experience underpinning scientific inquiry; both highlight and suppress different features of the (perceived) world.[59]

But James Bono's perspective best captures the various levels of significance of metaphors—cognitive, material, cultural, and political—by making the strong claim:

> Metaphorical aspects of language are essential to understanding the dynamic of conceptual change in science precisely because they ground complex scientific texts and discourses in other social, political, religious, or "cultural" texts and discourses. Rather than mirroring the "legible face" of a reality envisioned by scientists and "deciphered" within a single, dominant paradigm, complex scientific texts and discourses constitute themselves through their intersection with other multiple discourses.

He contends that, in science, metaphor functions as a medium of exchange between the *two* interconnected domains: *intrascientific* and *extrascientific*.[60] Thus, a look beyond the confines of theoretical and material practices of molecular biology reveals the multivalences of the information metaphor: on several fronts it worked as an interdisciplinary and cultural medium of exchange. In the 1950s it served as a means of demarcating the disciplinary boundary of the new molecular biology, especially from the traditions of biochemistry (though eventually both fields reshaped each other). Many molecular biologists deployed the idiom of information in the mid-1950s, but it was Francis Crick (with his superb expository skills) who formalized the information discourse as a way of imposing thematic order and rhetorical imperatives on the central problem of protein synthesis, reproduction, and the disciplinary turf of molecular biology. Echoing Wiener's writings of a decade earlier, in which Wiener set forth that representations of organisms were shifting from the materialistic and energetic to the informational, Crick asserted that the essence of protein synthesis was flux: flow of energy, flow of matter, and principally the unidirectional flow of information (meaning the specificity inherent in the sequence) from DNA to proteins. "Once 'information' has passed into protein it *cannot get out again* [my emphasis]," he proclaimed in the notorious "Central Dogma" (1958).[61]

In a single masterly stroke, Crick encapsulated the imperative logic of the genetic code and the ideology and experimental mandate of the new biology: genetic information, qua DNA, was both the origin and universal agent of all life (proteins)—the Aristotelian prime mover—according to Delbrück. By that time biochemist Marshall Nirenberg had already begun tracking that genetic information by envisioning protein synthesis as "the code of life." For biochemistry, information was the (dangerous) Derridian *supplement*, as Rheinberger noted, a tagalong term that innocuously smuggled the informational perspectives of molecular genetics into protein synthesis research, eventually requiring a reorientation of biochemistry and nearly merging the two fields.[62]

Beyond the disciplinary arenas, the information discourse served as a currency of shared historical sensibilities and cultural experiences. As Paul Fussell observes in his classic book, *The Great War and Modern Memory*, the memory of World War I became imprinted in British culture through a linguistic repertoire. "Nobody alive during the war, whether a combatant or not, ever got over its special diction and system of metaphor, its whole jargon of techniques and tactics and strategy."[63] Similarly, World War II and the cold war imprinted their own discursive traces on science and society, some of them captured in the information discourse. This discourse linked the biosemiotics of molecular biology to the imaginaries of postwar technoculture. Beyond the control of bodies and populations—in all their material messiness—the power over life was being envisioned within the new paradigms of communication. It was to be exercised on the pristine metalevel of controlling information flow, the sequence, the word, and the text.

I will now follow, episodically and sketchily to be sure, the aporic metaphor of the "Book of Life" and "Book of Nature" through the centuries. By highlighting vignettes along its turning points from antiquity to the twentieth century, I aim to provide some understanding of its diachronic and synchronic attributes, namely, its enduring powers as nature's scriptural sublime, as well as its technoepistemic meaning in the postwar era.

THE GENETIC CODE, THE BOOK OF LIFE, AND THE BOOK OF NATURE

> A thing is called a book because it has received writing. But a thing is said to be receptive in so far as it contains material potency, which cannot exist in God. Therefore, nothing uncreated is called the book of life. Since

> *book* means a kind of collection, it signifies distinction and difference. . . . In every book the writing is something other than the book.
>
> —*St. Thomas Aquinas*[64]

Synchronic representations of the genomic "Book of Life" in the 1960s were inextricably linked to the diachronic symbolism of "The Book" as natural, eternal, and universal writing. This metaphor, so pervasive throughout Judeo-Christian history, contains several paradoxical features that scholars through the ages have pondered. With the perceived linkages between the metaphor and information theory, that metaphor has been strained to the point of deconstruction. When Thomas Aquinas first posed numerous questions and difficulties inherent in the concept of the "Book of Life," he was responding not only to its various meanings in the Old and New Testaments but also to interpretations by previous commentators on the subject, notably St. Augustine. Plato's imagery of the human and world soul as eternal writing has long animated that metaphoric tradition. Although the Book of Life and the Book of Nature are not quite the same—the first represents the eternity and logos of human souls, the second the eternity and logos of all animate and inanimate things—in both the term *book* has always served as a metaphor for the material record of creation. "The invisible things of Him, from the creation of the world, are clearly seen, being understood by things that are made," decreed St. Paul's Epistle to the Romans (Romans 1:20). Within this tradition, the Book of Nature has functioned as scriptural exegesis, especially after the thirteenth century, with the demarcation of the boundary between knowledge in the light of grace (theology) and that attained through the light of nature (natural philosophy).[65]

But the aporia inherent in this simultaneously material and textual record of creation, as articulated in the Gospel of St. John (1:1–14), challenged believers even back then: "In the beginning was the Word and the Word was with God. . . . In Him was life; and the life was the light of man . . . and the Word was made flesh and dwelt among us. . . . " How can a word, a signifier, precede what is signified—thought or act? How can language represent the yet unthought? And if God is nonmaterial, then how can His Word carry material potency and be "made flesh"? As Aquinas pointed out, "Nothing uncreated is called the book of life." Thus both the Book of Life and Book of Nature pose the same age-old conundrum: What is this nonmaterial writing? Is it creation or revelation? In the beginning was the Wor(l)d? (In the postwar era the "word" would present additional conundrums: If not God, then what is the agency of this writing in the secular context of molecular biology? What meaning can this writing convey within the stochastic and syntactical logic of cybernetics and information theory? And how is language possible without human consciousness? This conundrum compelled Derrida to assert, "The idea of the book, which always refers to a natural totality, is profoundly alien to the sense of writing. It is the encyclopedic protection of theology and of logocentrism against the disruption of writing, against its aphoristic energy, and . . . against difference in general."[66]

Thus, from ancient times to recent years, as Hans Blumenberg documented, nature has always been textualized;[67] its aporias have challenged even religious practitioners of science. Apart from the theological conundrums, the synchronic meanings of writing and the Book of Nature also changed with time; they were reconfigured within the regimes of signification of changing epistemes and cultural experiences: ancient, medieval, Renaissance, scientific revolution, Enlightenment, romanticism, and twentieth-century modernism. As Mark Poster observed, "Every age employs forms of symbolic exchange which contain internal and external structures, means and relations of signification. . . . The shift from oral and print wrapped language to electronically wrapped language thus configures the subject's relation to the world."[68] Never monolithic, these epochal ruptures have always carried the instabilities and resonances of prior representations and hybrid constructions. Thus in the punctuated history of the Book of Nature the metaphor has had a remarkable endurance, but was always (re)historicized within complex and overlapping layers of old and new meanings.

Pondering the problem of Nature's logos and the diversity of life, Lucretius (50 B.C., *De Rerum Natura*) postulated, "The matter and first bodies of each thing must be inside the seed; because of this all things can not arise from anything. For each particular seed has, in itself, hidden inside, its own distinctive powers." And given the manifestation of this diversity, he concluded, "Isn't it reasonable to conclude that many things have tiny elements in common, just as different words may have letters the same?"[69] Textualizations of nature were reinvented within the scribal culture of literacy in twelfth-century Europe.

As Brian Stock has shown, the Book of Nature had a specific meaning in the high Middle Ages, a time when Naturalism bore its own historicity. With the growth of a more literate society, the science of nature had to accord with the inner logic of contemporary texts. Nature was then constituted as a book through interconnections between words, thoughts, and things. The medieval structures of knowledge—logic, grammar, rhetoric, and theology—were mobilized to unravel the "secrets of nature." For William of Conches, Nature consisted of many books corresponding to the number of controlled interpretations. For Alan of Lille unnatural deviance was a grammatical error. Hugh of St. Victor saw knowledge of Nature as inextricable from logic. Ideally signification through things was preferable, but the philosopher, limited to *scientia*, knew only the meanings of words. Words, texts, reason, and nature formed the seamless fabric of medieval natural knowledge.[70]

By 1500 the age of scribal culture had ended and the age of print culture had begun, ushering in a major transformation. According to Elizabeth Eisenstein the shift reconfigured boundaries of knowledge, beaux arts and belles lettres, church, nobility, and learned elites. The physical and social status of "book" had changed. The Book of Nature became a printed text; though the printed text too had multiple social valences and contextual meanings. In that history of Nature and/as text, a decisive moment came in the seventeenth century with the birth of modern science. With the rise of autonomous experimental tradition, representations of nature became intertwined with interventions, or to use Derrida's terms: (reading) the book of nature/life became inseparable from its writing.[71] In a vision reminiscent of the discourse on the genetic code, the Book of Nature in the seventeenth century awaited "decoding" by the experimental investigator equipped with an "ideal language." Bacon argues that one reads "God's natural truths with the alphabet of Nature." Both Descartes and Galileo spoke of the writing and reading of the great book of Nature; Leibniz searched for the ideal language, the "characteristica universalis," which would correspond exactly to Nature's written language; Bonnet presumed that "our earth is a book the God has given to intelligences far superior to ours to read."[72]

Giving voice to the German Enlightenment, Immanuel Kant seemed to regard a purely mechanistic command of language as a necessary but not sufficient condition for the acquisition of knowledge. "Nature," he wrote, "is a book, an epistle, a fable (in a philosophical sense) or however you want to call her. Suppose we know all its letters as well as possible, we can break all its words into syllables and pronounce them, even know the language in which it is written—is all this already enough to understand a book, moreover, to judge its character from it, or to extract its essence?"[73] Straddling the passage from the Enlightenment to romanticism, Goethe spoke of nature in terms of secret writing: "How readable the Book of Nature will be to me I cannot express to you, my high spelling has helped me," he wrote. Perhaps it is his preoccupation with the ambiguities of nature's writing that inspired Goethe to interrogate them through the figure of Faust, who, in his quest for epistemic mastery and worldly power interprets "In the beginning was the Word" as "In the beginning was the Act," thus submitting the "word" as signifier to the primacy of the (transcendental) signified. Images of nature as linguistic communication, as hieroglyphs and ciphers, were ubiquitous in the eighteenth century, guiding the studies of practitioners of *Naturphilosophie*, who cultivated poetic and aesthetic impulses as a means of reading and narrating nature's hidden poetry (here Novalis stands out). Images of nature's ciphers persisted into the nineteenth century, when modern notions of organic memory (engram or Mneme principle) became linked to biological and molecular knowledge (notably to the combinatorial nature of proteins); telegraphy supplied additional imagery for physiological phenomena as linguistic communication. Such concepts and imprinting imagery also informed Schrödinger's visions of a hereditary code-script into the 1940s.[74]

But in the 1950s, with the spread of information theory, electronic computers, communication and simulation technologies, the very notion of message, text, and language was transformed, once again, as was the textualization of nature. It now seemed technically legitimate to speak of molecules and organisms as texts, namely as information storage and transfer systems. "This [the human chromosomes] is the book of human life. In this book are instructions, in a curious and wonderful code, for making a human being," stated Sinsheimer.[75] Heredity became a programmed communication system governed by a code that transferred "linguistic information" through the cell and cycles of life. But aside from the paradoxes associated with a stochastic concept of information devoid of semantics there was also the problem of linguistic signification devoid of agency. Geneticist Philippe L'Héritier pointed out in 1967 that, "Being a symbolic language, human language presupposes an interlocutor and a comprehending brain but in genetic language we have nothing but information transfer between molecules [and even then 'information transfer' is just a metaphor]"; an objection later echoed by Florkin. Claude Lévi-Strauss put his finger on this fundamental philosophical conundrum when he wrote, "Can there be a prediscursive knowledge of language existing prior to its construction by humans? Could there be something, as biologists claim, which resembles the structure of language but which involves neither consciousness nor subject?" Jean Baudrillard has gone so far as to assert that the genetic code, as an icon of command and control, simulations, electronic programs and texts, can exist only through this crisis of (linguistic) representations. He observed, "End of the theater of representation, the space of signs, their conflict, their silence; only the black box of the code, the molecular emitter of signals form which we have been irradiated, crossed by answer/questions like signifying radiations, tested continuously by our own program inscribed in the cells."[76] Indeed, it was the information catachresis—the double metaphorical construction of information—that seemed to validate the representations of the genetic code as natural, eternal, and universal writing. It is the space created by these slippages, ambiguities, paradoxes, and loss of referentialities that served as a repository for the scientific imaginary of the genomic Book of Life.

Moreover, both diachronic and synchronic visions of nature's writings are problematized if one also questions the objectivist perspective on knowledge. This Platonic, or logocentric, standpoint takes it as given that genomic writing exited before humans' entry into the world, awaiting decoding, divinely inspired or otherwise. The material and theoretical tools of molecular biology—quantification and experimentation—have enabled, in principle, its unambiguous reading. Guided mostly by mechanistic ideals of language as transparent signification, by faith in the exact correspondence of words and things, of signifier and signified, this objectivist view endowed the initiated with access to positive knowledge of the Book of Life. Yet such absolute concepts of language (like the absolutes of mass, space, and time in the mechanical world picture) had already been challenged in the beginning of the twentieth century by admitting the contextualities within the knowledge-system into analyses of meaning, by accepting the contingencies of structuralism. Rather than precise correspondence between signifier and signified and absolute reference, the sign derived its meaning only through differences with other signs, from the context of the whole linguistic system. Meanings and words then become polysemic. Thus, the very concept of language was radically altered: Since the idea of the Book of Life first came into being as universal and absolute writings, the polysemic aspect of its so-called writing undermines the possibility of its absolute reading.[77]

Beyond difference and structuralism, how does one define "the system" itself (such as a genome), including its origins and boundaries? As Derrida (and poststructuralism more generally) has destabilized the notion of a linguistic "system," so too life scientists have been problematizing biological systems through the theory of autopoiesis ("self-production"). For what characterizes all living things is that they are continually self-producing according to their own internal rules and requirements, thus blurring a clear distinction between "inside" and "outside," "closed" and "open." In such a revision, information is not an independent, prespecified quantity functioning as input for the genomic system; rather, the "meaning" of information is continuously adjusted, not only by the contextualities *within* the system but also by the system's interaction with the *outside*. The distance between the genotype and phenotype is therefore considerably increased; it is a dynamic dialectic of preformation and epigenesis.[78]

Furthermore, the logocentric view of genomic writing has probed neither the construction process of the tools of science nor their physical and discursive fashioning. In short, it has not questioned the cognitive assumptions and technological imperatives behind nature's writing. These issues may be illuminated by critiques of technology (e.g., Heidegger) and of meaning (e.g., Derrida) that are grounded in the dialectic between *episteme* and *techne* as well as between intervention and representation (after Ian Hacking) in the workings of modern science, in general, and molecular biology, in particular. This dialectic challenges the objectivist viewpoint that one can have access to an unmediated nature—visible or submicroscopic—or to natural phenomena that are prediscursive and independent of the tools of representation. Even without an emersion in pre-Socratic debates of being and becoming, about essences and ontologies, one can hardly escape the ancient problem of being and knowing: Can *being* be separated from its manifestation, can an entity or phenomenon be known independently of the means—discursive and material—that form its representation? When *episteme* and *techne* are seen as intertwined (thus rejecting the Greek logocentric legacy), the time-honored dichotomies between theory and practice, discovery and invention, and observer and phenomenon are blurred. Technology and theory generate each other; epistemic things become technical things and vice versa, as Hans-Jörg Rheinberger has shown.[79]

Writing, from this vantage point, is then on the side of *techne*. It is the process of signification—ordering, naming, isolating, measuring, describing—by which knowledge of entities and phenomena become manifest; writing could be seen as a technology of representation, be it the surface of the earth, cells, or DNA. From this Derridean vantage point it is the writing itself (qua production of representation) that writes; it comes to possess a kind of agency. For once committed to describing and manipulating biological entities through the information discourse and its scriptural technologies, the scientists became part of the representational space within which techno-epistemic events of molecular biology take place. The actors' freedom of movement, from experimental design to data interpretation and presentation, is always already mediated through that discursive/material space.[80]

Finally there is the additional conundrum of the authorship of the Book of Life in a godless scientific universe. Perhaps, because of its subject matter of life and reproduction, molecular biology since the 1950s has been suffused with theistic images and religious icons; its practitioners transacting a kind of divine biopower. References to Delbrück's priesthood of the phage church; Caltech and Cold Spring Harbor as molecular biology's Mecca and Medina; disciples' pilgrimages to centers of enlightenment; Monod's college of cardinals issuing an encyclical to abolish the terminology of enzyme "adaptation"; Crick's Central Dogma of biology that information flowed only unidirectionally from DNA to RNA to protein; and classic molecular biology texts analogized to the Old and New Testaments are but a few (of many seriously humorous) perceptions of such transcendent authority.[81] There is little doubt that this biopower entailed mastery of the genomic Book of Life, first through secular exegesis and subsequently through secular (re)creation.

Edward Trifonov and Volker Brendel, authors of the book *Gnomic: A Dictionary of Genetic Codes* (1986) and pioneers of Chomskian DNA linguistics, which they christened "gnomic," might have unwittingly grasped the powers and limits of genomic writings, with all their theistic, epistemic, and deconstructive force. Situating their linguistic project in the genealogy of primal knowledge, they envisioned molecular biologists as struggling with the age-old difficulties of what they called "*lingua prima* of life."

> The nature of the beginning and the foundations of life are central issues in man's spiritual and scientific quest. Goethe had his Faust struggle with this in trying to interpret the first verse of St. John's gospel: "In the beginning was the Word. . . . " Was it really to mean "Word" in this context or had it to be translated as "Thought" or "Deed?" Whatever the answer, the literal domain of words—language—is surely associated at least with the beginning of man and with the understanding of man.[82]

Thus, if the genome stands for the origins of human life, then the Word—the DNA sequence—has brought molecular biologists as close to the act of creation as could be experienced, invoking supernatural, Faustian powers. This scriptural and material mastery was articulated by James Watson in the mandate for the Human Genome Project: "For the genetic dice will continue to inflict cruel fates on all too many individuals and their families who do not deserve this damnation. Decency demands that someone must rescue them from genetic hells. If we don't play God, who will?"[83]

CHAPTER TWO

Spaces of Specificity: The Discourse of Molecular Biology Before the Age of Information

In 1971 Max Delbrück, a Caltech physicist-turned-biologist, entertained his audience, as he often did, with a serious joke. Looking back at nearly two decades of DNA-based molecular biology and nearly two-score Nobelists in the field (including himself), Delbrück suggested it was Aristotle who had, in fact, discovered the DNA principle, and if the Nobel Committee were to award posthumous prizes, they should consider him.[1] Delbrück's sophisticated argument consisted of a subtle scientific subversion and historical inversion. Providing his own English translation of passages from *Generation of Animals* on organismic development—matter and form—he applauded Aristotle's prescience: "Put into modern language, what all of these quotations say is this. The form principle (DNA) is the information which is stored in the semen. After fertilization it is read out in a programmed way; the readout alters the matter upon which it acts, but it does not alter the stored information, which is not, properly speaking, part of the finished product." This logos, DNA's information, had even deeper significance, Delbrück seriously jested: it was Aristotle's principle of the *Primum Mobile*, the "unmoved mover," the source of all motions. (Georges Canguilhem had also noted that representations of life as information corresponded to Aristotle's notion of form and to the inscription of its logos.) Originating with Aristotle's biological works, the principle was then grafted onto physics, astronomy, and cosmological theology. Catastrophic as it was for Newtonian physics, he admitted, "the 'unmoved mover' perfectly describes DNA. It acts, creates form and development, and is not changed in the process."[2] It is therefore an ontological entity, a cosmological principle.

A master of the carnivalesque, Delbrück often challenged his audience by dissolving conventions and inverting the logic of temporalities and causalities. But ironically, his intentional play with the anachronism of Aristotle's "genetic information" exposes other, unintended, anachronisms: how molecular biologists in the second half of the twentieth century re-represented key concepts of heredity belonging to the first half of the century. Rather than displacing the Aristotelian discourse of generation, emergent descriptions of organisms and molecules as information transfer systems reveal a more recent rupture. A discursive shift and a reinvention of history, a reconfiguration of epistemic, experimental, and social structures, and a remaking of what had been the space of representations of molecular biology before the 1950s all transpired in the course of two decades.

Throughout the 1940s, in numerous studies of phage infection, multiplication, mutations, recombination, and resistance, Delbrück and his many colleagues in molecular biology had explained such phenomena in biological and physico-chemical terms, without any references to information transfer. But like others in the 1950s, Delbrück gave information theory in biology serious consideration. By the late 1950s when surveying the mechanisms of phage replication, Delbrück and Gunther Stent believed—as did many molecular biologists—that "it would be unwise not give some currency to 'information transfer' as a possible replication mechanism."[3]

While in the 1950s terms such as *information* and *code* in molecular biology sometimes appeared in quotation marks to acknowledge their metaphorical and heuristic dimensions, by the end of the decade the quotation marks had disappeared. The new biosemiotics and its linguistic tropes became naturalized within the scientific and cultural discourses of the postwar era to the point where it became virtually impossible to think of genetic mechanisms and organisms outside the discursive framework of information. (Note also that the meaning of "genes" had evolved since the beginning of the century: from Mendelian inferential constructs, to physical entities equated with proteins, then to nucleoproteins, and finally to DNA.) The age-old metaphor of the "Book of Life" was similarly resignified; the book now contained "information." *Information* and *book* came to define and validate each other.

But what did the metaphor of information denote? What biological property did "information" signify? What animate processes or physiological mechanisms did the notion "information transfer" represent? What did genes do before they stored and carried information? How did researchers who shaped the course of molecular biology, for example, Max Delbrück, Linus Pauling, George Beadle, Jacques Monod, Oswald Avery, Erwin Chargaff, and James Watson, represent the functional and structural attributes of genes in the 1940s? In short, what features—material, discursive, and social—distinguished the discourse of molecular biology before the age of information?

This chapter examines the cognitive commitments, semiotic tools, the discursive and material practices of molecular biology in the 1940s, and the historical knowledge/power nexus in which they functioned: namely, the Rockefeller Foundation's sponsorship of molecular life sciences in the interwar period. Such investigation ultimately brings into sharp relief the centrality of the concept of *biological specificity* in the contemporary explanatory accounts surrounding gene action, with its intrascientific and extrascientific linkages.

Canguilhem spotlighted the discontinuities in the conceptualization of life from antiquity to the present, tracing life as animation, life as mechanism, life as organization, and finally, life as information. But it was Michel Foucault (Canguilhem's protégé) who in *The Order of Things* elaborated on the significance of the discourse of organization in the emergence of a science of life in the nineteenth century. (The term *biology*—in contradistinction to natural history—was coined by at least four life scientists in the first decade of the nineteenth century.) Foucault also examined biology's linkages to contemporary cultural/discursive transformations in labor, wealth, and linguistics. Hidden organization displaced the visible form; comparative anatomy challenged natural history as a way of understanding life.

> Character [biological character starting in the nineteenth century] is not, then, established by a relation of the visible to itself; it's nothing in itself but the visible point of a complex and hierarchized organic structure in which function plays an essential governing and determining role. . . . The preeminence of one function over the others implies that the organism, in its visible arrangements, obeys a *plan* [emphasis added].

Karl Figlio elaborated on and refined Foucault's approach. His analysis of the historical role of the metaphor of organization in nineteenth-century biomedical sciences showed how it was a natural carrier of its cultural context.[4] François Jacob in *The Logic of Life* drew on Canguilhem and Foucault (and the general trend in French history of science) and traced the history of heredity to a series of epistemic ruptures. (Foucault considered Jacob's book to be "the most remarkable history of biology that has ever been written," because it demonstrated that the trajectory of science corresponded to that of human thought.) Jacob wrote:

> In the second half of the eighteenth century and the beginning of the nineteenth, the very nature of empirical knowledge was gradually transformed. . . . It was within living bodies themselves that the very cause of their existence had to be found. It was the interaction of the parts that gave meaning to the whole. Living bodies then became three-dimensional entities in which the structures were arranged in depth, according to an order prescribed by the working of the total organism. The surface properties of a living being were controlled by the inside, what is visible by what is hidden. Form, attributes and behavior all became expressions of organization. By its organization the living could be distinguished from the non-living. . . . Thus with the start of the nineteenth cen-

> tury, a new science was to appear. Its aim was no longer to classify organisms, but to study the processes of life; its object of investigation was no longer visible structure, but organization.

Organized, versus unorganized, matter marked the difference between the animate and inanimate. It is this notion of biological organization that guided life science into the mid-twentieth century and that would later be subsumed by an imaginary computer program of life (what Jacob called the "Integron").[5]

Within the discourse of organization specificity was a thematic thread running through all the life sciences: biochemistry, immunology, genetics, physiology, embryology, taxonomy, and evolution. Most notably in molecular biology, the concept and uses of "specificity" would often later be displaced by "information," because both specificity and information signified the complementarity of highly ordered biological structures. While there were good grounds for the interchange, specificity and information were not equivalent representations. Specificity was the Aristotelian material cause, grounded in three-dimensional molecular structures and in experimentally determined measures, linking molecules to organisms to species. Information was the Aristotelian form, abstracted as a one-dimensional tape, a transaction devoid of experimental measures and material linkages. Specificity corresponded to body, information to soul. Being historically situated, discourses, such as the discourse of organization and the discourse of information, do not map directly onto each other: certain features are highlighted, others suppressed; some articulations are silenced, others overflow with new meanings. Biological specificity (and its embodiments in various experimental practices) would be re-represented through the scriptural tropes of information—*message*, *alphabet*, *instructions*, *code*, *text*, *reading*, *program*. The narratives of heredity and life would be rewritten as programmed communication systems, aligning molecular biology with other contemporary scientific disciplines and cold-war technoculture, where the very idea of organization was reformulated within the algorithms of complex systems.[6]

I first examine the spaces of "specificity" in life science during the first half of the twentieth century via material and discursive structures. Briefly exploring its uses and explanatory functions in several research areas, I will follow its uses within the discourse of organization, or life's "grand design." I then trace notions of biological and chemical specificity within the changing paradigms of molecular biology—from proteins to nucleic acids as the prime movers of heredity—and within its contemporary institutional contexts. Looking at the Rockefeller Foundation's sponsored program of molecular biology, I focus on studies by Pauling, Beadle, and Monod, conducted within the protein paradigm.[7] Moving next to the role of specificity in researches linking nucleic acids to heredity, I will glance at the works of Avery, Chargaff, and finally Watson and Crick, who in 1953 presented their double helix structure of DNA in terms of an informational code, thus inspiring George Gamow to take on the coding problem.[8] I then reexamine Erwin Schrödinger's contributions to the genetic code. Ubiquitous "founding-father" genealogies have attributed the generative role in the work on the genetic code to Schrödinger's book, *What Is Life?* (1944). By analyzing these retrospective accounts, which invariably conflate Schrödinger's coding with later notions of information transfer, I will resituate Schrödinger's concerns within the discourse of organization of the interwar period.[9] Finally, I follow other so-called precursors of the genetic code (protocodes?): proposals by Kurt Stern, Cyril Hinshelwood, and Alexander Dounce to explain the biological specificity of nucleic acids and proteins before the age of information.

INSIDE THE "GRAND DESIGN" OF ORGANIZATION

François Jacob observed that biology, in failing to offer genuine mathematical theories as the exact sciences do, must function mostly through models (and, he might have added, metaphors). Furthermore, he noted, there exist a number of generalizations in biology, but few genuine theories.[10] Specificity, one such generalization, referred to the concept of biological and chemical specificity, which was a central preoccupation and unifying theme in early-twentieth-century life science. Key problems and experimental agendas that collectively contributed to the emergence of molecular biology in the 1930s and 1940s were articulated in terms of specificities.[11] Even before the compositions and structures of genes, enzymes, antibodies, bacteria, and viruses were elucidated, impressive knowledge of these entities and their precise material workings had accumulated based on their remarkable functional specificities. Experimental studies showed that genes were highly specific with respect to gene products; enzymes exhibited a high degree of specificity for their substrates; the binding of antigen and antibody became an index of specificity in immunology and related fields; bacteria and viruses were often characterized with respect to their host-range specificities; taxonomies of species (and human "races") were established based on experimentally measured serological differences.

Like "information," the word "specificity" was in generic use in science for centuries, but, as Arthur Silverstein has pointed out, it acquired a particular technical meaning, conceptual coherence, and material potency within the science of immunology at the beginning of the twentieth century. This potency was grounded in the idea of *stereocomplementarity*. From Paul Ehrlich's biological formulation of the side-chain theory in the 1890s, to its displacement in the 1920s by Karl Landsteiner's work on the chemical perspectives on antibody synthesis, stereocomplementarity supplied powerful physical and visual representations of specificity. Transported to immunology from organic chemistry—through the efforts to explain enzyme-substrate specificity—the idea and image of complementarity was captured by Emil Fischer's noted lock-and-key hypothesis (1894–98). The fit between reacting substances and the complementarity of their three-dimensional configurations determined specificity.[12] Across the life sciences, the lock-and-key metaphor served as an exchange medium, a conceptual and experimental bridge, which related form to function along the material continuum of biological specificity, from species to molecules.

As Scott Gilbert observed, in embryology, through the model-system of fertilization, the specificities of intercellular interactions during development were represented in immunological terms. Prominent Chicago embryologist Frank R. Lillie proposed in 1914 that sperm and egg were linked by stereocomplementary reactions at the cell surface, though Lillie apologized for "The terminology [which] has been largely adopted from immunology, because it seemed best suited to express the facts."[13] Rockefeller Institute physiologist Jacques Loeb objected to Lillie's biological model. Aspiring to elevate embryology to a status of an exact science, Loeb sought to replace the underlying "metaphorical" side-chain theory by mechanistic explanations grounded in physico-chemical immunology (promulgated by Swedish physical chemist Svante Arrhenius).[14] What went uncontested in that acrimonious controversy was the concept of specificity as a meaningful generalization in biology and its material embodiment as proteins.

For Loeb, as for most contemporary biochemists, the specificities determining the physiological differences within and between species resided in the compositional and constitutional differences in proteins. That blood serum proteins exhibited quantifiable species-specificities had been already established at the beginning of the twentieth century.[15] Organismic and species specificity were related, as Robert Olby pointed out. Physiologist Edward Reichert of the Carnegie Institution of Washington proposed in 1909 that if a definite relationship between differences in proteins and physiological differences between species could be demonstrated, then "a fundamental principle of the utmost importance would be established in the explanation of heredity, mutation, the influence of food and environment, the differentiation of sex, and other great problems of biology, normal and pathological."[16] Reichert, together with Amos Brown, examined hemoglobin crystals from about two hundred mammalian species, establishing a taxonomy of hemoglobins that paralleled traditional organismic classification.[17] Mammalian visible attributes were thus replaced by the properties hidden in their molecular structures. Specificity therefore served as a probe into evolutionary change, measuring phylogenetic distances from molecules to species through various methods (e.g., structural comparisons, cross-reactivity tests). Information did not do likewise.

By the time Loeb wrote his important treatise, *The Organism as a Whole* (1916), he could credit numerous studies that together warranted a broad generalization of the chemical basis of genus and species specificity. "What is the nature of the substances which are responsible for and transmit this specificity?" he asked rhetorically. "There can be no doubt that on the basis of our present knowledge proteins are in most or practically all cases the bearers of this specificity."[18] An avid admirer of genetics and a friend of its American doyen, Thomas Hunt Morgan, Loeb nevertheless doubted whether the constituents of the nucleus contributed to the determination of the species. As Jan Sapp has shown, Loeb was articulating some of the contemporary disagreements over cytoplasmic inheritance and the role of genetics in evolution. Loeb conjectured, "This in its ultimate consequences might lead to the idea that the Mendelian characters which are equally transmitted by egg and spermatozoon determine the individual or variety heredity, but not the genus or species heredity."[19]

Morgan strongly disagreed. There was no evidence in all the Mendelian studies that there was a "distinction between generic, versus *specific* characters or even 'specificity,'" he argued, responding to the challenge to his "nuclear monopoly";[20] though the material workings of specificity were not an urgent problem for Morganian genetics. Within the formalisms of Mendelian crosses, notably its linkage maps and inference structures which were oblivious to the physical existence of genes, the materialities of specificity had only little relevance. By the early 1920s, the consensus emerged that one-to-one correspondence between genes and characters did not exist: instead some traits were polygenic (determined by several genes), and others pleiotropic (one gene controlled several traits), further muddying up the relevance of specificity for genetics. (Though the intuitive idea of one gene–one trait continued to fuel eugenic commitments for years to come.)

As Morgan observed, "Each gene may have a specific effect on a particular organ, but this gene is by no means the sole representative of that organ, and it has also equally specific effects on other organs and, in extreme cases, perhaps on all the organs or characters of the body." Only on rare occasions, when prodded toward reflecting about the physical meaning of genes, did Morgan speculate on their organic nature; and like most of his contemporaries he envisioned genes as protein bodies.[21] The notion of specificity remained important, however, in several European traditions of genetics, where heredity was not reduced to Mendelian mechanisms but was broadly conceived to encompass problems of development, physiology, and evolution.[22]

Specificity did assume major importance in American genetics in the 1930s, with the reorientation of the field toward physico-chemical investigations under the aegis of the Rockefeller Foundation. Within its new agenda, termed "Science of Man," which was aimed at a science-based rationalization of the social order, the foundation targeted vast resources toward a new biology. Foundation president and physicist Max Mason announced that the "salients" of the concentration "are directed to the general problem of human behavior, with the aim of control through understanding. The Social sciences, for example, will concern themselves with the rationalization of social control; the Medical and Natural sciences propose a closely coordinated study of the sciences which underlie personal understanding and personal control."[23] Under the direction of his protégé, mathematical physicist Warren Weaver—who in 1938 coined the term molecular biology—the new interdisciplinary biology was grounded in theories and technologies of the physical sciences. Genetics occupied a key position in this cooperation between the social, medical, and biological sciences. With all its eugenic implications, the query "How are things inherited?" led Weaver's list of twoscore items for investigation, spanning physiology, sexuality, intelligence, and mental attributes. Molecular genetics was poised at the pivot of the axis linking control of bodies and control of populations; it was a historically specific mode of biopower.[24]

"Genetics is broadening its scope; the gross structural features of inheritance are today fairly well-known, and the workers are turning to the physiological aspects of heredity," announced Morgan, who also acted as adviser to the foundation in 1933.[25] It is within the new agenda of physiological (or physico-chemical) genetics that specificity gained direct relevance to gene action. And as in the other life sciences, the notion of specificity was transported into genetics from immunology through the nascent field of immunogenetics. Serological studies from animal systematics had demonstrated a direct relation between the formation of antibodies and heritable genetic markers, relations with immense promise for eugenic intervention. With the intensified programmatic development of physiological and biochemical genetics in the 1930s, the relations between genes and their products came to be conceptualized in terms of biological and chemical specificities.[26]

In many instances the term *specificity* (like the term *information* later on) possessed more of a metaphorical quality and heuristic value than operational force. Unless detailed through some kind of concrete structure, measure, mechanism, and experimental procedure, specificity was not really an explanation (*explanan*) but that which needed explaining (*explanandum*).[27] But in several lines of research it did have concrete meaning and practical value derived from experimental work. In immunology, bacteriology, enzymology, and taxonomy, for example, specificity served as a laboratory tool for ordering, predicting, and sometimes measuring activities and relations (cross-reactivities) of molecules, organisms, and species, or "races." (Henry Quastler would employ "information" as an alternative measure of specificity, a point taken up later in Chapter 3.) More important, specificity was primarily a *biological* concept, signifying animate phenomena, processes, and characteristics. As such, specificity related form to function by representing life within a three-dimensional biological space (or four-dimensional, when including ontogeny's time arrow). Within the epistemic, technical, and disciplinary configurations linking genes, antibodies, enzymes, organismic growth, and animal taxonomy, biological and chemical specificities were central. They accounted, often in material terms and concrete methods, for the myriad of spatial and temporal events from reproduction, fertilization, embryogenesis, maturation, to speciation; specificities dictated and governed the successive cycles of life.

Similar to information discourse in the 1950s, biological specificity in the early twentieth century found articulation in statements that came to be systematically configured together. The statements formed elements of an older biological world picture, which was marked (among other things) by the discourse of organization, as several scholars have suggested.[28] Although the term *organization* (like information) had been in circulation for centuries (derived, in fact, from organs and bodily structures), in the nineteenth and early twentieth centuries it came to signify for life scientists the hidden agency governing the visible body: the plan of life, the "grand design," demarcating the animate from the inanimate. Within that biological grand design, it was the harmonious interaction of hierarchies and parts that gave meaning to the whole and defined the order behind the bewildering complexities. "What makes an organism an organism," observed Paul Weiss in 1939, "is that the diverse portions are *definitely grouped and arranged*, maintain *specific mutual relationships*, and conform to a pattern which is *essentially identical* for all members of a species. . . . This order is called *organization*."[29] Organization was predicated on specificity. For life scientists of the early twentieth century, "organization," that is, the coordinated structures and activities among and within organs, tissues, cells, chromosomes, enzymes, and antibodies, framed the unity, stability, and specificity of organisms and species.

This organization, or hierarchical order of life, was predicated on specialization modeled after ideas of division of labor (which themselves drew on visions of society as a body). The body's parts and organic constituents were envisioned as highly specific or specialized, their structures and functions having coevolved over several million years. In the industrialized epoch of modernity organization and specialization were discursive formations of the human sciences, occupying the space between biology and labor.[30] Across Victorian doctrines of laissez-faire, Weberian theories of "rationalization," Durkheimian laments of "differentiation," and an interwar managerial culture of "Fordism" (and cooperative individualism), scientific representations and human experience were naturalized through analogies between the living body and the body politic, each signifying and validating the other through the circulation and economy of discourse.[31] Biological specificity was enmeshed within other sociotechnical constructions of modernity: organization, differentiation, specialization, cooperation, stability, and control. Configured together, these terms captured the premise of knowability and control of the grand design of nature and society, which provided coherence across diverse biological fields and linked specificity to the discourses of modernist culture.

"Are there not general principles of stabilization?" queried Walter Cannon in his noted essay on self-regulation. "Might it not be useful to examine other forms of organization—industrial, domestic or social—in light of the organization of the body?"

> The centrally important fact is that with the division of labor, which is implicit in the massing of cells in great multitudes and their arrangement in specific organs, most of the individual units become fixed in place so that they cannot forage for themselves. . . . Only when human beings are grouped in large aggregations, much as cells are grouped to form organisms, is there the opportunity of developing an internal organization which can offer mutual aid and the advantage, to many, of special individual ingenuity and skill. . . . Each one finds security in the general cooperation. Once more. Just as in the body physiologic, so in the body politic, the whole and its parts are mutually dependent; the welfare of the large community and the welfare of its individual members are reciprocal.[32]

Referring in a circular manner to physiological and social processes, these representations of organized specialization enabled the circulation of its meanings, historically constituted through these conjunctions and through congruences of the social, economic, and biological. These discursive formations formed the objects of biological research and shaped the macro- and microrepresentations of living bodies until the middle of the twentieth century. Different, fragmented representations of life would emerge out of the regimes of significations of the post–World War II era. The problems of organization would be reconfigured within the models of systems analysis cybernetics and the discourse of information, eventually decomposing the grand design and the organism.

PRIME MOVERS: PROTEINS AND NUCLEIC ACIDS

Of all the macro- and microelements conjoined in the organization and perpetuation of the body, proteins came first, privileged as the ontological substance of life. As the material representatives of heredity, at least until the early 1950s, they bore biological and chemical specificity. Conceptualizations of the material basis of life in terms of proteins reached back to the nineteenth century, deriving conviction and imagery from the enormous influence of Thomas H. Huxley's protoplasmic view of life (1864). His theory bestowed upon the protoplasm—simple or nucleated—all the physical and mental attributes of life, enshrining the gelatinous substance as the source of diversity and organization and the locus of cognitive and social control.[33] With the rise of eugenics and genetics in the early twentieth century, the "national protoplasm" became a key site for the management of biopower, for control of bodies and control of populations.[34]

By the 1930s, due to developments in enzymology most of the centralized attributes of the protoplasm had been fragmented and distributed among hundreds of constituent enzymes.[35] Some turned out to be autocatalytic, displaying an astonishing property of self-duplication, generating more of themselves by accelerating the reaction in which the enzyme itself was an end product. Autocatalysis, often analogized with crystal growth in the mother liquor, became a popular catchall term for a range of vital processes in cellular reproduction and organismic growth. The crystallization of the tobacco mosaic virus by Wendell M. Stanley and its characterization as a protein possessing autocatalytic properties (1935) supplied a sensational "proof" for the enzyme theory of life, which had far-reaching consequences for genetics. The work seemed to offer concrete evidence for the compositional and functional equivalence of proteins, viruses, enzymes, genes, and antibodies.[36]

This cognitive and cultural significance of proteins, within the discourse of organization, undergirded the Rockefeller Foundation's molecular biology program from the 1930s to the early 1950s. "All that association of phenomena which we term life is manifested only by matter made up to a large extent of proteins," argued Weaver in justifying the centrality of protein research in the foundation's new program:

> [Proteins] enter into nearly every vital process. They are the principal component of the chromosomes which govern our heredity; they are the basic building stuff for the protoplasm of each cell of every living thing. Our immunity to many diseases depends upon the mysterious ability of serum globulin. . . . Several of the hormones, including insulin, are protein in nature. . . . The invasion of certain huge protein molecules, otherwise known as viruses, gives us common cold, influenza. . . . *Enzymes, those strange chemical controllers of so many of the detailed processes of the body, those perfect executives which stimulate and organize all sorts of activities without using up any of their own substances or energy*—these enzymes are now believed to be protein in nature. Indeed many diverse scientists, each with his own special enthusiasm, would be willing to agree that these proteins deserve their names of "first substances" [my emphasis].[37]

Here are the workings of a managerial imagination in which proteins served as strategic sites for managing the molecularized body, a predicate of biological and social control.[38]

Backed by the foundation's enormous financial and institutional resources —about $25 million during the years 1932–59—the molecular biology program served to mutually validate the organization of the living body and the body politic. With the rise of molecular biology one observes a stabilization of sociotechnical meanings produced within the regimes of signification of the interwar era and how certain ideas in the 1930s and 1940s were systematically configured together. Organization governed molecules, bodies, and societies. "Cooperation" represented both a managerial, cognitive, and biological prescription. "Social control" aimed at rationalizing and controlling individual and group behavior. Behavior, both a medical and social domain, was partly biological. Biological organization was increasingly conflated with genetic determinism; the materiality and specificity of genes originated in protoplasmic endowment; and protein specificities resided on the molecular level, to be studied with molecular technologies that are at once representations and interventions. In this economy of discourse modes of representing bodies and the means in their intervention became two sides of the same coin, a purely material mode of control. There were neither messages, information, nor texts in the pre–World War II era.[39]

Under the aegis of the Rockefeller Foundation, one of the key contributors to molecular biology was Caltech chemist Linus Pauling. His landmark work on protein structure and immunochemistry underscored the centrality of proteins' specificity in reproduction (and related biological phenomena) as well as for the eventual rationalization of society through birth control and population control.[40] With Rockefeller Institute biochemist Alfred E. Mirsky, Pauling published a seminal paper, "On the Structure of Native, Denatured, and Coagulated Proteins" (1936), which outlined a general theory of protein structure. Hydrogen bonds—weak but flexible links between molecules—were assigned a major physiological role in determining specificity: a polypeptide chain was coiled in a specific configuration, stabilized mostly by hydrogen bonds; denaturation implied a loss of such configuration. In suggesting that hydrogen bonds determined the three-dimensional configuration of proteins, and thus their biological specificity, Pauling enunciated a fundamental link between molecular structure and biological function, adding a new perspective to the older concept of stereocomplementarity.[41]

Pauling's conception of protein specificity in terms of spatial folding (independent of the ordering of its constituent amino acids) became the basis for his influential program in immunochemistry, which formed one of the pillars of molecular biology in the 1940s. Challenged and inspired by Landsteiner's work and his recent book, *The Specificity of Serological Reactions*, Pauling targeted protein chemistry, notably stereocomplementarity, toward the solution of the long-standing problem of antibody formation: chemical specificity. An often-cited theoretical article Pauling authored with Max Delbrück (1940), entitled "The Nature of the Intermolecular Forces Operating in Biological Process," made complementarity a predicate of specificity, linking processes of antibody formation to enzyme synthesis, virus replication, and gene action. Arguing that the syntheses and folding of complex molecules in the living cell involved not only covalent bonds but also other weak molecular forces, Pauling and Delbrück concluded, "We accordingly feel that complementariness should be given primary consideration in the discussion of the specific attraction between molecules and the enzymatic synthesis of molecules."[42] The entire argument hinged on the primacy and specificity of proteins as templates of heredity, growth, and cellular regulation,[43] or on the ability of these proteins to store and transfer information, as Norbert Wiener, John von Neumann, Henry Quastler, and many others would put it about a decade later.

The notion of complementarity as template for auto- and heterocatalysis did not originate with Pauling. In 1936 John B. S. Haldane was one of the first to suggest a reciprocal complementary copying in gene replication, conjecturing, "We could conceive of a process analogous to the copying of a gramophone record by the intermediation of a 'negative' perhaps related to the original as an antibody to an antigen."[44] Rockefeller Institute biochemists Max Bergmann and Carl Niemann, and British-trained mathematician Dorothy Wrinch, envisioned genetic copying patterns, their specificities determined by amino acid sequences.[45] Other researchers in the 1930s deployed similar analogies and images. Pauling and Delbrück's joint contribution at once narrowed and broadened the template concept by pinpointing the underlying physical mechanisms and generalizing them to all biological phenomena.

These ideas provided the framework for Pauling's template hypothesis of antibody formation, fully articulated in 1940 in "The Theory of the Structure and Process of Formation of Antibodies." Combining Landsteiner's theory of antibody specificity and Pauling's model of protein folding, the theory outlined an elegant mechanism of antibody formation. Later called the *instructional theory*, it seemed to explain several chemical aspects of antibody activities and promised an effective mechanism for a commercially lucrative artificial production of antibodies. The theory's key roles in molecular biology, however, lay in promising to elucidate the action of antigens as templates for antibodies and as a mechanism for explaining gene action. The impact of the instructive theory lasted until the mid-1950s.[46]

Until then, the theory of antibody formation, including its images and metaphors, shaped the accounts of specificity in molecular biology: the replication and mutations of viruses and genes.

> The phenomenon is the same as the production of a coin by a die, or in general of a replica by the process of pressing a plastic material against a mould and permitting it to harden. The polypeptide chain, with its power of assuming alternative configuration is the plastic material, and the surface of the antigen serves as the die or mould. The process of hardening is the result of the operation of the weak forces between different portions of the polypeptide chain that find themselves in juxtaposition.[47]

This representation of the materiality of specificity spread beyond immunochemistry. At Caltech, embryologist Albert Tyler combined Pauling's immunochemistry with Lillie's ideas of specificity to explain fertilization, and Alfred H. Sturtevant applied serological concepts and techniques to *Drosophila* genetics.[48] Through its supposed ability to modify immunity and induce genetic mutations, the instructive theory also held enormous potential for biological engineering.[49] Throughout the 1940s the icons and metaphors of enzymes, antibodies, viruses, and genes as patterns, templates, molds, lock-and-key, die-and-coin, and photographic and phonographic negatives abounded in the scientific and popular literature of molecular biology. (In 1956 F. Macfarlane Burnet would attempt to recast the very same patterns of immunological specificity as an information-based communication system.)[50]

Geneticist George Beadle, too, conceptualized his research projects in terms of biological specificity. Having moved away from the formalisms of corn and *Drosophila* genetics, Beadle focused in the early 1940s on the thorny problem of the relation between genes and enzymes: Were genes enzymes, or did they only make enzymes? As geneticist Jack Schultz aptly put it, if one knew what a gene was, then one could probably find out how it worked; if one understood the mechanisms of gene action, then one could begin to predict what the gene was. The problem was to solve both puzzles at once, knowing the answer to neither.[51] Collaborating with Edward L. Tatum at Stanford University (and sponsored by the Rockefeller Foundation) Beadle used the fungus *Neurospora* as a probe into the gene. Linking biochemical methods with the techniques of Mendelian genetics the two researchers demonstrated that one gene controlled a single chemical reaction, which in turn was regulated by a specific enzyme; this finding was later crowned the "one gene–one enzyme hypothesis."[52]

For Beadle, like his Caltech partner Pauling, genetic specificities were embedded within the folds of protein molecules and assumed a central place in the conjunction of Mendelian crosses with material processes, physiology, and behavior. As master molecules, genes direct the configurations of protein molecules, determine organismic antigens, impose enzymatic specificities in a one-to-one relation, and correspond to specific kinds of psychological abnormalities and feeble-mindedness. Beadle further pointed out, "It should follow, indeed, that every enzymatically catalyzed reaction that goes on in an organism should depend directly on the gene responsible for the specificity of the enzyme concerned. Furthermore, for reasons of economy in the evolutionary process, one might expect that with few exceptions the final specificity of a particular enzyme would be imposed by only one gene."[53] Specificities evolved with material life. Inspired by Beadle's feats, his Caltech friend Sterling Emerson developed an experimental agenda in serological genetics using the elegant *Neurospora* system; it sought to show the antigenic relation between an enzyme and specific gene protein.[54] A mere decade later Beadle—like most geneticists—would accept as a "working hypothesis the view that primary genetic information in all organisms is carried in the form of DNA." The next obvious question was "how the information coded in DNA is used in the development and functioning of a complex organism." By the 1960s Beadle, like most molecular biologists, came to view genes as the universal language of life.[55] DNA did not merely assume the specificities attributed to proteins; instead, it would be elevated to the originator and sole bearer of biological information—the prime mover.

But as microbial geneticist Joshua Lederberg observed, there were some problems with the displacement of specificity by information. He wrote, "The hypothesis which obviously underlies the one-to-one theory is that a gene works as a unique template for 'stamping the specificity' on an enzyme. My philosophical reservation is against the implication that 'specificity' (or 'information' as it is called nowadays) is something apart from structure."[56] Noting such interchanges, George Gamow, Alexander Rich, and Martynas Yčas stated in 1955 that they used "the term 'information' in the sense of molecular specificity."[57] Although Lederberg thought that the concept of "biological specificity" was on several counts superior to "information"—especially when it came to laboratory practice—like the others, he eventually adopted the discourse of information.

While it is true that "information" was often interchanged with "specificity," the complex displacement was only partial. For studies of static structures—crystallography, three-dimensional molecular folding, or material compositions of proteins and nucleic acids—informational representations seem to have been of little conceptual relevance and experimental appeal (Quastler did try to apply it to such structural studies by quantifying information as measure of specificity). However, in studies of dynamic functions—molecular and cellular processes of transport and exchange—informational representations acquired the conceptual potency shaping experimental reasoning. Because molecular specificity was immobile and grounded in matter, information came to serve as its carrier beyond material bounds, as the body's transcendent soul, so to speak. Possessing motion, information could transcend the limits of structure. Specificity was mute; information communicated specificity's messages.

This transition from specificity and the discourse of organization to the discourse of information was especially striking in the 1950s works of Monod, who at the Pasteur Institute had been studying lactose metabolism in E. coli (specifically the formation of the enzyme β-galactosidase in response to lactose), or the problem of "enzyme adaptation." This problem was one of the most intriguing aspects of biological specificity because it asked: How is it that certain enzymes were formed selectively through a stimulation by the specific substrate of the enzyme? How is it that bacteria could synthesize new enzymes in response to changes in the medium? Enzymologists and bacteriologists had observed this phenomenon since the beginning of the twentieth century—the term *enzyme adaptation* was coined in the 1930s[58]—but it was only in the mid-1940s that genetics came to bear on this question.[59] Thereafter the problem of enzymatic adaptation, too, came to inhabit the epistemic/technical interface linking genes, enzymes, antibodies, and cellular development within the theme of specificity.

Having been introduced to genetics in 1936 as a Rockefeller Fellow at Caltech and through associations with George Beadle and French biologist Boris Ephrussi, Monod, unlike most French life scientists, appreciated quite early the relevance of genetics to enzyme adaptation.[60] He closely followed the developments in immunochemistry and *Neurospora* research, including

Emerson's recent work in serological genetics. Drawing on these studies, he assessed the scope and limits of the template concept (and the one-gene–one-enzyme hypothesis) in his masterful examination in 1947 of "The Phenomenon of Enzymatic Adaptation and Its Bearings on Problems of Genetics and Cellular Differentiation."[61] The essay brings into sharp relief the centrality of biological specificity within the discourse of organization and its relation to the epistemic and experimental commitments of Monod and his collaborators at the Pasteur Institute.

"One of the most characteristic tendencies of the present period in the development of biology may perhaps be seen in the focusing of attention on problems of *specificity*," Monod wrote, opening his essay. Quoting Paul Weiss on the concept of specificity, he emphasized that, in spite of the various connotations of the word, specificity had to be associated with a "specific" pattern in space or time or both.

> Thus it is generally recognized that one of the main problems of modern biology is the understanding of the physical basis of specificity, and of the mechanisms by which specific molecular configurations (or multi-molecular patterns) are developed, maintained, and differentiated. The means, the experimental tools for this study, are found in those experiments which result in inducing the formation, or suppressing the synthesis, or modifying the distribution of a specific substance or substances.[62]

This was his point of departure in an attempt to explain how cells with identical genomes could manufacture molecules with different specific patterns. For Monod in the 1940s, genetics was inextricably linked to problems of organismic development and biological organization.

He even rejected the brittle images conjured by the rigid "template" metaphor and its mold-cast processes in favor of "prototype" or "master pattern," seeking more fluid and contingent representations of cellular adaptation. He argued, "Hereditary factors would only determine a certain *range* of structural possibilities, within which specific configuration, i.e. activities, could be evoked by environmental factors." Within the discourse of organization, genetic specificity was only one form of many manifestations of specificity. The cell could also learn from experience; indeed, it had a kind of cellular memory. Based on Weiss, Monod also sought to represent cellular modulations and differentiations in terms of *molecular ecology*, "where the cell is viewed as a complex population of specific molecules and molecular groups, cellular organizations resulting from the interactions, competitions and regrouping of elementary units."[63] This stand is remarkably instructive and illuminates the representational repertoire through which organisms and biological phenomena were constituted before the age of information. Before the 1950s, Monod's semiotic and experimental tools created fluid representations of the cell. Then, heredity—coterminal with development—was interactive, open-ended, and contingent. This picture would contrast sharply with Monod's representations in the mid-1950s of cells as closed cybernetic systems, a point upon which he initiated a discursive turn from enzyme adaptation to enzyme induction. By the late 1950s, as the techniques of phage genetics and biochemistry converged in demonstrating the genetic control of enzyme induction, that phenomenon was further reformulated within the discourse of information: the operon, its repressors, and its messengers (see Chapter 5). Fully determined and controlled by genetic information, the cellular program of 1950s was no longer embedded in a molecular ecology. It did not learn from experience; its memory (information) instead was internal, stored and transferred across generations. The organism would be compressed into a sequence, its functions collapsed into a message inscribed in a one-dimensional DNA tape.

Monod's transition straddled the paradigm shift in molecular biology from proteins to nucleic acids. In the late 1940s, as Monod reviewed the state of knowledge in the protein-based molecular biology, new cognitive currents were already discernable. Recent findings led to the inescapable conclusion that nucleic acids, not proteins, were the carriers of genetic specificity (not yet information). The story has been told many times: in the 1930s, at the zenith of the protein supremacy and the nadir of nucleic acids, several cytologists and biochemists (notably Torbjorn O. Caspersson and Jean Brachet) provided telling glimpses into the role of nucleic acids (DNA and RNA) in genetic replication and protein synthesis.[64] But the tyranny of the "tetranucleotide hypothesis," enunciated in the mid-1920s by Rockefeller Institute biochemist Phoebus A. T. Levene, had denied nucleic acids a claim to biological specificity. Given the enormous investments in protein research, these prescient studies of nucleic acids had received scant attention, especially in the United States. Biological theories, laboratory technologies, institutional and cultural authority all had converged on proteins as agents of organization of bodies and society and were not easily displaced. A turning point occurred in 1944. Attention began shifting to DNA after the publication of the paper on type-transformation of pneumococcus bacteria by the Rockefeller Institute team of Oswald T. Avery, Colin M. MacLeod, and Maclyn McCarty, who argued that the transforming agent was a nucleic acid.[65]

Like most discovery stories, this one too can be narrated somewhat differently from the usual accounts in order to highlight another perspective: the central role of the biological specificity of nucleic acids in the web of theories and practices linking bacteriology, biochemistry, immunology, taxonomy, and heredity. The quiet and persistent Avery, known at the institute as

"Professor" (or "Fess"), sought alongside his contemporaries in life science to explain the biological specificity involved in the production of the type-specific polysaccharides in transformed bacteria. He wrote, "Thus, it is evident that the inducing substance and the substance produced in turn are chemically distinct and biologically specific in their action and both are requisite in determining the type specificity of the cell of which they form a part."[66] It was nucleic acids, "at least those of the deoxyribose type," that possessed different specificities. His argument flew against the prevailing proteins paradigm.

The genetic implications did not escape his notice. "Not only is the capsular material reproduced in successive generations," he pointed out, "but the primary factor, which controls the occurrence and specificity of capsular development, is also duplicated in the daughter cells." Reinforcing his findings with Theodosius Dobzhansky's view that "we are dealing with authentic cases of induction of specific mutations by specific treatments," Avery correlated inducer with gene and capsular antigen with gene product. If these findings were confirmed, Avery predicted, "then nucleic acids must be regarded as possessing biological specificity the chemical basis of which is yet undetermined."[67] Not everyone was convinced. Certainly, half a century of protein supremacy would not be easily overturned, not even by rigorous experiments and argumentation.

But at Columbia University, Erwin Chargaff, an Austrian-born biochemist (and polymath), was one of the first to grasp the significance of Avery's results and to redirect his research in the late 1940s toward the study of nucleic acids; Rockefeller support lagged several years behind these efforts. This story too, with its many heroes, villains, and twists of irony, has been narrated from various vantage points;[68] but it is particularly instructive for viewing the circulation of discourse in the transition from "specificity" to "information." A focus on Chargaff's changing significations of nucleic acids reveals not only the epistemic and linguistic shifts from chemical specificity to information transfer and his later retraction but also a reconfiguration of disciplinary authority in life science through the impact of molecular biology, especially on biochemical concepts and practices.

When Chargaff presented his preliminary studies at the 1947 Cold Spring Harbor symposium, he placed them along Avery's path. He reported, "If, as we may take for granted on the basis of the very convincing work of Avery and his associates, certain bacterial nucleic acids of the desoxypentose type are endowed with a specific biological activity, a quest for the chemical or physical causes of these specificities appears appropriate."[69] In collaboration with E. Vischer, they embarked on such a quest: the development of micro separation methods that would permit the separation and identification of minute amounts of nucleic acid constituents (purines and pyrimidines). Chargaff predicted differences in the proportions or in the sequence of the several nucleotides forming the nucleic acid chain would account for its specific biological effects.[70]

In the next three years, Chargaff's analyses of nucleotide base composition of DNA samples obtained from widely different organisms revealed that the molar proportions of the bases adenine [A], guanine [G], cytosine [C], and thymine [T] varied within wide limits, thus contradicting the tetranucleotide hypothesis, which had been grounded on the assumption of equimolar ratios of the bases. The findings were published in 1950 in Chargaff's often-cited article, "Chemical Specificity of Nucleic Acids and Mechanism of Their Enzymatic Degradation." There, he not only exploded the tetranucleotide hypothesis but also pointed out that the molar base ratios of [A] to [T] and [G] to [C] were not far from 1.[71] Known as Chargaff's rule, the finding formed a key insight in the elucidation of DNA's structure. Many felt Chargaff should have shared the Nobel Prize with Watson and Crick.

Equally important, specificity situated his investigations along the biological continuum from molecules to species. It would be gratifying to learn, he wishfully mused, "that just as desoxypentose nucleic acids of the nucleus [DNA] are species-specific and concerned with the maintenance of the species, the pentose nucleic acids [RNA] of the cytoplasm are organ-specific and involved in the important task of differentiation."[72] Within a few years, the very same regularities of combinatorial variations in DNA bases that governed biological specificity became the hallmark of genetic information and the genetic code. In 1955, as information theory reached a crescendo, Chargaff reminded his audience, "Regularities of this sort may be the best, or only, means of recognizing the existence of systems concerned with the preservation, or the transfer, of information: tasks that we should like to assign to the nucleic acids or, more probably, to the nucleoproteins."[73] Chargaff was then a member of the RNA Tie Club and in frequent communication with Gamow about the coding problem. A few years later he would suggest that these impressive informational properties referred to the very chemical specificities, which, in aggregate, maintained cell specificity. An astute cultural observer, Chargaff pondered these new modes of signification.

> When a science approaches the frontiers of its knowledge, it seeks refuge in allegory or in analogy. The latter attempt . . . has enlisted the support of modern disciplines, such as cybernetics or information theory. A subordinate analogy to the maintenance of cellular specificity, but one, perhaps, more easily grasped, is that of the manner in which communication is brought about through language.[74]

Chargaff soon came to object to and mock the information discourse and the theoretical representation of coding based on it. By 1962—at the peak of the euphoria around the biochemical "cracking of the genetic code"—Chargaff would back off, probably because of his bitterness about the lack of acknowledgment for his contributions and toward molecular biologists' growing dominance in biochemistry. He even confessed to his "share of guilt" in promoting "the concept of 'biological information' [which] raised its head and began to sport a multicolored beard which has become ever more luxurious despite numerous applications of Occam's razor." In time his caricatures of "information transfer," targeted at the brash young Turks of molecular biology, reintensified. He challenged young molecular biologists in an imaginary dialogue between the old and young scientific generation: "Is it not possible that the entire imposing terminological scaffold is nothing but a suitcase for the emperor's new clothes? Is not possible that there is no message, no messenger, that the entire question is asked, and therefore answered, wrongly?"[75]

Chargaff had communicated freely the results of his chemical analyses to Watson and Crick, who were then racing to determine DNA's structure in Cambridge, England, at the Unit for Molecular Structures of Biological Systems (later renamed Unit for Molecular Biology). Chargaff's rule (regularities of base ratios) immediately acquired for the two researchers a structural meaning, implying a complementary base-pairing scheme nested within the double helix interior, and a mechanism for genetic replication and mutation.[76] For Watson, 1953 marked a shift in representation. Prior to 1953, his papers had referred to replication and mutation in terms of genetic transfer, genetic specificity, and genetic continuity.[77] But in anticipation of the potential contributions of cybernetics and information theory to biology, Watson and Crick consciously began to represent their findings through an information discourse. In a letter to *Nature* (just before their announcement of the double helix), Watson, together with geneticist Boris Ephrussi and physicists Urs Leopold and J. J. Weigle, suggested new terminology in bacterial genetics. Seeking to impose rhetorical order on the proliferating semantic confusion in the field (transformation, recombination, induction, transduction, etc.), they proposed to replace these uses with the term *inter-bacterial information*. "It does not imply," they argued, "necessarily the transfer of material substances, and recognizes the possible future importance of cybernetics at the bacterial level."[78]

Elaborating on the genetic implications of the newly minted DNA structure in their second article in *Nature*, Watson and Crick used a metaphor of information to call attention to the molecular specificities inherent in the diverse permutations of DNA's four nucleotide bases: "The phosphate-sugar backbone of our model is completely regular, but any sequence of the pairs of bases can fit into the structure. It follows that in a long molecule many different permutations are possible, and it therefore seems likely that the precise sequence of the bases is the code which carries the genetical information."[79] Placing no quotation marks around code or information, the authors thus did not signal the metaphoricities of these terms (against which communication theorists would caution throughout the 1950s). With the new representations, the authors invested DNA—Delbrück's "prime mover" —with the cultural power of a new semiotics, then reshaping representations of the animate, inanimate, and the social. "The code," not "a code"—the preexisting *logos* preceding its experimental warrant—governed that emergent communication system by supposedly transferring genetic information unilaterally from the DNA master molecule to its now subordinate protein recipient.

THE RAPHAEL TAPESTRY OF ERWIN SCHRÖDINGER

The idea of a biological code did not arise out of the newly elucidated DNA structure. Indeed, the idea was not entirely new in 1953. The association of protein specificities with permutations of amino acids dates to the beginning of the century, while their association with a kind of cipher dates to the 1930s. Even representations of genetic specificity in terms of combinatorial properties that conveyed information through a code did not originate with Watson and Crick. Instead, such representations of heredity, still well within the protein paradigm, emerged in the late 1940s with the rise of cybernetics and information theory. Nevertheless, in retrospect, the representations have been commonly attributed to the singular impact of Erwin Schrödinger's little book, *What Is Life?* (1944), through which his historical role was reinvented. As Stent proclaimed in his classic textbook on molecular genetics (1970), "In that book, Schrödinger heralded the dawn of a new epoch in biological research."[80] "[Schrödinger's] ambition and interest were limited to a single problem: the physical basis of genetic information," pronounced François Jacob in 1970, reinforcing the canonization process.[81] (Schrödinger's generative role in molecular biology has been documented by several scholars, nearly all of whom serve to buttress this "founding-father" narrative.) Of all the insights and foresights attributed to Schrödinger, the crowning achievement bestowed upon him is that of progenitor of the genetic code and information prophet of molecular biology. As such, Schrödinger presumably supplied a new discourse for biology and new semiotics for representing heredity and life.[82]

FIGURE 2. Erwin Schrödinger, 1930s. Courtesy of the Emilio Segrè Visual Archive, American Institute of Physics.

Undoubtedly Schrödinger's book played an important role in molecular biology, and his essay contained perceptive—if by then already outdated—articulations of some of the principal problems in biology. But precisely what role did the book play and how did it become historically and historiographically significant? These questions have intrigued scholars, who have arrived at substantial agreement on the disciplinary and social impact of the book: the cognitive legitimacy it bestowed upon biology gave the book much of its legitimacy. By both aligning biology with and demarcating it from physics (even elevating it above physics), Schrödinger conveyed his blessings upon the postwar exodus of disillusioned physical scientists migrating into molecular biology. Biology, as Delbrück self-referentially noted, became a respectable playground for physical scientists. Or as Schrödinger's protégé Neville Symonds put it, "Biology stopped being a 'sissy' subject and came of age."[83]

Beyond such disciplinary and social effects, later decontextualization and reinvention of *What Is Life?* enhanced its prescience and durability. Schrödinger's musings about the protein gene and its coded design, for example, would be cleaved from his central preoccupation with the problem of biological organization, organismic development, thermodynamic order, and social order. Of course, it is true that Schrödinger himself evoked some of the conceptual constructs attributed to him. In fact, his two most celebrated ideas—the "aperiodic crystal" and the "code-script"—resulted from his pondering the enigma of how complex organisms develop from a tiny bundle of chromosomal material. But these ideas, even as he invoked them, remained firmly grounded in older discourses and representations of life. The concept of an *aperiodic crystal* derives from Schrödinger's analogy of life with crystal growth, a subject that had fascinated students of nature since the seventeenth century. Whereas the inanimate realm displayed merely repetitive, "periodic crystals," the chromosome fiber—the protean carrier of life—consisted of "an aperiodic crystal," Schrödinger observed. "The difference in structure is of the same kind as that between an ordinary wallpaper in which the same pattern is repeated again and again in regular periodicity and masterpiece of embroidery, say a Raphael tapestry, which shows no dull repetition, but an elaborate, coherent, meaningful design traced by the great master."[84] Schrödinger's poetic image of Raphael's tapestry was both grounded in the protein view of life and configured together with other 1930s idioms such as "pattern," "plan," and "design," which formed the discourse of organization. However, the imagery would be retrospectively interpreted by molecular biologists in the 1960s to convey a different meaning: it would refer to the combinatorial properties of DNA and to its informational properties.

Schrödinger's Raphaelesque pattern was four-dimensional, spatial as well as temporal, "meaning not only the structure and functioning of that organism . . . but the whole of its ontogenetic development."[85] In order to explain how such a grand design could unfold from a tiny speck of material, how "the egg would develop, under suitable conditions, into a black cock or into a speckled hen, into a fly or a maize plant, a rhododendron, a beetle, a mouse or a woman,"[86] Schrödinger invoked the idea of a miniature code-script. Analogized as a Morse-like code, it could account for biological complexity and specificity, for even a small number of molecular elements could generate a great diversity of effects through combinatorial rearrangements. "With the molecular picture of the [protein] gene," Schrödinger promised, "It is no longer inconceivable that the miniature code should precisely correspond with a highly complicated and specified plan of development and should somehow contain the means to put it into operation."[87] It was a plan grounded in organismic memory, the *engram*. Note that Schrödinger's code-script was based on permutations in proteins, not a *relation* (as codes technically are) between DNA and protein. It neither related one system of symbols (nucleic acids) to another (amino acids) as did genetic codes after 1953, nor, most importantly, did it claim to transfer information.

From a diachronic point of view, Schrödinger's scriptural representation of heredity belonged to the age-old tradition of textualizing nature by tacitly expressing Western tradition's representations of nature as cipher, though he made no connection to the idea of alphabet. From a synchronic perspective, Schrödinger's semiotic repertoire, including the telegraphic imagery of a Morse-like code, belonged to an older biological discourse of organization (and the philosophical debates of the interwar era), to a different circulation of meaning.[88] These and related statements would later be lifted out of their temporal and intellectual context, as Delbrück playfully did with Aristotle's generation. They would become the stuff of prophecy, but only as re-representations of what was articulated and what was silenced.

The notion of a cellular repository of organismic memory (engram) belongs to the nineteenth century. Indeed, Schrödinger's fascination with these ideas dates to his student years (1906–10) at the University of Vienna—the "golden age" he would always long for. There, together with his closest friend, biologist Franz Frimmel, he luxuriated in the influential book by Richard Semon, *Die Mneme als erhaltendes Prinzip* (*The Mneme as Conservative Principle*, 1904). A vibrant Germanic patchwork of Haeckelian, Lamarckian, and Darwinian themes, the book explained the enigma of material organization and psychic existence through the concept of *mneme*, a sort of cellular memory record governing all biological phenomena. As the sum of all organismic *engrams*, that is, imprints of innate and acquired stimuli, the *mneme* was believed to control heredity, differentiation, regeneration, development, instinct, behavior, movement, and consciousness. According to Schrödinger's biographer, Walter Moore, the book had a lasting impact upon Schrödinger because it helped shape his philosophical inquiries into the nature of life.[89]

Onto these fin-de-siècle preoccupations Schrödinger wove ideas (some of them outdated by 1944) from Mendelian and radiation genetics, quandaries and conundrums of thermodynamics, statistical and quantum mechanics, and Spenglerian gloom of cultural decline and social decay. Amid the dislocations of the 1930s and through communications with Niels Bohr, Pascual Jordan, Frederick G. Donnan, Max Born, and Max Delbrück, Schrödinger reflected, lectured, and wrote on the thermodynamics of inanimate versus living matter, the instability of human history and the stability of genes. History, he pronounced as if peering into the darkness beyond the twilight of an epoch, is the outcome of statistical fluctuations that cannot be altered at will.[90]

The discourse of organization, viewing the cell as a utopian society, provided Schrödinger as well as his predecessors (e.g., Rudolph Virchow, Emil Dubois-Reymond) and contemporaries (e.g., Walter Cannon, Paul Weiss) with an escape from entropic chaos.

> I like to compare the "cell-state" with a society which is organized on the principle that every official, every civil-servant, every person who has any duty at all within that organization is, at least in principle, given the same universal training and is so well informed about the plan of the whole, that every clerk could, in principle take over the duties of the prime-minister, every police-man that of a chief-surgeon, etc., etc. Thus I am not at all surprised, that one out of the two—or one out of the four cells should be just as capable of forming the whole individual when called upon to do so. And I am not so very astonished that in Spemann's experiments, the liver cells were able to produce something like an eye, when, by some odd treatment they were given the impression that this was the moment for them to perform their duty.[91]

Like the organismic plan, the plan of the body politic was determined by a finite number of universal principles, which under suitable conditions could generate the diversities and specificities of animate life. Human experience, often conveyed through metaphors, provides the structures for scientific reasoning. In this way, science naturalizes human experience.

These ideas, supplemented by a few additional sources, formed the backbone of Schrödinger's obligatory public lecture series he delivered in his capacity as director at the Institute of Advanced Studies in Dublin in 1943. Taken together, the extensively prepared lectures, which reveal only minor departures from earlier ideas, constituted *What Is Life?*[92] (Professor Frederick Donnan arranged for the book's publication with Cambridge University Press.) Among Schrödinger's vast sources, physiologist Charles Sherrington's lectures in the 1930s were particularly important. Like Haldane, Bergmann, Niemann, Wrinch, and others, Sherrington had tracked down biological specificity to the different possible arrangements (not relations) of amino acids in the chromosomal proteins, surmising, "With even only thirty amino-acids to ring the changes, different proteins are possible to a number requiring twenty-three *ciphers* after the third figure."[93] More metaphors of nature's hidden writings, ciphers, and decipherability. These newer semiotic tools helped refurbish the *mneme* into the more serviceable idiom of code-script as the "all penetrating mind" guiding the development of organisms.[94]

Schrödinger's interest in the gene was secondary and reflected his intrigue with the process of development, which, in turn, exemplified the conundrums of stability and change as subsets of his main preoccupation with the problem of order: he asked, How can the laws of thermodynamics account for the emergence and maintenance of order? The discussion of the gene was a fortuitous by-product of his concern with the problem of entropy as the key to explicating the problem of animate organization.[95] From this vantage point and with his expertise in statistical mechanics, Schrödinger formulated his much celebrated (albeit problematic) concept of *negative entropy* as the hallmark of life: $-(\text{entropy}) = k \log (I/D)$, where I/D is a measure of disorder and k is the Boltzmann constant. Schrödinger argued tautologically that organization was maintained by extracting "order" from the environment: the living organism delayed its decay into thermodynamic equilibrium (death) by feeding upon negative entropy. Chromosomes merely epitomized this property. He wrote, "An organism's astonishing gift of concentrating a 'stream of order' on itself and thus escaping the decay into atomic chaos—of 'drinking orderliness' from a suitable environment—seems to be connected with the presence of the 'aperiodic solids,' the chromosome molecules, which doubtless represent the highest degree of well-ordered atomic association we know of."[96]

This eloquent account of how negative entropy maintained organismic design was not the prescient anticipation of the information concept, as it was later acclaimed to be. At the moment Schrödinger's book went to press, the war-born fields of cybernetics and information theory were already imparting new meanings to negative entropy by integrating it with old and new concepts in communication sciences. Through the works of Norbert Wiener and Claude Shannon, communication as control was being recast as information transmission—and information defined as negative entropy—thus sweeping Schrödinger's representations into the cybernetic groundswell.

The information watershed also resignified other articulations, such as the 1929 paper of the Hungarian émigré physicist Leo Szilard, in which he had proposed a solution to the problem of "Maxwell's demon": how, in a closed system, more ordered structures could arise with no energy compensation and still obey the second law of thermodynamics (i.e., in a closed system, energy is always degraded and entropy, or disorder, increases). In an 1871 thought experiment physicist James Clerk Maxwell had conjured a noncorporeal being, a demon, who could traffic energy against the concentration gradient (namely, separate swift and slow molecules in a two-chambered container, thereby creating the impossible situation of raising the temperature in one chamber). Szilard met the challenge by showing that the answer lay in the relation between entropy and the demon's use of "some kind of memory," or "intelligence." By mentally recording the energy traffic, the demon's entropy gain exactly compensated for the system's entropy loss; thinking generated entropy according to Szilard's scheme. But he never used the term *information* (that mathematical concept was just then coined at Bell Laboratories). In the late 1940s, after information became a scientific entity defined as negative entropy, Szilard's "rediscovered" paper, like Schrödinger's essay, was reinterpreted as the earliest demonstration of the connection between entropy and information. Due mainly to French émigré physicist Leon Brillouin, who in the early 1950s was forging a grand synthesis of information sciences, Szilard too (to whom Brillouin was introduced by Weaver) was ushered into the information hall of fame.[97]

While Szilard welcomed these new attributions, Schrödinger rejected them. He did not see his ideas on negative entropy as part of the information discourse and considered the analogy between negative entropy and information unwarranted. His adoring colleague Brillouin, however, kept Schrödinger informed about the developments in information theory. Schrödinger, in a long letter to Brillouin, explained his objections and his position on the subject.

> I fully acknowledge the interesting *analogy* between information and 'negentropy,' but I considered it inadequate to identify the two. The information attained by measuring the length of a meter-rod with an accuracy of 10^{-3} mm corresponds to a negentropy of 6 k log 10, which is thermodynamically entirely negligible. . . . It is difficult to assess the negentropy represented by the *organization* of a steam-engine or a cat or an oak-tree, or the body of Max Born. But I believe that in all these cases there is a great discrepancy between the insignificant thermodynamic value of the negentropy and the significance these organizations have for us.[98]

Schrödinger reasoned within the older discourse of organization. He did not see his broader concept of negative entropy as equivalent to the narrower technical construct of information and rejected the notion of information as a valid representation of organization in the animal or machine.

My laborious examination and retracing of these scientific articulations are neither motivated by the quest for precursors or roots (invariably leading to Descartes or Aristotle) nor performed as a service to some diachronicity, the timelessness of "unit ideas." Rather, this backward glance underscores the impact of discursive ruptures and the roles of scientific and cultural synchronicities.[99] As a scientific-literary genre, Schrödinger's book bears its own historicity. It must be situated within the cognitive and material space of representations of the 1930s, the space defined by the intersection of the physical, biological, and social discourses of organization: the steam engine, the thermodynamic system, the living body, and the body politic. Wiener grasped the import of these discursive formations when noting that organismic representations were shifting from the material and energetic to the informational (a transition later underscored by Crick's Central Dogma). In analogizing cellular organization with social order, Schrödinger—like his predecessors and contemporaries—perpetuated the resonances of science and culture of the modernist era.

Only in retrospect do all the pieces appear to have been nearly in place for the grand synthesis—gene structure, function, coding, and information—attributed to Schrödinger. But he did not think of DNA or information; in fact, epochal distance separated his voice from its distorted echoes. Within the new space of representations defined by the emergent communication technosciences of the late 1940s, the mathematical formulation of negative entropy came to be identified with information; Schrödinger's four-dimensional Raphaelesque pattern was digitalized and collapsed into a one-dimensional Boolean message inscribed on a magnetic tape. His quaint code-script—and the few attempts in the subsequent decade to explain the physical nature of genetic specificity—were resignified in the 1960s within the information discourse and its new biosemiotics of coding (neither Crick nor Gamow seem to have cited Schrödinger in the 1950s). Early attempts by chemists Kurt G. Stern, Cyril N. Hinshelwood, and Alexander L. Dounce to explicate the physico-chemical mechanisms governing that specificity were also retrospectively reinterpreted as "decoding" and incorporated into the reinvented history of founders' narrative.[100]

PROTOCODES?

A number of available metaphors existed for visualizing the logic governing aperiodic patterns in genes. Since the early nineteenth century, more than a century before the notion of a stored program, the silk-weaving industry had been using punched cards for controlling complex patterns woven into cloth. By the mid-nineteenth century the Morse code—the alphabetical system of signals comprised of dots, dashes, and spaces, which guided Schrödinger's imagination—was in standard use in telegraphy. And the image (ca. 1904) of needles vibrating in time to sound waves and impressing their wavering tracks on gramophone records (utilizing "negative" patterns for copying) served as heuristic for genetic replication (e.g., John B. S. Haldane). But hidden below the surface of such analogies remained, unexplained, the imperatives of material structures and processes of the gene.

After the appearance of Schrödinger's book and before 1953, at least three scientists tried to explain the specificities inherent in the combinatorial properties of genes, each with somewhat different concerns. Kurt G. Stern at Brooklyn Polytechnic Institute was probably the first to act on and re-represent Schrödinger's notion of a code. Stern had distinguished himself in the 1930s for his theoretical and technical contributions to the physical chemistry of proteins.[101] But as evidence for nucleic acids' role in heredity mounted, he began pondering their underlying physico-chemical mechanisms. In 1946, before the acknowledged demise of the tetranucleotide hypothesis and the ascent of Chargaff's rule, Stern found it necessary

> to search for a principle of variation which does not affect the chemical stoichiometry of the nucleoprotein molecules and yet allows for a number of permutations sufficient to account for all genic constitutions (genotypes) which have been described in the past and which may be encountered in the future. . . . If one admits the possibility that the bases may vary in their sequence or orientation with reference to backbone, this would constitute a principle of *modulation* which does not affect the over-all stoichiometry of composition of the nucleoprotein chain, and which is fully capable of accounting for all possible genotypes.[102]

Unlike Schrödinger, who abstracted ideas from permutations in proteins, Stern reasoned within the framework of the nucleoprotein gene and fixed his gaze exclusively on nucleic acids.

Stern's proposed structure showed how "modulated" nucleic acid chains could incorporate many different "gene codes." But these inscriptions hardly related to codes of the 1950s, because they did not address the relational system of nucleic acids and proteins. And unlike Schrödinger's scriptural representation of a Morse-like code, Stern thought in concrete physical terms, analogizing the "coding" process to the material process of sound recording, where "Genes are modulations of a 'neutral' nucleoprotein structure, similar to the modulations impressed by audio frequency signals on a carrier wave radio transmission; or to the modulations impressed on a smooth surface by the stylus of a sound recorder."[103] His analogies were pictorial (see Fig. 3).

The appended pictures displayed photomicrographs of sound tracks, where the unmodulated grooves, or grooves modulated at a constant frequency, corresponded to "neutral" (i.e., "uncoded") nucleoprotein configuration, while the speech modulated tracks to "genic modulations." Deploying neither concepts of information nor formalisms of language and alphabet,

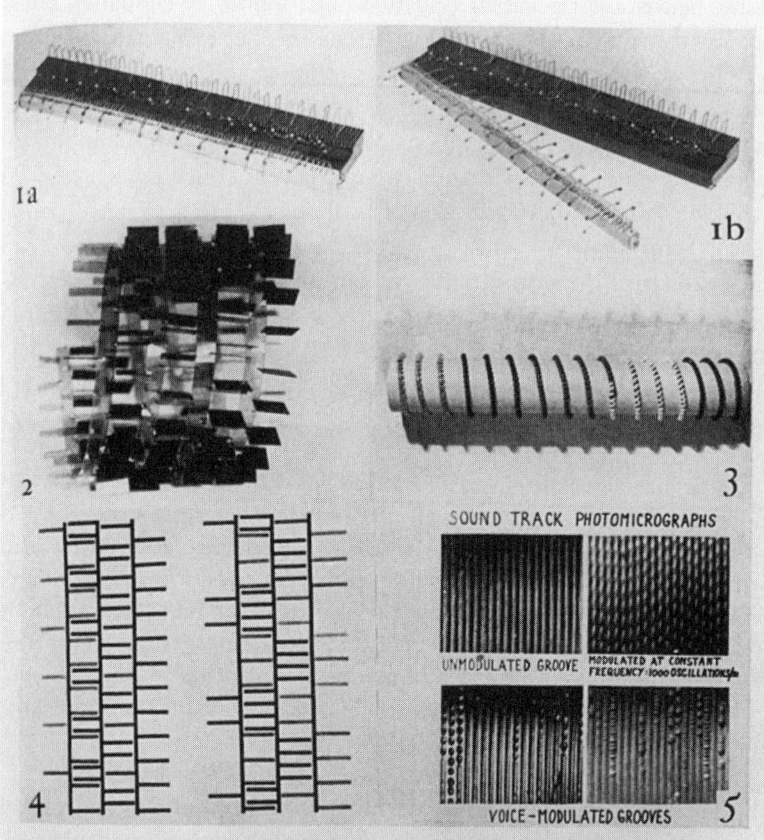

FIG. 1a. Model of a section of a nucleoprotein molecule, based chiefly on Astbury and Bell's x-ray diffraction data[2] for a synthetic thymonucleic acid-albumin complex. The amino acid residues of the polypeptide chain (in foreground) are projecting alternately right and left from the peptide backbone (Neurath[28]). The planes of the purine and pyrimidine bases of the polynucleotide chain (in background) are parallel to each other and have a spacing identical with that of the successive amino acid residues of the peptide chain (3.3 A). The heterocyclic bases are linked together by phosphate ester bonds (dark beads) between the sugar residues. In this arrangement, at least every other phosphate radical is available for binding a sodium atom (light beads). The bond between the basic side chains of the polypeptide (arginine residues) and the acidic phosphate ester groups is of a salt-like or electrostatic character.

FIG. 1b. Dissociation of the protein from the nucleic acid moiety of the complex is indicated by spatial separation. This dissociation is caused by the use of high-ionic strength solvents, e.g., *M* NaCl.

FIG. 2. Model of a nucleoprotein helix, formed by interlacing polynucleotide and polypeptide threads. The purine and pyrimidine bases of the nucleic acid chains (2nd and 4th turns) are represented by the black, large rectangles, the amino acid residues of the protein chains (1st and 3rd turns) by the light, narrow rectangles. The model is designed to illustrate the bilaterally functional character of both types of linear polymer molecules and their ability to form stable, condensed structures by coiling.

FIG. 3. Model of a chromosome, showing a more or less evenly spaced coiled nucleoprotein helix, embedded in or attached to a matrix. The differently shaded portions of the helix represent different genes or genic modulations of the chemically identical nucleoprotein spiral (see text).

FIG. 4. Hypothetical nucleoprotein lattices composed of interlaced polynucleotide and polypeptide chains. The two structures shown here are *chemically* identical but differ with regard to the *modulation* of the side chains: In the structure, illustrated at the left, the adjacent purine and pyrimidine base *pairs* of the nucleic acid chain (central strand) project alternately right and left from the phosphate ester backbone. In the structure on the right, blocks of *four* bases project to the same side of the backbone. The possible function of the polypeptide chains (outer strands) as locks of the specific genic configurations of the polynucleotide chains is also indicated.

FIG. 5. Photomicrographs of sound tracks, engraved into a wax surface by a recording stylus. They are intended to illustrate the principle of genic modulation of nucleoprotein spirals outlined in the text. (Courtesy of Dr. L. D. Norton, Director of Research, Dictaphone Corporation, Bridgeport, Conn.).

FIGURE 3. Kurt G. Stern, "Nucleoproteins and Gene Structure," *Yale Journal of Biology and Medicine* 19 (1947): 937–49. Stern's representations of "modulated" nucleic acid chains incorporating different "gene codes." Reprinted by permission.

Stern's representations aimed at visualizing the combinatorial effects of DNA's bases as a material process that accounted exclusively for its specificity. He did not address the problem of replication or protein synthesis and surely did not probe the correspondences between nucleic acids and proteins, as did genetic codes after the mid-1950s. His codes did not code a protean Raphaelesque tapestry but rather DNA's sound tracks.

The Oxford physical chemist Cyril N. Hinshelwood, on the other hand, tackled a different problem: autosynthesis. Having gained international acclaim in the 1930s for what became his Nobel-winning work (1956) on the kinetics of polymerization chain reactions, in the mid-1940s he began applying principles of chemical kinetics to mechanisms of bacterial replication. He probably knew Schrödinger at Oxford and had read his book,[104] but their concerns diverged. Reasoning within a nucleoprotein framework, Hinshelwood focused not on the protein code-script, which so intrigued Schrödinger, but on the dialectical reactions controlling the mutual coordination and synthesis of proteins and nucleic acids.

> In the synthesis of protein, the nucleic acid, by process analogous to crystallization, guides the order in which the various amino-acids are laid down; in the formation of nucleic acid the converse holds, the protein molecule governing the order in which the different nucleotides are arranged. . . . This suggests some sort of *correspondence* between the units in the two kinds of polymer. In a protein, about 23 different amino-acids may occur, whereas in a nucleic acid only 5 basic units are found [including DNA's and RNA's bases]. . . . Clearly there cannot be one-to-one correspondence between the position of an individual amino-acid in the protein part of the nucleoprotein and the position of an individual nucleotide in the nucleic acid part [my emphasis].

Correspondence, not code. He solved the problem by assuming that a newly synthesized amino acid chain was wedged between two adjacent nucleotide units at the surface of a nucleic acid polymer. Twenty-five [5^2] different internucleotide arrangements were then possible; just the right order of correspondence with the different possibilities in a protein chain, he argued.[105] Thus, within the nucleoprotein view of heredity—a compromise position between the protein monopoly and nucleic acids' embryonic ascent—Hinshelwood articulated not a genetic code, a term he never used, but rather a physical relation between (not function of) the two substances determining autosynthesis, a relation devoid of directionality, scriptural signification, or semantic import.

Like most biochemists and biologists, Alexander L. Dounce—often credited for his prescient contributions to what became the coding problem—either did not read Schrödinger's book or found it irrelevant. Nor was he tempted initially by the possibilities of the concept of information.[106] What did intrigue him was the much debated question about the biochemical action of templates in protein synthesis. That preoccupation had stayed with him since his Ph.D. oral examination at Cornell about a decade earlier when the Nobel laureate enzymologist John B. Sumner challenged him with the problem of protein synthesis: how to explain the paradox of infinite regress (proteins synthesized by proteinaceous enzymes, synthesized off other protein templates, and so ad infinitum).[107] By the time Dounce, then a professor of biochemistry at the University of Rochester Medical School, wrote his often-cited 1952 article, "Duplicating Mechanisms for Peptide Chain and Nucleic Acid Synthesis," the spell of the protein circularity seemed to be almost broken by nucleotide templates.[108] Dounce's goal (and Sumner's mandate) became to explain the coordination between nucleotide templates and the stringing together of amino acids in peptide synthesis.

Hypothesizing that there were at least as many specific nucleic acid combinations in the cell as there were specific peptide chain arrangements and postulating that specific arrangements of amino acid residues corresponded to arrangements of nucleotide residues, he proposed a detailed biochemical mechanism for protein synthesis. Without invoking metaphors of coding, information transfer, or even numerological logic, Dounce reasoned that if a particular base, say G, specified the selection of a particular amino acid, then that chemical reaction would be necessarily influenced by G's linkage to the two neighboring bases, for example, A and C. Hence, he reasoned, a minimum of three bases specified peptide synthesis. He arrived at what would later be known as a "triplet code" strictly through biochemical, rather than mathematical, considerations. His calculations confirmed that if three nucleotide bases influenced the attachment of an amino acid, there would be forty different directional combinations of such triads. Even if direction did not determine specificity, (e.g., C-A-G being equivalent to G-A-C) there would be more than enough triads to account for all known amino acids in proteins. He also visualized a "peeling off" transamination mechanism that would cleave newly synthesized peptide chains from the RNA template and free the nucleic acids for subsequent protein syntheses.[109] (See Fig. 4.)

Apparently conscious of the timeliness and priority of his ideas, Dounce did not let them remain buried in the specialized journal *Enzymologia*. He reiterated and elaborated on them a few months later (before Watson and Crick's publication of DNA structure) in a note to *Science*. His "Nucleic Acid Template Hypothesis" suggested that proteins could be synthesized from a DNA template, not directly, but via an RNA intermediate, an idea just beginning to gain currency among biochemists. Citing the growing evidence for the high turnover of cytoplasmic RNA and studies of viruses and plasma

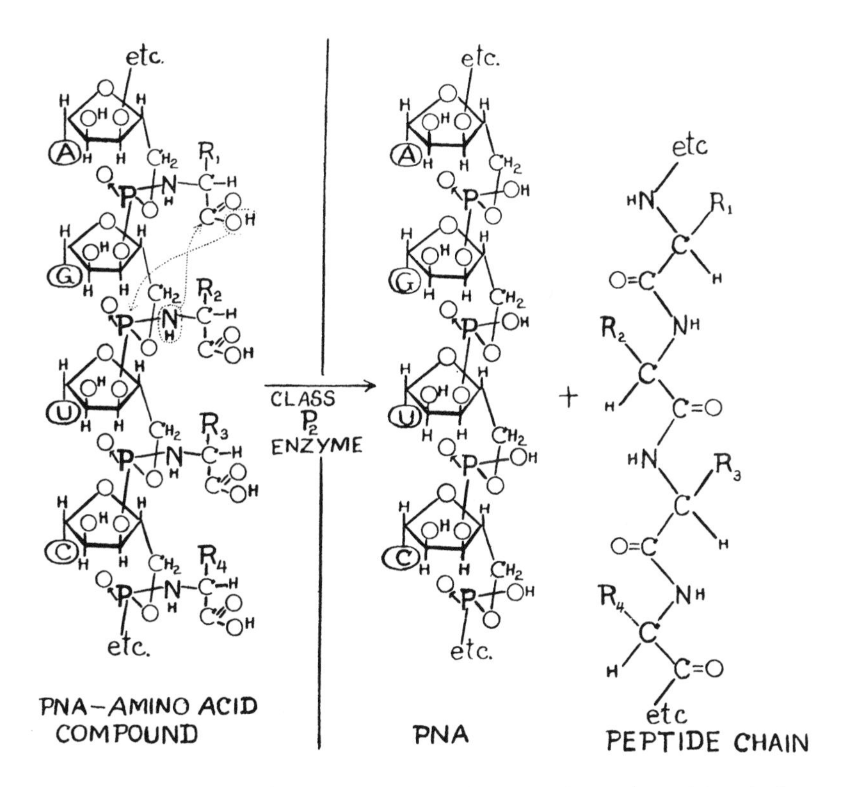

FIGURE 4. *Enzymologia* 15 (1952): 254. Dounce's scheme of protein synthesis via an RNA intermediate. Reprinted by permission.

genes, Dounce observed that these findings "could be explained by the template sequence deoxyribonucleic acid (DNA) → ribonucleic acid (RNA) → protein," a scheme that after 1958 would be ceremoniously enunciated in Crick's Central Dogma.[110] In any case, what mattered to Dounce was not to "crack *the* code"—a term he did not use—but to establish priority and stimulate interest in the template hypothesis. He stated, "The template hypothesis constructed by me was published mainly in the hope of promoting thinking and experimentation of a detailed nature concerned with the possibility that nucleic acids may participate in protein synthesis. It will be gratifying if this aim is achieved."[111] In 1955, Dounce, by then a member of the decoding group known as the RNA Tie Club, would recast the coding problem as information transfer and articulate his ideas through the information discourse.

"The code"—the hidden writing of the cryptogram inscribed in the chromosomes—did not yet exist in 1952. There was no coding problem. That icon, as invoked by Schrödinger, attempted to account for the stability and diversity of proteins based on the combinatorial effects of their building blocks, the twenty-some amino acids. When Stern referred to "gene codes" he imagined variations in the physical tracks impressed in the chromosomes by the permutations of the four bases in DNA with no relation to proteins. Neither picture addressed a *relation between two entities*, or two sets of symbols. Hinshelwood, on the other hand, pondered the correspondence between the syntheses of nucleic acids and proteins, without ever envisioning a code governing their nondirectional or *dialectical relations*. Dounce enunciated the nucleic acid template hypothesis, not the coding problem, to explicate the unidirectional (DNA to RNA to protein) coordination (not information flow) between the specific arrangements of the nucleic acid bases and the specificity of amino acid assembly. None of these supposed contributors of the history of the genetic code viewed their studies as the "coding problem." And neither Schrödinger nor Stern, Hinshelwood, or Dounce (before 1955) used the terms *information*, *program*, *instructions*, *alphabet*, *words*, *messages*, and *texts*. These were not yet in the lexical repertoire.

The picture would change by the mid-1950s. New codes—information codes—would come to bear on studies of heredity. With the emergence of new representations from the sciences of communication, electronic computing, guidance and control, and espionage technologies, the information discourse and its tropes would acquire currency and potency. Wiener's Cybernetics, Shannon's mathematical theory of communication, John von Neumann's automata studies, and Quastler's adaptation of these concepts to biological problems would furnish a new biosemiotics for genetic codes. Seen as both mediums and agents of information storage and transfer, codes as scriptural technologies relayed the messages of life's hidden writing in the form of a cryptogram through cells and cycles of life. These semiotics would be transported into studies of the genetic code by Gamow, the RNA Tie Club, and the physicists, chemists, mathematicians, cryptologists, computer analysts, and communication scientists intrigued by the coding problem. They would be formulated within new regimes of signification of cold-war culture and the industrial-military-academic complex.

CHAPTER THREE

Production of Discourse: Cybernetics, Information, Life

In 1947, amid the aftershocks of the atomic bombing of Hiroshima and Nagasaki and the spread of military power into American culture, MIT mathematician Norbert Wiener dispatched a note, "A Scientist Rebels," to the *Atlantic Monthly.* A jeremiad on the powers that be and the scientists' fall from grace, his pacifist manifesto declared his refusal to put his science at the service of weapons design.[1] An outpouring of congratulations followed. Among the accolades was a provocative response challenging Wiener's well-intentioned but naive stand. "To contribute to society with the realm of science, yet to actually repudiate one's role as a citizen of the world seems not only a remarkable stance, but one whose entire justification certainly needs more rigorous development," the writer (a scientist) pointed out, following with a penetrating critique of contemporary culture.[2]

He shared Wiener's dark view that "wartime technology, which has furnished from small bits of scientific truth great warehouses of potential threat and arsenals of destruction has much to thank scientists for—their intellectual prostitution has been significant." But he did not applaud Wiener's tokenism, countering, "Is it not a futile gesture to attempt to halt this onslaught in the particular manner which you propose, Dr. Wiener?"[3]

> Not only have scientists become intellectual prostitutes to militarism, but the rest of mankind has done a note-worthy job of keeping step in analogous ways. The tempo of life has skipped merrily over a Mozart sonata which illuminated the days of Galileo [*sic*]; it has trod heavily through the degraded ruins of the World Wars with a Tchaikovskian air of inexorable finality; and now, one may discern the prelude to a theme which will sublime before our very eyes to a note of Wagnerian clamor and Shostokovichian dissonance. Such a parallel as this could be drawn from any of the myriads of tiny paths of mortal endeavour and the result is relentlessly the same. Why then, Dr. Wiener, do you lament the downfall of a race of scientists? Is it not more obvious that this downfall extends to every nook and cranny of the pettiness and arrogance of mankind as a whole?

FIGURE 5. Norbert Wiener at the blackboard, MIT, 1950s. Courtesy of the MIT Museum.

His lament of the wholesale militarization of Western culture, that is, the seepage of national security into every nook and cranny of social life, was at once retrospective and prospective: an indictment of scientists' roles in World War II and a prescient critique of America's emergent identity as a military power at the threshold of the cold war. That perspective informs this chapter as it traces the formation of a politico-scientific hegemony and its multivalent discourses: namely, the new structures of science, as configured by the military, industry, and university in the 1940s and the emergence of a discourse of cybernetics and information within these new regimes of signification.[4]

World War II was a watershed in American science; hardly any aspect of science, including the cognitive, technological, disciplinary, and organizational, remained unaffected. The war created new institutions and reconfigured older relations, recasting them in new forms.[5] The Office of Scientific Research and Development (OSRD) conducted wartime military research. OSRD, endowed with resources and power beyond any previous coalition of science, industry, and the military, was headed by Vannevar Bush, former MIT dean, cofounder of Raytheon, director of AT&T, and developer of analog computers for military ballistics. By 1944, under his leadership, the government was spending $700 million per year on research—ten times the amount in 1938. In 1939 only 1 percent of the total research expenditure of the Bell Laboratories of AT&T was for government contract work; by 1943 it had risen to 83 percent. By the end of the war, OSRD had spent $450 million on weapons research and development and played a key role in most of the war's technical feats: radar, the proximity fuse, solid fuel rockets, and the atomic bomb. OSRD also sustained the less visible but crucial biomedical technologies linked to neuropsychology, aviation medicine, drugs, vaccines, blood substitutes, and germ warfare. Four elite universities—MIT, Caltech, Columbia, and Harvard—and three leading industries—Bell Labs, General Motors, and General Electric—spearheaded the war mobilization. In procuring the bulk of contracts, they set the pattern for the trajectory of science and technology in the postwar era.[6]

As scholars have observed, wartime experience supplied an academic and industrial elite with new visions of postwar technoscientific order. Scientists had gotten used to wartime efficiency, priority considerations of projects, and massive support for research. Reluctant to relinquish the spoils of war, in 1944 they already began to look for ways to support their programs in the style to which they had become accustomed.[7] Perhaps no document better captured these expectations than Vannevar Bush's *Science: The Endless Frontier* (1945), calling for a strong federally funded science geared toward the nation's political and economic goals. Chief architect of wartime science policy, Bush became a visionary for a new technocratic world order in which science, in particular physical science wedded to engineering and biological science focused on medicine, would justify its own endless expansion.[8]

This expansion of science was enabled by a massive buildup of a military technology and bureaucracy in response to actual or perceived global threats. Prodded by Harry Truman, amid the conflicts over Poland and Iran in 1946, Winston Churchill proclaimed the peril to Christian civilization as an iron curtain descended across the European continent from the Baltic to the Adriatic. Placing the United States at the pinnacle of Western power, Churchill thus supplied a potent metaphor for a bipolar world riveted by American and Soviet spheres of domination.[9] The Marshall Plan—a softer version of the Truman Doctrine's containment—was announced in mid-1947 and launched in 1948. Governed by a logic that a stable free-market Europe would be resistant to the virulence of communism, the plan, notably reinforced by ubiquitous military installations, financed a massive recovery program of Western Europe ($12.4 billion, or 1.2 percent of the United States Gross National Product).[10] By 1949, Mao Zedong won control of mainland China ("the loss of China") and the Soviet Union ran its first atomic bomb test following failed attempts at international control of atomic energy. The cold war, national security, and the military moved into center stage of international life. The anticommunist crusade abroad was matched by the fervor on the home front: massive screening programs of federal employees, the institution of the loyalty oath, and the rise of McCarthyism. Although popular pressure forced cutbacks in military forces from their wartime peak, Truman retained the largest peacetime military establishment in U.S. history.[11]

Unwilling to relinquish its central place in the laboratory and the academy, the still-fluid military establishment lobbied successfully to prolong the wartime pattern of cooperative research. Before 1947 the army and the navy were separate services, each with its own cabinet-level representative. During World War II the Army Air Force (AAF) had achieved such a significant strategic role that it sought a place as a third service. In 1947 the National Defense Act elevated the air force to the status of an independent service. The National Security Act (1947) replaced the three separate military administrations with the Joint Chiefs of Staff, under a unified Department of Defense. In 1949 the air force joined the navy as a major patron of scientific research (mainly in the United States but also abroad), supporting basic and applied research in material, human resources, and medical fields; these encompassed mainly studies relevant to computer-based command, control, and communications systems, airplane and missile design, guidance systems, and industrial automation.[12] The sponsorship of these agencies, boosted by the vast resources of the military-controlled Atomic Energy Commission (the AEC, founded in 1946), spanned diverse fields in the physical, life, and social sciences. The AEC, in fact, became a principal supporter of genetics and cellular biology in the United States (with significant contributions to European biology). By 1949 the Department of Defense's research activities accounted for 65 percent of all research and development (R&D) expenditures, including more than 60 percent of federal grants to universities.[13] Budgets rose even more steeply in the 1950s. General Dwight Eisenhower generated elaborate designs to extend the partnership of science, industry, and the military into the postwar era.[14]

As Paul Forman has demonstrated in his study of postwar physics and as Stuart Leslie has shown in his study of MIT and Stanford, instead of demobilizing the wartime academic laboratories, the military establishment either picked them up under new contractual arrangements or kept them afloat until more permanent arrangements could be made. Studies of the trajectories of other research laboratories have reinforced these findings.[15] These revitalized prewar institutions were joined by others that mushroomed in the immediate postwar years: the AEC-sponsored national laboratories, especially Argonne, Brookhaven, and Oak Ridge, with a strong focus on biological and medical research. With the exception of a minor lull during the stringencies of the Truman administration, by 1950 defense R&D budgets reached wartime levels.[16]

These institutional conjunctions of science and the military influenced both partners, altering in both the military and in science modus vivandi, policies, and attitude toward research. What was the impact of these new modes of patronage on the organization and content of knowledge? How did this massive and systematic sponsorship of the military shape scientific research, especially in the universities? As several scholars have argued persuasively, military patronage was not simply a resource external to the scientists' projects; patronage power did not reside outside the organization and practice of science. Rather the web of military institutions sponsoring scientific research defined the conditions of possibility for the production of particular forms of knowledge. As Ian Hacking has suggested, beyond overt gross military manifestations, such as federal, academic, and industrial budgets, number of fields and scientists funded, military power has extended into the world of the mind.[17] The capillary workings of power permeated the material and symbolic space in which information, cybernetics, and life were constituted; in turn, these new modes of signification served to naturalize and sustain the circulation of power.

This chapter examines the emergence, scope, and limits, from the 1940s to the mid-1950s, of this new field of representation. Some of the terms and ideas in cybernetics and information theory date back at least to the nineteenth century, but they were reconstituted during the postwar period. The word *information*, for example, was in common usage for centuries, meaning *in-forming*: formation of mind and character, instruction, and communicated knowledge. As such it was deployed in a generic sense in physics, mathematical logic, electrical engineering, psychology, and biology since the turn of the century. But as Bill Aspray has convincingly argued, it was only in the decade following World War II that information became for the first time a physical parameter and a precisely defined concept amenable to scientific study.[18] As such, it was wedged at the intersection of several lines of military-sponsored research on both machines and living organisms: the mathematical theory of communication, modeling of the brain, artificial intelligence, command and control systems, cybernetics, automata theory, genetics, and behaviorism. In the late 1940s it was this intersection that defined a new discursive and semiotic space in which the meanings of cybernetics and information could be constituted.

The production of the information discourse was a complex development embedded within the postwar industrial-military-academic complex. While it cannot be reduced purely to the influences of powerful individuals, it likewise cannot be understood without them. This chapter follows the emergence of cybernetics by focusing primarily on four subjects: Norbert Wiener and cybernetics; the mathematical theory of communication and Claude Shannon; computing and automata studies as articulated by John von Neumann; and Henry Quastler's attempts to apply some of these concepts to biology, recasting it as an information science. Historically these efforts to represent heredity and life in terms of information and communication were not grounded in DNA genetics, nor did they follow the elucidation of the architecture of the double helix in 1953. These discursive practices were instead deployed to account for processes of heredity still within the protein paradigm of the gene in the late 1940s and were linked to material and social practices that defined the interlocking space of representations and interventions in life science.[19]

THE ORGANISM AS THE WORD: WIENER AND THE BIRTH OF THE CYBORG

Norbert Wiener (1896–1964) had known for a long time that in case of a national emergency his function would be determined by two things: computing machines and electrical networks. In 1940 Wiener began working with Vannevar Bush on computing machinery for solving partial differential equations. Out of this collaboration he crafted a vision for future computing machines with features such as numerical instead of analog modes of computing; electronic tubes instead of mechanical relays for switching; binary instead of decimal representation; built-in logical facilities which obviated human manipulation of data; and a memory capable of rapid storage, recall, and erasure. According to Wiener, these ideas sprung from his general interest in the analogy between the nervous system and electronic computing, though the technological mandate of war shaped the specific form these ideas took.[20]

In November 1942 Vannevar Bush established the Applied Mathematics Panel (AMP) within the National Defense Research Committee (NDRC) of the OSRD to coordinate the services of mathematicians and act as a clearinghouse for their war-related projects. Under the leadership of Warren Weaver, director of the Rockefeller Foundation's Natural Sciences Division since 1932 and now chief of the fire control analysis section D-2, the AMP grew in scope and size. Beyond its concerns with analyses of bombing accuracy, AMP activities included studies of rocket accuracy and numerous problems relating to gunfire. These projects affected not only weapons design but also the development of statistics, numerical analysis and computation, theory of shock waves, command and control systems, and the emerging field of operations research. By the end of the war, the AMP had spent nearly $3 million and had coordinated the efforts of about three hundred researchers and consultants for both the U.S. and British services. Among them were world-class mathematicians such as John von Neumann, Richard Courant, Jerzy Neyman, Garrett Birkhoff, Harold Hotelling, Oswald Veblen, and Norbert Wiener.[21]

Like some of his illustrious cohorts—"the dreamy moonchildren, the prima donnas, the a-social geniuses," as Weaver put it—Wiener was not an ideal candidate for wartime team projects. Weaver bent over backward to apply Wiener's talents to unglamorous cooperative tasks, to no avail. Affiliated with the Statistical Research Group and Operational Research Laboratory at Columbia, Wiener was also part of an interdisciplinary team at MIT studying the mathematical aspects of guidance and control of antiaircraft fire, working on the design of fire control apparatus for antiaircraft guns with MIT engineer Julian Bigelow. According to Bigelow, Wiener was just not interested in problem solving, "simply on the basis of utility, particularly if it lacks the qualities suggestive of an elegant, general, formal solution." Bigelow suggested that to get Wiener to comply Weaver should

> Get a stack of paper, pencils and erasers, hire a hotel room in N.Y., Wash. D.C., or wherever else convenient, send him [Wiener] an emergency telegram preferably to reach him at his house at 2:00 A.M. stressing an emergency of catastrophic consequences requiring his decision immediately; when he arrives rush him to the room and lock him in, with request for a written report to be printed and published immediately, then come around again in about 1/10th the time you think it will take him to read it, and your report will be ready for you.

Indeed, Weaver's most challenging task as AMP director was to manage the eccentricities of genius.[22]

Wiener, despite his difficult character, did produce outstanding results in areas that captured his scientific imagination, especially the problem of guns tracking the curving course of a plane and predicting its future position. This work, in collaboration with Bigelow, formed the beginnings of his statistical theory of prediction and the cornerstone of cybernetics. Wiener and Bigelow quickly reached the conclusion that any solution of the self-correcting tracking problem was predicated on the feedback principle, operating not only in the apparatus but also in the human operators of the gun and the plane. Wiener had long-standing interests in Walter Cannon's neurological work at Harvard Medical School. He now consulted with his Mexican friend at Cannon's laboratory, Arturo Rosenblueth, who was an expert in electronic apparatus and nerve and muscle physiology, about problems of oscillations in the nervous system. The analysis of muscle tremors reinforced Wiener's conceptualization of voluntary action around the principle of feedback control. It would be another five years before Wiener coined the term *cybernetics*, but the gestalt switch had already occurred. In defining gunnery processes as systems of communication and behavior, and a system in terms of both human and machine components, the conception of *cyborg* had emerged. The cybernetic organism—a heterogeneous construction, part living and part machine—germinated within the wartime academic-military matrix and matured within the national security practices of the cold war.[23]

In his characteristic ebullient style, Wiener conveyed excitement about his success at "the edge of some biological work" to his admiring friend, British geneticist John B. S. Haldane. "Fundamentally the matter is this," Wiener explained:

> Behaviorism as we all know is an established method of biological and psychological study but I have nowhere seen an adequate attempt to analyze the intrinsic possibilities of types of behavior. This has become necessary to me in connection with the design of apparatus to accomplish specific purposes in the way of the repetition and modification of time patterns. . . . The problem of examining the behavior of an instrument from this point of view is fundamental in communication engineering and in related fields where we often have to specify what the apparatus between four terminals in a box is to do. . . . We are quite clear on the fact that no behaviorist has ever really understood the possibilities of behavior.[24]

While geography and war impeded frequent exchanges between Wiener and Haldane, they helped to intensify the value of local communications. Wiener, whose interests and collegial relations in biology, linguistics, and philosophy dated back to the 1930s, now shared and refined his protocybernetic ideas through informal and quasi-formal interdisciplinary gatherings in Cambridge. These gatherings of what Gerald Holton has termed "the Vienna circle in exile"—the Unity of Science Movement devoted to nurturing Ernst Mach's positivist legacy in the United States—provided Wiener with an inspiring and critical audience that spanned the physical, biological, and social sciences. Besides Wiener, regular participants included P. W. Bridgman, W. V. Quine, Phillip Frank, Leon Brillouin, Karl Deutch, Gorgio de Santillana, Roman Jakobson, Gyorgy Kepes, Philippe LeCorbeiller, Wassily Leontief, George Uhlenbeck, Joseph Schumpeter, Laszlo Tisza, Henry Aiken, George D. Birkhoff, E. G. Boring, Talcott Parsons, B. F. Skinner, L. J. Henderson, and K. S. Lashley, many of whom, like Wiener, wove the contingencies of interdisciplinary war experience into the transcendent discourse of the unity of science.[25]

This dual mode of intervening and representing, of linking applied mathematics of guidance weapons to a new scientific epistemology was clearly displayed in Wiener's publications in 1943. That year he completed his top secret, highly technical report on control theory, "The Extrapolation, Interpolation, and Smoothing of Stationary Time Series with Engineering Applications," which was circulated to several research centers in the United States and Britain. "At the time it was felt that copies should have been distributed to the enemy so that they would have to devote such time to it and enable us to get on with winning the war," reminisced one electrical engineer about the obtuse mathematical language of stochastic processes.[26] Yet Wiener was able to convey the general epistemological import of this new control theory to a wider audience. In his famous joint paper entitled "Behavior, Purpose and Teleology," coauthored with Bigelow and Rosenblueth, Wiener articulated the new representation of control systems as a conjunction of physiological homeostasis, servomechanisms, and behavioral processes.[27]

The essay, which spoke to practitioners and patrons of the physical and biomedical sciences, sought to define the behavioristic study of natural events, to classify behavior, and to underscore the importance of the concept of purpose. Its point of departure was that "Given any object, relatively abstracted from its surroundings for study, the behavioristic approach consists in the examination of the output of the object and of the relations of this output to the input. By output is meant any change produced in the surroundings by the object. By input, conversely, is meant any event external to the object that modifies this object in any manner."[28] In the essay, Wiener distinguished purposeful from nonpurposeful systems and subdivided purposeful ones into feedback (teleological) and nonfeedback (nonteleological) systems. He further delineated positive feedback, where the fraction of the output that reenters the object has the same sign as the original input signal, from negative feedback, where the signals from the goal are used to restrict outputs. Predictive from nonpredictive negative feedback systems were also demarcated. The term *servomechanisms*, or self-correcting negative feedback, came to designate machines—living or inanimate—with intrinsic purposeful behavior.[29]

The method of studying organisms and machines was similar, Wiener reasoned. As in living organisms, servomechanistic machines involved a continuous negative feedback from the goal that modified and guided the behaving object. He concluded, "Whether they [organisms and machines] should always be the same may depend on whether or not there are one or more qualitatively distinct, unique characteristics present in one group and absent in the other. Such qualitative differences have not appeared so far."[30] A kind of manifesto for a new form of knowledge, this military-inspired theory was a cognitive implosion, conflating categories and hierarchies that until then had determined the analyses of behavior. Within the constant movement between binary oppositions (such as stimulus and response, input and output, purpose and outcome, organism and machine), each binary signified its opposite. The ambiguities, aporias, and circularities inherent in this construction would be challenged only after the war, as Wiener began to promote his servomechanistic vision among colleagues and patrons.

Even before the war ended, Wiener, like many of his academic colleagues, began planning for the opportunities in a postwar technocratic order. In 1944 he and John von Neumann of the Institute for Advanced Study at Princeton began to formulate their grand vision and to organize meetings linking the complex field of control and communication engineering to several areas in biomedicine.[31] Early in 1945 Wiener announced to Rosenblueth, who had by then returned to Mexico, plans to organize a society (tentatively named the Teleological Society) and a journal (perhaps to be entitled *Teleologia*), and to found a research center in "our new field" (yet unnamed). The bottleneck was the declassification of all the relevant war work, but resources appeared to be abundant.

> In this matter Moe [Henry Allen Moe, Rockefeller trustee and head of the Guggenheim Foundation] is giving his good will and expects to be able to help us with fellowships. We are also getting a good backing from Warren Weaver, and he has said to me that this is just the sort of thing that the Rockefeller should consider pushing. In addition McCulloch [Warren McCulloch, the Chicago neurophysiologist] and von Neumann are very slick organizers, and I have heard from von Neumann mysterious words concerning some thirty megabucks which is likely to be available for scientific research. von Neumann is quite confident that he can siphon some of it off. When this scheme really gets going, I for one will not be content unless we can bring you and Bigelow directly into it.[32]

Von Neumann agreed "that we ought to interest anybody and everybody and then see what happens. . . . The best way to get 'something' done is to propagandize everybody who is a reasonable potential support."[33] By September 1945, in addition to federal and institutional backing, the fellowship program in applied (biological) mathematics was in place. In 1946 the Josiah Macy Foundation began promoting the application of protocybernetics to the biological and social sciences, and funding from the Rockefeller Foundation was secured for Wiener's future collaboration with Rosenblueth.[34]

During the period of grant accountability the conceptual flaws and logical circularities in Wiener and Rosenblueth's biological project (grist to philosophers' mill in subsequent decades) came to the forefront.[35] Five years of Rockefeller support notwithstanding, on rereading "Behavior, Purpose and Teleology," Robert S. Morison, director of the Rockefeller Foundation's Medical Sciences Division, urged Wiener and Rosenblueth to address the ambiguities in their program. His criticisms centered on the flaws in the cybernetic project, especially as applied to biology. Several items troubled him: their reducing the distinction between teleological and nonteleological purposeful mechanisms to feedback; their failure to distinguish between feedback from various stages of the apparatus and reception of signals from a goal; and the murky relation between these concepts and their supposed technical correlates of statistical mechanics, Poisson distributions, and so on. A more significant problem, Morison noted, was their tendency to lump together equilibrium and goal-seeking mechanisms.[36] Indeed, as William Wimsatt later showed, it is not possible to define the concept of feedback solely in terms of the externally observable behavior of a purported feedback system, and all known criteria of analysis of self-regulating and goal-directed systems fail to distinguish between negative feedback system (servomechanism) and any open biochemical system tending to steady-state equilibrium.[37] Signifying homeostasis as negative feedback and then resignifying such servomechanisms as organismic homeostasis amounted to a circularity.

The incongruities and tautologies inherent in Wiener and Rosenblueth's conceptual scheme were not lost on keen observers. Weaver put his finger on the problem when he later criticized "Behavior, Purpose and Teleology." He complained to Wiener, "I want to read this article but so far I have not succeeded in getting beyond the first four paragraphs."

> At the outset you exclude any interest in the "specific structure and intrinsic organization" of the object but in the second paragraph you have to talk about events external to the object which "modify the object." What is the meaning of this phrase? Since we are not permitted, so to speak, to look inside the object, the only way in which we can tell whether or not it has been modified is to observe a modified aspect of its external behavior. But behavior is defined in terms of output and input. Thus behavior is defined in terms of behavior. What is wrong?[38]

Having followed the tautological construction of Wiener's heterogeneous phenomenology to its underlying premise, Weaver was led to the annulment of the project's rationale.

Neither Rosenblueth nor Wiener took these criticisms seriously. In fact, after a rather arrogant attempt to brush them aside, they offered cosmetic and terminological tokens of clarification in response to these serious epistemic challenges. Rosenblueth perhaps came more to the point when he suggested that any judgment of their program should be based more on their experimental accomplishments in nerve and muscle physiology and their forays into theoretical biology, and less on the admittedly important, but nevertheless secondary philosophical issues.[39] In any case, in light of Wiener's acknowledged genius and the enormous technoscientific potential involved, these philosophical incongruities hardly threatened the project's institutional support. Gazing into the future, Wiener was euphoric. In mid-1947 (with the book on cybernetics already in press) he stopped by, "completely unannounced as usual," Weaver's office on his way to Europe. "In something like ten minutes W. sprayed WW with a perfectly amazing and totally confusing assortment of half formed sentences, almost any one of which would sound perfectly insane if taken alone," Weaver wrote. He was convinced, however, that all of this—the salient concepts of Wiener's upcoming book—added up to something important and exciting.[40]

Cybernetics: or Control and Communication in the Animal and the Machine was published in 1948, almost simultaneously in France and the United States. "We have decided," Wiener explained, "to call the entire field of control and communication theory, whether in the machine or in the animal, by the name of cybernetics which we form from the Greek (κυβερνήτης) meaning 'steerman.'"[41] As it later turned out, the word had been coined by a Polish philosopher in the middle of the nineteenth century and belonged to the tradition of Comptian positivism;[42] but the novelty did not reside in the word but rather in its culturally specific meaning and discursive valence.

As a synthesis of Wiener's technical and mathematical work on control systems, the book aimed well beyond communication engineering. Cybernetics, as Wiener saw it, was a metadiscipline, something akin to what Michel Foucault would later term "episteme." For Wiener the evolution of thought and technique manifested itself in punctuated leaps: from a preoccupation with matter, to the study of energy, to the formulation of information. "If the seventeenth and early eighteenth centuries are the age of clocks, and the later eighteenth and the nineteenth centuries constitute the age of steam engines, the present time is the age of communication and control," he proclaimed. Defining epochs by technological markers—the Stone Age, Bronze Age, automobile age—was hardly historically novel. Wiener's argument, however, went deeper. He understood the dialectical relations between doing and knowing and saw technologies as coextensive with the forms of knowledge they engender.[43]

The book expounded a new form of technological epistemology. Its central notion was that problems of control and communication engineering were inseparable—communication and control being two faces of the same coin—and centered on the fundamental notion of the message, a discrete or continuous sequence of measurable events distributed in time. Control qua feedback was nothing but the sending of messages that affect behavior. And the measure of information content of a message was defined as negative entropy. Inspired by the potentialities of Schrödinger's formulation of organismic order in terms of negative entropy, Wiener transformed Schrödinger's statistical mechanical arguments into an information discourse encompassing all self-regulating systems. "Just as the amount of information in a system is a measure of its degree of organization, so the entropy of a system is a measure of its degree of disorganization; and one is simply the negative of the other," Wiener argued, generously crediting others, notably Claude Shannon, with the simultaneous formulation of these ideas.[44] (Recall that Schrödinger disagreed.)

But the conceptual and semiotic impact of cybernetics did not derive so much from its constitutive technical features—feedback, control, message, or information—as from their synchronic meaning, namely, from their particular configuration within postwar technoculture. As several scholars have documented, the knowledge and means of controlling technologies through feedback mechanisms reaches back to antiquity. The displays of automata and pneumatica in Heron's Alexendria, the float-valve regulators of ancient and medieval water clocks, temperature and pressure regulators in the seventeenth and eighteenth centuries, feedback control on mills and speed regulation of steam engines in the nineteenth century were all based on closed-loop control mechanisms.[45] Such mechanisms were not confined to hardware; they also served as software for social technologies. Philosophers and social theorists since the eighteenth century have visualized political, economic, and physiological stabilizations and correctives through models of closed cycles (reminding us of the diachronic and synchronic nature of metaphors).[46] By the early 1930s, moving beyond specific recreational gadgets, industrial and military technologies, and their attendant mathematical analyses, scientists in several countries had published theories of feedback control.[47]

Thus many features of Wiener's cybernetics were not new. The potency of his project, however, stemmed from the ways in which these terms—*feedback*, *control*, *message*, and *information*—became resignified. Configured together, they acquired new meanings within a new space of representation formed by the intersection of researches in the physical, biological, and social sciences. Within that space, control was abstracted and diffused: it was not a thing but a manifestation; not a mode of decision making but a process pervading the whole system.[48] Information and message, formerly deployed in a generic sense, not only became physical parameters but also gained new meanings through their discursive equivalence to control and through the erasure of differences between the animate and inanimate. By the late 1940s—with the buildup of guidance and control weapons systems—information processing and feedback control had emerged as a new way of thinking. Beyond its status as a new academic specialty within electrical engineering, control systems was redefining the meanings of social and biological phenomena.

As Wiener was well aware, the chain of signification linking feedback with control, messages, and information carried profound implications for the applications of the second law of thermodynamics to animate and inanimate processes. These linkages were particularly intriguing for studies of the possibility of Maxwell's demons—the imaginary beings invented by Maxwell in 1871 to account for phenomena (notably life) where a natural system increased its orderliness in apparent contradiction of the second law of thermodynamics.[49] Historically self-conscious, Wiener's concept of information was densely interwoven from the start with his historicization of biological phenomenology: the twentieth-century body was sharply demarcated from that of the nineteenth century. "In the nineteenth century," he observed:

> The automata which are humanly constructed and those other natural automata, the animal and plants of the materialist are studied from a very different aspect. The conservation and the degradation of energy are the ruling principles of the day. The living organism is above all a heat engine. . . . The engineering of the body is a branch of power engineering. . . . The newer study of automata, whether in the metal or in the flesh, is a branch of communication engineering, and its cardinal notions are those of the message, amount of disturbance or "noise," . . . quantity of information, coding technique, and so on.[50]

According to Wiener, this cybernetic phenomenology of the body obtained also on the cellular and subcellular levels of life. Partly through his ongoing dialogue with Haldane, an enthusiastic convert to cybernetics, Wiener prophesied a cybernetics of heredity by invoking the then-dominant view of the primacy of proteins. Like his contemporaries in life science, he reasoned that the combinatorial mechanisms by which mixtures of amino acids organized themselves into protein chains, which, in turn, formed stable associations with their likes, could be the same mechanism by which genes and viruses reproduced themselves. As in all transmissions of messages, such a protein-based genetic transmission could be ultimately explained by information theory.[51]

Wiener's technical argument was not easy to follow.[52] But despite the challenging concepts and difficult mathematics, *Cybernetics* received instant international recognition in the academy, as well as military-industrial attention. Haldane, who invited Wiener to lecture at University College, London, was probably one of the handful of biologists who could actually follow Wiener's mathematical arguments. He embraced cybernetics and information concepts enthusiastically and uncritically. "I am gradually learning to think in terms of messages and noise," he reported in 1948, offering sketchy calculations of information content in nerve fibers. He liked Shannon's notion of redundancy: "I suspect that a large amount of an animal or plant is redundant because it has to take some trouble to get accurately reproduced, and there is a lot of noise around. A mutation seems to be a bit of noise which gets incorporated into a message. If I could see heredity in terms of message and noise I could get somewhere."[53] Haldane continued to develop his new thoughts. By 1950 he had written a manuscript on "population cybernetics," and by 1952 he reported having prepared a paper where he "worked out the total amount of control (information=instruction) in a fertilized egg, and various other similar points."[54]

Haldane was not alone in rethinking genetic transmission and mutations in terms of messages and information bits. In 1950, Haldane's colleague, geneticist H. Kalmus, in a thought-piece called "A Cybernetical Aspect of Genetics," made similar attempts. Any geneticist reading Wiener's *Cybernetics*, he noted, will find that this new way of looking at life offered unifying principles and a powerful interpretive framework. Reasoning within the protein theory of the gene, he speculated that genes could be described as messages, or sources of messages, thus conflating cause and effect. He granted that the analogy might be problematic, since messages conveyed by genes were neither numerical nor electrical (as required by information theory). Nevertheless, he concluded that genes formed the basic element of control within the organism's integrated control systems.[55]

In Paris, after proofreading Wiener's manuscript, mathematician Benoit Mandelbrot immediately arranged for Wiener to lecture at the College de France on the new field.[56] "It isn't every day that a new science is born and christened," announced *Science Service* upon *Cybernetics*' release on "the American market," predicting its wide-ranging impact on society.[57] Anthropologist Julian M. Sturtevant immediately invited Wiener to explain cybernetics to a nonmathematical audience at Yale. "Wherever you go across the Cornell campus you find someone reading *Cybernetics*," wrote mathematician Will Feller. Warren McCulloch reported that everyone at the University of Illinois (a major research center for control systems and computing) is "all steamed up about your book" and added that his own copies had been stolen.[58]

The French newspaper *Le Monde* carried an article on cybernetics, and the book was creating a great impression in Sweden, reported a friend from Gothenburg, who advertised *Cybernetics* to his Indian colleague. (Wiener later lectured in India.) "I do not like to use cheap superlatives such as 'epoch making,' etc.," disclaimed the director of the Institute of General Semantics, as he proceeded to acclaim this "astonishing book" a "turning-leaf in the history of human evolution and socio-cultural adjustment."[59] Roman Jakobson, the linguist who a decade later would apply cybernetically informed notions of language to linguistic analyses of the genetic code (see Chapter 7 of this book) also believed that *Cybernetics* was an epoch-making book. "At every step I was again and again surprised at the extreme parallelism between the problems of modern linguistic analysis and the fascinating problems which you discuss. The linguistic pattern fits excellently into the structures you analyze and it is becoming still clearer how great are the outlooks for a consistent co-operation between modern linguistics and the exact sciences." No doubt, anticipating his move to Harvard (which would later include a joint appointment at MIT's new communications center), Jakobson looked forward to closer working contact with Wiener.[60]

The Rockefeller patrons, however, were not enthusiastic. Weaver considered *Cybernetics* a "profound, stimulating, and baffling book," the last because the interrelationships it postulated remained "largely in Professor Wiener's own remarkable head." Then he countered, "As long as there are persons who can write books like this I will not lose my respect for central nervous systems, nor be prepared to substitute, for them, machines."[61] After an "unsatisfactory conversation" with Wiener, Robert Morison learned that, following the enormous success of *Cybernetics*, various publishers approached Wiener about writing a popular version. Morison hoped Wiener would drop the project, because he felt the hurried popularized treatment of "the vexed subject of communication" was likely to draw much criticism.[62]

Wiener, however, did not drop the project. Through his remarkably appealing book, *The Human Use of Human Beings*, Wiener's ideas of cybernetic phenomenology and representation of animate phenomena as information flow reached many thousands of readers, even if his technical treatment of information-based communication and control had been inaccessible to researchers in the biological and social sciences.[63] Published in 1950, this popularization of *Cybernetics* forcefully expounded the coming of the new information epoch. Here, Wiener's thesis was that contemporary society could be understood only through a study of messages and communication facilities; the human individual and the living organism must be recast in terms of information.

In a chapter entitled "The Individual as the Word," Wiener elaborated this concept and its corollary: the technical possibility of transmitting the (genetic) essence of organisms. Information transfer enabled writing the Book of Life. "The earlier accounts of individuality were associated with some sort of identity of matter, whether of the material substance of the animal or the spiritual substance of the human soul. We are forced nowadays to recognize individuality as something which has to do with continuity of pattern, and consequently with something that shares the nature of communication." Noting the age-old commitment to notions of soul, form, or monad, Wiener observed that the biological individuality of an organism lay in a certain continuity and memory of process. It is not matter, he insisted, but the memory of the form that is perpetuated during cell division and genetic transmission.[64]

Wiener, whose youthful imagination had been captured by R. J. Kipling's old science fiction of a world physically united through coordinated air travel (via an Aerial Control Board), now saw himself as extending this vision beyond physical travel to the transmission of organismic information. The information content of a germ cell is enormous; he conceded that its coded message would dwarf that of a complete set of *Encyclopaedia Britannica.* But he concluded, "There is no fundamental absolute line between the type of transmission which we use for sending a telegram and the types of transmission which are theoretically possible for a living organism such as a human being." Wiener did not wish to engage in science fiction, he cautioned, but surmised that our ability to telegraph the pattern of a man from one place to another is probably due to technical difficulties: "keeping an organism in being during such a radical reconstruction. It is not due to any impossibility of the idea."[65] He believed that one could, in principle, transmit the coded messages that compose a human being; that it should be possible to write (and therefore control) the Book of Life.

Shorn of technical analyses and mathematical notations, it was *The Human Use of Human Beings* that spread the cybernetic vision in a culture enchanted with the nascent communication technosciences: electronic computers, systems analysis, operations research, industrial and military automation. As David Noble has written, taken together, these preoccupations gave rise to an unwieldy new metatheory of systems in quest for total control.[66] The laudatory responses from lay circles and the many invitations extended to Wiener from industrial and military organizations to lecture, consult, and conduct seminars all attest to the multivalence of Wiener's cybernetics discourse. One witnesses here the remarkable resonance of the new form of representation, a resonance derived from the epistemic, technological, and semiotic links of information technosciences to many different strata of postwar culture.

These new regimes of signification also encompassed the academic disciplines. Workers in fields as diverse as engineering, psychology, neurology, physiology, endocrinology, political science, economics, anthropology, linguistics, and architecture wrote enthusiastic letters to Wiener conveying a remarkable eagerness to represent old and new research problems in cybernetic terms, to which Wiener generally responded warmly.[67] An invitation to Wiener from New York University's Department of Administrative Engineering urged his participation in a meeting on "Cybernetics and Management Control" and stressed, "The newest managerial field, that of Management Control—the planning and checking phases of Management—is seemingly developing along lines some of which have astounding similarity to Cybernetics concepts [in *Human Use of Human Beings*]."[68] "I am . . . interested in exploring the planning and administrative significance of servomechanistic systems for processing large amounts of data. . . . I am wondering if you can be of any help to me in tracing this down?" queried a Johns Hopkins political scientist. Indeed, Wiener's frequent communications with Harvard political scientist Karl W. Deutsch had a strong impact on the field through Deutsch's cybernetic explanations of political patterns. Harvard sociologist Talcott Parsons, too, came to conceptualize mechanisms of social control as cybernetic systems.[69] University of Michigan economist Kenneth E. Boulding, who was organizing a large interdisciplinary seminar on the theory of information and communication, invited Wiener to participate in this "missionary work" on the campus. And his colleague Gyorgy Kepes, professor of visual design at MIT, extolled Wiener's contribution about the "three different attitudes toward nature and pattern" (the third being cybernetics) in his book on *The New Landscape in Art and Science*.[70]

To encourage information thinking in life science, the prominent British applied mathematician W. Ross Ashby reported having started writing an "Introduction to Cybernetics" intended for biologists. Its premise was the principle that "regulation" was the basis for all living things.[71] "About three years ago I first became aware of the incapacity of our present language to express adequately certain endocrine relationships," reported an endocrinologist from George Washington University. Eager to engage in some kind of a collaboration with Wiener, he suggested that, as a feedback system, the endocrine portion of homeostasis was more instructive than the nervous system. Wiener confirmed, "It is perfectly clear to me that feedback chains exist in the hormonal system, and that the excessive quantity of one hormone will have a back reaction on the amount of hormone determining the secretion of the first. They are thus systems with an intrinsic capacity to carry large amount of information."[72]

Wiener was merely one (albeit very effective) of numerous producers of the cybernetic discourse. Kurt Vonnegut captured the subject's pervasive materiality in *Player Piano* (1952), a chilling critique in which he depicts a society economically and spiritually polarized by automation. (Wiener appears as the prophet of "the second industrial revolution"; von Neumann is a minor character.)[73] True, in a valiant display of virtue Wiener might have condemned the military "abuses" of such knowledge while condoning its industrial "benefits." But the Gordian knot of military, industrial, and academic interests could not be so easily disentangled. As a constituent of that tripartite power structure, he unwittingly and even well-intentionally helped buttress the very forms of knowledge that sustained it. Wiener, indeed, seemed to have completely missed the point of Vonnegut's novel. Categorizing it as ordinary "science fiction" he merely questioned the uses of names of living persons (himself and von Neumann). But the story was never intended as science fiction. "In the book itself you will find an indictment of science as it is being run today," Vonnegut responded, emphatically correcting Wiener's trivialization of his penetrating message.[74]

Since his denunciation of military science in 1947, Wiener had responded in a characteristically erratic and idiosyncratic manner to the many solicitations of his cybernetic expertise. He did ceremoniously turn down a couple of invitations to cooperate with military authorities and with Rand Corporation. Yet in the early 1950s he also accepted several invitations to teach various aspects of automated control technology to military-based groups.[75] He refused to partake in the activities of the *Bulletin of the Atomic Scientists* because he was "sick of trying to pick up the pieces which atomic scientists have so carefully dropped" and tired of all their "wind under the diaphragm." He also remained unswayed by the *Bulletin* editor's perceptive counterargument that Wiener belonged "to the same Satanic breed of intellectual subversives who gave God's children toys too dangerous for them to play with."[76] The chain of signification linking feedback, control, message, and information also linked them to guidance and control systems, national security, and postwar politics. From the vantage point of history Wiener's contributions to the cognitive armamentarium of the cold war were far more effective than his protests. Naively, if passionately, committed to pacifist ideals and intellectual esthetics, Wiener missed the deeper significance of military pervasiveness: its impact on the world of the mind.

THE ECLIPSE OF SEMANTICS: SHANNON'S THEORY OF COMMUNICATION

The impact of the military on the world of the mind, particularly in the realms of science and engineering, was not confined to academia; instead, it extended to the industrial laboratories, most notable among them Bell Labs, which led the industrial sector in volume of military contracts. Guided by the managerial philosophy of its first president, Frank B. Jewett, who set a high premium on the integration of science and engineering, Bell Labs had fostered basic and applied research in electronics, physics, chemistry, magnetics, radio, and mathematics since its incorporation in 1925. With the advent of World War II, much of that expertise was adapted toward war projects. Bell Labs's war effort was diverse and extensive, including military work in telecommunications, radar, gun-director systems, electronic computer systems, cryptography, and teaching services for the School for War Training.[77] Shannon's mathematical theory of communication, which complemented and extended Wiener's work on cybernetics, developed at Bell Labs as part of the war effort, where military imperatives simultaneously guided technological design and theory construction. These communications technologies shaped not only the form, scope, and limits of Shannon's information theory but also its peculiar feature as a communication devoid of semantics.

An extraordinarily promising MIT Ph.D. in mathematics, Claude Shannon (b. 1906) joined Bell Labs in 1941, after having served, as did Wiener, as consultant with the NDRC in fire-control work. (As a graduate student Shannon even produced a mathematical theory of population genetics, apparently on the encouragement of Vannevar Bush, then director of the Carnegie Institution of Washington.[78]) An expert tightrope walker and unicycle rider, Shannon relished games and puzzles of all kinds. He could be often seen riding his unicycle up and down the halls of Bell Labs. "During World War II," Shannon recalled:

> Bell Labs were working on secrecy systems. I'd worked on communication systems and I was appointed to some of the committees studying cryptanalytic techniques. The work on both the mathematical theory of communications and the cryptology went forward concurrently from about 1941. I worked on both of them together and I had some of the ideas while working on the other. I wouldn't say one came before the other—they were so close together you couldn't separate them.[79]

The work on top secret "Project X" entailed developing a speech-encoding system, which quantized the speech waveform and added to it a digital code before transmission. To decode a signal, an eavesdropper would have to possess not only the equipment for converting the digitized information into sound waves but also the code pattern, a formidable challenge.[80] The work on both the secrecy codes and communication theory was substantially complete by about 1944, but Shannon kept refining them until their publication as separate papers in 1948 and 1949. "Communication Theory of Secrecy

FIGURE 6. Claude E. Shannon holding a toy mouse in a mouse maze, 1950s. Courtesy of AT&T.

Systems" treated cryptology in information-theoretical terms. Introducing concepts such as mathematical redundancy and binary coding into communication, it furnished fundamental tools to both information theory and cryptanalysis.[81] "The Mathematical Theory of Communication" linked these new features with older concepts of information transmission to form a general theory applicable to any system in which information can be quantified and transmitted.

Shannon's "The Mathematical Theory of Communication" appeared at the same time as Wiener's *Cybernetics.* Although Shannon and Wiener worked independently, their conceptual frameworks turned out to be remarkably close (probably due to the earlier background of Wiener's work and his influence on Shannon's training at MIT). Shannon praised Wiener for having written another classic; yet he found it interesting "how closely your work has been paralleling mine in a number of directions." He pointed to an obvious difference that seemed confusing: Wiener defined information as negative entropy, while Shannon used the regular entropy formula with no change in sign. Shannon did not believe that difference had real significance, "but is due to our taking somewhat complementary views of information." He continued, "I consider how much information is *produced* when a choice is made from a set—the larger the set the *more* information. You consider the larger uncertainty in the case of a larger set to mean less knowledge of the situation and hence *less* information. The difference in viewpoint is partially a mathematical pun. We would obtain the same numerical answers in any particular question."[82] Perhaps so. But the pun spotlighted a fundamental difference: Shannon represented phenomena positively, while Wiener traced their negative space. He invited Wiener's comments on his papers. Meanwhile, Warren Weaver had just read both works and dispatched a curious note to Wiener. Praising both authors he wondered about the similarity of their products. "He [Shannon] is so loyal an admirer of yours that I find it difficult to decide how much of this was really inspired by you, and how much he deserves individual credit for." He then confessed, "That is probably a very bad question to put to you."[83] This was not the last time the uncanny similarity of the two theories was questioned. Wiener expressed the "highest regard for Dr. Shannon both as to his scientific accomplishment and to his personal integrity." Nevertheless, he claimed priority to many of the features of the new science, pointing out how the different institutional contexts—MIT and Bell Labs—shaped his and Shannon's perspectives on the communication problem.

> Dr. Shannon is an employee of Bell Telephone Company, and is committed to a career of developing communication notions within a certain rather limited range confronting the interests of the company. Within this range, he must work much more definitely towards immediately usable results than I do, and he has been both industrious and prolific in ideas in this work. On the other hand, I am a college professor, and I have always interpreted my position, with the consent and encouragement of my school, as that of being a free-lance. I have found the new realm of communication ideas a fertile source of new concepts not only in communication theory, but in the study of the living organism and in many related problems.

It seems that cybernetics' emphasis on continuous, rather than discreet, modes of communication was directly related to the biological grounding of the theory. In any case Wiener did not wish to be forced into a personal competition. "I would prefer," he stressed, "that the theory be called by the names of the two of us or objectively by the names of neither of us, but if it comes to a matter of names, I have the right by historical priority to have my name first."[84] For a while the new science became known as the Wiener-Shannon theory of communication, but eventually it was Shannon's positive and powerful formulation that endured. In 1957 Shannon became a permanent member of the MIT faculty while continuing his association with Bell Labs as a mathematical consultant.

Shannon began developing key features of information theory at MIT before joining Bell Labs. But the institutional setting of his work in the 1940s guided his investigations in directions relevant to industrial-military priorities. In constructing his theory, Shannon—in contrast to Wiener—relied heavily on the important studies of telegraphic transmission by Harry Nyquist and R. V. Hartley, his predecessors at Bell Labs. Working on improving transmission speeds over telegraph wires in the 1920s, Nyquist had written an article on the transmission of "intelligence."[85] Apart from one theoretical section, "Theoretical Possibilities of Using Codes with Different Numbers of Current Values," the paper focused mainly on practical engineering problems. But in that one section Nyquist presented two key points. He formulated the first logarithmic rule governing the transmission of information: $W = k \log m$, where W is the speed of transmission of intelligence, m is the number of current values that can be transmitted, and k is a constant. (By "speed of transmission of intelligence," he meant "the number of characters, representing different letters, figures, etc.," transmitted in a given length of time.) He also offered the first analysis of the theoretical bounds for ideal transmission codes. Nyquist's law later became a specific instance of Shannon's definition of information.[86]

"Intelligence," however, smacked of anthropomorphic and psychological attributes, a problem R. V. Hartley, a research engineer at Bell Labs, sought to circumvent. Hoping to develop a theory of sufficient generality to include all modes of electronic communication—telegraphy, telephony, radio, television, and cinema—he provided in 1928 a preliminary analysis of the theoretical limits of information transmission under idealized conditions. His often-cited paper, "Transmission of Information," stressed that the capacity of a system to transmit any sequence of symbols depended solely on distinguishing at the receiving end between the results of various selections made at the sending end—not on the meanings of these sequences. He viewed information as "logical instructions to select," since any such scientifically usable definition of information had be grounded in what he called "physical," rather than "psychological," considerations. "Information" and even its precursor, "intelligence," were used metaphorically. Information—defined as the number of possible messages—was thus demarcated from meaning. He used this definition to derive a logarithmic law for information transmission:

$H = K \log s^n$, where H is the amount of information, K is a constant, n is the number of symbols in the message, s is the size of the set of symbols, and s^n the number of possible symbolic sequences of specified length n.[87]

Thus Hartley had established key concepts of a mathematical theory of communication: the difference between information and meaning, information as physical quantity, the logarithmic rule of information transmission. He also formulated the concept of noise as an impediment to information transmission. But it was Shannon who generalized these aspects of telegraph transmission into a general theory of communication spanning the spectrum from telegraphs to computers[88] and from cryptology to command and control. The synergy of his concurrent projects—improving the fidelity of information transmission and the design of secret coding systems—seemed to generate a theory applicable, in principle, to any system, physical or biological, in which information can be properly coded, quantified, and manipulated through time and space.

The cleavage of information from semantics, the predicate of information theory, was enunciated in the introduction to Shannon's article:

> The fundamental problem of communication is that of reproducing at one point either exactly or approximately a message selected at another point. Frequently the messages have *meaning*; that is they refer to or are correlated according to some system with certain physical or conceptual entities. These semantic aspects of communication are irrelevant to the engineering problem. The significant aspect is that the actual message is one *selected from a set* of possible messages.[89]

In this system, the stochastic source of information selects symbols according to probabilities. Shannon, like Wiener and others, described information in statistical terms: the information conveyed by a symbol increased as its probability of occurrence increased. He thus generalized Hartley's logarithmic rule as: $H_n = -\Sigma_i p_i \log p_i$, where H_n is the information content and $p_1 \, p_2 \ldots p_i$ are the probabilities attached to the n symbols of the message. (This expression, which happens to be similar to the entropy of a thermodynamic system in Boltzmann's statistical mechanics, was thus implicitly assumed to signify similar underlying physical phenomena; redundancy, then, was defined as one minus the relative entropy.)[90] Shannon assigned the base of the logarithm the value of 2. Having years earlier applied Boolean algebra to relay circuits, where switches have only two positions, yes (1) and no (0), Shannon now adapted these concepts to information theory: any message expressed in language may be written in such a binary code and is said to be logically communicable. He thus defined the amount of information as:

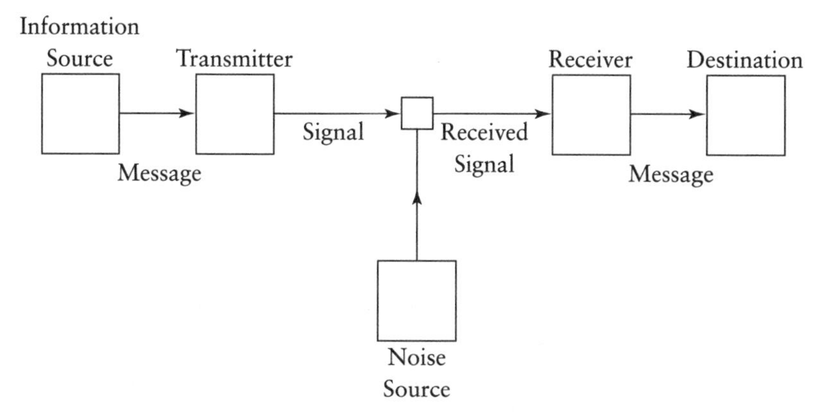

FIGURE 7. Claude Shannon, "The Mathematical Theory of Communication," *Bell System Technical Journal* 28, no. 4 (1949): 656–715. Schematic diagram of a general communication system. Reprinted by permission.

$H_n = -\Sigma_i p_i \log_2 p_i$, and introduced the new unit "bit" (from "binary digit") as a measure of information.[91]

The generality of Shannon's conceptual framework was displayed diagrammatically with seductive austerity (see Fig. 7).

1. The *information source* produces a message or sequence of messages to be communicated to the receiving terminal . . .
2. The *Transmitter* operates on (encodes) the message in some way to produce a signal suitable for transmission over the channel . . .
3. The *Channel* is merely the medium used to transmit the signal from transmitter to receiver . . .
4. The *Receiver* reverses the operation of the transmitter, reconstructing (decoding) the message from the signal . . .
5. The *Destination* is the person (or thing) for whom the message is intended . . .

This was a scheme designed for communications between machines, where information was conceptualized in a manner divorced from content, subject matter, or nature of the channel. Although Shannon did not extrapolate beyond electronic information, his scheme encouraged such projections. (In fact, Shannon himself was intrigued by its implications for human communications.) Provided the components were properly interpreted, communication between muscle and brain and chromosomes and cells could be, in principle, analogized to machine communications between target and guided missiles, provided the components and parameters were mathematically interpreted.[92]

In its original form, as a technical piece in a specialized journal, Shannon's article, unlike Wiener's second book, did not aim for a wide audience. Neither did it look like a work destined to achieve popularity among psychologists, anthropologists, linguists, biologists, economists, philosophers, or historians. Yet this is what happened, partly because in 1949 the University of Illinois Press issued a book entitled *The Mathematical Theory of Communication*, written jointly by Shannon and Weaver. Perhaps as a reaction to the thunderous accolades to Wiener's *Cybernetics* and clearly an enthusiastic endorsement, Weaver applied his talent for science management and exposition to making Shannon's technical work accessible to a broad audience.[93]

The word "communications," Weaver explained, is used in a very broad sense. Used metaphorically, it included all of the representations by which one mind may affect another, not only in written and oral speech but also in music, the pictorial arts, theater, ballet, and, in fact, all human behavior. He distinguished three levels of the communication problem: level A, the technical problem, asks, How accurately can symbols of communication be transmitted?; level B, the semantic problem, asks, How precisely do transmitted symbols convey the desired meaning?; and level C, the effectiveness problem, asks, How does received meaning affect conduct in the desired way? These three levels correspond to the tripartite hierarchy of human communications: syntax, semantics, and pragmatics. But for important reasons the book deals largely with level A, the technical level. The kind of questions relevant to level A generally are:

1. How does one measure the amount of information?
2. How does one measure the capacity of a communication channel?
3. Questions about the coding process when changing a message into a signal; the code's characteristics and efficiency.
4. What are the characteristics of noise? Its relation to accuracy of received messages? How can we minimize noise?
5. If the transmitted signal is continuous (as in oral speech or music) rather than discrete symbols (as written speech, or telegraphy) how does this affect the communication problem?[94]

The reason for bracketing out the semantic level and focusing on the narrow A domain is key. As Weaver explained, "The word 'information' in this theory is used in a special sense that *must not* [my emphasis] be confused with its ordinary usage. In particular, information must not be confused with meaning." In fact, he pointed out, two messages, one heavily loaded with meaning, the other pure nonsense—a Shakespearean sonnet and a random collection of letters, for example—can be exactly equivalent from an information theory standpoint, because in the technical sense, "information" is used metaphorically. It is a measure of a freedom of choice when one selects a message. The amount of information is measured by the logarithm of the number of available choices (typically the logarithm of 2 because in a digital switching system based on Boolean algebra, 0 and 1 are taken symbolically to represent any two choices); the bit is then associated with a two-choice situation having a unit information. But this measure of information does not supply semantic value. Weaver even had the feeling that information and meaning may be related as the pair of canonically conjugate variables as in Heisenberg's uncertainty principle, thus condemning a person to sacrifice one when insisting on having much of the other.[95]

Semiotics (and language), like information theory, deals with the manipulation and study of signs. But unlike information processing, semiotics has generally been studied on three different but interrelated levels, representing different types of abstracting: syntactics (study of signs and of the relations between signs); semantics (study of the relations between signs and designata); and pragmatics (study of signs in relation to their users). The problem of semantics in information theory has engaged students of semiotics and language since 1950. Shannon himself has pondered the relations between information theory and human communications when, as a participant in the Seventh Macy Conference, he presented a paper on "The Redundancy of English," which scrupulously circumvented semantic questions. He also shunned any extrapolation of information theory to genetics, an area with which he was familiar.[96] Wiener, too, attempted superficially to reconcile cybernetics and semantics by considering words as types of deeds, thus being caught in the Derridian paradox, the catachresis, of a metaphor of a metaphor.[97] But the Wiener-Shannon theory of communication concerns only signs. If it were considered as a contribution to semiotics it would lie on the syntactic level alone, as several specialists pointed out.[98]

True, the noted logician Yehoshua Bar-Hillel, elaborating on Rudolph Carnap's theory of language as logical syntax, has attempted to establish a measure for semantic information content. But these logic-driven linguistic analyses were concerned only with language systems amenable to machine communications and not with ordinary human language. They were the analytics of communication patterns geared toward automation problems in which electronic computing, linguistics, and psychology were often configured together as tools of command and control systems.[99] Bar-Hillel, in fact, was a vociferous critic of colloquial misuses of information theory. He wrote, "Unfortunately, however, it often turned out that impatient scientists in various fields applied the terminology and the theorems of the statistical (communication) theory to fields in which the term 'information' was used, presystematically, in a semantic sense . . . or even in a pragmatic sense."[100] The Wiener-Shannon measure of selective information rate concerns the statistical rarity of signals; what these signals signify or mean, or what their value or truth is, cannot be gleaned from their theory of communication. As such, information theory embodied the historical conditions of its formation. While several of its earlier features were shaped by industrial innovations in communication during the interwar era, its final shape was determined by the technological momentum of World War II. The form of the theory, that is, information as physical quantity, binary coding, and the cleavage of information from meaning, as well as the theory's scope, or the applicability to all situations obeying Shannon's schema, and its limits, which were strictly a mathematical concept designed primarily for machine communications without regard for meaning, were all reflected some of the technological imperatives of the period.

Although in its generic sense the information concept promises broad application, when it is used technically the domain of applicability of the Wiener-Shannon theory is restricted. But an adherence to that technical level circumvents the critical difficulties that crop up when Shannon's analysis of communication extends beyond electrical engineering, for example, to biological domains where information cannot be readily measured, and where the materiality of the channel, context, and semantics do matter. As the leading British information theorist, Colin Cherry, cautioned:

> It is in telecommunication that a really hard core of mathematical theory has developed. . . . In such technical systems, the commodity which is bought and sold, called *information capacity*, may be defined strictly on a mathematical basis, without any of the vagueness which arises when human beings or biological organisms are regarded as "communication systems." . . . This is not to say that the mathematical concepts or technique are completely forbidden elsewhere, but if so used, this must not be regarded as simple application of existing "theory of (tele)communication" by extrapolation from its legitimate domain of applicability.[101]

Ultimately these technical strictures severely limit the conceptualization of heredity as an information system and destabilize the representations of organisms as words and texts and the meaning of the genetic code as the key to the Book of Life.

Based on Shannon's theory, the knowledge qua "measure of information" of, say, a genetic message or a sentence in the Book of Life cannot imply a knowledge of its meaning. Cherry cautioned against such trivializations by the diverse researchers who jumped on the information bandwagon.

> How much can we know from a particular set of observations or experiments? The experimenter is really not forming a "communication-link" with Mother Nature. He is not receiving signs or signals, which are physical embodiments of messages, not words, picture, or symbols. The stimuli received from Nature—the sights and sounds—are not pictures of reality but are the evidence from which we build our personal models, or impressions, of reality.[102]

Of course one could conjure vague analogies between such narrow communications systems as telegraphy and electronic servomechanisms on one hand, and the complex processes in the organism on the other. And, as such, they have offered productive models and powerful imagery for conceptualizing central phenomena in molecular biology. But such analogies, which restore semantic value to the information concept, have had little in common with information theory. At best—when analyzed quantitatively—they might illuminate notions of information as *a measure* of organization and control.

So how *does* one get from communication engineering systems to biological systems, *and for what purpose*? *How and why* did scientists employ the semiotics of information to represent organisms and to write biology in the language of machine communication? A partial answer, at least for the champions (scientists and patrons) of information technosciences, lay in the enormous promise to the study and construction of automata. The Wiener-Shannon theory offered new representations for the heterogeneous construction of a simulated reality composed of animate and inanimate components. Within these technological regimes of signification, relay circuits of electronic computers, syntactic rules of communication, cognitive operations, and biological mechanisms of regulation and reproduction became elements in the conceptual armamentarium for the study and construction of automata as organisms.

Like Wiener, Shannon had been interested in automata since at least the mid-1940s. And though the roots of computationally based automata reach back to the nineteenth century (as with the cases of feedback and telegraphy), these technosciences were reconstituted during World War II. Shannon built a maze-solving mechanical mouse (named Theseus), developed circuitry for logic machines, and worked on a chess-playing machine. (Shannon's interests were shared by Warren McCulloch, a pioneer in the design of logical automata [see next section].) At the Eighth Macy Conference on Cybernetics, Shannon presented his maze-solving machine. Such war-game technologies functioned in a dual mode: as interventions and as representations. On the one hand, they may be regarded as the first step in the construction of computers for evaluating military situations and deciding the best move. On the other hand, governed by "logical choice" and "decision theory," these machines came to serve as heuristics for analyzing biological and social processes; they began to generate a new (hyper)reality, where simulations became models for understanding life, death, and human experience.[103]

JOHN VON NEUMANN'S GENETIC SIMULACRA

Hungarian émigré mathematician John von Neumann (1903–57) is unrivaled in his promotion of logical automata (including computers) at the hub of power that linked military control systems to cold-war strategies. Having been a research professor of mathematics at the Institute for Advanced Study (IAS) at Princeton since his arrival in the United States in 1933, von Neumann began his lifelong collaboration with the military in 1940 as a member of the Scientific Advisory Committee at the Ballistic Research Laboratories, Aberdeen Proving Ground, Maryland. Driven by profound loathing of communism and the Soviet Union, he had since served as consultant to nearly two dozen government and industrial organizations, among them the Navy Bureau of Ordnance, Los Alamos Scientific Laboratory, Rand Corporation, IBM, Oak Ridge National Laboratory, Armed Forces Special Weapons Project, Scientific Advisory Board, U.S. Air Force, the Central Intelligence Agency, the Radiation Laboratory at Livermore, the Strategic Missiles Evaluation Committee of the U.S. Air Force, National Security Agency Advisory Board, the Nuclear Weapons Panel of the U.S. Air Force, and the General Advisory Committee of the U.S. Atomic Energy Commission. In 1955 he became an AEC commissioner. From the mid-1940s on, he steadily curtailed his scientific research in pursuit of ever more advanced weapons systems.[104]

The development of computers, including von Neumann's contributions to the field, was central to cold-war strategic and technological objectives. From its inception in 1943 as the Electronic Numerical Integrator and Calculator (ENIAC) Project at the Moore School of the University of Pennsylvania, the construction of the high-speed digital computer was supported at a price of nearly half a million dollars—all of it in the service of weapons automation. (Harvard's Mark I, designed by Howard Aiken in cooperation with IBM, did not have the power and flexibility required by Los Ala-

FIGURE 8. John von Neumann in front of IAS computer, ca. 1952. Courtesy of the Library of Congress.

mos physicists.) Von Neumann joined the ENIAC Project in 1944. By the end of the war, with codirector Herman Goldstine and Moore School engineers John Mauchly and J. Presper Eckert, he had planned an improved machine, introducing the first written account of the stored program computer. The Electronic Discrete Variable Arithmetic Computer (EDVAC), the first machine to incorporate an internally stored program, thus became the first true computer in the modern sense. Its logical design—often called "von Neumann architecture"—became not only a blueprint for his enormous computer project at IAS but also for general computer design well into the 1980s.[105]

With the war's end, corporate funding for computer research became available; however, in various forms, military agencies continued to serve as the principal and enduring source of support. What function did computers serve in the military projects of the bipolar world of the cold war? In the feedback loop linking computers to weapons systems, computers served as both input and output, origin and outcome. The commitment to complex high-speed and high-technology warfare generated demands for control, communications, and information analysis surpassing natural human capacities. Computers could automate and accelerate military operations, which led to intensified commitment to faster and more powerful computers.[106]

As Paul Edwards has written, the pursuit of automated, centralized command and control—a kind of information panopticon for remote battle "management"—that began in the mid-1940s would reach its palpable articulation in 1969, with General William Westmoreland's call for an "electronic battlefield." As Edwards observed:

> To a very large degree *the Cold War was actually prosecuted through* simulations. Each side based its weapons purchases, force deployments, technological R&D, and negotiation postures on its models of strategic conflict and its projections about the future choices of the other. This is why the Cold War must be understood in terms of *discourses* that connect technology, strategy, and culture. The Cold War was quite literally fought inside a quintessentially semiotic space, existing in models, language, iconography, and metaphor, embodied in technologies which lent to these semiotic dimensions their heavily inertial mass. In turn, this technological embodiment allowed closed-world discourse to ramify, proliferate, and entwine new strands, in a self-elaborating process.[107]

Although discourse, as a culturally grounded system of representations, cannot be reduced purely to the role of powerful individuals, individuals serve as its agents. Influential figures such as von Neumann shaped and amplified the production of the information discourse—and its place in biology—by the scope and intensity of their action and ideas.

Von Neumann, strongly committed to the arms race, was an outspoken advocate of a preventive war and feigned no leniency or sentimentality. "If you say why not bomb them tomorrow, I say why not today? If you say today at five o'clock, I say why not one o'clock?" he remarked in 1950. A champion of "quantum jump" technologies, he quickly shifted his gaze from the atom bomb to the hydrogen bomb and from the bomber as delivery system to the nuclear-tipped intercontinental missile. For von Neumann, no weapon was too large. Under the guise of pure technical advice, the charming and politic strategist accrued the quiet power of a decision-making elite at the inner sanctum of the cold war. Unlike his colleague Wiener, who invited public attention to cybernetics qua *Lebens philosophy* and popularized his vision of a new information epoch, von Neumann shunned publicity, preferring to work behind the scenes and declining participation in public activities devoid of technical content. As Steve Heims observed, as a disinterested expert von Neumann was that much more valuable to the military, industry, and government.[108]

Through the multitude of his government consultancies, the fast-thinking, fast-talking von Neumann promoted the design and funding of ever-faster and more powerful computers. For example, as a consultant for IBM in 1951 he participated in the plans to develop reliable and cost-effective memory systems for the NORC, a computer being built for the navy.[109] At the same time, as consultant to the Navy Bureau of Ordnance, he argued for the indispensability of the new computer for military operations, a solution in search of a problem. "I have no doubt that the Bureau of Ordnance and the activities that it controls have an immediate, as well as a lasting need for an automatic high speed computing machine, and specifically for one of the most advanced design," he argued, proceeding to enumerate the areas in need for advanced computing: aerodynamics, hydrodynamics, elasticity and plasticity, high explosives, missile design, atomic weapons, and motors. "The characteristics of the proposed IBM machine are very bold, but not at all impossible. The speed characteristics exceed by a considerable amount those of all machines about which I am informed, whose completion in the nearer future is at all likely."[110] Such upward spiraling feedback loops of computers and weapons systems rendered both hardware and software of these technologies inextricably bound. As Frank Rose observed, "The computerization of society . . . has essentially been a side effect of the computerization of war."[111]

Von Neumann's general interests in biological automata were aroused in the early 1940s, primarily through his camaraderie with Wiener. Captivated by the 1943 paper of Warren McCulloch and Walter Pitts, "A Logical Calculus of the Ideas Immanent in Nervous Activity," which he read on Wiener's and Bigelow's urgings, von Neumann immediately saw its potential for treating biological phenomena as information systems. He helped popularize the paper even before the Macy conferences lavished attention on McCulloch. McCulloch, a psychiatrist and philosopher by training then working at the University of Illinois, and Pitts, who was trained under Rashevsky and Carnap in mathematical logic and then employed at MIT's Research Laboratory of Electronics, joined together with a unique combination of skills and temperaments to construct a mathematical model of the neural networks of the brain based on Carnap's logical calculus and on Alan Turing's work on theoretical machines. Neurons were treated as black boxes, obeying mathematical rules governing the input and output of signals. A unit psychic event, or "psychon," was based on all-or-none (1,0; yes,no) impulses of neurons combined by convergence upon the next neuron to yield complexes of propositional events.[112]

These logic-driven representations of biological mechanisms served as epistemic modules in the design of military cybernetic systems of the 1940s and

1950s, a point McCulloch appreciated quite early. F. C. S. Northrop, McCulloch's mentor at Yale, wrote to Wiener with some concern in 1947: "As McCulloch maintains,"

> All this ideological business [attention to social implications of cybernetics] is nothing but window-dressing and a rationalization after the event for rough-and-tumble scrapping in the power politics field. What this amounts to, if one acquiesces in it, is the continuation of a situation in which the power politics boys, the militarists, instead of scientifically informed people, give the machine its major teleological instructions.[113]

Like von Neumann, McCulloch cultivated close working relations with the navy, air force, and army. He thought of his intellectual labor as contiguous with military technologies: "As the industrial revolution concludes in bigger and better bombs, an intellectual revolution opens with bigger and better robots."[114] By the early 1960s the military establishment would come to honor McCulloch as "the High Priest of Cybernetics, Bionics, or Self-Organizing Systems, whatever you call it."[115]

Since his captivation by the promise of the applicability of McCulloch's work to the design of biological automata and concurrent with his emersion in computing, von Neumann began cultivating contacts with the biomedical community. He also started to participate in various interdisciplinary conferences—including the Macy conferences—that linked areas of life science to physics, mathematics, and cybernetics.[116] His interest in biology, in general, and genetics, in particular, became closely linked to his mission of developing self-reproducing machines. That project aimed at simultaneously explaining and simulating living systems, thus opening up a space of representation in which heredity was to be modeled on simulacra.

The Ninth Washington Conference on Theoretical Physics (1946), entitled "The Physics of Living Matter" and sponsored by the Carnegie Institution of Washington (CIW), seems to have been of particular significance for von Neumann's involvement in biological automata, particularly the self-reproducing kind. Organized by physicist George Gamow of George Washington University and chemist Philip H. Abelson of the CIW's Department of Terrestrial Magnetism, the conference brought "together leaders in biology and physics giving them ample opportunity to know each other and to exchange ideas, with a view to the possibility of future closer collaboration." The list of participants included representatives from twenty-four universities, research centers, and government agencies, among them, George W. Beadle, Niels Bohr, Carl F. Cori, Max Delbrück, Milislav Demerec, John Edsall, James Franck, H. J. Muller, Francis O. Schmitt, Sol Spiegelman, Wendell Stanley, Leo Szilard, Edward Teller, and John von Neumann. Conference topics included the action and reduplication of genes and chromosomes; the tobacco mosaic virus; the mutant strains of bacteriophage, energy-rich phosphate bonds; extranuclear hereditary factors as links in the genetic control of enzymatic action; and photosynthesis. Von Neumann lectured on the possible role of servomechanisms in chromosome pairing.[117]

As the organizers had hoped, the exchange of ideas continued well beyond the conference. Through his discussions with Delbrück (who had just moved back to Caltech) von Neumann came to appreciate the nuances of phage as the simplest replication model. Delbrück, in turn, became intrigued by information theory and by certain aspects of cybernetics. Von Neumann was particularly interested in the gene-enzyme relation as explicated by microbiologist Sol Spiegelman of Washington University, St. Louis. Spiegelman, then working on "adaptive fermentation" in yeast, had just proposed "the plasmagene theory of gene action." (His work paralleled Jacques Monod's studies on "bacterial adaptation"; cybernetic and informational representations would guide his subsequent studies of enzyme systems and messenger RNA.) Postulating that genes continually produced, at different rates, partial replicas of themselves that entered the cytoplasm, Spiegelman's theory had the unique feature of linking nuclear genes to cellular differentiation.[118] Spiegelman was much taken with von Neumann's model. He wrote:

> I have been thinking about the results of the Washington Conference and, in particular, about some of the suggestions you made. I have been preparing a paper on the problem of self-duplication and, of course, your suggestions are quite pertinent. It seems to me that your remarks should be communicated to the biological public in some way so that they can think about it. I have discussed the desirability of this with Dr. Muller, who heartily agrees. I wonder whether you would be willing to entertain the idea of writing up the possible involvement of "servo mechanisms" in the phenomenon of chromosome pairing as a possible solution to the specific long range force paradox [how two homologous chromosomes lined up together during mitosis, seemingly defying the law that like repels like]. I also think that a brief description of your self-duplicating machine would certainly be pertinent. I would be happy to do anything I could to expedite your getting these ideas before biological readers.[119]

Spiegelman's enthusiasm and the interest of other life scientists helped validate von Neumann's concepts. Von Neumann now gained confidence that his self-duplicating automata might not be merely a formalism but represented actual "mechanisms essential in nature." He was gratified, he wrote to Spiegelman, that his "unorganized and amateurish remarks on the subject of 'servo-mechanisms' or amplification on one hand and of self-duplication on the other deserves a future discussion in your opinion as well as in the opinion of Dr. Muller."[120] Von Neumann and Spiegelman exchanged relevant sets of literature—Spiegelman supplying recent studies in biochemical genetics, von Neumann citing recent publications in information theory—and planned future meetings to explore their new common ground. Von Neumann promised he was working on the first draft of self-duplicating automata.[121]

Von Neumann participated enthusiastically in the biomedical project to link computers to neural networks and brain functions, efforts spearheaded by Wiener, Rosenblueth, McCulloch, and Pitts. But it appears that his contact with molecular biologists, notably Delbrück, influenced his perception of the problem of biological automata. Shortly after the Washington Conference he expressed doubts that a system as complex and poorly understood as the human nervous system was the best way to explore and simulate the salient processes of life. "I feel that we have to turn to simpler systems," he wrote to Wiener. Going down the phylogenetic ladder, von Neumann fixed his attention on viruses. These minimalist organisms ("units of life" as some referred to them) displayed orientation, reproduction, and mutation. "The less-than-cellular" organisms of the virus or bacteriophage type possessed the decisive traits of any living organism, especially self-reproduction, he pointed out, thus revealing obvious influence by the rationale of Delbrück's phage model. Von Neumann estimated that out of about six million atoms constituting a phage particle only a few hundred thousand were likely to be "mechanical elements."[122]

Outlining an alternative strategy, which he followed, von Neumann recommended five objectives: (1) study viruses and phage and all that is known about the gene-enzyme relationship, since genes are probably much like viruses and phages; (2) learn about the present state of knowledge of protein structure; (3) study methods of X-ray diffraction and Fourier analysis; (4) study methods of electron microscopy; and (5) gain knowledge of the relevant literature and scientific community.[123] Clearly, in linking viruses to information processing, von Neumann, like most researchers (especially in the United States), operated within the dominant paradigm in life science. Conceptualizing reproduction within the protein view of heredity, he looked to autocatalytic enzymatic mechanisms as explanations of gene and virus replication.[124]

Through his interest in protein self-replication von Neumann became interested in X-ray crystallography of proteins. Proteins were then an intellectually contested area, including hypotheses of both linear and ring structures. Von Neumann's discussions with physical chemist Irving Langmuir, associate director of General Electric's Research Laboratory, focused his attention on the (later aborted) cyclol theory of protein structure, and its attendant proposal for genetic replication, championed by British-trained mathematician Dorothy Wrench. He intended to scale up the linear and cyclol models by a factor of 10^8 and deploy similarly scaled-up electromagnetic centimeter waves to simulate X rays in order to study medium- and large-sized proteins.[125] He also held dialogues about the protein problem with Harvard physical chemist John T. Edsall and with biochemist Albert Szent-Gyorgyi, visited Woods Hole meetings, attended conferences on proteins, and disseminated the information discourse in life science.[126] Although von Neumann never did write up his 1946 lecture on genetic servomechanisms, in 1947 he committed himself to a presentation and publication of these concepts. He also agreed to participate in the 1948 Hixon symposium on "Cerebral Mechanisms in Behavior," to be held at Caltech. This symposium is generally regarded as a turning point in the history of the behavioral sciences because it initiated a focus away from behaviorism toward cognition studies.[127] But the symposium's significance also derives from the formal conceptual bridges that von Neumann began to build there between automata and molecular genetics. Von Neumann urged both Caltech's Hixon Committee, composed of George Beadle, James Bonner, Henry Borsook, Max Delbrück, Linus Pauling, Alfred H. Sturtevant, Cornelius A. G. Wiersma, and Anthony van Harreveld, and the symposium's organizer, psychologist Lloyd Jeffress of the University of Texas, to invite the participation of Muller, Spiegelman, and Salvador Luria.[128]

In broaching biological automata von Neumann pared down the problem of living organisms into two parts: (1) the functioning of their elementary units; and (2) the system (logical, organizational, or other) by which these parts are assembled. Focusing on two cases, the system of neurons and the system of genes, he believed that the functioning of these systems as a whole (part 2) was a problem independent of the knowledge of their constitutive elements (part 1). (This antireductionist, system-based model turned out to be a major stumbling block for molecular biologists committed to abstracting the minimum unit of replicative function.) With the McCulloch-Pitts model as a point of departure, the task of researchers was "to explain the correlating and analogizing principles by which the central nervous system analyzes and organizes the material furnished by the sense organs." Believing that this applied equally to the genetic system, von Neumann aimed toward explaining its most conspicuous feature: the ability of self-reproduction, coupled with the transmission of mutated traits.[129]

Unleashing his scientific imagination, von Neumann's proposed paper called for a theory of biological automata based on mathematical logic. Accordingly, he envisioned automata constructed on a conventionalized system of relatively few and simple elements. In fact, after a few plausible assumptions have been made, he argued, one would be forced toward constructions of automata that exhibited some of the principal traits of genetics. All these constructions and the choice of the basic conventions and elements would ultimately rely on refining the mathematical theory of automata based on information theory.

> Such a theory remains to be developed, but some of its major characteristics can already be predicted today. I think that it will have as one of its bases the modern theory of communications, and that some of its most essential techniques will have a character very near to Boltzmann-ian thermodynamics [referring to the homology between Shannon's statistical formula for information and the entropy equation derived from statistical mechanics; see the following section]. It should give a mathematical basis for concepts like the "degree of complication" and of "logical efficiency" of an automata (or of a procedure).[130]

By all accounts von Neumann's opening lecture at the symposium, based only on a handful of notes, was inspirational. Typically, however, his free-flowing imagination resisted the constraints of formal write-ups, and he stalled the publication process of the symposium's collected papers for more than two years.[131]

The published version of his talk, entitled "The General and Logical Theory of Automata," was von Neumann's first written account on the subject and a cornerstone for subsequent lectures. It described the general traits of computing machines, detailed the comparisons between computing machines and living organisms, speculated about future logical theory of automata, explained the principles of digitalization, related these to formal neural networks, and concluded with the problem of self-reproduction. With Turing's theory as a point of departure, von Neumann's theory of automata transposed the trait of self-replication into the concept of complication. His automaton was essentially a universal Turing machine, constructed to read a description and then to imitate the object described. In order to make an automaton duplicate an operation, which any other automaton could perform, it needed only a description of the relevant automaton and the required operational instructions. [132]

In one respect Turing's procedure was too narrow: his automata were purely computing machines, outputting only tapes studded with zeros and ones, not other automata. And there was another, trickier dilemma: in the biological realm species increased in complexity through evolution, but how could a Turing machine produce anything more complicated than itself? Von Neumann thus sought to derive an equivalent of Turing's theory for self-reproducing and evolving biological automata. He sketched plans for a kinematic model, where an automaton floating in a reservoir containing a dozen or so elementary parts sorted and assembled them into progeny according to instructions. He hypothesized that a specific set of instructions could roughly affect the functions of a gene, while a copying mechanism performed the fundamental act of reproduction: the duplication of the genetic material. Genes became essentially an "information tape."[133]

As von Neumann saw it, his model was flexible enough to encompass not only mechanisms of autocatalysis but also mutations and heterocatalysis. By altering a set of instructions one "can exhibit certain typical traits which appear in connection with mutation, lethally as a rule, but with a possibility of continuing reproduction with modification of traits." An additional set of instructions supposedly could s(t)imulate the production of gene-specific enzymes. Thus the solution to his evolutionary dilemma was that below a certain level of complexity automata were degenerative (producing less complex automata) but above which they could reproduce themselves and higher entities. The construction of such automata required several simplifying assumptions, von Neumann conceded, but these simplifications had great heuristic value: they were predicates for understanding the more complex natural processes.[134] Such simulations opened a new category of representations of life. Discursive and semiotic structure promised to be epistemically and technically compatible with the burgeoning cybernetic systems of the 1950s: organisms as computers and computers as organisms.

The Hixon presentation formed the foundation for von Neumann's subsequent lectures on automata. He elaborated on these principles in a series of five lectures, entitled "The Theory and Organization of Complicated Automata," delivered at the University of Illinois in 1949, and again in a series of five lectures, entitled "Probabilistic Logics and the Syntheses of Reliable Organisms from Unreliable Components," delivered at Caltech in 1952. He later developed a theory of cellular automata, which was more amenable to mathematical analysis and could circumvent the difficulties inherent in fusing together physical components. Aspects of this theory were presented in the 1953 Vanuxem Lectures at Princeton. Though none of von Neumann's presentations yielded a complete manuscript, John Kemeny, von Neumann's Princeton associate, assembled the written fragments and lecture notes to produce an article for *Scientific American* (1955), thus introducing von Neumann's ideas to a broader audience.[135]

Machines, like organisms, so the analogy went, could utilize raw materials from their surroundings and transform them into complex and specific matter that made up their parts. Thus a reproducing machine had to be endowed with the ability to transform pieces of matter into machine parts—rolls of tape, vacuum tubes, dials, photoelectric cells, motors, shafts, wire batteries—and to organize them into a new machine. Comprising three conceptual units, the von Neumann machine possessed a "nervous system," a "brain," and "neurons," to provide logical control; body "muscles," cells which could change surrounding cells, lowering and raising their levels of complexity; and a genetic "tail," transmission cells to carry messages from control centers. The genetic tail, the machine's crucial unit, was viewed as the set of chromosomes; the machine always copying its tail for the new machine via coded instructions, switching cells "on" and "off." Such machines could presumably undergo an evolutionary process, the article explained. By introducing random changes (mutations) in the machine code and limiting the supply of raw materials, one could simulate the dynamics of selective pressures. Machines would have to compete for *Lebensraum*, even to the extent of killing one another.[136] The premise was a functionalist one: "life," "brain," or "chromosomes" were not used as foundational entities or essentialist categories. Instead, they were what they did. Genes were copiers, mutators, and instructors, their function stated in terms of operational outcomes. In von Neumann's scheme, epistemology and technology defined and folded into each other.

It is difficult to assess the specific impact of von Neumann's automata on researchers in robotics or life science, or the shape that impact would have taken had he not essentially abandoned these scientific interests to become an AEC commissioner in 1955, two years before his death. There is no doubt, however, that his technological visions engaged engineers, mathematicians, and geneticists. At Bell Labs, where Shannon's mouse drew inspiration from von Neumann's automata, mathematician E. F. Moore conjured a possible use for self-reproducing machines. He produced a feasibility study for a large artificial plant that would extract minerals from the ocean to obtain materials from which it would build copies of itself.[137] Homer Jacobson of Brooklyn College took up the mechanical challenge of constructing von Neumann's machine, producing an intricate trainlike apparatus that generated a kind of primitive self-replication. And British geneticist Lionel S. Penrose devoted a few years to the pursuit of von Neumann's ideas. With crystal growth as organic analogy, he built a clunky mechanical "crystal" as a primitive replicating model, hoping it would mimic reproduction in living cells but ultimately conceding that it was more likely to assist in understanding the evolution of simple prebiotic molecules.[138] With Manfred Eigen's work on molecular evolution in the 1970s, however, von Neumann's dream of self-reproducing and evolving automata came to fruition.

Geneticist Joshua Lederberg at the University of Wisconsin, then blazing through the mapping of the E. coli genome, became intrigued by von Neumann's replication model. He had read and also heard from several friends, among them Sol Spiegelman, about von Neumann's contributions to "the problem of the minimum information required for 'self-reproduction'" and related aspects of genetic theory. As a geneticist, Lederberg explained, he too was challenged by that problem: What did it mean for a nucleus, chromosome, or gene to "reproduce"?

> I have occasion to feel rather uncomfortable at the adequacy of the genetical concept of "self-reproduction," or rather at the operations by which this is inferred for any sub-cellular particle. . . . Since most intracellular particles will not reproduce at all outside the context of a given kind of cell, the criteria for the capacity of "self-reproduction" of individual genes, etc., may be grossly deficient.[139]

He requested von Neumann's reprints and citations on the topic. Von Neumann referred him to the Hixon symposium paper. Lederberg also read Kemeny's *Scientific American* article. The lively exchange that ensued underscores not only the similarities and incongruities between von Neumann's self-reproducing automata and molecular biologists' representations of the genetic process but also brings into sharp relief the problematic uses of information theory in biology.

Lederberg took issue with the principal premise of von Neumann's theory: that, since the entire assembly alone possessed the self-replicating property, the functioning of the system as a whole was a problem independent of the constitutive elements. He too had been groping for the same inference, "simply on the basis that genes, or even nuclei, are incapable of producing anything, much less copies of themselves, when isolated from the whole machine." Nevertheless the geneticist wished to abstract from the entire organism "the least structure that will perpetuate the genetic function" (for example, which components of the tobacco mosaic virus could be removed without impairing the generation of progeny). He wondered about the apparent contradiction between digital codes used by automatons and nature's reliance on what he termed "non-digital coding": "The linear chromosome must be one of the most elegantly coded sequences, having baffled even Gamow's cryptography." He also challenged von Neumann's analogy between genes and "information tape," suspecting that such a conceptual construct had no structural representation. On these issues, he wrote a paper and sent a draft to von Neumann.[140]

The correspondence reveals how their debate proceeded at cognitive and disciplinary cross-purposes, the bottleneck forming around the problematic use of the term *information*. Was information to be construed as rules of behavior, instructions, or material content? If it meant all three, then, as Lederberg saw it, one was really dealing with the well-established and experimentally more useful concept of *biological specificity*. Lederberg wrote to von Neumann, "In this event, then, 'biological specificity' would have an even broader content than information, and I would merely substitute the terms in my account. I would also be more discouraged about the possibility of learning something that could be put to good use in the laboratory."[141] (As we shall see, the problem with Quastler's project was that it could not be put to good use in the laboratory.) Von Neumann, for his part, did not follow the technical meaning of Lederberg's notion of "information" as specificity, nor his other terms, such as "self-sufficiency" and "indifference."[142]

Finally, after consultation with informed colleagues, notably Leo Szilard, Lederberg acquired sufficient technical knowledge to conclude, "I think the root of our trouble is that we are working at very different planes." Molecular biologists were concerned with *realistic* working models of reproduction. Perhaps they could design a chemical analog to a punched-card reproducer, and thus go "a long ways toward the experimental initiation of life," if only they possessed an equal knowledge of the model's parts. They could simulate an autocatalytic system by punched-card reproduction with more than one or a few bits on it, Lederberg pointed out, but none of the chemical machines they could devise would produce results anywhere near the complexity of an organism. He noted, "I can see that you have been looking for the foundation of an axiomatic theory of reproduction, and that I had been needlessly reading my own mechanical interpretations into it."[143] Hoping to remove the constraints of correspondence, Lederberg pursued these discussions face-to-face in Washington, D.C., von Neumann's new center of operations. But nothing more concrete (e.g., experimental projects) came out of their meeting. Similarly, no tangible technical innovations seem to spring out of the Spiegelman–von Neumann exchange.

Nevertheless, the discussions around von Neumann's genetic simulacra served to transport the information discourse into biology. For example, by the early 1950s Spiegelman (by then at the University of Illinois) had come to conceptualize the "enzyme forming system"—template, enzyme, and inducer—as a cybernetic feedback model; the DNA template signified as the locus of information storage and transmission. Willy-nilly, his rigorous critique notwithstanding, Lederberg too adopted these discursive practices.[144] A new way of thinking and speaking began to permeate molecular genetics. Living entities were increasingly conceptualized as programmed communications systems, in which, as Lederberg astutely sensed, instructions and material content were collapsed into a single amorphous fabric of information. The "medium was the message," as Marshall McLuhan later observed in his critique of the cybernetic society. "There is no longer any medium in the literal sense: it is now intangible, diffuse and diffracted in the real, and it can no longer even be said that the latter is distorted by it," added Baudrillard years later.[145] The information discourse and its modes of signification bestowed upon the biological sciences—long beleaguered by Comtean inferiority—some of the high status and promise of command and control fields. Of course, as Cherry and Bar-Hillel complained, biologists, with a few exceptions, tended to deploy the concept of information very loosely—in ways quite unrelated to its intended technical meaning and quantitative usage in the Wiener-Shannon theory. (But, of course, the theory had metaphorized the "information" in the first place.) One such exception was radiation biologist and physician Henry Quastler, whose mission in the 1950s was to transform biology into a technical information science.

QUASTLER'S QUEST: FROM BIOLOGICAL SPECIFICITY TO INFORMATION

Histories of molecular biology have generally privileged "winners'" accounts, while "losers" have gotten short shrift and, in the process of the canonization of others, stripped of recognition. Yet works written outside canonical history afford equally instructive lessons. Quastler's efforts are illuminating examples of a well-reasoned epistemic quest and a curious disciplinary failure. They also provide an opportunity for situating his project within the new scientific space traced by the intersection of cybernetics, information theory, and molecular life sciences. That new space was generally characterized by the grassroot forays into cybernetics of life scientists, among them Haldane, Kalmus, Penrose, Spiegelman, Lederberg, Delbrück, Sinsheimer, Yčas, Chargaff, and Burnet, certainly conveyed an attraction; their work did not constitute formal institutional or disciplinary ventures (as would have been indicated by symposia, proceedings, and funding).

Quastler embarked on building a new subdiscipline—an information-based biology—in a technically proper form. His output, in the form of articles, reports, symposia, and edited volumes, was prolific. He achieved a measure of acclaim in the 1950s as a pioneer in a highly technical and intellectually challenging branch of biology. *Science Citation Index* (1955–63) contains nearly four hundred references to his work. Many people were impressed with his biomathematical prowess and valued his theoretical framework, but his work was plagued by problems: outdated data, unwarranted assumptions, some dubious numerology, and, importantly, an inability to generate an experimental agenda. These weaknesses, coupled with Quastler's early death, led to the eventual eclipse of his quantitative studies. Yet his discursive framework survived and flourished, as did the cybernetic imagery he propagated. Its semiotic impact endured long after his project fell into obscurity.

Like many European refugee scientists, Henry Quastler (1908–63) arrived in New York in 1939 from Vienna with impressive credentials and uncertain prospects. He had received a medical degree from the University of Vienna in 1932, with a focus on histology and radiology, buttressed by training in physics, chemistry, and mathematics. With his chances for a medical career in Austria rapidly dwindling after 1933, Quastler moved to Tirana, Albania, where his five-year stay led to some unusual advances. Charming, cultured, and skilled, he allegedly won the confidence of Albania's King Zog (Ahmed Bey Zogu) and became court physician. As chief of radiology at the Tirana General Hospital he managed to carry out clinical research in radiology, as well as experimental studies on malaria, the latter earning him a position in 1939 with the local bureau of Rockefeller's International Health Board. That year, with Mussolini pressing on Albania's border, the unpopular Zog fled to Greece, while Henry Quastler, rescued by Rockefeller Foundation officer Marston Bates, sailed to America. Bates had argued that Quastler possessed "a first class scientific mind—apparently a rather rare phenomenon" that would be a pity to waste. Within a year Quastler was employed as assistant radiologist at New Rochelle Hospital in New York. In 1942 he relocated to Urbana, Illinois, where he became chief radiologist of the prestigious Carle Hospital Clinic, an offshoot of the Mayo Clinic.[146]

With a penchant for research and a score of publications to his credit, Quastler soon associated himself with the University of Illinois. He made friends exceptionally easily, especially through the "Viennese mafia," as Heinz von Foerster put it. He participated informally in research, gaining a half-time appointment in radiobiology in 1947, while pursuing full-time medical practice. The University of Illinois was then modernizing and expanding its research programs, building up computing and cybernetics, molecular biology, and high-energy physics. Part of the Argonne National Laboratory Midwestern consortium, the university was then operating Betatron, a powerful electron accelerator, as a centerpiece of its high-energy physics and biomedical research. Quastler's studies of the effects of X rays on organisms and the role of the Betatron in cancer therapy fit squarely within that postwar agenda. After giving up his medical practice in 1949, he was appointed assistant professor of physiology at the new Control Systems Laboratory. The study of the therapeutic effects of radiation became his venue for "healing the world's wounds" inflicted by the atom bomb. While maintaining an active experimental program (and secret war work) in radiobiology, Quastler's scientific imagination, social skills, and relentless energies turned to applications of information theory to biology. Warren Weaver was "very well impressed by Q.," by his command of physics and by his close familiarity with the work in England.[147]

FIGURE 9. Henry Quastler with his wife, Gertrud, 1950s.
Courtesy of the Library of Congress.

Several years in advance of a DNA-based molecular biology that featured the genome as a "text" and protein synthesis as translation of "instructions," Quastler had used these concepts to establish measures for genetic information based on knowledge of protein structure, specificity, and function. Surrounded in the late 1940s by friends such as Warren McCulloch and Heinz von Foerster (a leader of cybernetics) and captivated by the Wiener-Shannon theory, the Macy conferences, and von Neumann's lectures, Quastler joined the champions of the cybernetic vision of life. Teaming up with physicists and chemists, he combined his knowledge of cytology and genetics with his newly upgraded mathematical skills to erect an applied framework for quantifying the information flow in biological control systems. His close collaboration with Sydney M. Dancoff (1913–51) was especially significant, initiating them both into the terra incognita of cybernetics. A leading theoretical physicist of his generation, Dancoff, as participant with Quastler in the Betatron group, was branching away from quantum electrodynamics and nuclear physics to pursue his interest in biological growth, biological complexity, and reproduction as a problem of stability.[148] Their collaboration led to the first technical application of the Wiener-Shannon theory in genetics.

Quastler and Dancoff's paper, "The Information Content and Error Rate of Living Things," written in 1949, benefited from critical consultations with Salvador Luria, Sol Spiegelman, Barbara McClintock, Tracy Sonneborn, and Aaron Novick and required major revisions that delayed its publication by four years. These exchanges engaged life scientists in new ways of seeing and speaking about organisms. In a bold departure from the biological canon that natural selection serves to preserve the accuracy of replication by disposing of errors, Quastler and Dancoff proposed a cybernetic system: a chromosomal error control. Viewing replication as a high-error (mutability) process, they hypothesized a checking device—a purely statistical process—within the elements that received the messages from the chromosomes. Brushing off Luria's objections, Quastler wrote Dancoff on the fourth of July, "I believe we do like the American Constitution and stick to the system of independent checks and balances. Our next problem ought to be: how can an organism go about evolving an independent check?"[149] Nature and society, he mused, were governed by similar control mechanisms.

They envisioned the complexity of living systems in terms of high information content, and the chromosomal thread as being

> A linear coded tape of instructions. The entire thread constitutes a "message." This message can be broken down into sub-units which may be called "paragraphs," "words," etc. The smallest message unit is perhaps some flip-flop

> which can make a yes-no decision. If the result of this yes-no decision is evident in the grown organism, we can call this smallest message unit a gene (Note that genetic alleles are *two* in number for each character—not three, or some other number [*sic*]).[150]

And if a gene was to a chromosome as a vacuum tube was to a radio, or a neuron to the nervous system, then the property of such an element would be simply that of a switch, or a relay, or an amplifier, they reasoned. As such, the statistically distributed binary decisions in the self-correcting chromosomal system seemed truly amenable to a Wiener-Shannon analysis.[151]

With that analysis, the final paper derived an information content based on hypothetical "instruction to build an organism" and proceeded to use that formula to estimate the information content of a human being in terms of atoms: 2×10^{28} bits, and of molecules: 5×10^{25} bits. The information content of single printed page is about 10^4 bits; thus the description of a human in terms of molecules would take up to 5×10^{21} pages—many orders of magnitude greater than the content of the largest library, the authors concluded. They approximated the information content of a germ cell as 10^{11} bits and that of a "gene catalog" as 10^5. They then used these figures to ascertain that the generally accepted theoretical range of error rate per generation (10^{-4} and 10^{-12}) agreed with the estimated theoretical range derived from Dancoff's proposed checking mechanism. Their calculations remind one that numerical similitudes and mathematical correspondences have a long history of epistemic seduction. Succumbing to what biochemist Joseph Fruton has aptly characterized as the "hypnotic power of numerology" seems to be reenacted in new forms with every scientific generation.[152]

After the blow of Dancoff's death, Quastler mobilized other colleagues for his new venture through personal appeal and local gatherings. Inspired by British and French information theorists such as Denis Gabor, Ross Ashby, Colin Cherry, Benoit Mandelbrot, and Leon Brillouin, Quastler entered into collaboration with some of them, hoping to bring a similar kind of leadership to biology. The symposium on "Information Theory in Biology," which Quastler organized in 1952 under the auspices of the Control Systems Laboratory with funding from the Office of Naval Research (ONR), was intended to be a first step in extending the "new movement" of information to the life sciences, even if it featured primarily local researchers. The often-cited symposium proceedings represented the earliest authentic effort to rewrite biology as an information science.[153]

The symposium focused on four areas: definition and measurement of information; structural analysis; functional analysis; and biosystems. Of the papers in the first area only Quastler's centered on the salient feature of information theory in biology: the mathematical interchangeability of information with specificity. His paper, "The Measure of Specificity," devised a framework and case study that supplied a persuasive argument for linking information theory to biology through the central concept of specificity. Acknowledging the stricture that "information" in the Wiener-Shannon sense did not convey meaning, he nevertheless expected to construct powerful representations in the new biosemiotic space. Information, he reasoned, related to activities of living systems—designing, deciding, messaging, differentiating, ordering, restricting, selecting, specializing, specifying, and systematizing. It could be used in operations aimed at decreasing quantities such as disorder, entropy, generality, noise, randomness, and uncertainty, and increasing the degree of design, orderliness, regularity, differentiation, and specificity. Since specificity was the governing principle in life science, Quastler reasoned that, when properly qualified, an information content would supply the exact measure of biological specificity. Establishing an estimate of "enzyme specificity" based on binary choices of "right" and "wrong" substrates, he showed that the expression resembled Shannon's entropy equation. His point was that the amount of specificity could be determined without reference to causal mechanisms or the nature of "reactive gadgetry." Thus Shannon's principle that the "quantity of information in a message could be defined independently of its meaning" also obtained in biology, according to Quastler.[154]

Of course, structure mattered; it imposed definite limits on the functional range of molecules. But according to information theory, when properly applied, *any* organized structure displaying biospecificity in a form of feedback or communication—enzyme, hormone, antibody, gene—was an information carrier. Proteins epitomized biological specificity. Thus Howard University physicist Herman R. Branson expected that an information theoretical analysis of protein structure would lead to discoveries of new biological properties. Proteins possessed especially attractive properties from the standpoint of information theory, he argued:

> They are constructed much as a message, since they consist of some definite arrangement of about 20 different amino acid residues. Thus, the protein molecule could be looked upon as the message and amino acid residues as the alphabet. We do not know if the letters of this alphabet (the amino acids) are arranged in words within the message or not, that is we know nothing of the redundancy characteristics of the protein molecule.[155]

But assuming that the message was one of the many possible arrangements of the letters and spaces (i.e., neglecting intersymbol influence), one could then use the equivalence of information and negative entropy to calculate the information content in proteins, based on the content of their amino acid residues. Branson produced such calculations for nearly thirty proteins (e.g., insulin, 3.55 bits/residue; salmine, 1.43 bits/residue) and even observed regularities in the distribution of information, which he thought might represent fundamental attributes of biological systems. Specifically, he intuited that these patterns held insights into the antigenic components of proteins and into the origin and evolution of life.

In remarkable resemblance to the reasoning and symbolic representations of genetic codes only a couple of years later, Branson and his colleagues from the Control Systems Laboratory embarked on a search for intersymbol influences in protein structure, in quest of regularities in the amino acid pattern. Applying Shannon's theory, where intersymbol influences in the English language had been measured, they chose twenty paragraphs from diverse sources (want ads, textbooks, newspapers, and magazines) that matched the proteins in their sample in symbol length, treating letters as amino acids and paragraphs as proteins. (Gamow, Rich, and Yčas preferred the more dignified Milton's *Paradise Lost* as their analytic sample.) That method failed to reveal a type of intersymbol influence known to exist in language. (In hindsight their result makes sense, since the assembly of proteins is determined by nucleic acid codons, with no logical restriction on their position.) The authors concluded with the inconclusiveness of negative results but conceded to the possibility of real differences between the laws of language-construction and those governing the construction of proteins.[156]

Information based on specificity of protein structure, Quastler argued, was a necessary but insufficient condition for biological function. Structural specificity constituted an upper bound to functional specificity, with only a fraction of structural specificity finding functional expression. Focusing on antibody and gene action as forms of highly specific communication, he invited to the symposium immunochemist Felix Haurowitz from Indiana University and immunogeneticist M. R. Irwin from the University of Wisconsin, with the intent of estimating the specificities associated with the function of genes and antigens. Quastler also invited Joshua Lederberg, but Lederberg remembers being "pretty quizzical" about the symposium's aims and ONR's sponsorship. Lederberg worried that recorded discussions might spill over to topics subject to security classification. The virulence of McCarthyism and the sting of the loyalty oath in the academy deterred him from such potential security entanglements.[157]

Neither Haurowitz nor Irwin possessed mathematical skills for performing information theoretical analyses. Haurowitz merely described the specificities of immunity and antibody formation based on the protein template model, and Irwin lectured on genetic specificities in human, avian, and cattle blood groups. It was Quastler who fed these qualitative accounts into his information mill to spew data on measures of functional specificities of genes and antibodies. His mathematical manipulations seemed to yield an astonishing result: functional specificity could be measured in multiples of 9 bits. Succumbing to the hypnotic power of numerology, Quastler tentatively proposed that "biological specificity is quantized in units of approximately 9 bits," suggesting a selective mechanism at work. The degree of selectivity, he hypothesized, "could be achieved by nine binary decisions under optimum conditions of maximum efficiency and error-free operation."[158]

Moving on to communication studies of "bio-systems," several participants examined the hormonal control of blood-sugar level, as a scheme where the "address of the target organ" and the "instruction" it received were encoded into the hormone at the place of production and decoded at the place of action. Another participant, Kenyon Tweedell, analyzing the development of zygotes and identical twinning, praised information theory for providing a compromise to the age-old debate of preformation versus epigenesis: specificity corresponded to preformation, epigenesis to nonspecificity. He wrote, "The information content is a set of instructions coded into the fertilized egg as dictated by genetic constitution; if a section of the instructions happens to lie in the zone which will give rise to the part to which this section refers, the part will behave as if preformed."[159] This argument preceded François Jacob's analysis of the "genetic program" and Delbrück's informational reinterpretation of Aristotle's theory of generation by more than a decade.

Only Henry Linschitz, a physical chemist from Syracuse University, kept a skeptical posture, offering a perceptive critique of the application of information theory to biology. Physical entropy did not provide a proper measure for the *functional* information contained within an organism, he countered. Examining the relation between structural entropy and functional organization, he questioned the applicability of information derived from entropy considerations in inert systems to living ones. Further, he believed that various problems complicated such applications, all of which related to the difficulty of defining the entropy of a system composed of *functionally interdependent* units. In a cell containing numerous protein molecules, entropy might be closely related to functional information content only if all the cell's constituents where chemically linked, so that the organized entity was truly molecular. If, on the other hand, cellular interactions occurred simply by proximity, by virtue of the existence of a common enclosing membrane, then physical entropy did not measure the essential organization of the functioning complex.[160]

One might calculate, as Linschitz did, the *physical entropy* of the cell, or its *structural organization*—the amount of negative physical entropy the cell needs to make itself—but these calculations said nothing about the way this entropy was channeled to produce a functioning cell. Functional organization was the result of nonstructural coupling between functioning units constituting the organism, he argued. (Recall that Schrödinger rejected even the equivalence of information and negentropy as a measure of structural organization.) Scientists must have also objected to the misapplication of a thermodynamics of closed systems to nonequilibrium (animate) processes.

Quastler was not deterred by these weaknesses. Shuttling back and forth along the provocative similitude linking the Wiener-Shannon measure of information with the Boltzmann-based entropy equation (a mathematical homology which did not necessarily prove the existence of an underlying common physical mechanism), he kept transporting new conceptual models and analytical tools into biology. Based on rough-and-ready estimates and piling assumption on top of hypothesis, Quastler built a semiotic house of cards: symbolic representations, quantitative relations, and numerological patterns. His pioneering investigations were not necessarily wrong, but they were devoid of predictive capability, means of theory-testing, and experimental agenda. Nevertheless they contributed to the construction of a new epistemic space, in which "biological control systems" could be represented axiomatically, independent of the knowledge of their materiality and components. By explaining precisely how cryptanalytic and mathematical principles of information theory could be applied to living systems, Quastler and his collaborators seemed to actualize the potentialities in Wiener's cybernetic vision, Shannon's communication scheme, and von Neumann's self-reproducing machines. Von Foerster's invitation to present these ideas at the Ninth Macy Conference on Cybernetics (1953) attested to Quastler's growing authority in the field.[161]

Quastler, by now, was receiving considerable scientific recognition. "If a man is 46 and famous, of fame newly confirmed on a trip in the big world, shouldn't he be showered with gifts?" Quastler asked his wife, Gertrud (a successful artist), on his birthday in 1954. "In the few months that he spent at Argonne before he came to Brookhaven he influenced more people than I would have thought possible in a few years or even in a life time," noted his colleague at Quastler's memorial service. (Quastler committed suicide the day after Gertrud's death.)[162]

Quastler refined his ideas on information theory in biology (and psychology) while pursuing his primary research on radiation biology and planning a major career move. After a brief interlude at the Argonne National Laboratory, which sponsored a large research program on clinical applications of radioisotopes and high-energy radiation, in 1955 Quastler joined the biology department of the Brookhaven National Laboratory in Upton, New York. All three national laboratories, Argonne, Oak Ridge, and Brookhaven, had their roots in the Manhattan Project and were greatly expanding in the postwar decade to form a principal research base for the AEC. These centers offered luxurious research facilities for those who could embrace the ethos of secrecy in science.[163]

The national laboratories were central to the tripartite agenda of the AEC, declared commissioner Henry D. Smythe in 1949:

> The first is to make more and better weapons. The second is to develop possible peacetime uses of atomic energy, and the third is to develop such scientific strength in the country as is needed in the long run to support the other two. . . . There are good reasons why the National Laboratories are needed and why their functions cannot be discharged by already existing research or development organizations. One of these, of course, is secrecy. . . . Another reason is . . . the advantage of gathering together a large group of men from the various divisions of science to work in close cooperation.[164]

As the only national laboratory in 1955 with a large research reactor and a proton synchrotron in the billion-electron-volt range, Brookhaven offered the most expansive setting for research in radiation sciences. It also had the advantage of being formally founded in 1946 after the war, meaning it could bypass policies that constrained other national laboratories. Of the three laboratories, Brookhaven came closest to realizing the original model of a regional, cooperative research center.[165] In the 1950s, the Brookhaven symposia, sponsored by the university's biology department, centered on topics such as "Mutation" (1955), "Genetics in Plant Breeding" (1956), and "Structure and Function of Genetic Elements" (1959). These became prestigious gatherings in molecular biology.[166]

Quastler flourished at Brookhaven. While pursuing his two projects—radiation biology and information biology—his circle of collaborators and students grew steadily. In 1956 he helped organize a sequel symposium on information theory and biology, which was far more extensive than the one in 1952. Sponsored by Oak Ridge National Laboratory, Tennessee (which employed scores of life scientists in its health physics and biology divisions), the symposium focused on storage and transfer of biological information, information measures, information and ionizing radiation, aging and radiation damage, and information networks. McCulloch was invited to participate but declined, expressing his reservations about the project. He felt that information theory might not yet be suitable for probing biological complexities, and, in any case, its application hinged on cracking the genetic code. He wrote:

> I doubt whether information theory is yet properly attuned to the complexities of biological problems. To apply the theory in its present state except in a most rudimentary fashion we need to crack the code in genetics as surely as we do in the Central Nervous System. I presume nature has somewhat optimized the code for the kind of noise it encounters, but even when we know something of the noise we can not yet devise the optimum code enough to put a significant lower bound on the information capacity of a biological channel. . . . I have talked with Dr. Wiesner and several other members of the department and they are inclined to agree with me.

Indeed, the most illuminating aspects of the symposium—from the standpoint of the history of molecular biology—were the introduction, the papers on the genetic code, and the conclusion.[167] The introduction included an elaborate primer (with exercises) on information theory, originally prepared by Quastler as a technical report for the Office of Ordnance Research and, according to von Foerster, one of the best pieces ever written on the subject. Designed to supply biologists and psychologists with the tools for recasting their problems in informational terms, this primer gave considerable attention to principles of coding as an aspect of representing information. Building on earlier studies of "protein coding" and obviously responding to the growing enchantment with the DNA "code," the primer explained symbols, alphabet, and "words" as units of representation and examined the status of different types of codes in information theory. It also probed the problem of organization in terms of systems analysis and game theory. Several exercises demonstrated how to apply these techniques to biological problems.[168] As part of the introduction, Oak Ridge biophysicist Hubert P. Yockey drew on the latest knowledge in molecular genetics, including the Watson-Crick DNA model, to devise an information-theoretical model of protein synthesis in the cell. His main argument, based on Dancoff's principle, hinged on the role of noise in the genome. He underscored the principal contribution of information theory in biology, emphasizing its ability to quantify the two key concepts of organization and specificity.[169]

George Gamow and microbiologist Martynas Yčas, engaged in an intense collaboration to decipher what by then was increasingly referred to as "the genetic code," presented a cryptographic approach to protein synthesis. Assuming a "transfer of information from nucleic acids to proteins" and performing a cryptanalysis of the possible distribution of amino acids in two proteins, they showed that these sequences were subject to fewer strictures than language-based cryptograms; thus they expected protein decoding to be far more challenging than ordinary codes.[170] Yčas examined various "protein texts" (partial protein sequences), analyzed the frequency of occurrence of amino acids, and via the Wiener-Shannon relation, applied these findings to the problem of information storage and transfer in RNA and protein, all of it with, as he readily admitted, inconclusive results.[171]

In fact, a cloud of inconclusiveness seemed to hang over the entire 1956 symposium. The promotions of information theory in biology notwithstanding, the organizers conveyed an overanxious tone of those seeking recognition. Perhaps due to their quasi-academic status, or quizzical insecurity about their hypotheses, or just a sense of frustration, the concluding roundtable, led by Quastler, lacked the upbeat tone of earlier discussions.

> Information theory is very strong on the negative side, i.e., in demonstrating what cannot be done; on the positive side its application to the study of living things has not produced many results so far; it has not yet led to the discovery of new facts, nor has its application to known facts been tested in critical experiments. To date, a definitive and valid judgment of the value of information theory in biology is not possible.[172]

As participants put it, information theory furnished a sort of thread that enabled them to sense a continuum in the order of the universe, a means of relating existence to the nonexistence of life, a quest of regularities in irregular phenomena. It also encouraged analysis in terms of system sciences: the whole rather than its parts, the general instead of the particular, patterns rather than specific mechanisms. Information theory in biology was here to stay, they believed, but perhaps at a price of a compromise, as a discursive rather than mathematical tool. Given "the irreducibly relative nature of information measures" and the difficulties with quantitative applications, it might be "preferable to use information theory only in a semi-quantitative fashion," namely metaphorically.[173] By 1961 Yčas would reinforce this assessment. "Workers have been aware of information theory and have made qualitative use of some of its concepts," he conceded, then continued, "no explicit, and especially no quantitative use of information theory has, however, been made in practice."[174]

Nearly a decade after the emergence of the Wiener-Shannon theory and von Neumann's models of self-reproducing automata, with their promise of bringing operational techniques to "genetic communications," the technical status of information theory and cybernetics in molecular biology was dubious. By the late 1950s Shannon's skepticism regarding the applicability of information theory beyond the engineering domain seemed abundantly clear. As Cherry repeatedly stressed: information is not a commodity. He noted, "Signals do not convey information as railway trucks carry coal. Rather we should say: signals have an information content by virtue of their *potential for making selections*." Von Foerster too voiced similar objections, arguing that it was the military context of command and control that created that confusing epistemology.[175] Yet despite the acknowledged technical impotence of information theory in molecular biology, its discursive potency intensified by compromising its technical structures. In the theory's proper form, and indeed in Quastler's mathematical analyses, all organized entities—carbohydrates, proteins, nucleic acids—contained information. Molecular geneticists (and biochemists in the late 1950s) singled out nucleic acids as the unique carriers of informational attributes. Information—as meaning and commodity—came to signify the privileged status of DNA as "master molecule." Emptied of its technical content, it actually became a metaphor of a metaphor, a signification without a referent. This, however, did not diminish its scientific and cultural potency. The discourse of information linked biology to other postwar discourses of automated communication systems as a way of conceptualizing and managing nature and society. And it provided discursive, epistemic, and, occasionally, technical frameworks for the scriptural representations of genetic codes in the 1950s.

CHAPTER FOUR

Scriptural Technologies: Genetic Codes in the 1950s

BLACK CHAMBERS: THE RISE AND FALL OF OVERLAPPING CODES

Scientists have generally tended to regard the theoretical work on the genetic code in the 1950s as naively optimistic at best, erroneous and unfruitful at worst. These attempts seem to represent the demise of the Pythagorean ideal by the world of matter. To explicate the significance of this important work, I aim to situate it within the wider network of code workers, particularly George Gamow and the RNA Tie Club, and also within the cultural and military context of the cold-war era. By placing Francis Crick's work in these contexts, I make apparent the other narratives and other histories of the genetic code. The dominant code narratives built around Crick are potent, as implied in what one critic observed: "One could hardly escape the conclusion that after the Eighth Day, Francis Crick rested."[1] These narratives have served to underrate, even displace, the work of others, notably George Gamow, a Russian émigré physicist, cartoonist, science popularizer, and military strategist, who has been duly credited as the first to define what became "the coding problem" and to inspire researchers to engage in its solution. Yet his enduring marks on molecular biology have become diminished in the shadow of Crick's history.

Not only did Gamow define, articulate, and attempt to solve the coding problem but also he brought the powerful culture of postwar physics with its various military linkages to bear on representations of heredity and life. And although his participation in molecular biology proved to be temporary, he left an enduring legacy: his approach supplied the potent imagery and discursive software with which that mythical object—the genetic code—became constituted. Gamow, deploying the discourse of information and representations of heredity promulgated in the works of Norbert Wiener, Claude Shannon, John von Neumann, and Henry Quastler, visualized heredity as a process of information transfer, operating via a code, much like an enemy code or a cryptogram. Meanwhile, Gamow's intense attempts to break that extraordinarily recalcitrant code became a seductive challenge for a significant number of his colleagues.

Soon, eminent physicists, biophysicists, physical chemists, mathematicians, communication engineers, and computer analysts—whose projects, like Gamow's, situated them near the hub of weapons design, operations research, and cryptology—joined in the effort to "crack the code." During their five-year research into the code, Gamow and his collaborators brought to molecular biology the tropes of communication sciences, namely, information theory, linguistics, and computer-based cryptanalysis, coupled with their simultaneous involvement in physics and defense. Gamow thus extended and amplified the discursive space that emerged with Norbert Wiener and John von Neumann in the late 1940s and had been elaborated upon by Henry Quastler in the early 1950s: heredity was conceptualized as information transfer; organisms and genes were represented in terms of messages, words, letters, instructions, and texts. Gamow's contribution of semiotic tools, soon followed by other linguistic tropes, such as commas, dictionaries, sense, nonsense, and missense, as deployed by Crick, Delbrück, and their collaborators, helped fix the image of the genome as a codebook. As Carl R. Woese, one of the code breakers and a user of information theory, reflected, "What has not been generally appreciated is that the subsequent spectacular advances in the field, occurring in the second period [1961–67], were interpreted and assimilated with ease, their values appreciated, and new experiments readily designed, precisely because of the conceptual framework that had already been laid."[2]

Not merely the conceptual framework had been laid down; rather, a knowledge-power nexus had formed within which molecular biology reconfigured itself as information science and represented its objects in terms of electronic communication systems, including linguistic communication. Scholars have either celebrated or lamented the cognitive and disciplinary impact of physicists on molecular biology, as well as the cultural impact of cold-war technosciences on the life sciences. As Evelyn Keller commented, it is on the social and material dimensions of that knowledge/power nexus that we must focus our attention. Physics served as a social resource for biology, which borrowed physicslike agendas, language, attitudes, and even names of physicists, thus eventually reframing the character and goals of biology.[3] But physics was not the only resource. By examining the work on the genetic code in the 1950s, including its cognitive strategies and discursive, material, and social practices, this chapter shows how the old problem of genetic specificity was reframed, or recast, through scriptural technologies

FIGURE 10. George Gamow, ca. 1955. Collection of Martynas Yčas.

poised at the interface of several interlocking postwar discourses: information theory, cryptoanalysis, and linguistics.[4]

The Diamond Code

Like most physicists, George Gamow (1904–68) enjoyed posing as a neophyte in biology, a mere dilettante, who had stumbled on an interesting problem. However, his abundant and cheerfully illustrated paper trails indicate otherwise. Already in the early 1940s, while at George Washington University in Washington, D.C., working on theories of the origins of chemical elements and the birth of the universe, he planned a biology book, entitled *The Dance of the Chromosomes*, in collaboration with Max Delbrück, his friend since their postdoctoral year at Niels Bohr's institute in Copenhagen (1932). The book's plans and outlines convey Gamow's up-to-date knowledge of and childlike enthusiasm for the subject matter. (The book was never completed.) He also organized the important 1946 Washington conference, "The Physics of Living Matter," designed to engage leading physical scientists in biological problems (addressed in Chapter 3). By spring of 1953, at the time of Watson and Crick's publication on the structure of DNA in *Nature*, Gamow's popular biology book, *Mr. Tompkins Learns the Facts of Life*, found wide circulation.[5]

Watson and Crick's report excited Gamow even beyond his normal Slavic exuberance.

> I remember very well this day, I was for some reason visiting Berkeley and I was walking through the corridor in the Radiation Lab, and there was Luis Alvarez going with *Nature* in his hand (Luis Alvarez was interested at this time in biology) and he said, "Look, what a wonderful article Watson and Crick have written." This was the first time I saw it. And then I returned to Washington and started thinking about it.[6]

He soon dispatched a letter to Watson and Crick.

> Dear Drs. Watson and Crick,
>
> I am a physicist, not a biologist. . . . But I am very much excited by your article in May 30th *Nature*, and think that brings Biology over into the group of "exact" sciences. I plan to be in England through most of September, and hope to have a chance to talk to you about all that, but I would like to ask a few questions now. If your point of view is correct each organism will be characterized by a long number written in quadrucal (?) system with figures 1,2,3,4 standing for different bases . . . [the "number of the beast" as quantitative indicator of the species, he suggested]. This would open a very exciting possibility of theo-

> retical research based on combinatorix [*sic*] and the theory of numbers! . . . I have a feeling this can be done. What do you think?

Gamow perceived the relationship between DNA structure and protein synthesis as a numerical cryptanalytic problem: How can a long sequence of four nucleotides determine the assignment of long protein sequences comprised of twenty amino acids? "And the question was to find, is it possible?"[7]

> And at this time I was consultant in the Navy and I knew some people in this top secret business in the Navy basement who were deciphering and broke the Japanese code and so on. So I talked to the admiral, the head of the Bureau of Ordnance. . . . So I told them the problem, gave them the protein things [list of amino acids], and they put it in a machine [computer] and after two weeks they informed me there is no solution. Ha!

The year Gamow plunged into deciphering the genetic code, 1953, was a key time in the cold war. Significantly for Gamow, cryptanalytic theories and technologies were just then transforming within a new knowledge/power nexus configured by computer-based linguistic analysis and information theory. That year too—the year of Stalin's death, the end of the Korean War, and the testing of the first Soviet hydrogen bomb—marked new heights in the escalation of national security. As the rage of McCarthyism intensified at home, newly inaugurated President Eisenhower substantially broadened and deepened Truman's loyalty program. Security clearance for the Manhattan Project's hero J. Robert Oppenheimer, who now opposed the work on the hydrogen bomb, was suspended, an event that polarized the physics community. Even scientists who had been relatively accommodating to the loyalty program now mounted resistance against the government's relentless pressure, which had even spread into areas of unclassified research, notably the life sciences. By 1954, America's 15-megaton H-bomb would be tested on the Pacific's Bikini Island.[8] Geneticist George Beadle, chairman of Caltech's biology division and member of the Atomic Energy Commission advisory committee in biology and medicine, warned, "Like many others, I see in our present security system a machine so complex in its operation that, once having been set in motion, it is almost impossible either to control or to stop. And I sense a real danger that it may destroy our way of life if we do not find a way to control it."[9] Science, indeed, played a pivotal role within that national security state. Within the newly launched geopolitics of "brinkmanship" and the "domino theory" of totalitarianism, as outlined by President Eisenhower and Secretary of State John Foster Dulles, the Defense Department's budget soared to more than $50 billion, more than doubling since 1951. Much of that money was spent on (nuclear) weapons research, space research, and the development of ever-faster electronic computers, thus placing scientific research—especially physics, mathematics, computer science, and cryptology—at the center of cold-war knowledge production.[10]

Cryptography, the methods of rendering a secret message (plaintext) unintelligible to outsiders through transformations based on letter transposition, substitution, or both, is as old as human civilization. Ciphers and codes have long occupied important political roles in antiquity: the Spartans established the first system of military cryptography, while Roman Caesars employed cryptography regularly. Monks, too, used ciphers throughout history. The use of political cryptology reemerged in the Middle Ages and underwent remarkable innovations in the Renaissance through the introduction of polyalphabetic substitutions, autokeys (use of the message itself as a key), enciphered codes, and use of frequency analysis. Analyzing letter frequency (how often a letter occurs in the text) and contacts (which letter it touches, and how many different ones) have remained the most universal and basic of cryptanalytic procedures in all languages (in postwar communication science it would relate to the measure of informational redundancy). Beyond technical innovations, the institutionalization of cryptology occurred in the seventeenth century, when the great Antoine Rossignol, who introduced into cryptography a two-part nomenclator, or a kind of bilingual dictionary, became France's cryptologist at the court of Louis XIV. Throughout the eighteenth century cryptanalytic activities of states came to be housed in central black chambers, called "Cabinet Noir," which was also a label for later centers of political cryptology in Europe and America.[11]

Clearly not all codes were devised for political or military ends. Commercial codes proliferated in the nineteenth and early twentieth centuries. In 1843, the first public electric telegraph line was laid in England, and in 1844 Samuel F. B. Morse (inventor of the Morse code) established the first public telegraph line in the United States. This new technology quickly displaced the Chappe semaphore. The Atlantic cable gave immense impetus to commercial codes. By the beginning of the twentieth century virtually every nonlocal industry had compiled commercial codes; however, their use began to decline in the 1920s and 1930s.[12] During the First World War, cryptanalysis played only a limited role. By war's end its two core areas—cryptographic operations (encoding) and cryptanalytic techniques (decoding)—were left inadequate and depleted. Manual encipherment could hardly cope with the message load, and brute frequency analysis challenged the greatest of masters. For cryptology, the war marked an end to a long era; afterward political codes flourished. In the 1920s and 1930s cryptography underwent fundamental institutional, mechanical, and theoretical transformations.

In Britain the government's cryptanalytic agency was greatly expanded within the country's Foreign Office; in 1939, its euphemistically named "Department of Communication" moved to the famous Bletchley Park, probably the world's leading center of cryptanalysis in the 1940s, which employed some of the greatest mathematicians and physicists of the period (among them, Alan M. Turing). In the United States the organization that became known as the American Black Chamber was ensconced in New York (1919), under the leadership of the noted cryptologist Herbert Osborne Yardley. Newly invented cipher machines replaced pencil-and-paper methods in the late 1920s and mathematical methods were introduced in cryptology. The New York mathematician Lester S. Hill deployed algebraic solutions (linear equations and their matrices) for encipherment, and the Russian-born Wolfe Friedman, a Cornell-trained geneticist, pioneered statistical methods in cryptanalysis (perhaps inspired by population genetics). For the first time in cryptology, letter frequency distribution was treated as a mathematical parameter (yielding a curve whose several points were causally linked), signaling a conceptual leap that inspired diverse statistical tools in cryptology. Friedman served as the first director of the Signal Intelligence Service (SIS, founded in 1930 as the successor to the American Black Chambers and renamed Army Security Agency after 1945), which within a few decades grew to be America's cryptologic empire at the service of the National Security Agency (NSA, signed into existence by President Truman in 1952). The institutional, mechanical, and mathematical trends of the 1930s culminated in a major transformation by 1945. World War II mechanized cryptography and mathematized cryptanalysis, while cryptology became a nation's prime source of intelligence.[13]

By the time Gamow consulted with the navy cryptologists, those practices were already being transformed by the mathematical theory of communication. Information theory raised cryptanalysis to a new technical level through the application of the concept of redundancy to theory of codes, as formulated by Claude Shannon in the late 1940s and early 1950s.[14] Recall that Shannon's statistical theory of communication aimed, among other things, to set up a measure of redundancy in codes (e.g., telegraph codes), counted in bits. (Redundancy was defined as one minus the relative entropy; see previous chapter.) With coded messages expressed in signs (i.e., as physical signals), the amount of redundancy added or subtracted when code structures were modified could be formulated quantitatively.

Shannon applied this idea to linguistics and cryptanalysis, thus simultaneously influencing both. Redundancy meant that more symbols were transmitted in a message than were actually needed to bear the information. In English the *u* of *qu* is redundant since *q* is always followed by *u*, Shannon argued; many *the*'s are redundant, as perfectly readable telegrams attest. This kind of redundancy usually arises from an excess of linguistic rules and restrictions; Shannon calculated that English is 50 to 75 percent redundant. By pointing out that redundancy (backed by letter frequency counts) furnished the ground for cryptanalysis, Shannon's information theory showed how to raise the difficulty of cryptanalysis and how much ciphertext was needed to reach a valid solution.

These new approaches were coupled with the use of electronic computers. As such, they did not merely raise cryptanalysis in the 1950s to new levels of sophistication and efficiency, they reconstituted it within a new science of electronic communication. Combined with symbolic logic, linguistics as logical syntax, and mechanical translation, cryptanalysis was resignified within the compelling logic of guidance and control systems. The NSA possessed more computer equipment than any other installation in the world.[15] But other defense centers, such as the Ordnance Bureau in Washington, D.C., or Los Alamos Scientific Laboratory, also possessed state-of-the-art capabilities for electronic cryptanalysis, to which Gamow and his code-breaking team turned in the 1950s.

At that time Gamow's fame extended well beyond theoretical physics or even cosmology; in fact, he had captured the scientific imagination of 1950s America. He showily drove around in a large white convertible with red seats. Crick recalled, "He told me that a third of his income came from his academic salary, a third from writing, and a third from consulting, which partly explained his somewhat expensive car."[16] Gamow's activities, namely his defense consultantships with the U.S. Navy Bureau of Ordnance, Air Force Scientific Advisory Board, Army Office of Operations Research, Los Alamos Scientific Laboratory, Convair (San Diego, California), and Rand, epitomized the capillary workings of the military-industrial-academic complex.[17] His career as a military strategist resembled that of his friend, John von Neumann ("Johnny," Gamow called him), with whom he shared not only cultural and political bonds but also intellectual interests.[18]

While a professor of physics at George Washington University (1934–56) Gamow held several visiting professorships, among them at the University of Michigan, Ohio State University, the University of California at Berkeley and at Santa Barbara, as well as faculty fellowships in Japan, India, and Australia. He spent the last twelve years of his life as professor of physics at the University of Colorado at Boulder.[19] He published more than twenty popular books on physics, cosmology, mathematical puzzles, and biology, including the acclaimed series of Mr. Tompkins's—Gamow's Everyman—adventures in science. Owing to his status and his exuberance and penchant for storytelling, he was in demand by the media and especially enjoyed invitations for television appearances.[20] Between his numerous coast-to-coast academic commitments, consulting engagements, and popularization activities, Gamow, like his colleagues von Neumann and Leo Szilard, led a peripatetic existence. His picturesque letters, often composed on different university, company, or hotel stationery, mark his meteoric tracks.

Gamow sent his preliminary formulation and solution of what later became the "coding problem"—"Possible Relation Between Deoxyribonucleic Acid and Protein Structures"—as a short note to *Nature* in October 1953, which he followed with a full-length exposition. As he had described to Watson and Crick, the hereditary properties of an organism could be represented as "a long number written in a four-digital system" (adenine, thymine, guanine, cytosine), which fully determined the long peptide chains formed by about twenty different amino acids. As those before him, particularly the contributors to Henry Quastler's 1952 symposium, "Information Theory in Biology," Gamow viewed these peptides as "Long 'words' based on a 20-letter alphabet. The question arises about the way in which four digital numbers can be translated into such 'words.'" Gamow thus extended and reinforced the discursive trend of representing cells and molecules as texts, or, as Richard Doyle put it, as sites of lexical, textual problematic.[21]

Gamow's proposed solution, the so-called diamond code, was based on a correlation with specific properties (see Fig. 11). It was an overlapping triplet code—though it was not called so until a couple of years later—based on a combinatorial scheme where four nucleotides, arranged three at the time ($4\times4\times4=64$), were more than sufficient to specify twenty amino acids. (A nucleotide doublet, $4\times4=16$, was clearly inadequate for the job.) Accordingly, the sequence AGCTGAACT . . . would consist of the overlapping combinations AGC, GCT, CTG, TGA, and so forth (specifying amino acids A_1, A_2, A_3, A_4, respectively), so that every nucleotide triplet shared two bases with its adjacent triplet. Somewhat like a language, this was a highly restrictive scheme in terms of intersymbol correlation. It was also based strictly on DNA-protein interactions and on mechanisms predicated on a simultaneous "translation" of both DNA chains. As Gamow visualized it, "such translation procedure can be easily established by considering the 'key-and-lock' relation between various amino-acids [he obviously had read Emil Fischer's works], and the rhomb-shaped 'holes' [the diamonds] formed by various nucleotides in the deoxyribonucleic acid chain."[22] As it happened, elementary combinatorial analysis decreed that there were exactly twenty such diamond-shaped configurations, corresponding exactly to the twenty amino acids in the "alphabet"; the "magic twenty," as they came to be called.

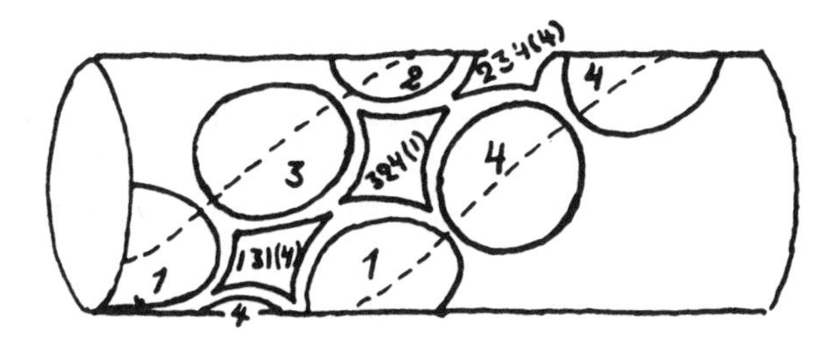

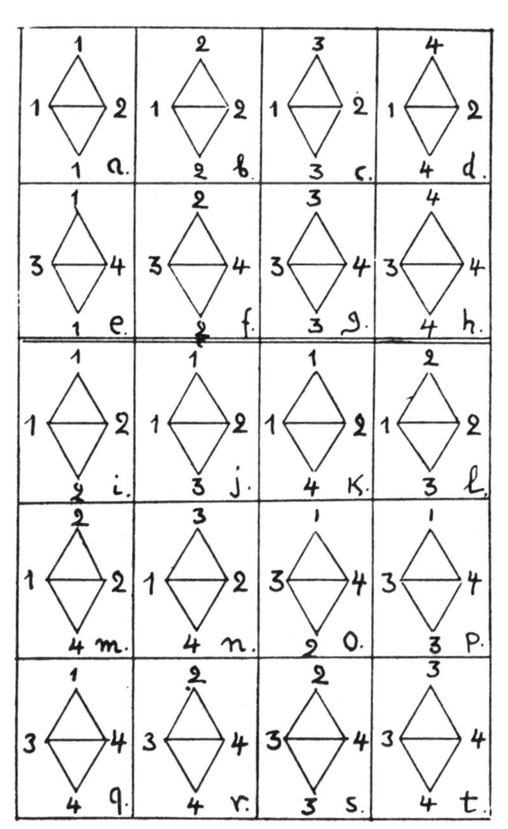

FIGURE 11. George Gamow, "Possible Relation," *Nature* 173 (1954): 318. *Bottom*, the twenty different types of diamonds; *top*, diamonds arranged along a schematic representation of the double helix. © 1954 Macmillan Magazines Ltd.

On the same day that Gamow sent his note to *Nature* he also forwarded his scheme to Linus Pauling at Caltech. "What do you think about it?" he asked the molecular grandmaster. "It would be wonderful if it could be true!" But Pauling did not think much of the scheme. The gist of Pauling's response was: what was good was not new, and what was new was not good.[23] Crick remembered that winter (1953–54), while at Brooklyn Polytechnic, he met with Gamow and managed to disprove all possible versions of Gamow's code by using the small amount of sequence data then available [Fred Sanger's insulin sequence which was just then being completed], and by tacitly assuming that "the code" was colinear (that the DNA and protein strands were coextensive) and that it was "universal" (the same in all living organisms).[24] Colinearity and universality continued to be the guiding assumptions well into the 1960s.

The expanded version of Gamow's idea, "Possible Mathematical Relation Between Deoxynucleic Acid and Proteins," fared no better. Newly elected to the National Academy of Sciences (NAS), Gamow submitted the paper as an inaugural contribution to the Academy's *Proceedings* (*PNAS*) in biology. Apparently the paper caused a furor; Gamow recalled the biologists were quite unhappy about it. He withdrew the paper, sent a copy to Crick and Chargaff, and then he published it in the proceedings of the Royal Danish Academy, of which he was a member. The paper circulated widely, it seems, since Gamow sent reprints to all members of NAS.[25]

It is in this paper that Gamow represented the problem of DNA-protein specificity in terms of information transfer, cryptanalysis, and linguistics. An epigram (in German) by the great organic chemist Emil Fischer served to link the old notions of biochemical specificity to the idea of organisms as texts, as well as to "mathematically possible" forms of life, "the number of beast." "The methods of polypeptide synthesis consist of building long chains with numerous variations in the sequence. It is thus not merely a play with numbers when one calculates the specified probabilities," Gamow argued through Fischer's epitaph.[26] In the chromosomes, these numerical permutations were responsible for carrying all the hereditary information for building the proteins forming the organisms, Gamow explained, thus tacitly linking the old idea of specificity, as inherent in the combinatorial arrangements of amino acids, to the new notion of information content and, in turn, to heredity as secret linguistic communication.

The overlap was a key feature for cryptanalysis. Gamow reasoned there should be a mathematical correspondence between DNA and proteins. It would be based on the "*expected* intersymbol correlation" of the overlapping nucleotides of the diamonds and the "*observed* intersymbol correlation" between the amino acids in polypeptide chains. By using the Rosetta stone of ciphers, the recently published insulin sequence, it would be then possible to "decipher" part of the molecule in terms of the corresponding diamonds (i.e., nucleotide triplets). Although the results of his "decipherment" did not match the insulin sequence data, and Crick's criticisms notwithstanding, Gamow was optimistic. Checking his analysis against related calculations of intersymbol restrictions—reported in Quastler's volume—Gamow believed he was generally on the right track.[27]

Apparently not all life scientists opposed his ideas. Biochemist Erwin Chargaff, at Columbia University, welcomed Gamow's contributions, though eventually Chargaff became antagonistic to theoretical decoding and its informational tropes. He reminded Gamow that in 1949 he had pointed to the specific nature of base pairing but only recently succeeded in engaging physicists and crystallographers in the problem, "and I still have to wait for a mathematician to get hot about it."[28] He did have objections to Gamow's scheme, especially after receiving the prepublication draft of the Danish version: it was too simplistic; and it might be that proteins were not synthesized directly off the DNA. "Perhaps DNA makes RNA and RNA makes proteins," Chargaff suggested, reflecting a view first championed by Jean Brachet and "Rouge-Clôitre Group" and by Hubert Chantrenne's group in Brussels, which was increasingly accepted by molecular biologists (this was also the basis for Dounce's model). But Chargaff hoped to continue to communicate about the "decoding" attempts. "They are most interesting and I wish you much luck. Fun I don't have to wish you; you seem to be having it anyway."[29]

Gamow's project received a major boost when the Russian-born biologist Martynas Yčas (b. 1917) was seduced by Gamow's ideas and served for nearly four years as his principal code-mate. Trained in jurisprudence, Yčas switched over to biology after immigrating to United States in 1941 and completing his BA degree in zoology at the University of Wisconsin. After three years of military service he went to Caltech, where his graduate work in embryology and marine biology (under Albert Tyler) culminated in a 1950 dissertation on respiratory enzymes of sea urchin eggs. From 1951 to 1956—during the peak of his "decoding" work with Gamow—Yčas was employed as a biologist at the Pioneering Research Division, Quartermaster Research and Development Command, U.S. Army, first located in Philadelphia and then in their new expanded quarters in Natick, Massachusetts. In 1956, with Gamow's and George Beadle's endorsement, Yčas became professor of microbiology at the State University of New York at Syracuse.[30]

In spite of, or perhaps because of, his extensive biological training, Yčas, like many biologists, harbored an intellectual and disciplinary inferiority vis-à-vis physicists. He considered most biologists to be rather limited, revered the physicists' analytic powers and mathematical magic wand, and enjoyed participating in the culture of physics in biology, the "fine playground for serious children who ask ambitious questions," as Delbrück once put it.[31] Gamow, the acclaimed landsman, became Yčas's alter ego; Yčas, in turn, provided Gamow with intellectual and disciplinary links to biology. As Gamow wrote a couple of years later, "During the last two years we have been working as a team. . . . Being a biologist, Dr. Yčas takes care of the biological part of the problems we are trying to attack, while I handle the more mathematical part of the picture."[32]

Yčas initiated a correspondence with Gamow soon after the publication of Gamow's note in *Nature*, informing him that the sequence of amino acids in insulin could not be determined by his diamond code. (He later recalled that Gamow admitted that two naval cryptologists in Washington had come to the same conclusion.) Yčas had a more biological view of the problem. As he saw it, "Using current jargon, the information which completely determines the organism resides in various degrees in the whole organism, rather than exclusively in one part," but he was convinced that DNA carried information according to some version of Gamow's code. By then Gamow had already understood that the diamond code was theoretically much too simplistic; he was now attacking the problem empirically by constructing Fischer atomic models of DNA and amino acids.[33]

Gamow welcomed Yčas's criticisms. "I do not think your letters are of the "wet blanket" type, he reassured Yčas; "They sound more like those from a 'buddy' from another company along the line in a difficult advance against the entrenched enemy. As the consultant to the Army in the line of nuclear weapons and operational research (battle theory) I am quite thrilled to know that Quartermaster Corps take such vital interest in DNA and protein synthesis."[34] In addition to their intellectual interests and Russian bonds, the two shared in the cultural values of their military patrons. At that time Gamow was a visiting professor at Berkeley, teaching graduate courses on "Relativity and Cosmology" and on the "Evolution of Stars." He used his West Coast sojourn to visit Caltech, discussing the coding problem with Delbrück and his colleagues, among them molecular biophysicist Alexander Rich, a research fellow in Pauling's chemistry division, and physicist Richard Feynman.[35]

The diamond code met with skepticism, but Gamow was already excited about the new "purely formal scheme (which may not work for DNA, but may [be] the informative [message] for RNA) in which such substitutions can take place." He thought that decoding along these new lines (the so-called major-minor code) would require an electronic computer, and he intended to put the problem on the MANIAC in Los Alamos in July. "We have seen a lot of Gamow since then," reported Gunther Stent, the phage geneticist at Wendell Stanley's Virus Laboratory at Berkeley, to his friend Delbrück. "He devotes one day a week to 'Biology' these days. He has built the DNA structure with Herschfelder atomic models, and is very proud of this imposing edifice. But I am afraid nothing will fit into the 'diamonds.'"[36]

Even if no amino acids would fit into Gamow's diamonds, his colleagues were absorbed by them. Gamow and his coding scheme generated infectious enthusiasm, resulting in a growing chain of prominent physicists who were challenged by the mathematical properties of the code. Given the rising interest in the problem and in RNA as a possible code for the assembly of proteins, Gamow decided to formalize the coding network by establishing the RNA Tie Club, or the "RNA-Tie brotherhood," as Gamow sometimes referred to it. "We were just drinking California wine and we got the idea," Gamow recalled. Wine, beer, and whiskey seem to have been critical lubricants for the physicists' imaginations. "If you had three or four scientists and you had three or four bottles of beer, you either had a new reactor or a new bomb come out of the evening," reflected Sumner T. Pike, acting chairman of the Atomic Energy Commission; "Scientists like to talk and drink beer, I guess." Gamow, sometimes nicknamed "Whiskey-twisty," was by then on the slippery slope of alcoholism.[37]

Instead of, or rather in addition to, a bomb or a reactor, Gamow and his collaborators reveled in the genetic "enemy" code; life was re-represented within the regimes of signification of cold-war military imagination. Gamow wrote, "As in the breaking of enemy messages during the war (hot, or cold!), the success depends on the available *length* of the coded text. As every intelligence officer will tell you, the work is very hard, and the success depends mostly on luck. There are $20! = 10^{17}$ possible assignments of aa's [amino acids] to base triplets! . . . I am afraid that the problem cannot be solved without the help of electronic computer."[38] The discourse of genetic decoding was now being formulated within the operational space of electronic technology.

The aim of the RNA Tie Club was to foster communication and camaraderie by circulating notes and manuscripts on the coding problem; its hope was to locate funding for regular meetings. It was a select boys' club of twenty picked by Gamow, corresponding to the "magic twenty" amino acids: G. Gamow, ALA (he always wanted to be deity, he explained); A. Rich, ARG; P. Doty, ASP; R. Ledley, ASN; M. Yčas, CYS; R. Williams, GLU; A. Dounce, GLN; R. Feynman, GLY; M. Calvin, HIS; N. Simons, ISO; E. Teller, LEU; E. Chargaff, LYS; N. Metropolis, MET; G. Stent, PHE; J. Watson, PRO; H. Gordon, SER; L. Orgel, THR; M. Delbrück, TRY; F. Crick, TYR; and

FIGURE 12. Members of the RNA Tie Club, ca. 1955. *Front (left to right)*, Alexander Rich, James Watson. *Back (left to right)*, Francis Crick, Leslie Orgel. Collection of Alexander Rich.

S. Brenner, VAL. F. Lipmann and A. Szent-Gyorgyi were honorary members. Thirteen members were physical scientists—chemists, physicists, and mathematicians. Delbrück's motto "Do or die, or don't try" graced the club's stationery, along with an officers' list: Gamow, synthesizer; Watson, optimist; Crick, pessimist; Yčas, archivist; Rich, lord privy seal. Each member had a diagrammed tie (proposed by British chemist Leslie Orgel) made to Gamow's design by a Los Angeles haberdasher and a tiepin with an abbreviation of the amino acids corresponding to the member.[39] (See Fig. 13.)

Gamow, a perpetual optimist, had high hopes for his cryptologic brotherhood, as well as for some wealthy, worthy patronage. And he tirelessly maintained the club's epistolary and organizational chores; Yčas, RNA Tie Club archivist, scrupulously documented every turn in the decoding work, includ-

FIGURE 13. Gamow's drawing of the "card game of life," played by members of the RNA Tie Club. The four suits—diamonds, hearts, clubs, and spades—are analogized to the four nucleotide bases, A, T, C, and G. Gamow Papers, Manuscript Division, Library of Congress.

ing his own considerable contributions. Yčas even offered a solution for the patronage problem: there was a strong possibility for military funding to cover the cost of RNA Tie Club biannual conferences, which Gamow enthusiastically endorsed. Yčas and Gamow's campaign for sponsorship by the Quartermaster Corps of the U.S. Army lasted for about three months and almost came to fruition. "We in the Quartermaster are very interested in the problems which you folks are considering and are most willing to serve in any capacity that may contribute to the success of your meetings," the division chief assured Gamow. But the RNA Tie Club never convened as an official group and the sponsorship eventually vanished in the bureaucratic morgue of aborted schemes (perhaps also due to security clearance problems).[40] A few members—notably Delbrück, Crick, and molecular biologist Sydney Brenner—sent significant contributions. But mostly, according to Gamow (and Yčas concurred), "it was just a result of good California wine. And it very quickly came to nothing."[41] Too hastily and harshly spoken. For beyond the handful of significant cognitive contributions, the club, despite, or rather because of, its geographic dispersion, transported the coding problem into biology with all its discursive and operational resources, which were reshaping the representations of heredity and life.

Military Codes? Logic, Statistics, Linguistics

By early May 1954, Gamow fully accepted the evidence that his "diamond system" was impossible; yet he was far from discouraged. "Playing with diamonds was still quite useful because this heartbreaking exercise indicated that such kind of decoding is not a hopeless undertaking." Having just returned from a visit to Caltech and "with the southern branch of 'RNA Club' (as we call ourselves)," he was excited about a new possible code. With evidence mounting that it was not the double-stranded DNA but the single-stranded RNA that is responsible for protein synthesis, Gamow, Rich, Feynman, and Orgel proposed various coding schemes; all based on the principle of matching polypeptide fragments with the overlapping nucleotide triplets along a single-stranded nucleic acid. All such decoding schemes were extraordinarily complicated and necessitated the use of computerized cryptanalysis, to be executed during Gamow's July engagement at Los Alamos.[42]

There was the triangular code and its two variants: compact and loose. (See Fig. 14.) That code was based on a correspondence between twenty different possible triangles (due to the pitch of the helix) formed by consecutive bases: 1,2,4,4,2,3,1,1,4 . . . and the twenty amino acids, A,B,C,D,E,F,G,H, I,K,L,M,N,O,P,R,S,T,U,V. In the compact triangular code the amino acid assembly along the template would produce the protein: I-P-P-A-O-F-G-I (the staggered sequence shown in Fig. 14, *top*), rendering the intersymbol combination rules even more restrictive than the diamond code. In the loose triangular code a single template would synthesize simultaneously two proteins complementary to one another: I-P-O-G and P-A-F-I (the top and bottom linear sequences in Fig. 14, *top*). The loose version of the triangular code was less restrictive than the compact version, thus significantly raising the difficulty of decoding known polypeptide sequences (the more restrictions, the more clues).[43]

Feynman and Orgel proposed another version of an overlapping triplet code: the major-minor code, where the central base of each triplet was called "major" and the two neighboring bases "minors." In that code, each amino acid had a particular nucleotide acting as a major determining factor in the positioning of the amino acid, and several different amino acids could possess the same major determinant, but adjoining nucleotides might also influence the positioning of the amino acid (as Dounce's 1952 model suggested).[44]

After the major-minor code collapsed under the weight of contradictions accumulated from the growing roster of known peptide sequences, nuclear physicist Edward Teller—"busy as he was with H bomb, and Oppenhei-

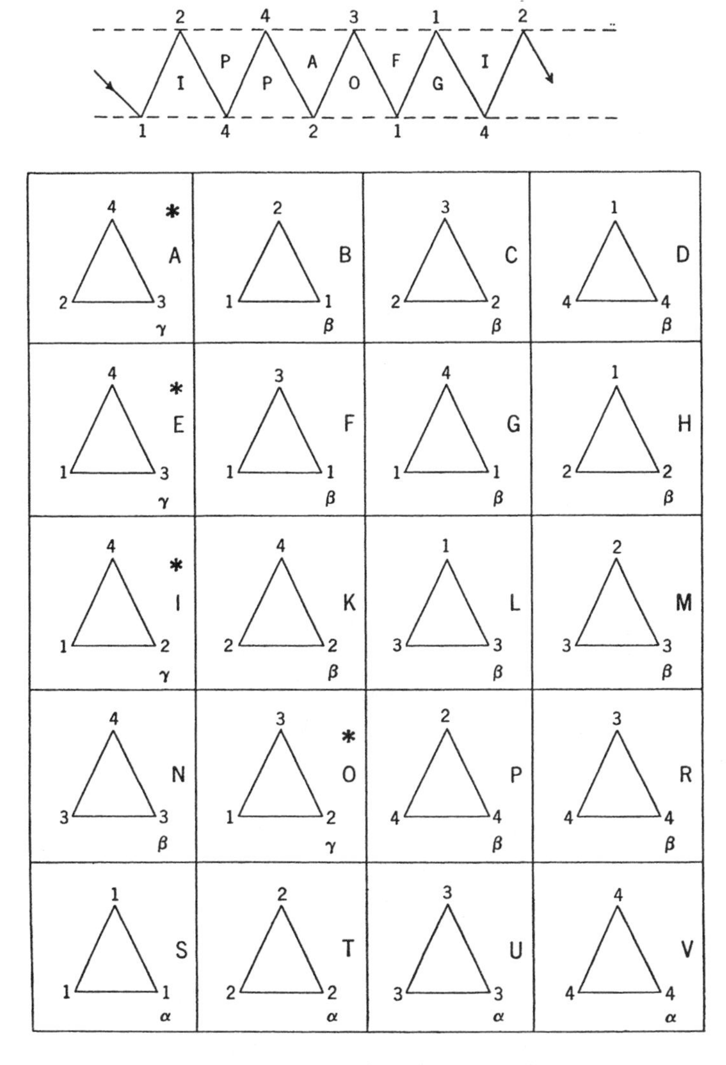

FIGURE 14. George Gamow, Alexander Rich, and Martynas Yčas, "The Problem of Information Transfer from Nucleic Acids to Proteins," *Advances in Biological and Medical Physics* 4 (New York: Academic Press, 1956), pp. 48–49. *Bottom*, twenty possible triads of triangular code; *top*, schematic diagram of triangular code. Reprinted by permission.

mer"—proposed a new kind of code: the sequential code. (Teller's testimony played a key role in the suspension of Oppenheimer's security clearance in 1954, making Teller a pariah within the scientific community.) The idea of Teller's code, nicknamed by Gamow the Russian-bath code (*bana*), was that an amino acid was defined by two bases and the previous amino acid. His

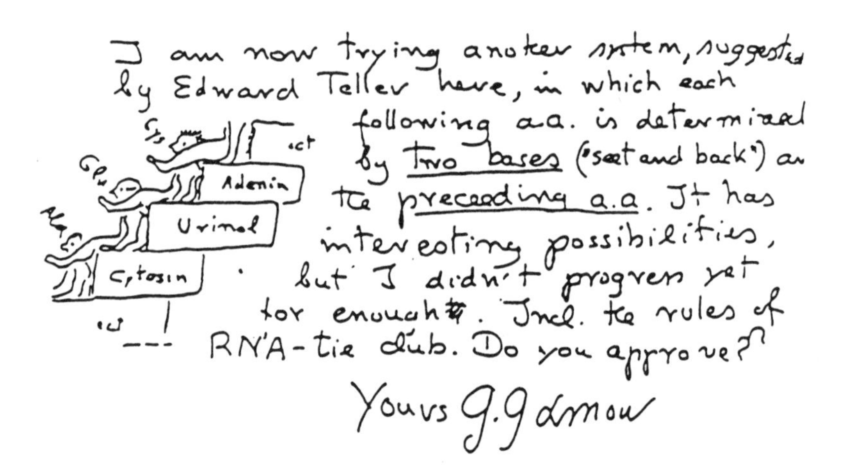
I am now trying another system, suggested by Edward Teller here, in which each following a.a. is determined by two bases ("seat and back") and the preceding a.a. It has interesting possibilities, but I didn't progress yet far enough. Incl. the rules of RNA-tie club. Do you approve?

Yours G. Gamow

FIGURE 15. Gamow to Yčas, 27 May 1954, Gamow Papers, Yčas fld. (1954), Library of Congress. Gamow's illustration of Teller's sequential (Russian bath) code.

provocative scheme implied that there were some preferred sequences and that certain intersymbol relations were more important than others. The decoding procedure of the sequential code consisted of analyzing experimental data of peptides for any systematic patterns.[45] (See Fig. 15.)

In the summer of 1954, Gamow collaborated with his colleague at Los Alamos Nicholas Metropolis, a theoretical physicist and expert in electronic computing and logic design, to test the various overlapping (restrictive) codes on the MANIAC computer, using the cutting-edge technology of Monte Carlo simulations. Using random numbers (as if in a roulette game) the Monte Carlo method was designed to simulate stochastic processes too difficult to calculate analytically and too remote for the laboratory. As Peter Galison has shown, it was in 1954 that this method came into its own, blurring the boundary between theory, experiment, and simulation. The MANIAC had become the site of an artificial reality, not only in physics but also in biology. As with frequency cryptanalysis, they compared the different proposed codes by studying statistically the dependence between the frequency of occurrence of different amino acids and the number of their different neighbors in natural sequences, as well as in artificial sequences based on the proposed coding procedures. The stronger the restrictions governing the associations between the amino acids in the sequence, the smaller the number of different neighbors as compared with random distribution. The results came in at the end of September. *Science Service* reported, "The Maniac, an electronic computer at Los Alamos Scientific Laboratory is being used to learn more about protein structure. . . . The computer 'builds' artificial proteins molecules at random from any of the twenty amino acids building blocks, according to a specified code. The proteins resulting from the machine's computation are compared with those found in nature, Prof. George Gamow . . . and Dr. N. Metropolis reported [at a Columbia University meeting]." But the observed and artificially constructed protein sequences, compared through neighbor saturation curves, yielded negative results.[46]

The theoretical data (based on coding specifications) and the empirical data (from experimentally known peptide sequences) suggested that the distribution of amino acids in the known protein sequence was purely random. In spite of the strong intersymbol restrictions in the proposed codes, no noticeable difference between observed and calculated results was discerned. Paradoxically, the restrictive codes still produced random amino acid distribution. But the decoders' commitment was unfailing. Rather than question their guiding premise that the code was overlapping, or more fundamentally, that the scheme was, in fact, a code (namely, a systematic manipulation of language), Gamow and Metropolis stuck to their premises. They inferred that the neighbor-saturation method was not sensitive enough to distinguish between random distributions and those subject to certain intersymbol restrictions.[47]

The deliberations over the various codes—compact and loose triangular codes, major-minor code, Teller's sequential code—continued in August at Woods Hole, the waterfront cottage of Albert Szent-Gyorgyi (the Hungarian Nobel laureate biochemist), where Gamow and his wife stayed and which became a social and intellectual locus for RNA Tie Club members Crick, Watson, Yčas, and Brenner. Yčas and Brenner stayed only for a week or so. Crick recalled that "[on] most afternoons Jim and I went out to the cottage and sat on the shore with Gamow, discussing all the different aspects of the coding problem, idly chatting or just watching Gamow showing some of his card tricks to any pretty girl who happened to be around." This was Sydney Brenner's first visit to the United States and after a four-month trip he left quite an impression. "Boy, do we miss you here!" wrote Gunther Stent, beginning their long collaboration. "You simply must come back to Berkeley." Brenner (b. 1927) had by then earned his medical degree in Johannesburg and had just completed his doctoral training under Sir Cyril Hinshelwood at the Department of Physical Chemistry in Oxford, studying the nature of bacteriophage mutations. He had been scrupulously collecting the growing roster of amino acid sequences in various proteins since 1954 and from it compiling a table of distribution of all known dipeptides. With this impressive table, he had outlined schemes refuting most of Gamow's triangular codes and buttressing his own argument that the code was not overlapping. Gamow did not much care for Brenner's code but made extensive use of his data. Brenner spent several months in the United States sponsored by a Carnegie Fellowship. He visited Woods Hole, Cold Spring Harbor, Caltech,

Berkeley, and Washington, D.C., conferring with Gamow before returning to South Africa. "I haven't found a code as good as yours," Crick assured Brenner. "I would suggest you write up yours, have it mimeographed and send it to the RNA Tie Club. This seems to be the only way that isolated members like yourself can keep in touch." Brenner remained in Johannesburg for two years, mounting phage research and in 1956, through Crick's efforts, was offered a position at the Medical Research Council Unit for the Molecular Structure of Biological Systems (soon after expanded and renamed Unit for Molecular Biology).[48]

That September in 1954, Gamow, together with Yčas and Alex Rich, now chief of the Section of Physical Chemistry at NIH, undertook the challenging and enormous project of consolidating extant knowledge by preparing a comprehensive review of the coding problem for *Advances in Biological and Medical Physics*. Not only did this task entail assembling previously unpublished materials and reconstructing fragments of dispersed communications but also it required integrating the writing with the avalanche of nearly daily information that admittedly offered no solutions to the coding problem. More than a year elapsed before the appearance of the article that also required security clearance by Yčas's army employer. Although by the time of its publication in 1956 all the reviewed codes had been disproved, the article has remained a significant historical document, because it reveals in detail how the genetic code was textualized within the space of representations configured by the discourses of information, cryptanalysis, and linguistics.

While working on the manuscript, Gamow added yet another line of attack to the coding problem: the use of symbolic logic. During his stay at the Office of Operations Research at Johns Hopkins in October, he managed to engage the mathematical biophysicist Robert Ledley to apply his expertise in digital computational methods of symbolic logic (and the related Boolean algebra) to the coding problem. Ledley viewed the task as an opportunity to broaden the scope of operations research. He noted, "Besides the use of logical propositional methods in problems concerned with sentences, such as analysis of military information reports and legal and insurance documents, there appear to be even more important applications to fields of operations research, biology, medicine, design of experiments, etc."[49] The electronic sciences of information and secrecy could be now applied to the secrets of life. Since all overlapping translation codes display stronger or weaker intersymbol correlation, Ledley explained, one could set up equations that would lead to the solutions of the various overlapping triplet codes. He readily demonstrated the technique on the limited case of $3!=6$ for three amino acids and their three corresponding hypothetical nucleotide triplets. This calculation took only a hundred hours with the MANIAC. However, to solve the case of $20! = 2.3 \times 10^{17}$ (twenty amino acids and their corresponding triplets), "a computer put to work in the days of the Roman Empire, at a rate of one million solutions per second, 24 hours a day, all year round, would not yet be close to finishing the job," Ledley estimated;[50] such analysis would be impotent for nonoverlapping codes.

Two other decoding systems were devised while the coding resume was being written: statistical analysis of the correlation between nucleic acid and protein composition in viruses and neighbor distribution plots based on Brenner's table of dipeptides. Brenner gave that table, which implied that overlapping codes were impossible, to Gamow in Washington at the end of 1954. Relying on well-established data (e.g., Chargaff's studies) of base composition as well as protein composition data of RNA plant viruses (tobacco mosaic virus, or TMV, and turnip yellow virus, or TYV, supplied by Stanley's Berkeley group), the authors reasoned that if the two viruses differed in their base composition, then they had to differ correspondingly in their amino acid compositions (tacit, yet unproven, premise of colinearity). Statistical analysis verified the correlation between amino acids and base triplets, but disappointingly, the TMV and TYV data led to contradictory assignments between amino acids and the triplets.[51] Having the viral protein sequences, and not just compositions, might have facilitated the decoding. TMV would remain an important biological system for subsequent sequencing and decoding efforts.

The other method of cryptanalysis, called *neighbor distribution plots*, deployed statistical analysis (akin to the neighbor saturation study by Gamow and Metropolis) of the experimentally known protein sequences: How would the observed frequency distributions in known proteins correlate with those expected from the various diamond codes and triangles? Gamow, Rich, and Yčas made extensive use of Brenner's table (the "South African" graph, as Gamow called it), which displayed the neighbor distribution in a grid comprised of all four hundred (20×20) possible amino acid pairs.

Subjecting the dipeptide grid to computerized statistical analysis, the team calculated the average density of points in it and compared (via the χ^2 test) the observed density distribution with those predicted by the codes. "Seems to be very little intersymbol correlations when one applies Poisson distribution to Brenner's table (Fishy story!)," Gamow admitted to Watson. Brenner had already pointed out that the observed frequencies from natural protein sequences followed a Poisson (random) distribution. Gamow agreed, supplying his own imagery.

> When Germans bombed London by V1's and V2's, British Operational Annalists tried to find out whether or not the Germans were *aiming* at some partic-

> ular spots in the city. To do this they overlapped on the map of London a square lattice, and counted the number of hits within each square [drawing of lattice and hits included]. I do not know what was the result of Poisson analysis in that case, but our problem is exactly of the same nature . . . protein synthesis "aims" for definite amigo-pairs.[52]

Alas, the frequencies in the proteins predicted by the codes deviated noticeably from the Poisson distribution and thus from their protein sample. Another nail driven into the coffin of overlapping codes.

"It is my impression now that we shall shortly be in a position to demonstrate rather clearly that all types of interlocking codes do not correspond to reality," Yčas confessed to Rich, as their manuscript went to press. "This leaves only the possibility of codes, the bases of which do not overlap, and codes on which the RNA forms only part such as nuclear proteins." He relayed the same verdict to Crick. "All coding schemes so far fail. . . . We are forced to the conclusion that there is no trace of any preferred neighbors relation [as in language]. . . . Thus if RNA is a template, the aa [amino acid] sites must be nonoverlapping, or interact with other sites more than two residues removed." When he mailed the manuscript to Crick a month later Yčas referred to their paper as "Why Gamow, Rich, and Yčas failed to solve the problem."[53] Although the authors' conclusions were mainly negative, their work was historically significant on several levels. They had defined "the coding problem," opened up a path toward analyses of nonoverlapping codes, and suggested novel and subsequently fruitful approaches to decoding (such as amino acid replacements). Equally important, their work elaborated the discursivities, semiotics, and imagery that resignified heredity as a communication system. Its code was the key to the scriptural technology governing the secret linguistic transactions—the transfer of information.

Code, Language, and Discourse of Information

In titling their article "The Problem of Information Transfer from the Nucleic Acids to Proteins," Gamow, Rich, and Yčas were not deploying the idiom of information unconsciously or casually, as did many life scientists in subsequent years. They used information—with all its connections to mathematics, logic, cryptanalysis, linguistics, computers, operations research, and weapons systems—as a means of framing the "coding problem." The trope of information served to integrate mechanisms of molecular specificity, structural considerations, mathematical relations, and linguistic attributes within a single explanatory framework. It reshaped the problem of genetic specificity through a discourse that resonated with the technosciences of command and control.

The use of the code idiom and its corollary—language—was extremely problematic, laden with aporias and tautologies. While from the start the idea of code was implicit in Gamow's analysis of the specific relation between the four bases and the twenty amino acids, he had not explicitly used the term. It was only in the review article by Gamow, Rich, and Yčas that the code idiom was used deliberately and defined, albeit ambiguously, for the first time since Watson and Crick had casually mentioned the "code which carries the genetical information" in their 1953 note to *Nature*. Codes were "information-transfer models," the authors declared and then clarified, using "the term 'information' in the sense of molecular specificity."[54]

The problem, they noted, was really one of translation from a four-letter code to a twenty-letter code. Namely, the protein "text" (e.g., insulin, hemoglobin, lysozyme) was seen as coded by twenty amino acids, which in turn, were coded by nucleotide triplets. "In its fullest detail, this is the problem of specifying the detailed interactions between amino acids and nucleic acids which determine site specificity," Gamow, Rich, and Yčas explained.[55] They thus articulated the problem of chemical specificity in terms of *two* types of codes: the DNA (or RNA) code, and the amino acid code (Schrödinger's code-script). Herein lies an aporia: the "translation" process, as understood by the authors, would thus operate between two codes rather than, as is customary, between a cryptogram and its plaintext. While ordinarily two different codes would correspond to two different texts, in their model the nucleic acid triplets did not "code" a true message (a plaintext) but yet another "coded" amino acid message, which only then referred to the protein "text." They were engaged in the logically confounding task of decoding a code of codes (were it a code to begin with).

Moreover, the application of code to the problem of genetic specificity was merely analogical, and as such it was marred by confusion between two very different devices: code and cipher. In modern practice of cryptology, the differences between code and cipher are quite marked. Ciphers operate on plaintext units of regular length (all single letters or all groups of, say, three letters, as in the triplet DNA scheme), whereas codes operate on plaintext groups of variable length (words, phrases, individual letters, and so on). A more penetrating distinction—and a crucial one for this analysis of the "genetic code"—is that the *code operates on linguistic entities*, dividing its raw material into meaningful elements such as words and syllables (while ciphers will split the "t" from the "h" in "the," for example).[56]

Thus even if one grants for a second the problematic analogy between combinatorial elements in a molecule and the alphabetical elements of a language, the "translation" between the four nucleic acid bases and the twenty amino acids would obey the rules of cipher. "It always makes cryptographers indignant to hear the terms 'code' and 'cipher' used synonymously."[57] In fact, years later Crick granted, "The proper technical term for such a translation rule is, strictly speaking, not a code but a cipher. In the same way the Morse code should really be called the Morse cipher. I did not know this at the time, which was fortunate because 'genetic code' sounds a lot more intriguing than 'genetic cipher.'"[58] This is the poetics of technoscience. More fundamentally, whether code or cipher, the tacit operating assumption behind all of the so-called deciphering or decoding attempts was that the code operated on language-like entities. Once the analogy between combinatorial elements in nucleic acids or proteins and the letters of the alphabet took root, the comparison with language assumed a life of its own. Eventually the metaphor of language, with all its ambiguities, aporias, and tautologies, began to be taken literally. How could it be otherwise? "The Code"—that logocentric icon preceding empirical warrant demanded the existence of language as its predicate. Only through signification of molecules as linguistic entities could there be a code to manipulate them.

First there was the analogy. In defining the "coding problem," Gamow, Rich, and Yčas explained that if one distinguished between four different bases by means of numerals (1,2,3,4), and used the reduced English alphabet (a,b,c,d,e,f,g,h,i,k,l,m,n,o,p,r,s,t,u,v) for twenty different amino acids, then one might represent the correlation between the structure of the template and the structure of the corresponding protein by a long word written under a long number formed by only four different digits. Thus the order of amino acids in the known polypeptide sequences had to be far from arbitrary, and there had to be a strong correlation between the neighboring amino acids. They analogized the situation to that of any language. Just as short (twenty-one-word) sentences like "Amino acids form proteins" or "Read *Alice in Wonderland*" display only a negligible fraction of all possible sequences of the same length formed by the letters of the alphabet, the known polypeptide fragments (only a few hundred amino acids long) represent minuscule snippets of all possible arrangements (2.1×10^{27}). "We here face difficulties similar to those encountered by an Armed Force's Intelligence Office trying to break an enemy code on the basis of a single message less than two printed lines long," they estimated with some pride.[59]

The analogy did not remain an external aid to the scientific imagination but rather became constitutive of the decoding analyses and interpretations. When Brenner's table showed that the frequency of dipeptides from known proteins followed a Poisson (random) distribution, while the frequencies in proteins predicted by the various codes deviated noticeably from Poisson, Gamow, Rich, and Yčas turned to cryptanalysis. It was well known that in a grid comprised of twenty letters ($20 \times 20 = 400$) neighbor frequencies would follow a Poisson distribution, while in a language that introduces preferred neighbors (e.g., qu, th, be, an) the resulting restrictions bring about deviations from the Poisson distribution. "As an example of this, we determined the neighbor frequency in the English language, using the beginning of Milton's *Paradise Lost*," the authors explained; "Even though repeated words were deleted from the tabulation, the distribution nonetheless deviates widely from Poisson. . . . Neighbor frequency distributions have been listed for several languages and are employed extensively in various cryptographic texts [they cited Pratt's book *Secret and Urgent*]. In no written language, however, does the letter neighbor distribution follow that of Poisson." [60]

They then compared that deviation against the frequency distribution of the artificially constructed amino acid sequences based on the various codes. They even used Shannon's innovations in cryptanalysis to determine the information content of these sequences ("177 bits of sequential information"), only to learn that, once again, their theoretical sequences deviated markedly from Poisson and their experimental sample. It was disheartening. Most codes express themselves by restrictions and neighbor preferences; without them, the authors lamented, decoding is a forbidding task.[61] Herein lies another aporia. Clearly the amino acid "text" did not obey the rules of any known language (as Quastler's colleagues had already demonstrated three years earlier); either the code overlaps and, thus, contradicts the experimental evidence, or it does not overlap and contradicts the meaning of language. If the code were nonoverlapping, as the evidence implied, then cryptanalysis could no longer be applied.

At this point it would have seemed reasonable to abandon, or at least reconsider, the linguistic representation of the nucleic acid–protein relations and discard their signification as codes, since codes operate strictly on linguistic structures, and the protein "text" did not seem to obey the rules of language. But precisely because language and code made each other necessary in the authors' representation, the tautology survived intact: the code was conceptualized through linguistic significations of molecules; linguistic attributes, in turn, justified the notion of code. Instead, finessing the negative results, they shifted their focus to the significance of the random distributions of amino acids in nature. They proceeded to valorize randomness in biology.

Drawing on information theory, Gamow, Rich, and Yčas extracted some poignant lessons about the nature of so-called information storage in proteins. In their *Mathematical Theory of Communication*, Shannon and Weaver developed quantitatively the concept of the entropy of a given informational source or language, as the authors reminded their readers. In systems in which certain sequences are preferred, such as th, or perhaps Ala-Leu (amino acids, Alanine and Leucine), there is a loss of potential entropy (or increase in redundancy). Those languages that are the most flexible and economical are those with few restrictions. "In this case, it may be that Nature uses a polypeptide 'language' in which the entropy is high (few restrictions), and the concomitant potential specificity for a given polypeptide chain is very great (negative entropy is a measure of specificity, as Quastler had shown)."[62] In any case, the unwavering commitment to an information-bearing code survived all empirical refutations.

While the review by Gamow, Rich, and Yčas was addressed to researchers in the physical and life sciences, its message and imagery circulated widely after the publication of Gamow's article, "Information Transfer in the Living Cell," in *Scientific American* in 1955. That article, which raised its gaze from the microsanctum of nucleic acids–proteins transactions to the panoramic vistas of cells and organismic life, appeared a year earlier than its specialized counterpart. To appeal to a wider audience, Gamow compared the four bases of the genetic material to playing-card suits (diamonds, clubs, spades, hearts) and showed how through this card game cells store their identities in the form of chemical codes used to build replicas of themselves. That image was also captured in Gamow's drawing of the Coding Club (RNA Tie Club) playing the card game of life (see Fig. 13).

A master of popularization, Gamow knew how to convey the drama of life. And even though he harbored contempt for Shannon's information theory, he still used it metaphorically, borrowing its imagery and discourse to represent living cells as information systems. He wrote, "The nucleus of a cell is a storehouse of information. It is also something more remarkable: a self-activating transmitter which passes on very precise messages that direct the construction of identical new cells. The continuity of all life on our planet depends on this information system contained in the tiny cell nucleus. Let us examine what is known about the language of cells."[63] While visualizing heredity as an electronic communication technology, Gamow deployed the idiom of information in its nontechnical, metaphoric sense, thus tacitly imparting unwarranted semantic value and meaning to the language of cells.

The problem of explaining cell communication had two parts, he explained: how the information is stored in the chromosomes, and how it is transmitted from the chromosomes in the nucleus to the enzymes in the cytoplasm. The first part, he remarked, had been mostly answered by Watson and Crick; the second part—the subject of the article—traced the rise and fall of overlapping codes. Had the code been overlapping there would have been intersymbol restrictions in the DNA (and protein) sequences, but with nonoverlapping codes the sequence could be entirely random.

Like its specialized counterpart (in *Advances in Biological and Medical Physics*), Gamow's article finessed its defeat with a meditation on randomness. The first DNA sequence could possess randomness as exquisite as that of the number π: there is no discernable pattern in the sequence 3.14158265 . . . yet it is essential for geometrical solutions. He mused, "In a living or-

FIGURE 16. *Left to right*, Alexander Rich, Francis Crick, Sydney Brenner, ca. 1955. Collection of Alexander Rich.

ganism over eons of time the random mutations may once in a while produce a sequence of nucleotides which blueprints a new and helpful enzyme. A lizard living in the age of reptiles may thus have acquired an enzyme which catalyzed the production of milk—and so taken the first step toward the age of mammals."[64] Gamow's Big Bang theory of textual genesis: in the beginning was the word, the random sequence. This view on the origins of life and molecular evolution would later be operationalized in Manfred Eigen's works in the 1970s. There might be purpose in randomness, or higher wisdom in higher entropy, thought Gamow, and so an entry into life's sublime secrets locked in nonoverlapping codes, after all.

PROLIFERATION OF WRITING: NONOVERLAPPING CODES

Formalizing Degeneracy

By the winter of 1955 all overlapping codes were dead. As the manuscript, "The Problem of Information Transfer from Nucleic Acids to Proteins," by Gamow, Rich, and Yčas went to press, Gamow and Yčas were already pursuing alternative coding schemes and were now engaged in sophisticated statistical analyses of amino acids in various proteins in order to test their new nonoverlapping triplet code. In that so-called combination code only the *combination* of bases in a triplet mattered, not their *order*.

Visiting Caltech on his "trip west sponsored by Rand-USAF conference," Gamow reported that he and his Los Alamos colleagues, mathematician Stan Ulam and physicist Nicholas Metropolis, who was an expert in electronic computing and logic design, had made some progress on decoding. Indeed, it looked like they might have found a way out of the problem of randomness of amino acids in proteins (random distribution blocked any clues about the arrangement of corresponding nucleotide triplets),[65] an escape from the Poisson catastrophe.

Once again, the coding spirit was high, even though Crick had just dispatched an insightful and remarkably discouraging note to the RNA Tie Club, entitled "On Degenerate Templates and the Adaptor Hypothesis." Crick granted Gamow's important contributions: defining coding as an abstract problem independent of biochemical considerations, and introducing the key feature that several different base sequences could code for one amino acid. Crick called this feature "degeneracy" (after the analogous situation in physics when an equation has more than one solution), which, he thought, could be generalized in the analyses of all other codes. One could develop a logical method of degenerating that might yield clues about the relation between nucleotide triplets and amino acids. But as Crick saw it, there was not much beyond these two contributions.[66]

Skillfully weaving spatial considerations (e.g., how amino acids fit in the template via what Brenner called "adaptors") with formalistic arguments (e.g., directional properties and neighbor occurrences) Crick, the club's pessimist, pronounced all of Gamow's codes—including Gamow and Yčas's embryonic "combination Code"—impossible. "It is better to use one's head for a few minutes than a computing machine for a few days!" he snickered.[67] The most fundamental objection to Gamow's schemes was that they did not distinguish between directions of a sequence (e.g., between Thr. Pro. Lys. Ala, and Ala. Lys. Pro. Thr.). Crick thought there was little doubt that Nature made that distinction.

He pointed out that in the DNA double helix one chain runs up while the other runs down, as Delbrück had emphasized, implying that "a base sequence read one way makes sense, and read the other way makes nonsense." Molecular directionality was no longer merely analogized to the directionality of language, it was conflated with it through the act of reading and sense making. Once again one witnesses how the tropes of language—reading and sense—became constitutive of the decoders' modes of reasoning, and how the information discourse (despite the warnings of information theorists) bestowed semantic properties on the (syntactic) arrangement of molecular symbols (recall that syntax is a necessary but not sufficient condition for human language; however, it is a sufficient condition for formal or machine languages). We see the process by which Nature was continuously being textualized. Crick reported with frustration about his unsuccessful attempts to construct a new class of codes—"directional codes"—that, when read backward, made nonsense. The epigram by Kai Kā'us ibn Iskandar on the cover of the paper, "Is there anyone so utterly lost as he that seeks a way where there is no way?" captured Crick's gloom.[68] "Iskandar" became the club's intermittent invocation of coding frustration.

Crick's gloom may have been reinforced by the relatively limited resources for scientific research in Britain. Conditions there differed markedly from those in the United States. Having recently recovered from the ravages of war, Britain leaned heavily on American dollars, goods, technologies, and, above all, defense, for its reconstruction. Enormous defense expenditures, rendering it Europe's leading military power, placed a heavy burden on its economy. British science operated on a much smaller scale than its American counterpart. Laboring in "the comparative isolation of Cambridge," Crick sometimes felt that he "had no stomach for decoding." Gamow and Yčas did not share ibn Iskandar's despair. Absorbed in their analysis of nonoverlapping code, they perceived a sense of community and progress. "Decoding here is in little better shape than in Cambridge. I have set myself up as a center for collecting a card file on sequences, replacements, etc.," Yčas reported to Crick. In the spring of 1955 Yčas wrote, "For the first time I am somewhat hopeful that a solution to the formal problem of decoding may be at hand." He also observed that success brought its own set of problems. "Chemically, of course, this [combination code] makes no *obvious* sense. Since the triplets are non-overlapping, we have a 'punctuation mark' problem. Also the 'degeneracy' to get magic 20 [number of nucleotide triplets corresponding exactly to the twenty amino acids] raises stereochemical problems." For Crick the "combination code" made little physical sense. "But it is so hard to think of a structural basis, and the supporting evidence is so weak, that I cannot really take it seriously," he wrote to Brenner. "My own view is that 'coding,' etc, should be put on the shelf for a bit. . . . "[69] Yet with the problem of "punctuation marks," Yčas articulated what would soon become a key preoccupation for analysts of nonoverlapping codes: how to distinguish between consecutive triplets along the nucleic acid sequence.

Most of the work on the combination code was done in the summer of 1955, while Gamow was at Woods Hole, conferring with biologists (Yčas and cytologist Daniel Mazia), finishing more than half of his new book, and spinning social limericks.[70]

> King Dan-el Mazia of Sheba
> Was in love with tiny amoeba
> This Whee bit of Jelly
> Had crowled over his belly,

And metabowhisper "Ich liebe."
Queen Gertrude Mazia of Sheba
Was jealous of tiny amoeba
Said she: "Aint it odd
That this damned pseudopod
Entführte der Man dem ich liebe!"

Play aside, Gamow and Yčas hoped to publish their statistical analysis soon. Gamow was in steady communication with colleagues in Los Alamos who were trying to find a solution to the problem of randomness of amino acid distribution in relation to nucleotide triplets arrangements. Gamow and Yčas considered the relative amount of different amino acids found in about a dozen proteins derived from disparate sources. They reasoned that if nature selected amino acids for protein in a random way, then the most abundant amino acid, the second most abundant, and so on to n, should occur in similar proportions to a stick broken at random into twenty pieces. Writing to John von Neumann from Woods Hole, Gamow requested help in supplying an analytical solution to such a problem for the case of n=20 (twenty amino acids). Von Neumann complied. The results posed a contradiction: the distribution of amino acid residues in the protein samples deviated markedly from von Neumann's random model.[71]

But the lessons could be instructive, Gamow explained to von Neumann after receiving the results. If the code were indeed nonoverlapping, then the deviation from randomness was due either to the nonrandomness in the composition of the nucleotide template, or due to the translation process operating on a random template. Von Neumann replied, "I still somewhat shudder at the thought that highly efficient, purposive, organizational elements, like the proteins, should originate in a random process. Yet many efficient (?) and purposive (??) media, e.g., language, or the national economy, also look statistically controlled, when viewed from a suitably limited aspect. On balance, I would therefore say that your argument is quite strong."[72] To test the first hypothesis, Nicholas Metropolis and physicist Giulio Fermi applied von Neumann's formula to the case of N=4 (four bases) and after 3,000 runs of the Los Alamos MANIAC, the resulting curve deviated significantly from the observed amino acid distribution. The deviation from randomness of the amino acid distribution, Gamow then inferred, had to arise from the nonrandom distribution of the nucleotide triplets in the template.[73]

This result seemed to support recent observations by Chargaff and his collaborators that the distribution of bases in RNA was not quite random; instead, the total of the bases adenine plus cytosine tended to equal the sum of bases guanine plus uracil. This meant that only half of the base composition was random (the other half determined by the first). This finding prompted

Ulam to modify von Neumann's formula (breaking the stick in half, and then each half in two, and so on) and to rerun the analysis on MANIAC. The results indicated that the distribution of the RNA, from plant viruses, as well as from nonviral sources, deviated from random and in the same direction as the protein distribution.[74]

How to interpret these findings? Gamow was exasperated. From a consulting engagement at Convair (San Diego), a major site for developing intercontinental ballistic missiles (ICBMs), he wrote, "I am again completely mixed up about the randomness problem. I wonder what Johnny and Stan will tell about it. Cannot write the article before this question will be cleared up. What do *you* think about this 'new' trouble?" he prodded Yčas. Yčas had been having some reservations about the whole approach. He weighed the pros and cons of randomness versus nonrandomness of nucleotide triplets in RNA and concluded, "It would be best to say that when such eminent gray matter as von Neumann and Ulam are working at high speed it is best for myself to remain blank."[75]

In September, when they submitted their paper, "Statistical Correlation of Protein and Ribonucleic Acid Composition," to the *Proceedings of the National Academy of Sciences* (*PNAS*), they were relatively satisfied with their conclusions. The proportions of amino acids in proteins are not random; this nonrandomness was not due to the application of the triplet translation procedure to a random RNA constitution; and the application of the same translation procedure to the actual RNA composition led to excellent agreement with the observed amino acid distribution. These results, they felt, constituted a strong argument in favor of the assumption that each amino acid in a protein sequence was defined by a nonoverlapping triplet of nucleotides located in the RNA chain. Most important, their model degenerated the sixty-four base triplets without regard to their order—only to their combination—so as to save the "magic twenty" (namely, the number of nucleotide triplets corresponding exactly with the twenty amino acids).[76]

"Geo [Gamow] is about to submit another MS to PNAS about protein coding and this time there can be no doubt that he is not only on the right track but almost doing something trivial," remarked Delbrück in his playful acerbic style. Crick was less generous, finding a mere couple of interesting points in the paper and granting that it showed nonuniform degeneracy of coding. Unless it offered even sketchy structural justifications, it certainly did not make him believe in the combination code itself. Crick found errors in Ulam's proof and, in any case, could not see the point in their random model, because it did not correspond to any known random model in molecular terms. He quipped, "though no doubt it is fine to have von Neumann's name in the paper!" As Crick saw it, their model of an overlapping code was completely wrong.[77]

At least three other researchers tried their hand at "decoding" around 1955: geneticist Drew Schwartz; biochemist Alexander Dounce; and artificial intelligence expert Herbert Simon. Objecting to Gamow's diamond code on structural grounds, Schwartz proposed a complicated and restrictive scheme by which aromatic amino acids (those containing benzene ring structures) were differentially bound in cavities formed by adjacent bases.[78] Alexander Dounce, at the University of Rochester School of Medicine and Dentistry and member of the RNA Tie Club, had been the first to propose a prescient scheme in 1952 that could account for the assembly of amino acids along an RNA strand. He now updated his views in a form of an overlapping directional code: rather than mere triplets, he proposed a scheme by which sixty-four nucleotide diads, taken three at the time, were assigned to an amino acid. (Diads, or doublet codes, remained viable alternatives well into the 1960s.)

But the most striking feature of Dounce's paper is its reconceptualization of the same problem of molecular specificity in terms of both information storage capabilities of the diads and the transfer of information from polynucleotide to polypeptide chains. Even though Dounce did not deploy the tropes of language (alphabet, text, and so forth) he now tacitly represented protein synthesis as a unidirectional communication system. He acknowledged "Drs. G. Gamow, M. Yčas and A. Rich for the free communication of much of their thinking, which has been extremely helpful in bringing him up to date on the problem of the transfer of information."[79]

Herbert Simon, head of the Department of Industrial Management at Carnegie Institute of Technology, had been intrigued by the coding problem for a while; he had even communicated his thoughts to Gamow a year earlier, in 1954. Now responding to Gamow and Yčas's 1955 article on statistical correlation of frequency distributions, Simon was inspired to construct a sextuplet code based on permutations of four pairs of bases, which he claimed fit well with the distribution of amino acids in their table. Apologizing for his amateurish venture, he confessed to Yčas, "I continue to be intrigued by some of the analogies between the kinds of complex systems I am accustomed to consider in my research—large-scale human organizations—and the complex system you are studying." Like his view of language and intelligence, for Simon the genetic code was strictly an algorithmic process: an execution of a designated program.[80]

Gamow's 1955 paper with Yčas was essentially his last original contribution to the coding problem, though he continued to take an active interest in and to write thought pieces and popularizations on the subject. In fact, while a visiting professor at Berkeley's physics department in the fall of 1956, he also taught a course entitled "Biology on the Molecular Level."[81] That fall, he and Yčas presented papers at the symposium on "Information The-

	B	*C*	*A*	*C*	*D*	*D*	*A*	*B*	*A*	*B*	*D*	*C*
Overlapping code	*B*	*C*	*A*									
		C	*A*	*C*								
			A	*C*	*D*							
				C	*D*	*D*						
Partial overlapping code	*B*	*C*	*A*									
			A	*C*	*D*							
					D	*D*	*A*					
							A	*B*	*A*			
Nonoverlapping code	*B*	*C*	*A*									
				C	*D*	*D*						
							A	*B*	*A*			
										B	*D*	*C*

—The letters *A*, *B*, *C*, and *D* stand for the four bases of the four common nucleotides. The top row of letters represents an imaginary sequence of them. In the codes illustrated here each set of three letters represents an amino acid. The diagram shows how the first four amino acids of a sequence are coded in the three classes of codes.

FIGURE 17. Schematic representation of fully overlapping, partially overlapping, and nonoverlapping triplet codes. Francis H. C. Crick, John S. Griffith, Leslie E. Orgel, "Codes Without Commas," *PNAS* 43 (1957). Reprinted by permission.

ory in Biology," at Oak Ridge, Tennessee, organized by biophysicist Hubert Yockey with the assistance of Henry Quastler (see Chapter 3). By that time two important contributions from the Cambridge group had reinforced the potency of nonoverlapping codes: Brenner's proof of the impossibility of overlapping codes; and the construction of comma-free codes by Crick, Leslie Orgel, and John S. Griffith.

In the isolation of Johannesburg, Brenner finally consolidated his analyses of dipeptides, and in September 1956, he dispatched a proof, "On the Impossibility of All Overlapping Triplet Codes," to the RNA Tie Club.[82]

Strictly speaking, Brenner's proof pertained only to fully overlapping triplet codes (see Fig. 17) in which sixty-four triplets were degenerated into twenty sets, corresponding exactly to the twenty amino acids (the so-called magic twenty). His analysis critiqued codes like the diamond code and the compact triangular code, in which each of the triplets shared two nucleotides with the triplet succeeding it in a sequence. The sequence ABCDA, for example, coded for three amino acids: ABC for the first, BCD for the second, and CDA for the third. But his analysis did not examine partially overlapping codes, where the triplets overlapped by a single nucleotide. Nevertheless, his proof was dazzling in its directness and generality, a validation of the superiority of the mind over a computing machine, as Crick might have put it.

Brenner's proof was based on the analysis of dipeptides derived from amino acid sequences of about twenty-five proteins and consisted of showing that sixty-four triplets were clearly insufficient to specify the known sequences. Since any dipeptide sequence was represented by a sequence of four nucleotides (ABCD yielded ABC and BCD, each triplet specifying one amino acid), there could not be more than 256 different dipeptides ($4^4=256$). On the other hand, if all dipeptide sequences were possible, then 400 (20×20) would be expected. In a fully overlapping code, for every four different amino acids known to be adjacent to a particular amino acid on either side of it, that amino acid had to be assigned one triplet. Based on his dipeptide table of amino acid neighbors, Brenner enumerated how many different amino acids immediately preceded or succeeded each of the twenty amino acids and, from the larger of these two numbers, determined for each amino acid the minimum number of triplets that had to be assigned to it. In this manner Brenner showed that, in a fully overlapping code, at least seventy triplets (not sixty-four) would be required to accommodate the tabulated dipeptide sequences.

"It seems clear that non-overlapping equivalence between nucleic acids and proteins must exist," Brenner concluded, "and some of the problems arising from [the] conclusion will be discussed in a later note."[83] Here, he referred to the conundrum of "punctuation" in nonoverlapping codes, a problem which seemed to be nearing solution with the construction of comma-free codes by the Cambridge group. When Brenner's paper appeared a year later in *PNAS*, it was reframed within the increasingly accepted terms of information discourse: the impossibility of overlapping triplet codes now referred to their inability to transfer information from nucleic acids to proteins; and degeneracy meant "excess of information."[84] Brenner's and Crick's recent communications to the RNA Tie Club informed Gamow's and Yčas's presentations at the symposium, "Information Theory in Biology," at Oak Ridge National Laboratory (Gatlinburg) in the fall of 1956.

It was not a memorable event. Gamow did not care much for the information theorists though he shared their discourse. And Yčas recalled that "to the dismay of Gamow and many other participants, Gatlinburg was dry, but the day was saved through contacts with the local elks or some such benevolent organization." Gamow delivered a brief precis of the coding problem (information transfer), "The Cryptographic Approach to the Problem of Protein Synthesis," ending his paper with the thorny problem of "punctuation" in nonoverlapping codes. Yčas, by then a professor at the Department of Microbiology at the University of Syracuse, provided an extensive and up-to-date treatment of what he called "The Protein Text."[85] Representing the RNA molecule as a text written in a four-symbol alphabet that encoded the protein text written with a twenty-symbol alphabet, Yčas too analyzed in detail the various cryptanalytic endeavors. "Cryptography must be based on a study of texts, and I shall therefore attempt an examination of protein molecules from this point of view."[86]

If life is written, if DNA, RNA, and proteins are texts, as these scientists came to articulate them, then who wrote them? What was the source of this writing? How did texts precede the modes of representation that brought them into being? Could there be tacit theistic investments in the sequence as the primal word of genesis? While promoting these scriptural representations of molecules, Yčas somehow intuitively grasped their deconstructive dimensions: the conflation of the writer with the written. He prefaced his paper with this insightful witty quote:

> *And strange to tell, among that earthen Lot*
> *Some could articulate, while others not:*
> *And suddenly one more impatient cried—*
> *"Who is the Potter, pray, and who the Pot?"*
> (from) The Book of Pots

Having examined the protein text, Yčas moved on to the coding problem, namely to the problem of "storage, transfer, and the replication of the information contained in the protein molecule." Within that discourse, information transfer and coding (textualizing) were two sides of the same coin and increasingly visualized as the reading of instructions off a magnetic tape. Viewing the "informational content of the RNA forming system [as] highly redundant with respect to the end product formed," he used Shannon's formula to calculate the information content of the recently accepted model of an RNA as a template in order to show that it could indeed store the requisite information and thus could be, theoretically, self-replicating. Like Gamow, Yčas concluded his paper with the problem of punctuation in nonoverlapping codes, including a critique of a new type of code—the "comma-free code"—that had recently constructed by Crick and his Cambridge collaborators.[87]

Syntax to Semantics? Comma-free and Other Codes

As the analogy of a DNA (or RNA) strand to a text fixed itself in the minds of the code workers, the tropes of textuality instantiated themselves in their genetic phenomenology. Since 1955, when it looked likely that the code was nonoverlapping, the problem of how and in which direction to distinguish the order of nucleotide triplets along the strand became the key conundrum of that textuality. Much discussion ensued about the "punctuation mark problem," namely, how the "reading mechanism" recognized the start

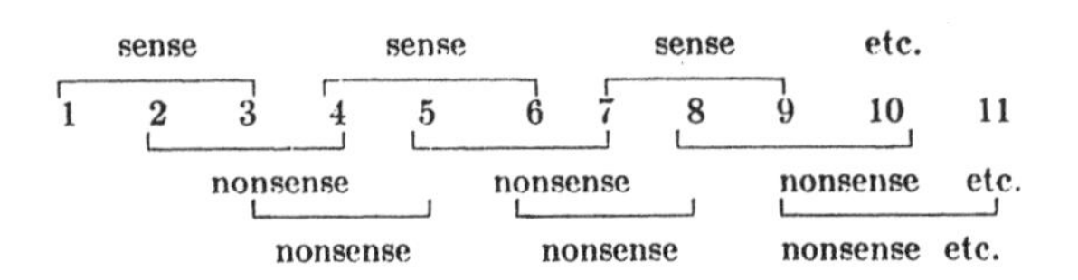

The numbers represent the positions occupied by the four letters *A*, *B*, *C*, and *D*. It is shown which triplets make sense and which nonsense.

FIGURE 18. Francis H. C. Crick, John S. Griffith, Leslie E. Orgel, "Codes Without Commas," *PNAS* 43 (1957). Reprinted by permission.

and end of a triplet, as Yčas recalled. Without a marker ("punctuation"), a sequence of bases could be read ambiguously, for example, ABC, DCC, BDA, . . . or, A, BCD, CCB, DA. . . .[88]

Three ways out of this ambiguity emerged: there could be an initial configuration, a "punctuation," marking the beginning of the count (as it was later shown); the "coding" triplets could be separated by some kind of "noncoding" (nonspecific) arrangement of bases; or the nature of what soon became known as the "dictionary" might itself prevent ambiguities. Only those particular configurations of triplets which specified amino acids would have "meaning" and count as "words." Crick, together with the chemist Leslie Orgel and biophysicist John Griffith, chose to elaborate the third option in their ingenious comma-free triplet code. Yčas observed, "This type of code was, of course, already well established in communication engineering practice."[89]

For a brief moment the authors did acknowledge the metaphoricity of coding and textuality. Since there were sixty-four possible base triplets and only twenty amino acids, certain triplets would be associated with amino acids, whereas others would not. Crick and his colleagues wrote, "Using the metaphors of coding, we say that some of the 64 triplets make sense while some make nonsense. We further assume that all possible sequences of the *amino acids* may occur (that is, can be coded) and that at every point in the string of letters one can only read 'sense' in the correct way."[90] (See Fig. 18.) The mathematical problem was to find the maximum number that could be coded; the burden of the paper was to show two things: first, that maximum could not exceed (magic) twenty, and, second, that a solution for twenty could be given.

From then on, the metaphor of language became constitutive of their reasoning and modes of signification. Combining logical argumentation with detailed enumeration, Crick, Orgel, and Griffith ruled out the four sequences of the form AAA, BBB, CCC, DDD, since those offered no way of distinguishing sense from nonsense (as I discuss in Chapter 5, it was exactly this kind of triplet, UUU, that turned out to be the first so-called code word, which specified the amino acid phenylalanine). The remaining sixty triplets could be grouped into twenty sets of three, each set of three being a cyclic permutation of another (ABC, BCA, CAB). If ABC were "sense" (namely, specifying an amino acid), then the other permutations would be nonsense. Thus only one triplet from each set could be chosen, yielding a total of no more than twenty.

Almost miraculously it turned out to be possible to select twenty "sense" triplets in such a way that juxtaposition of any two produced overlapping "nonsense" triplets. One such solution was:

```
               A
         A  A
   A  A        B
AB      CB  BD
   B  B        C
         C  C
               D
```

where A—B signifies ABA, ABB, and so on. There were many other possible solutions (two hundred and eighty-eight, to be exact) that yielded twenty "sense" triplets. All such commaless codes were nondegenerate and saved the "magic number twenty."[91]

That immaculately logical model had an additional advantage: it offered a physical interpretation of the assembly of polypeptides through what became known in the RNA Tie Club as the "adaptor hypothesis." In his 1955 note to the club, "On Degenerate Templates and the Adaptor Hypothesis," Crick, aided by Brenner's imagery, had proposed that each amino acid could be transported to its proper place via a small molecule (e.g., trinucleotide) capable of hydrogen bonding specifically with the nucleic acid template. (Around 1957 that "adaptor" was correlated with soluble RNA and soon after renamed "transfer RNA.")[92] Combining that hypothetical transport mechanism with the hypothetical commaless code, Crick, Griffith, and Orgel now suggested a way by which such a trinucleotide adaptor, loaded with its amino acid, would diffuse away from incorrect (nonsense) positions and bond only to the correct (sense) triplet along the template. Their model served to concretize the implicit idea of information transfer in protein synthesis.

Crick, Orgel, and Griffith were excited by the idea of a comma-free code. "It seemed so pretty, almost elegant," Crick recalled. Nevertheless they were rather circumspect in print. "The arguments and assumptions which we have had to employ to deduce this code are too precarious for us to feel much confidence in it on purely theoretical grounds," they wrote. "We put it forward because it gives the magic number—20—in a neat manner and from reasonable physical postulates." They granted that other codes, notably the "combination code" of Gamow and Yčas, also attained the goal of "magic twenty," but Crick thought theirs was grounded in implausible physical premises.[93]

Yčas, on the other hand, countered that a commaless code contradicted the observations: for example, it imposed intersymbol restrictions on the protein text where none had been observed. A mutation could result in a change at one site only, leaving the rest of the sequence unchanged. In any case, he insisted, the "punctuation mark" problem could be solved without resurrecting the ghosts of overlapping triplets if amino acids were selected in a sequential manner starting from one end of the template. In spite of reservations and objections the commaless code became remarkably popular, and after four people had asked if they could quote the paper, the authors decided to write it up for *PNAS*, where it appeared in May 1957. Temporarily, it became a kind of dogma.[94] By that time yet another physicist, the Hungarian émigré Leo Szilard, had sharpened his wits on the coding problem, reaching similar conclusions to those of the Cambridge group.

Szilard by then had been well ensconced in biology. Like several other atomic scientists who were disillusioned with the advent of nuclear power and the militarization of physics, he migrated from the science of death to the science of life and worked on halting the arms race. Max Delbrück's congenial summer community at Cold Spring Harbor served as the initiation rite and the bridge to biology. In the fall of 1946 Szilard joined the faculty of the University of Chicago as professor of biophysics in the Institute of Radiobiology and Biophysics. Teaming up with the physical chemist Aaron Novick, his junior colleague from their wartime work on the Atomic Energy Project at the University of Chicago, he trained in biology through Delbrück's summer phage courses at Cold Spring Harbor and Cornelius Van Niel's summer course in bacterial biochemistry at Pacific Grove.[95]

Szilard and Novick went on to work on light-reactivation in bacteria, phenotypic mixing in viruses, on experiments with the chemostat (a continuous-culture device, one of Szilard's several patented inventions), on chemically induced mutations and amino acid synthesis in bacteria, and with geneticist Maurice Fox on a device for growing bacterial populations under steady-state conditions. But more important, Szilard served as an intellectual bumblebee, as François Jacob put it, cross-pollinating ideas across the molecular biology international community. Like Gamow and von Neumann, Szilard had a peripatetic style that took him to some of the most active research centers across the country and across the Atlantic. At the time of his interest in the coding problem in 1957, Szilard had been a frequent visitor at the Pasteur Institute (see Chapter 5), the Medical Research Council molecular biology unit in Cambridge, Caltech, the Rockefeller Institute, New York University, and the University of Colorado (where his wife, Trude, taught public health at the Medical School, where Gamow was now professor of physics, and phage researcher Theodore Puck was a professor of biophysics).[96]

Szilard enjoyed that freewheeling transinstitutional mode of doing science. When the Institute of Radiobiology and Biophysics was discontinued in July 1956, rather than join the Institute of Nuclear Studies at the University of Chicago, Szilard sought to formalize his academic rovings as a senior scientist-at-large. He applied to the National Science Foundation (NSF) for funding of a "Roving Professorship." Presenting himself as a theoretical biologist who facilitated communications in key research areas in molecular biology, he pointed out to NSF officers:

> At present certain branches of biology in which I am interested are in rapid progress. The problems of protein synthesis, the role of RNA and DNA, and the general problem of self-reproduction, differentiation and aging are rapidly becoming open to attack by means of new techniques. . . . As a Senior Scientist-At-Large it should be possible for me to acquire intimate knowledge of experiments conducted with a great variety of biological material and diverse techniques, and thereby to be in a position to try to function as "theoretical biologist."[97]

Biology had not quite reached the stage of theoretical physics half a century earlier, Szilard explained, echoing Delbrück and Schrödinger. But it could be very well on the verge. That meant that it could be fruitful for a few scientists to "put less emphasis on their own experiments and spend more time trying to keep in close touch with the experiments of others in the hope of being able to recognize new patterns and to gain insights into some general biological laws that have so far not clearly emerged." Several leading biologists who strongly supported his roving professorship concurred. They acknowledged the peculiarity of the arrangement, but, they noted, Szilard was a peculiar person.

> In addition to producing a number of discoveries of the highest rank, Szilard has been an unusually effective catalyst in science. In recent years he has engaged extensively in visiting other laboratories, where he has been most generous in giving thoughtful and deep attention to the work in progress. He has a unique ability to grasp instantly the most varied problems, to seize upon their significant aspects, and to apply to them unusual imaginative and critical powers. These visits have often led to valuable new experiments and have given to many young biologists a much enhanced sense of the distinction between significant and trivial problems.[98]

Although this particular institutional scheme (and a few others) did not materialize, Szilard continued his global intellectual rovings.

In June 1957 he dispatched a letter to Crick and enclosed the manuscript "How May Amino Acids Read the Nucleotide Code?" which he was submitting to *PNAS* (he also sent a copy to Delbrück at Caltech). He had missed the latest issue of *PNAS* containing the article "Codes Without Commas," he admitted, but he had come independently to similar conclusions and had since incorporated some of their points. "If you see anything basically wrong with my views, please drop me a line or send me a cable. For another ten days I can still withdraw this paper without causing anyone any appreciable inconvenience," he wrote, as he often did, and indeed, often abstained from publishing. He would be in Europe in the fall and wished to spend some time with the Cambridge group (Seymour Benzer, Renato Dulbecco, George Streisinger, and Sidney Brenner), particularly if they were "fully assembled."[99]

Szilard had been following the contours of the coding problem in all its embeddedness within the information discourse. Though he had been retrospectively credited as one of founders of information theory (see my discussion in Chapter 2), he apparently did not take a strong interest in it, nor did he think much of its application to biology. Nevertheless, like Gamow, he readily adopted its discursive practices, which had come to constitute the code as the governing component in the genetic communication system: information content, messages, words, commas, and reading.

Relatively up-to-date, Szilard's scheme built on the premise that it was not necessarily the gene (DNA) but what he called the "paragene" (RNA)—carrying "the same information"—that directed protein synthesis. It incorporated ideas similar to the "adaptor hypothesis," addressed the issue of directionality in nonoverlapping codes (though he also envisioned codes comprised of four, or a higher number of nucleotides), offered a solution to the "punctuation problem," and examined in detail the reaction kinetics of amino acid assembly in protein synthesis in bacteria.

> Accordingly, sequences of three nucleotides along the paragene represent the code words, and the trinucleotides which carry the amino acids represent the anti-code words. We assume that these anti-code words are complementary to the code words. . . . If the sequences of bases along one strand of DNA represent a coded message which consists in three letter-words, then . . . the code on the paragene must be read consecutively from one end—say, from the "head" of the paragene downward. In these circumstances, the code would be misread if the trinucleotides, which represent the anti-code words, assemble alongside the paragene simultaneously, rather than—from one end on—consecutively.[100]

Some of these features had already been proposed, as he belatedly learned. But the most novel aspect of Szilard's paper was its second part: a detailed account of the kinetics of amino acid assembly in protein synthesis, which he envisioned as a chemical chain reaction starting from the head of the paragene and moving along downward. Using known rates of enzyme synthesis, he computed the rates at which the paragene synthesized the enzymes required to load the amino acids onto the anticode nucleotides; the equilibrium conditions for the formation of gaps along the paragene into which these loaded units could fit; and the time needed for amino acid assembly. He concluded that his theory explained the high rate of enzyme synthesis in bacteria when that rate was maximally enhanced by an inducer. His work on enzymatic feedback inhibition in E. coli (with Aaron Novick) greatly overlapped with Jacques Monod's studies at the Pasteur Institute (see Chapter 5).[101]

Crick did not read the second part. "We have only had time to digest the part of your paper dealing with coding," Crick promptly responded, "and I had better say straight away that we don't like it. The type of code you suggest is extremely restrictive," he countered, proceeding to outline the incongruities in Szilard's proposal. He had just sent him Brenner's *PNAS* preprint, "On the Impossibility of All Overlapping Codes," where all the known sequence data had been collected. "As you can see," Crick pointed out, "quite a modest selection from this will disprove all possible versions of your code. . . . I have shown your paper to Brenner and Orgel, and the above criticisms are the joint work of the three of us." By then Szilard had already met with Gamow at the University of Colorado and obtained from him Brenner's data. Szilard never published the manuscript, though he continued to analyze the issues during his September visit in Cambridge and, in general, maintained his participation in the coding problem in his informal episodic manner.[102]

Comma-free codes attracted considerable attention. While their novelty, pristine logic, and seductive elegance intrigued molecular biologists, this class of codes was being used in advanced communication systems, as Yčas pointed out. It seemed natural, therefore, to draw on the expertise of mathematicians and communication engineers to analyze and elaborate on the general properties of comma-free codes. Such expertise was readily available at Caltech's Jet Propulsion Laboratory (JPL), which sat only a stone's throw from Delbrück's group and which since the end of World War II had been sponsored by the United States Army.

Founded in 1944, JPL served as a major scientific resource for California-based giants such as Lockheed and Convair, both leaders in the development of the ICBM program and its aerospace technologies, as well as for the growing missile research project National Advisory Committee on Aeronautics (NACA), reconstituted in 1958 as National Aeronautics and Space Administration (NASA). JPL researchers developed some of the key technologies of the missile age, such as electronics and guidance and control. And their role in the arms race intensified after the passage of the National

Defense Security Act (1958) and the militarization of NASA, both reactions to the launching of *Sputnik I* on October 4, 1957.[103]

At Caltech, Delbrück cultivated close working relations with physicists and mathematicians, in and outside JPL. His effect as a man of action and the charismatic leader of a burgeoning phage school was matched by his reputation as a "think-man," whose theoretical grasp of problems in physics, chemistry, and biology endowed him with an unusual ability to integrate knowledge across scientific boundaries, an approach of special importance to Caltech's interdisciplinary agenda. A cultivated eccentric, Delbrück's iconoclasm was balanced by his talent for group cooperation, a prized asset at Caltech. He had set up an interdisciplinary seminar soon after he moved there. "Max has a biophysics seminar that has theoretical chemists and physicists not only attending but actually presenting papers!" George Beadle, head of the biology division exclaimed incredulously in 1947.[104] Ten years later, Delbrück's Tuesday seminar was still going strong, its participants persistently challenged by the recalcitrant genetic code, now represented as a comma-free code.

Three JPL scientists responded to the challenge: Solomon W. Golomb, a young mathematician and communication engineer; mathematician Basil Gordon, specialist in combinatorial analysis; and mathematician and electrical engineer Lloyd R. Welch, who two years later joined the Institute for Defense Analysis. In the summer of 1957 they tackled the challenge of providing a mathematical generalization of the coding problem: What was the maximum size of a comma-free dictionary in the case of an arbitrary number of symbols and an arbitrary length of the words?

They analyzed the comma-free codes as a general combinatorial problem. Namely, they examined an alphabet (note the metaphor) consisting of the numbers 1, 2, . . . , n, with which it was possible to form all k-letter words (a_1 a_2 . . . a_k), where both n and k were fixed positive integers, and n^k the number of possible words. And in order to obey the strictures of comma-free codes—that certain nucleotide triplets made "sense" (words) while others made "nonsense" (nonwords)—they defined the set D of k-letter words ("sense") as a "comma-free dictionary" in such a way as to exclude all overlapping triplets ("nonsense").[105]

Their solutions, though partial, demonstrated that it was indeed possible to construct such codes, where twenty sequences of nucleotides corresponding to the twenty amino acids ("magic twenty") formed a comma-free dictionary. The term *dictionary* added yet another linguistic trope to the discursive and semiotic space configured by information theory, electronic computing, and linguistics, within which the genetic code was being constituted as a scriptural technology.

Golomb, Gordon, and Welch grounded their analysis in information theory. One could think of "the sequence of nucleotides as an infinite message, written without punctuation, from which any finite portion must be decodable into a sequence of amino acids by suitable insertion of commas. If the manner of inserting commas were not unique, genetic chaos could result," they predicted. Rather unexpectedly, the authors further argued, comma-free codes could even contribute to the mathematical theory of coding, since they were a subset of a larger class of related codes.

In their search for optimum coding techniques, Claude Shannon and Brockway McMillan had studied theoretical codes where the entire message was available. But in actual communication applications—as in the problem with the genetic code—only disjointed portions of a message were likely to be received. This is where comma-free codes could indeed be useful, they concluded. These genetic codes were now becoming "boundary objects," migrating along the two-way traffic between molecular genetics on the one hand, and the world of mathematics, communication engineering, and the military on the other.[106]

Golomb, Gordon, and Welch's treatment was purely mathematical. But soon after, Delbrück teamed up with Golomb and Welch to address comma-free codes as both a mathematical and genetic problem. On the mathematical end, Golomb and Welch now analyzed classes of dictionaries for the specific case of triplet "words" ($k=3$), examining the reversible portions of such dictionaries and characterizing messages composed of a maximum of twenty triplets generated by the four nucleotide bases ($k=3$, $n=4$).[107]

They then showed that there were five classes of such codes (a total of 408 codes) and that no message written with any of these codes ever contained a fourfold repeat of any base. In some of the codes certain threefold repeats were excluded. They even considered quadruplet codes. Together with Delbrück, they also expanded the range of possible types of coding errors—"misprints" as they called them—to include what they termed "missense." They explained, "Certain misprints in the coded message will produce nonsense (the resulting triplet does not code for any amino acid), other misprints will produce missense (the resulting triplet codes for a different amino acid). The codes were studied with respect to missense/nonsense ratios produced by various classes of misprints."[108]

From a biological standpoint, the most thoughtful feature of their model underscored that such genetic codes had to meet another requirement: transposability (directionality relative to a DNA strand). As Delbrück had stressed for several years, DNA has no intrinsic sense of direction; its two helices run in opposite directions. And since there were no known cases of one genetic locus coding for two different proteins, it seemed likely that only one DNA strand was read. Thus, Delbrück argued, the dictionary had to be not just comma-free but transposable as well. Its reversed complement had to be all nonsense.

> We wish to emphasize that we consider the postulate of comma-freedom and the postulate of transposability to be almost on the same footing. Indeed the principal virtue of comma-freedom is that any message can be read unambiguously starting at any point, with the proviso, however, that one must know in advance *in which direction to proceed*. Since the equivalence of the two opposite directions in a structural sense seems to be one of the more firmly established features regarding the DNA molecule the advance knowledge as to the direction in which to read cannot come from the basic structure. Comma-freedom would therefore seem to be a worthless virtue unless it is coupled with transposability.[109]

As he had been doing for a couple of years, Delbrück framed his arguments within the discourse of information. The discovery that genetic information in many organisms was transmitted from parent to offspring through DNA had raised the problem as to the nature of the code used to carry this information and as to the mechanism by which the code was read, his paper, coauthored with Golomb and Welch, began.[110] Delbrück felt that in view of the mounting evidence that phage DNA did not replicate itself directly in the bacterial cell but did so through an intermediary substance (RNA), "it would be unwise not to give some currency to 'information transfer' as a possible replication mechanism."[111] Delbrück did not invoke the information idiom idly. Though, like most molecular biologists, he tended to use the term generically (rather than in its mathematical sense), unlike them, he was well versed in information theory. "I am teaching information theory this term," he wrote in the fall of 1955 to his colleague Robert Sinsheimer, a biophysicist at the physics department at Iowa State College, "and have been using your DNA data and other people's RNA data to illustrate the notions of inter-symbol influence and the statistical properties of information sources."[112] Clearly, his close interactions with the mathematicians and communication engineers at JPL shaped his thinking, thus reinforcing the growing trend of representing heredity as an electronic communication system; even if by the conflation of technical and generic usage of information he actually undermined the validity of DNA, RNA, and proteins as cryptographic texts.

Delbrück's colleague, Golomb, would continue his attempts to decipher the genetic code mathematically. As late as the spring of 1961 (just weeks before the code was "broken" biochemically) his approach still seemed to hold out great promise. "New Way To Read Life's Code Found," announced the *New York Times*. "A 'dictionary' of 24 Words Appears Able to Describe Inheritance Mechanism."

> The scientist, 29-year-old Dr. Solomon W. Golomb of the Jet Propulsion Laboratory of Pasadena Calif., told a meeting of the American Mathematical Society at the New Yorker Hotel of building what might be called a "dictionary of life". . . . Biologists who have examined the new theory say it fits all of the facts now known about the way nucleic acids and proteins work in transmitting hereditary information. . . . Scientists from several fields have been trying to crack the code of life since the middle Nineteen Fifties. . . . Dr. Golomb decided that this [nature's alphabetic] extravagance probably produced the redundancy in the coded messages that would be necessary to minimize mistakes and to assure a high degree of reliability in transmitting genetic information. . . . he is satisfied that his code is a good and workable one and could be found in nature—if only on another inhabited planet.[113]

In 1958 Francis Crick formalized information as a fundamental property of biological systems and the governing concept in the discipline of molecular biology. Crick was by now a celebrity; his meteoric rise from near obscurity in 1952 to the center stage of science had endowed him with considerable authority. He was a visible member of a prestigious group of molecular geneticists and protein crystallographers, including John Kendrew, Vernon Ingram, and Sydney Brenner, working in the MRC Unit for the Study of the Molecular Structure of Biological Systems under the direction of Max Perutz. As Soraya de Chadarevian has pointed out, this was a small group of researchers backed by some potent patrons, but without a clear location in the disciplinary landscape in the university. In the late 1950s, the group began negotiating with the MRC and the University of Cambridge about construction of the Laboratory of Molecular Biology.

Their lab plans represented not merely physical expansion or change to a trendy name but also a consolidation and formalization of the new discipline. As de Chadarevian has shown, the political negotiations and those on the bench level happened at the same time, as did the founding of the *Journal of Molecular Biology* in 1959 (with John Kendrew as its first editor).[114] Pnina Abir-Am observed further that while in the 1930s and 1940s the term *molecular biology* had been used only modestly in the scientific literature, by the 1950s it was gaining wide circulation.[115] The concept of information, or rather its various tropes of information flow, storage, transfer, and retrieval, not only served as a source of potent models and analogies but also as rhetorical software, both demarcating the disciplinary turf of molecular biology and embedding it within the discourses of postwar culture.

Crick formalized the role of information as a way of imposing thematic order and discursive imperatives on the multilayered problem of protein synthesis. In his paper on "The Biological Replication of Macromolecules," presented at the 1957 symposium of the Society for Experimental Biology, he proclaimed that the essence of the problem of protein synthesis was flux:

flow of energy, flow of matter, and flow of information. His categorization was remarkably similar to that of Norbert Wiener's a decade earlier: Wiener had predicted that representations of organisms had been shifting from the materialistic and energetic to the informational. Like Wiener, Crick too focused on the third category—the flow of information. "By information I mean the specification of the amino acid sequence of the protein," Crick explained. Amino acids had to be joined in the right order. He stressed, "It is this problem, the problem of 'sequentialization,' which is the crux of the matter."[116]

Crick thus echoed the definition of information as a measure of biological specificity as formulated by Henry Quastler and his colleagues five years earlier. "Information . . . is related to such diverse activities as arranging . . . determining, ordering, organizing . . . specifying . . . ," Quastler had stated as a general proposition, following up with a quantification of the measure of information of enzyme specificity. "Proteins have been treated as linear messages, with amino acids as symbols," elaborated Leroy Augenstine and his collaborators. They calculated the informational properties (intersymbol influence) of twenty-five proteins based on the relative frequencies of their amino acids and readily admitted that having the actual amino acid sequences would be crucial to the task. Insulin had just been sequenced.[117] Their accounts clearly recognized the specification of the amino acid sequence as key to the informational properties of proteins. Whereas Augenstine and his collaborators attempted to use information in its mathematical form (and defined it as a measure of specificity for *all* biological entities), Crick deployed it merely qualitatively, selectively (privileging nucleic acids).

With protein synthesis defined as information flow, Crick enunciated its two general principles: the "sequence hypothesis" and the "Central Dogma," rules which soon came to be regarded as the pillars of molecular biology. In its simplest form, the sequence hypothesis assumes that "The specificity of a piece of nucleic acid is expressed solely by the sequence of its bases, and that this sequence is a (simple) code for the amino acid sequence of a particular protein." The Central Dogma postulates:

> Once "information" has passed into protein it *cannot get out again*. In more detail, the transfer of information from nucleic acid to nucleic acid, or from nucleic acid to protein may be possible, but transfer from protein to protein, or from protein to nucleic acid is impossible. Information means here the *precise* determination of sequence, either of bases in the nucleic acid or of the amino acid residues in the protein.

In a single masterly stroke Crick encapsulated the ideology and experimental mandate of molecular genetics: genetic information, qua DNA, was both the origin and universal agent of all life (proteins)—the Aristotelian prime mover—as Delbrück suggested.[118]

Crick did this by subtly reworking the old concept of biological specificity. A three-dimensional configuration devoid of directionality was collapsed and propelled along the unidirectional, unidimensional flow of information. Specificity was bounded in matter, while information was mobile, transporting the memory of form beyond material bounds. Information was the body's soul and logos. Information was relayed from sender to receiver; specificity was solitary and mute. But these biosemiotics entailed a deconstruction. In order to privilege nucleic acids as the sole source of biological information this transformation had to tacitly subvert information from its scientific definition. From being a purely stochastic process independent of both material and semantic contexts, information was now re-represented as a medium that carried its own meaning, conveying messages across cellular spaces and the cycles of life. The still-unbroken genetic code governed that metaphoric process of information transfer. Whether the latest decoding attempts had any validity, "only time will show," Crick concluded.[119]

There was much to be pensive about. A recent report in *Nature* by two Russian biochemists, A. N. Belozersky and A. S. Spirin, threw the coding problem into cognitive turmoil. They showed that DNA of different microorganisms ranged widely in their base ratios. Namely, the ratio (G+C)/(A+T) was as low as 0.5 for some organisms and higher than 2.5 for others (A-adenine, G-guanine, T-thymine, C-cytosine). Their report drew attention to earlier findings at the Pasteur Institute, where in sixty bacterial strains these ratios varied between 0.4 and 2.7. Surprisingly, the base composition of the total RNA of these organisms (where C is replaced by U, uracil) hardly varied at all. (This incongruity would be resolved in a few years, when total cellular RNA was differentiated into three distinct species: messenger, or mRNA; transfer, or tRNA; ribosomal, or rRNA.) This huge variation of DNA composition was disquieting since, as far as was known, the abundance of various amino acids did not vary much from organism to organism. The incongruity seemed to be a turn back to Gamow and Yčas's conundrums of randomness in their paper on statistical correlations of protein and RNA composition.[120]

These findings provoked Robert Sinsheimer (then at Caltech's biology division) to take an ingenious and bold leap: he proposed a binary code. Drawing on information theory, he suggested that such a code could accommodate the data with maximal efficiency. "There is an elementary theorem in information theory (Shannon & Weaver, 1949) that if a message is to be written in a code of T symbols, the message can be written most efficiently (i.e., make use of the least quantity of symbols) if each symbol is used to an equal extent. In our case the message would be the protein content of a cell;

this is to be expressed in a two-symbol (N, K) RNA code."[121] Sinsheimer had become familiar with feedback controls, cybernetics, and communication theory as a graduate student, when he worked at MIT's Radiation Laboratory during the Second World War. "It therefore seemed very natural to apply the concepts of information content, stability to thermal and other interference (noise), etc. to the issues of genetic inheritance, mutation, et al.," he recalled. At Iowa State University he gave a series of lectures on these subjects in John W. Gowen's genetics seminar, lectures based on Wiener's *Cybernetics*, Shannon and Weaver's *The Mathematical Theory of Communication*, and Quastler's *Essays on the Use of Information Theory in Biology*.[122]

In his informationally efficient two-letter code (N, K) the only requisite structural features for coding were the presence of two configurations: either a 6-amino group (N), or a 6-keto group (K). Keto groups (carbon double-bonded to oxygen, C=O) characterized the DNA bases G and T (or RNA bases G and U), while bases A and C were characterized by amino groups (carbon singly bonded to an amine, NH_2). As Chargaff had demonstrated four years earlier, in RNA the sum of the 6-amino groups (N) closely equals the sum of 6-keto groups (K). Thus Sinsheimer proposed that, when reading the DNA message, C equaled A, and T equaled G. In such an alphabet the quantity of the two letters (N, K) was always equal—thus satisfying Shannon's theorem—and the effective composition of DNA was invariant. Support for his model seemed forthcoming from Geoffrey Zubay, Quastler's colleague, who proposed a mechanism for transferring the genetic code from DNA to RNA.[123]

With such a proliferation of codes, several hundreds of them, the optimism surrounding the coding problem had dwindled markedly by the close of the 1950s. And when Brenner and Crick attempted a scheme of a quadruplet code, their minor contribution in December 1959—an elaboration on the amino acid "adaptors" (soluble RNA, soon renamed tRNA)—terminated the sedentary life of the RNA Tie Club. Meanwhile, the club's president, Gamow, preoccupied with his new life in Boulder and beleaguered by alcohol-related ills, had become primarily a spectator and commentator on these researches. "I didn't get any new bright ideas on coding recently," he wrote to Brenner, "apparently the problem is stuck just as the theory of elementary particles."[124]

"Decoding" was not much of a spectacle. Assessing "The Present Situation of the Coding Problem" at the 1959 summer symposium at Brookhaven National Laboratory, Crick was glum. Given the numerous codes and the unexpectedly wide variation of DNA base ratios in microorganisms, it was now time to reevaluate some of the fundamental premises of the coding problem. Maybe only part of the DNA coded for proteins; possibly the DNA-to-RNA translation mechanism varied; perhaps the code was *not* universal (i.e., not the same for all life); the nucleic acid code might have fewer than four letters (after Sinsheimer); it could be that the amino acid composition of the protein varied, a possibility that led Crick to caution, "unfortunately a small variation would not do."[125]

Brenner offered a similar view on the coding problem at the 1959 CIBA symposium, where he questioned some of its fundamental features. This assessment came despite his progress: an ingenius blend of phage manipulation, genetic analysis, and deductive reasoning, based on Seymour Benzer's recent breakthrough of mapping the fine-structure of the rII region of the bacteriophage system T_4. (That region refers to the location of specific mutations on the phage genome map, identified by their peculiar plaques on a petri dish of E. coli K12.) A physicist-turned-phage geneticist, Benzer did work that necessitated a major revision of the classical gene as unit of recombination, mutation, and function toward a distinction between the three (Benzer gave each physicslike names—"recon," "muton," and "cistron," respectively—though the terms disappeared within a few years). And though he did not work directly on the coding problem his studies advanced it considerably. With Benzer's amplification of the gene, each mutation could be correlated with changes in the DNA sequence at a defined position and with the corresponding changes in the amino acid sequence that rendered the protein defective. Assuming DNA-protein colinearity, possession of such a protein would provide a direct link to the coding sequence, especially in combination with analyses based on several newly studied mutagens. Racing against Cyrus Levinthal's group at MIT, Brenner's team concentrated their efforts on the phage-head proteins—composing nearly 90 percent of the virus—relatively accessible through osmotic shock treatment and enzyme digests studies. In fact, they had just identified the "O gene," which seemed to control the assembly of the head protein. This was an important finding, but it was a long way from breaking the code, as Brenner himself emphasized.[126]

"I cannot comment on Dr. Brenner's speculations on protein synthesis," responded McGill's geneticist H. Kalmus in the discussion session, "but . . . the analogy between printed and genetical information seems at the moment to fire the imagination of many people and some general considerations seem timely." Offering a bird's-eye view of the evolution of language, Kalmus drew analogies for the genetic code. Originating in a spoken form and shaped also by chance, natural languages evolved from ideographic to phonetic transcriptions, representations having little to do with the objects they denote, he pointed out. It might therefore be profitable to analyze hierarchical orders and sequences of the DNA script, Kalmus suggested, rather than look for correspondences between the DNA and protein codes, since, as in language, there might not be a direct relation between the two. To which Brenner responded, "This is very difficult; [genetic] coding is not really like ciphering." He continued, "We have the problem texts but we do not know what the original language looks like. Suppose one was given a cipher and was not told whether this was captured from the Russian, American, Chinese, or Italian Army and asked to decipher it. This would be impossible if one did not know the original language." In response to Kalmus's assertion that the genetic situation was analogous to the deciphering of the Rosetta stone (elaborated upon below), Brenner argued that the genetic code was even more challenging than the recent (1952) decipherment of "Linear B," one of the most elegant and surprising code-breaking stories.[127]

The obscure script, identified as "Linear B" in Minoan tablets excavated in 1900 in Knossos, Crete, defied deciphering for five decades, primarily because of heated disagreements among scholars about the language; the script differed from classical Greek and appeared to be related to Semitic, Etruscan, or Hittite languages. At stake was the dating of Minoan civilization, the ascent of Hellenism, and the veracity of the *Iliad* and the *Odyssey*. It turned out to be a primitive form of Greek, after all. Not only did the finding (by British architect Michael Ventris, based on the work of American archaeologist Alice B. Kober) settle epic historical questions but elevated decipherment methods to new heights. "It shines with a clean Euclidean beauty," David Kahn put it.[128] But the genetic case was different, as Brenner was quick to point out.

> On reading the book by Chadwick, "The Decipherment of Linear B," one realizes quite quickly why the nucleic acid protein problem is different. When scholars were presented with linear B, they could recognize the semantic context of some of the symbols—they could recognize horses and women. We are presented with the proteins. Now almost certainly the semantic context of a polypeptide chain of an amino acid sequence has got nothing to do with the original nucleic acid code, but it has to do with the folding of the protein, the structure of the active centres, and so on. In other words, inspection of amino acid sequences can probably tell us nothing about the code, but everything about protein structure. That is what breaking of the protein code could mean in terms of deciphering linear B. We are, however, trying to find the letter congruences between language of an as yet unknown system, and a system the semantic context of which we do not fully understand.[129]

Given the ambiguities of what "code" signified and the incongruences acknowledged between natural language and DNA sequences, the leaps from syntax to semantics, and the fundamental differences between linguistic and genetic decoding, the power of language and coding metaphors becomes all the more astonishing. In 1959 there was little to hold on to but the cardinal belief in the existence of "The Code," the governor of the Central Dogma, an object prefixed in the mind and preceding its experimental warrant. In five years, despite contradictory data and dubious results, and with some of the most "eminent gray matter working at high speed," as Yčas had put it, there were several interesting theories but few concrete returns. Gamow's colleague, the admiral at the Bureau of Ordnance in Washington, D.C., turned out to be right after all. There was no cryptanalytic solution to the genetic code (since, linguistically speaking, it is not code). Yet the champions of the genetic code clung tenaciously to their logos. Crick could vouch for only three safe anchors supporting the foundations of the Central Dogma: the RNA of tobacco mosaic virus (TMV) controls (at least in part) the amino acid sequences of viral proteins; DNA determines genetic effects; and (based on Vernon Ingram's work) DNA controls (at least in part) the amino acid sequence of hemoglobin.[130] Hopes were now pinned on TMV as the Rosetta stone of the genetic code and on the Virus Laboratory at Berkeley, where biochemist Heinz Fraenkel-Conrat had been excavating clues at a staggering rate.

Rosetta Stone: TMV as Information Code

Ever since its crystallization by Wendell Meredith Stanley (1904–71) at the Rockefeller Institute in 1935, TMV had been perceived as a key to the age-old riddle: "What Is Life?" By successfully isolating TMV, and working within the protein paradigm of life science, Stanley seemed to have demonstrated that the mysterious submicroscopic organisms were merely protein molecules, a special class of self-reproducing, or autocatalytic, enzymes. These crystalline molecules, which in their inert state could be stored indefinitely on the laboratory shelf but in the tobacco plant sprang to life, reproduced, and mutated, became a symbolic victory for the champions of the chemical view of life. With Stanley's masterful command of the news media, the lessons of his TMV work reached a remarkably wide audience.[131]

Stanley's discovery, for which he won the 1946 Nobel Prize in chemistry, had come to be viewed as the symbolic beginning of molecular biology, having inspired several scientists, especially Max Delbrück, to treat a virus as a gene analogue and approach it on a molecular level. Like most biochemists, and reflecting the general outlook at the Rockefeller Institute, Stanley had little appreciation for the logic and methods of genetics. Nevertheless, throughout the 1940s, he engaged in building cognitive and disciplinary bridges between viruses and genes. He participated in genetics meetings and symposia at Cold Spring Harbor and Woods Hole and directed several of his staff, notably C. Arthur Knight, toward problems of protein changes and mutations in relation to their biological differences.[132]

By 1947 Knight had amassed impressive results on the chemical differences in proteins of various virus strains. Reasoning within the protein paradigm (where protein self-replicated), he hoped to elucidate not only their mode of replication but also "the relationship between chemical structure and biological specificity." Relying on chemical and microbiological assays for nineteen amino acids, he examined highly purified preparations of eight strains of TMV, looking for changes in their amino acid content. The results were dramatic. In general, those strains that, from a biological (immunological) standpoint, diverged the most from ordinary type TMV displayed marked differences in protein composition. Knight added parenthetically that there was "little direct chemical evidence for the existence of differences among the nucleic acid components of strains of tobacco mosaic virus," thus buttressing Stanley's staunch opposition to nucleic acids as carriers of genetic specificity. Knight's work strongly suggested "mutation among tobacco mosaic virus involves stepwise changes in amino acid content." [133] In the ensuing decade, Knight's analyses of the proteins of TMV and other viruses, combined with Chargaff's studies of RNA, supplied much of the data for Gamow and Yčas's statistical correlations of RNA and protein composition.

At the end of 1947, when the thriving Princeton branch of the Rockefeller Institute had recently been dissolved, Stanley's group and Knight moved to the University of California at Berkeley. Stanley was by then an international scientific celebrity and a powerful political figure in American science. Recipient of honors and prizes, well connected to editorial boards and the Nobel Committee, and deeply involved in the administration of postwar life science, Stanley was in a strong position to shape the field. Capitalizing on California's postwar boom and backed by abundant resources—state appropriations and Rockefeller Foundation funds—Stanley embarked on building a world-class research center: an innovative department of biochemistry, qua molecular biology, which was inaugurated in 1952.[134]

As Angela Creager has pointed out, Stanley's scheme was paradoxical. Caught in the persistent dilemma between the autonomy of basic research versus its economical and political justifications, Stanley advocated both simultaneously. He envisioned a free-standing department, loosed from the traditional service roles, which, at the same time, would integrate local medical and agricultural research and guard California's public health. Virus research framed this California dream. Stanley posited virus reproduction as the central problem for his Virus Laboratory, the field of biochemistry as a whole, and beyond it, for medicine and public health. From the mid-1950s Stanley became one of the early champions of virus research as part of the

FIGURE 19. *Left to right*, Heinz Fraenkel-Conrat and Wendell M. Stanley, ca. 1957. Courtesy of the Bancroft Library, University of California at Berkeley.

campaign against cancer, ultimately helping to channel massive resources into molecular biology.[135]

In keeping with his long-standing bias for a purely chemical approach to biology (his recognition of the importance of phage research notwithstanding), Stanley hired few researchers concerned with virus genetics. Dean Frazer and Gunther Stent, both devotees of Delbrück, were in fact the only geneticists in the building. And, as Creager observed, it did not help the cause of phage genetics at Berkeley that Stent tended to view his laboratory as an outpost of Caltech. Consequently, Creager argued, most of the virus researchers Stanley assembled adopted a biochemical, rather than genetic, viewpoint on problems of heredity and reproduction.[136] Heinz Fraenkel-Conrat (1910–99) epitomized the biochemical approach of the Virus Lab.

Fraenkel-Conrat's scientific career spanned a wide cognitive and geographical range. He had earned his medical degree in 1933 in his native town of Breslau, Germany, and, three years later, his doctorate in biochemistry in Edinburgh, Scotland. Fraenkel-Conrat was partly Jewish, and with his career prospects in Germany foreclosed, he moved to America in 1936. He worked for a year with the noted protein chemist Max Bergmann at the Rockefeller Institute, spent another year at the Instituto Butantan, São Paulo, Brazil, and the following years (1938–42) as a research associate at the Institute of Experimental Biology at the University of California at Berkeley. Finally he settled in Albany, California (1942–50), as an associate chemist at the Western Regional Research Laboratory, USDA. His research interests have spanned a wide range of physiological proteins—enzymes, toxins, hormones, and viruses.[137]

Fraenkel-Conrat had not been satisfied with his job at the Western Regional Research Laboratory. He had been interested in the chemistry of TMV since the mid-1940s, but, as he stressed to Stanley, such fundamental work was not encouraged at the lab. He wrote to Stanley:

> The restrictions and limitations imposed upon us by the rigid system of burocracy [*sic*] make me wish ever so often that I might be associated with a free research institution, similar to the Rockefeller, or with a University. If you should know of any openings of that nature, and paying not too much less than what I am getting now ($5000) [then a standard academic salary], would you be good enough to think of me?[138]

Understandably, Fraenkel-Conrat was elated to learn from Stanley about "the bright new future of biochemistry at the West Coast," news which held out hope for an academic career.

> You will understand that these developments must affect the planning or wishful thinking of someone as deeply interested in academic research in biochemistry as I am, particularly in view of your great interest in a field in which I happen to have specialized in recent years. Since an appointment as assistant professor in your department or institute would represent the fulfillment of all my hopes for the future.[139]

After a year in Europe as a Rockefeller fellow, where he studied protein sequencing in the laboratories of Frederick Sanger, R. R. Porter, and K. V. Linderstrom-Lang, Fraenkel-Conrat, now in his forties, joined Stanley's new Virus Lab in 1952. But the position in Stanley's lab only partly fulfilled Fraenkel-Conrat's hopes: he was hired as a research biochemist (staff, rather than faculty) with a promise of a professorship. For the next six years, while producing some of his most significant work—namely, the classic recon-

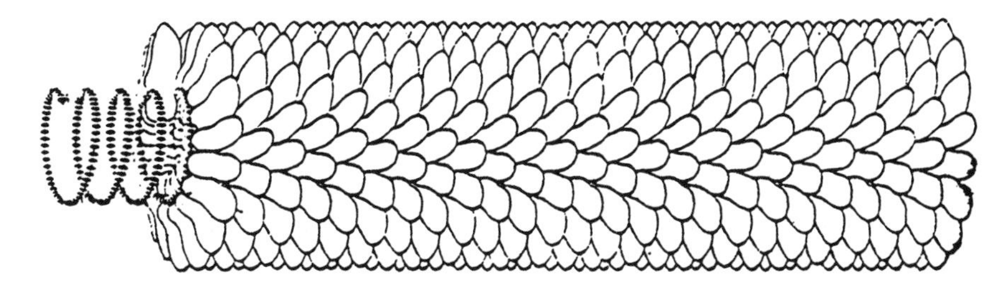

FIGURE 20. Heinz Fraenkel-Conrat, "The Genetic Code of a Virus," *Scientific American* 211 (1964): 47. Reprinted by permission.

stitution studies of TMV and the preliminary experiments toward the sequencing of TMV protein—he was funded by grants awarded to Stanley from the National Foundation, Rockefeller Foundation, and the United States Public Health Service.[140]

In 1954, when Fraenkel-Conrat embarked on his reconstitution experiments, the rod-shaped TMV was known to be comprised of two components: a tightly coiled inner core of RNA and an outer protein coat comprised of 2,200 repeating subunits (see Fig. 20). The architecture of these proteins had been established in the early 1950s by Knight and his collaborators at the Virus Lab.[141] While the role of DNA as the bearer of genetic specificity was by then accepted and its replication demonstrated in DNA viruses (e.g., bacteriophage), the function of RNA was uncertain. Evidence was mounting that RNA possessed genetic specificity and was an essential intermediary in protein synthesis, as Chargaff wrote to Gamow in 1954, when Gamow was a visiting professor at Berkeley. But between Stanley's general resistance to the genetic primacy of nucleic acids and Knight's specific demonstration that most TMV mutants differed in their amino acid composition but hardly in their nucleotide composition, TMV remained immune to the changing paradigm. Fraenkel-Conrat's experiments were designed to settle precisely that question: Was the infectivity of TMV a genetic property of its RNA or its protein?

Isolating native protein and RNA from TMV, Fraenkel-Conrat determined that, while each component separately was noninfective, the reconstituted virus was fully infective, producing typical TMV lesions in the tobacco plant. He extended his experiments to a hybrid TMV in which the protein coat came from TMV while the RNA core came from another viral strain (HR). From an immunological standpoint—due to the action of its protein coat—the hybrid behaved just like TMV, but the symptoms in the tobacco plant were only those of the HR strain, which supplied the RNA. And the progeny of the hybrid were nearly identical to HR the strain. "Thus, the ribonucleic acid seems to represent the main genetic determinant even for the progeny protein in the TMV strain," he concluded, just a couple of weeks ahead of their toughest competitors at the Max-Planck Institute in Tübingen.[142]

This was not exactly startling news in 1955 but it was decisive: a confirmation that RNA specified (at least in part) the viral proteins. The completeness and elegance of these experiments rendered them intellectually and aesthetically so satisfying that they were ushered into the hall of fame of classic experiments. Gamow exclaimed, "Alex [Rich] writes that Frenkel Conrad took RNA molecules from TMV, and a short protein. By plymerising he grew protein molecules in cylindrical form with RNA as the axis. These preparations acted properly on tobacco plant. Thus one can now synthesize live *virus* (i.e. life)!!!" Meanwhile, the *New York Times* announced, "Reconstruction of Virus in Laboratory Reopens the Question: What Is Life?"[143] As Stanley reflected, probably the most significant aspect of the reconstitution studies was, once again, its symbolic value: a reaffirmation of the chemical view of life. He wrote to the Nobel Committee, "It illustrated that there is nothing sacred about the integrity of the natural virus particle. Just as I had shown twenty-five years earlier that the active agent could be isolated and crystallized by chemical methods, so Fraenkel-Conrat now showed that it could be degraded into its components and reconstituted from them without loss of its biological function."[144] A cut-and-paste history, to be sure: when in 1935 Stanley had isolated the so-called active agent protein (purifying away the RNA), he erroneously validated the genetic role of viral proteins. In fact, from a technical standpoint Fraenkel-Conrat's experiments could have already been performed in the 1940s. The fact that they were not is a prime example of how paradigmatic commitments can foreclose certain lines of scientific inquiry.

With RNA's genetic role confirmed, Fraenkel-Conrat and his collaborator, C.-I. Niu, initiated in 1955 the studies that would lead by 1960 to the complete sequencing of TMV protein. These studies could potentially supply the most substantial clue for deciphering the genetic code and were being closely watched by Gamow and Yčas. It was known since 1952 that TMV's protein subunits consisted of about 2,200 chains, each with the amino acid threonine at its C-terminus (a polypeptide is flanked by C [carboxyl] terminus on one end and N [amino] terminus on the other). They elaborated on the nature of that C-terminus and isolated a tiny end fragment of the viral protein (hexapeptide thr-ser-gly-pro-ala-thr). "Where there is a C-terminus there must be an N-terminus," Fraenkel-Conrat playfully reasoned. Having spent a year (1951) with the sequencing experts Sanger, Porter, and K. Linderstrom-Lang, he had acquired "N-terminal techniques." By the end of 1957, while working with K. Narita, Fraenkel-Conrat isolated the

N-terminus fragment (acetyl-ser-tyr) for the first time. With both ends of the protein chain defined, the full sequencing of TMV protein was now in sight.[145]

Having amply distinguished himself, Fraenkel-Conrat now demanded his long-overdue professorship. When Stanley continued to renege on his promise, Fraenkel-Conrat decided to leave, come what may. "Heinz Fraenkel-Conrat visited NIH recently. He is being interviewed for the position which I am leaving. Apparently he has become restive with events at the virus lab and especially with Stanley, so that he states he is looking around for a position elsewhere," Rich reported to Crick.[146] Joining the chorus of disenchantment with Stanley's modus operandi at the Virus Lab, Stent appealed to Delbrück for help in the matter concerning Heinz Fraenkel-Conrat.

> The principal sufferer at the moment is F.-C., who has really been getting a raw deal for the past six years. Stanley has promised him a professorship year after year, without doing anything much about it . . . he is still only paid by one of the Virus Lab grants. Stanley's attitude seems to be that there is no hurry; after all, Heinz has waited already for so many years, why can't he wait a little longer? . . . So in trying to look out for our pal, we thought maybe it would be worth asking Beadle whether something couldn't be fixed up for Heinz at CalTech.[147]

Delbrück's discussions with Beadle and Sinsheimer on the "F. C. matter" came to naught, because F. C. did not possess much of a feeling for biology (e.g., taking a virus apart and putting it back together again), as Stent concurred, and F. C.'s insulting behavior toward his German competitor Gerhard Schramm did not reflect well on him. (He was not even aware of Schramm's Nazi past.) Furthermore, as Stent admitted, "F. C. is strictly an Einzelgänger [loner], rather difficult to get along with as a collaborator," a serious drawback, given Caltech's valorization of cooperative research. But "F. C. was pleasant to have around if you don't happen to work on exactly the same thing," Stent pointed out, and in any case he deserved a professorship. Now that the Caltech idea had fallen through, Stent and his colleagues would "try to go over Stanley's head directly to the Chancellor and see whether there isn't something that can be done for F. C. in Berkeley after all."[148] Finally, as a result of the chancellor's intercession, in 1958 Fraenkel-Conrat was made professor of virology.

As Fraenkel-Conrat was determining the genetic specificity of viral RNA and initiating the sequencing of TMV protein, the RNA Tie Club (and collaborators) had been recasting heredity as a communication system. Viruses and genes were becoming agents of information storage and transfer as well as texts governed by codes—combination codes, commaless codes, transposable commaless codes, codes comprised of two, three, four, six (or more) letters. Although studies of TMV (qua RNA) had direct relevance for the club's analyses and deciphering efforts—as its name signified—in 1958 neither Fraenkel-Conrat nor his colleagues at the Virus Lab thought in such terms. They did not represent their findings in terms of coding or information. The bearer of genetic specificity, their viral RNA, was mute and immobile; it did not yet transport any messages. Neither did it embody the code of life, as it did for the RNA Tie Club.

Gamow had been in contact with Stanley since 1946, when Gamow organized the 1946 conference "Physics of Living Matter" in Washington, D.C. "I have learned so much on our recent conference, that I have decided to include a chapter on genes, viruses, etc. in my lyman's book," he wrote to Stanley soon after, requesting photographs of TMV.[149] "There seems to be an epidemic among the physicists, 'maladia biologica' you may call it," Gamow informed Stanley a year later, explaining the economic significance of that condition. The solid-state subdivision of the physics section of the Office of Naval Research became interested in "aperiodic crystals" (in the sense of Schrödinger's book, *What Is Life?*). On the eve of Stanley's move to California, Gamow proposed, "The moral: may I and Mr. Mackenzi (in charge of ONR's solid state) come to Princeton sometime the end of March to talk to you about the way to spend few hundred thousand dollars? It isn't joke, it is serious!"[150]

Gamow spent considerable time at the new Virus Lab in 1954, devoting one day a week to biology, as Stent informed Delbrück. Gamow often consulted with Stanley while working with Yčas on the thorny problem of statistical correlation of RNA and protein in viruses in 1955. And in the fall of 1956, while a visiting professor at Berkeley's physics department, he taught a course, "Biology on a Molecular Level." What was the impact of Gamow's agenda on the biochemists of the Virus Lab? How did his representations of codes with their new discursive software affect the biochemists' epistemic commitments or technical practices? Hardly at all, according to Fraenkel-Conrat. He recalled that they did not follow Gamow's theoretical arguments, which, in any case, they did not take seriously; after all, Gamow did not even know the names of amino acids. They were vaguely aware of Crick's coding ventures, but these too appeared ethereal. The concerns of the RNA Tie Club seemed of little practical relevance to the biochemists' goal: to determine in detail the entire sequence of amino acids of TMV—not as a carrier of genetic information or some abstract code—but as a material (i.e., chemical) substance with genetic specificity.[151]

Their attitude began to change at the end of 1958, after the Tübingen group reported the sensational news of their production of TMV mutants by the chemical alteration of their RNA. The chemical modification of viral proteins had long been an active field (notably by Stanley's group), and many agents were known to act as mutagens on the living cell. But within the protein paradigm of heredity, there had been few incentives to study mutagenic action in nucleic acids. In the summer of 1958 Heinz Schüster and Gerhard Schramm used nitrous acid to modify specific bases of RNA (while the viral RNA strand remained intact), demonstrating that an alteration of one out of three thousand nucleotides (TMV has about six thousand) was lethal. Within a month, their colleagues Alfred Gierer and Karl-Wolfgang Mundry followed up by studying the effects of the mutagen (necrotic lesions in the plant), pointing out that a variety of different mutations could be produced that way.[152]

Controlled mutations are one of the most powerful molecular probes. If mutants could be artificially produced by chemical modification of the viral RNA core, then, assuming colinearity, one could begin to map out the corresponding changes in the amino acid composition of their protein coats. And with the most up-to-date methods of amino acid analysis, the detection of changes as minute as a replacement of a single amino acid now became possible. "Mutation Agent Held Clue to Life," announced the *New York Times*. The article reported:

> A method that may help scientists break the most important "code" in the world was described here at a symposium on virus research. The method is a chemical way of altering the hereditary material in viruses . . . seen as a promising tool for the giant task of unraveling the "code" by which chemicals called nucleic acids dictate the conditions of life, form and function to all living things.[153]

Fraenkel-Conrat and A. Tsugita were already racing. Preliminary survey of several TMV mutants produced by nitrous acid treatment showed that three amino acids (proline, aspartic acid, and threonine) were replaced by three others (leucine, alanine, and serine).[154]

The implications of their survey were profound even though a recent study, showing that not all changes in viral RNA altered its amino acid sequence, complicated the picture somewhat. For the first time (early 1960), Fraenkel-Conrat began to reconfigure his research within the framework of the coding problem. He concluded, "The implication of the fact that some chemically produced mutants differ in their protein composition, while others may not do so, as reported by Wittmann recently, appears of considerable interest in connection with the mechanism of coding of genetic properties by the RNA." Soon he would also begin to resignify "genetic properties" as "genetic information." It now became clear that knowledge of the entire TMV amino acid sequence, coupled with a chemical technology for amino acid replacement, could serve as the most powerful tool for deciphering the code.[155]

1 5 NH_2 10 15
Acetyl N-Ser→Tyr→Ser→Ileu→Thr→Thr→Pro→Ser→Glu→Phe→Val→Phe→Leu→Ser→Ser

30 NH_2 25 NH_2 20
Ala←Asp←Thr←CySH←Asp←Leu←Ileu←Leu←Glu←*Ileu*←*Pro*←*Asp*←*Ala*←Try←Ala

NH_2 NH_2 35 NH_2 NH_2 NH_2 40 NH_2 45
Leu→Gly→Asp→Glu→Phe→Glu→Thr→Glu→Glu→Ala→(Arg)→Thr→Val→Glu→Val

60 NH_2 55 50 NH_2 NH_2
Val←Thr←Val←Glu←Pro←Ser←Pro←Lys←Try←Val←Glu←Ser←Phe←Glu←(Arg)

65 70 NH_2 75
(Arg)→Phe→Pro→Asp→Ser→Asp→Phe→(Lys)→Val→Tyr→(Arg)→Tyr→Asp→Ala→Val

90 85 80
(Arg)←Thr←Asp←Phe←Ala←Gly←Leu←Leu←Ala←Thr←Val←Leu←Pro←Asp←Leu

NH_2 95 NH_2 NH_2 NH_2 100 NH_2 105
Asp→(Arg)→Ileu→Ileu→Glu→Val→Glu→Asp→Glu→Ala→Asp→Pro→Thr→Thr→Ala

120 115 110
Ala←Val←Thr←Ala←Asp←Asp←Val←Arg←(Arg)←Thr←Ala←Asp←Leu←Thr←Glu

125 NH_2 130 135
Ileu→(Arg)→Ser→Ala→Asp→Ileu→*Asp*→*Ileu*→*Ileu*→*Val*→*Glu*→Leu→Ileu→(Arg)→Gly

150 145 140 NH_2
Leu←Gly←Ser←Ser←Ser←Glu←Phe←Ser←Ser←(Arg)←Asp←Tyr←Ser←Gly←Thr

155 158
Val→Try→Thr→Ser→Gly→Pro→Ala→Thr

Sequence of the 158 amino acid residues in the protein subunit of tobacco mosaic virus. The encircled residues indicate the points of splitting by trypsin.

FIGURE 21. A. Tsugita, D. T. Gish, J. Young, H. Fraenkel-Conrat, C. A. Knight, and W. M. Stanley, "The Complete Amino Acid Sequence of the Protein of Tobacco Mosaic Virus," *PNAS* 46 (1960): 1463–69. Reprinted by permission.

"Scientists Find Clue to Heredity's Code," reported the *New York Times* on Fraenkel-Conrat and Tsugita's accomplishment.[156] The reporter recounted, "The discovery [is] the first definite link between mutation, or change in the inheritance code, and a specific alteration in the composition of a molecule manufactured according to that code. . . . The new finding was described as "one of those rare breakthroughs" by Dr. Wendell M. Stanley, Nobel laureate and director of the University's Virus Laboratory in Berkeley where the work was done."

Within a few months, well before Schramm's group, Stanley's laboratory announced the complete amino acid sequence of the protein of TMV (see Fig. 21). TMV protein became the first viral protein with a completely determined amino acid sequence. It was then also the largest protein elucidated; 158 amino acids in contrast to the 124 in insulin and 51 in ribonuclease. Its immense significance for the coding problem was now articulated within the discourse of information: the proteins of simple viruses were "produced as a result of the information carried by the nucleic acid." The epistemic and technical foundations were now in place for a variety of studies, the authors predicted.

> The next step will be to relate the structure of viral nucleic acid to its specific protein in a point-to-point manner. Although progress is also being made in this direction, it will probably be some time before the methodology of nucleic acid chemistry will catch up with that of protein chemistry and the code relating one structure to the other will have become unraveled. This is, of course, the problem of greatest importance in biology and medicine, and it appears that the viruses will provide a unique experimental approach.[157]

Stanley's laboratory was at the threshold of a breakthrough. "Please accept my congratulations for the beautiful 158 jewel necklace. Some six or seven years ago, I would spend some sleepless nights attempting to decode it using an overlapping code," Gamow reminisced. "Hope that somebody will get the RNA sequence in TMV soon." Stanley responded, "You have certainly placed an onerous obligation by expressing the hope that the RNA sequence in TMV will be obtained soon." But he thought it would be a long, long time, unless something unexpected happened.[158] Just to be on the safe side, Stanley quickly nominated Fraenkel-Conrat for a Nobel Prize (conceding that "there would be no miscarriage of justice if the Prize should be divided" [between Fraenkel-Conrat and Schramm]). The nomination was not only for the reconstitution experiments that demonstrated that RNA carried "all the genetic information" but also for the complete sequencing of TMV protein. Stanley predicted ceremoniously, "This type of work undoubtedly will play a crucial role in the deciphering of the code which relates nucleotide sequence to amino acid sequence, one of the key problems of molecular biology. This is, in fact, the basic problem of life itself," thereby elevating TMV to the code of life.[159]

And like a quarter of a century earlier with the crystallization of TMV, Stanley's mastery of the news media served to dramatize the findings at the Virus Lab. *Time* magazine reported:

> Dr. Stanley thinks that the techniques used by Tsugita and Fraenkel-Conrat may be developed to the point of proving "a *Rosetta Stone* for the language of life." If applied to many mutant viruses, they may break entire genetic codes, telling which groups of bases are responsible for what characteristics. The next step, perhaps years away, will be to do the same with the more complicated molecules of DNA that govern the heredity of higher animals.[160]

From being a symbol for the chemical view of life, TMV now came to signify the textualization of life. Postwar science confronted the world's most ancient "text" with the compelling, if far-flung, imagery of a genetic Rosetta stone.

For centuries Egyptian hieroglyphics had resisted decipherment. But the interest did not diminish with the passage of time. In 1799 a basalt slab covered with three bands of writings—hieroglyphs, something thought to be Syriac, and Greek—was discovered near the town of Rashid in the Nile Delta, a town known to Europeans as Rosetta. While the stone's potential was instantly recognized it took more than twenty years to decipher its hieroglyphics. As with the genetic code in the mid-1950s, there were only fragments to work with. The Greek consisted of fifty-four lines; of the hieroglyphic there remain only fourteen lines, corresponding to the last twenty-eight of the Greek, and the central band proved to be a simplified colloquial form of hieroglyphics (written in demotic, a language related to Coptic).[161]

The most eminent Orientalists and ancient-language experts who confronted the Rosetta stone worked mainly with the Greek and demotic texts. But the break came with the unconventional approach of a young French prodigy, Jean-François Champollion, whose lifelong passion had been to solve the hieroglyphs' mystery. His preliminary analysis proved a key point: hieroglyphs were phonetics, thus dispelling the age-old theory that they were idiographic. In 1822 he produced an near-complete translation of the code's relatively accessible part—the hieroglyphic names of Egypt's rulers from the Greco-Roman period. Rather than matching the plaintext (Greek) to the codetext (demotic and hieroglyphic) symbol by symbol, Champollion derived the sound-values of the other phonetic hieroglyphs by the cryptanalytic method of substituting known values, guessing the names, and testing the presumed values elsewhere. This kind of method might be envisioned with amino acid replacement analysis in TMV.

Upon Champollion's acquisition that year of some Egyptian inscriptions antedating the Greco-Roman era, the spell was finally broken with the word "Ramses." With his new understanding of the writing system he could now penetrate the language, while his knowledge of Coptic enabled him to approximate Egyptian. Shuttling back and forth, refining and correcting the language by the script and the script by the language, three years later he could accurately translate other early Egyptian inscriptions.[162] The Rosetta stone had finally yielded the secrets of ancient texts. There were some vague similarities to TMV, to be sure (ignoring the role of the third demotic text): the RNA "hieroglyphs" of a sequence could be matched with the known "Greek" protein text; and amino acid replacement analysis could be fancied as cryptanalytic substitutions. But the incongruities were significant. As before, these problematic analogies were already predicated on a metaphor, the metaphoricity of nucleic acids as linguistic entities. The Rosetta stone added another potent semiotic prop—images of ancient secret texts—to the representation of the genetic code as a scriptural technology.

From 1960 to 1961 TMV served as the main decoding apparatus; a long-awaited resource, especially for Yčas. He had been steadfastly pursuing the rationale laid out in his 1955 studies with Gamow: statistical correlations between nucleotides and amino acids based on comparisons of their relative amounts in different organismic sources. In a desperate effort to find a "Rosetta stone" he had even traveled all the way to Central Africa (1957) to hunt giant silk-producing caterpillars, a native delicacy. Silk proteins consist of only two amino acids (glycine and alanine). That simplicity, he hoped, would reflect in the DNA of the caterpillars' silk-producing glands, offering direct clues to the code. No difference in DNA composition was detected.[163]

With the TMV protein fully sequenced and with the growing inventory of nitrous acid mutations, amino acid replacement data now served as a prime cryptanalytic tool and Yčas expanded considerable efforts exploiting it. Assuming a nonoverlapping triplet code, if a mutation responsible for the replacement of amino acid A by B had been induced by a single mutagenic base transition, then the two nucleotide triplets representing A and B had to share two nucleotides. For example, hypothetically, if a base transition from GAC to GAU specified amino acid leucine instead of proline, then leucine and proline would share two bases G, A. And if an amino acid C had replaced A in another mutant protein induced by a single base mutation, then the triplet representing B and C would have only one nucleotide in common. Thus by surveying the many known step-wise replacements in mutant viral proteins and by assessing the nucleotide interchanges that might have produced the RNA mutation, one could construct a network of amino acid base triplet correlations and eventually—and very laboriously—break the entire genetic code.[164]

Yčas's industrious amino acid replacement analyses prompted others to follow his path, among them Richard V. Eck at NIH and the biophysicist Carl R. Woese at General Electric Research Laboratory (who was, after 1964, professor of physics and microbiology at the University of Illinois). Woese, who in 1967 published a major textbook on the genetic code, representing it as a communication system, actually devised a remarkably successful triplet code. It correctly predicted the nucleotide composition of six RNA viruses from their amino acid composition; an impressively large number of his amino acid–triplet assignments turned out to agree with those established biochemically a couple of years later.[165]

Thus by 1961 TMV came to embody the "code of life" that governed the communication system of heredity, with Stanley's laboratory at the vanguard. While in 1957 there was no trace of these scriptural icons in virus research (and hardly any in biochemistry) the situation had changed radically in a matter of three years. Even staunch materialists like Stanley and Fraenkel-Conrat reconfigured their object of study within the information discourse and re-represented them as a scriptural technology. Stanley was now giving lectures on "The Regulation and Transfer of Biological Information by Viruses"; Fraenkel-Conrat's discursive shift was equally striking. Reviewing the state of virus research for the monograph *Viruses and the Nature of Life* (1961), he wrote, "Now we come to the big question: how can the nucleic acid, a long thin, threadlike molecule, possess such unique and specific biological activity? How can it carry and communicate all the information necessary to make new viruses of one and only one specific type?"[166]

Explaining the nature of the nucleotide building blocks of viruses, he now pointed out, "You can see very well that this is a code," though just a couple of years earlier he did not see it that way. His reconstitution experiments had interpreted that same material in terms of genetic specificity, devoid of linguistic attributes, where now, he wrote:

> The sequence AGUACUCAGUCGUCGCAGUCUCAAGU in our model is like a sentence or a paragraph in the microscopic language of nucleic acids. And a sequence of 6,500 letters obviously can carry a good bit of information. The fact that there are only four different symbols to work with—a four-letter alphabet—does not worry us much. The Morse code, after all, is composed only of three symbols, a dot, a dash, and a gap, and with this international three-letter alphabet you can write everything that has ever been written or will ever be written.[167]

The next important question was: How was this information translated? Fraenkel-Conrat inquired, "Our problem is: how do we translate a language of four symbols into a language of 16 or 20 symbols? How can the four-letter code of nucleic acid be converted into a 20-letter protein code?" (He was clearly oblivious to the problematic intertextuality leading to a code of codes.) This was not a rhetorical question, but an urgent task. Recounting the recent decoding work based on amino acids replacement in mutant TMV he observed, "The chemical production of mutant plant virus, however, is only the very beginning of a lot of work that is before us. We are taking the first small steps. We are beginning to unravel the code that relates protein structure to the structure of nucleic acid. Gradually, in the course of many years, the way in which genetic information is transferred will become clear."[168] By the time the book reached the shelf, "the unexpected something" did happen. The genetic code was broken in the summer of 1961 by two young obscure biochemists at NIH, Marshall Nirenberg and his postdoctoral fellow Heinrich Matthaei, who broached the problem from a radically different vantage point. Inspired, in part, by the work of the Pasteur group in the late 1950s on genetic regulation and by their notions of information transfer from DNA to proteins via messenger RNA, the biochemical studies at NIH would completely reorient the approach to the coding problem in the 1960s.

CHAPTER FIVE

The Pasteur Connection: *Cybernétique Enzymatique*, *Gène Informateur*, and Messenger RNA

At the 1961 Cold Spring Harbor symposium, "Cellular Regulatory Mechanisms," Sydney Brenner offered an up-to-date overview of protein synthesis and the genetic code. As he put it, the dominant picture in the coalescing field of molecular biology was that DNA carried information in the form of a code (specific sequences of nucleotides determined the amino acid sequences of proteins). Exactly how such a specification was accomplished formed the greatest challenge of the field; it was a self-defining mandate. The solution, or rather "cracking the genetic code," was then the main ambition of many researchers, with several groups in leading laboratories in the United States and Europe racing one another to decode "the secret of life."

Broadly viewed, there existed two kinds of attacks on the coding problem, as Brenner explained. One approach entailed ignoring the internal biochemical machinery to view protein synthesis as a black box, with DNA information going in at one end and the polypeptide chain coming out the other. Deducing the code from the cryptic messages of amino acid sequences of known proteins (efforts exemplified by the RNA Tie Club) represented one such line of inquiry. Another "black box" decoding strategy (exemplified by the work on the tobacco mosaic virus, TMV, of Gerhard Schramm's group in Tübingen and Heinz Fraenkel-Conrat at Berkeley) consisted of analysis of alterations in amino acid compositions of viral proteins, caused through specific mutagenesis along the viral RNA strand. A related strategy, informed by genetic sensibility and techniques, consisted of deploying either naturally occurring viral and bacterial mutants, or ones produced by specific mutagenesis, and mapping the genetic crosses. (Representative of this line of inquiry are the E. coli studies of Charles Yanofsky's group at Stanford; Crick and Brenner's analyses of patterns of phage mating at Cambridge; and François Jacob's studies of genetic transfer between two different genera of bacteria at the Pasteur Institute.) Assuming that a gene and its protein product were colinear (a feature demonstrated simultaneously by Brenner and Yanofsky later in 1965), researchers could correlate the mapped genetic structure with the amino acid changes produced by mutations, hoping to deduce the code from exhaustive sets of such changes and exchanges. These black box approaches generally exemplified the studies of molecular biologists and theorists from the physical sciences lured by the coding problem.

The other, much more difficult approach entailed probing the actual biochemical machinery of protein synthesis and enzyme action by tracking its pathways and flagging down its components. Spanning a wide research spectrum, this approach relied far more heavily on traditional biochemistry, which until then was generally distant from genetics and the formalisms of decoding. Examples of this type of work are the investigations of polynucleotide synthesis by Arthur Kornberg at Washington University and by Severo Ochoa at New York University; studies of protein synthesis and RNA's role by Paul Zamecnik at the Huntington Memorial Hospital and by Mahlon Hoagland at Harvard; and Sol Spiegelman's work at the University of Illinois on nucleic acids and enzyme-forming systems. Genetic tools and coding theories did, however, inform several biochemical studies of protein synthesis: most importantly, the RNA studies of James Watson and colleagues at Harvard; the work on the operon—the coordinated genetic regulation of protein synthesis—by Jacques Monod (1910–76), François Jacob (b. 1927), and their collaborators at the Pasteur Institute; and its extension to the function of messenger RNA.[1] When Brenner summarized the state of the field, only weeks before Marshall Nirenberg and Heinrich Matthaei announced the breaking of the code, the coding problem was explicit in the first approach (the code as key) and implicit in the second (protein synthesis as key).

Whether researchers' efforts focused directly on the genetic code as the key to the problem of protein synthesis, or conversely, primarily on solving the problem of protein synthesis and enzymatic regulation and thus eventually "cracking the code of life," their paths converged by the end of the 1950s. They came to form a community of overlapping goals in "trading zones" of different though complementary experimental techniques and theoretical models, as they increasingly shared material, discursive, and even social practices. Indeed, as Jean-Paul Gaudillière and Richard Burian have each shown, the dynamics of disciplinary formation, such as molecular biology, must be understood in terms of the interactions between laboratory groups and local scientific cultures, where questions are refracted and transformed by exchanges of materials, methods, institutional niches, and modes of representation and communication.[2] Representations of the objects and mechanisms of molecular genetics were now increasingly constituted within the information discourse, as configured since the early 1950s through tropes and images from cybernetics, communications, and information theory. In fact, the information metaphor became a linguistic currency that increasingly linked biochemistry with molecular biology.

The Pasteur group formed a prominent site within that representational space. Occupying a privileged position within the international research community of protein synthesis and regulation, it played a major role in shaping the disciplinary identity and institutional form of molecular biology.[3] Monod's laboratory of cellular biochemistry focused on enzymatic synthesis and regulation in bacteria (it also hosted many European and American researchers, some who formed close collaborations, romances, marriages, as well as lasting amities and enmities). In André Lwoff's Department of Bacterial Physiology, Jacob investigated the genetic mechanism of phage lysogeny and bacterial sexuality. Together the two laboratories supplied some of the key pieces to the puzzle of protein synthesis and its genetic control and to the structure of genetic code (Lwoff, Monod, and Jacob shared the 1965 Nobel Prize). They also contributed to the linguistic software of molecular biology, constructing their accounts of enzyme induction around cybernetic models and of phage replication through metaphors of information flow; their concept of messenger RNA coalesced around the scriptural tropes of communication: "text," "coded message," its mobile "transcript," and the "translation system."

This chapter examines the transition to the information discourse at the Pasteur Institute. First, I briefly follow the paradigm shift from "enzymatic adaptation" to "enzymatic induction" (i.e., from Lamarckian and teleological to Darwinian and chance-based explanations of enzyme synthesis in bacteria) as a discursive turn designed to exorcise final causes from molecular biology. I also highlight the concomitant institutional changes. Next, I follow the merging of genetic and biochemical explanations and practices in the famous PaJaMa experiments (named after the three researchers, Arthur Pardee, Jacob, and Monod). These studies demonstrated the coordinated genetic control in protein synthesis, operating within a unified regulatory system (goal-directed, negative feedback). Swayed by the steady stream of findings on RNA's role in protein synthesis, especially from the Belgian group at Rouge-Clôitre, the Pasteur team was led to the idea of a messenger RNA. Finally, I follow the conceptual refinements of that general notion and the experimental identification of messenger RNA, which was a turning point in the history of the genetic code.

In following these local and international developments I aim to convey an appreciation for the important role of the new biosemiotics. Just as it did for the RNA Tie Club and in the late phase of the Berkeley work on the tobacco mosaic virus, the information discourse also played a pivotal and highly specific role in the Pasteur studies by providing the interpretive framework for their experimental results.[4] Indeed, it was adopted only in the summer of 1958, shortly after the completion of the PaJaMa experiment(s). Only then, when the different experimental strands where being linked, did Jacob and Monod began to use the communication tropes of information theory (information flow, information transfer, transcription, translation, messages, punctuations, and text) and deploy physical models derived from cybernetics and electronics (regulatory circuits, yes-no switches, magnetic tape, memory, transmitters, receivers, negative feedback systems, guidance systems, automated factories, and computer programs) in order to make sense of and give meaning to their experimental findings.

While between 1953 and 1958 Monod's mission had been to expunge all vestiges of teleological accounts from molecular biology by 1958, based on the interpretations of the PaJaMa experiments he substantially revised his position. The cybernetic model of goal-directed, negative feedback systems provided for Monod a legitimation of teleology sanctioned through the new biological concept of *teleonomy*, based on the goal-directedness of computer programs. (Teleology, which accounted for the open-ended adaptation of organisms to their environment, was replaced by teleonomy, the notion of a finite storage of genetic information; adaptation then became merely the activation of preexisting information.) The trope of "information transfer" supplied a key notion for delineating the difference between structural genes and regulatory genes in enzyme-synthesizing systems. The operon model, formulated in 1960 to consolidate these wide-ranging findings and interpretations, was conceptualized in terms of information flow in such regulatory negative feedback.

The results of the PaJaMa experiments also implied the existence of a cytoplasmic "messenger": a short-lived (unstable) RNA template acting as a message transmitted from the genome to the cytoplasm. That prediction initiated an international hunt in the form of a delicate balance of competition and cooperation, which culminated early in 1961 in the identification of messenger RNA (mRNA) by two teams. Working at Caltech, Jacob, Sydney Brenner, and Matthew Meselson examined phage-infected bacteria, while at Watson's laboratory, François Gros (a collaboration with H. Hiatt, W. Gilbert, C. G. Kurland, and R. W. Risebrough) led the studies on uninfected bacteria. Here too the messenger, as an element of the protein-synthesizing system, was signified through the information discourse. Beyond the nuts and bolts of the biochemical manipulations, the models of information transfer, cybernetic analogies, and tropes of communication were transported into the laboratory during the crucial conceptual phases of interpretation of results and, in turn, guided subsequent experimental design. The messenger was not viewed merely as a material or chemical entity (such as hormones were in the 1920s) but as informational. Namely, it was constructed as a *predicate* for specific chemical action by virtue of its possessing form: structural information. Information was the synecdoche for both chemical specificity and cellular memory. As such it was the currency of exchange between form and matter. It served as a communication link between genotypic potentialities and phenotypic actualities—being and becoming—across biological space and time.

Thus Monod's 1970 monograph, *Chance and Necessity*, which explained life as microscopic cybernetics of the cellular machinery, and Jacob's 1970 history, *The Logic of Life*, which recast molecular genetic mechanisms as cybernetic communication and information transfer systems, were not ex post facto constructions of earlier scientific experiences. These informational and cybernetic representations were neither dramatic devices and journalistic embellishments on prior, sober experimental facts, nor merely exemplars of sophisticated science popularization external to the science. True, the books clearly belong to the scientific literary genre of French public intellectuals and had enormous national and international impact among the intelligentsia.[5] But, as we shall see, these tropes, images, and models were never far from the laboratory bench and aided the scientific imagination in the quest of meaning making, an imagination shaped by shared scientific experiences, contemporary technoculture, and their regimes of signification.

EXORCISING FINAL CAUSES

The year 1953 was a pivotal one for molecular biology and for the Pasteur Institute. Watson and Crick elucidated the structure of DNA; the phenomenon of enzymatic adaptation underwent a major paradigm shift to enzymatic induction, and Monod became the director of an expanding Department of Cellular Biochemistry at the Pasteur Institute. Together, these changes shaped the material, discursive, and social dimensions of molecular biology at the Pasteur Institute in subsequent years.

Toward the end of 1953, about six months prior to the announcement in *Nature* of the DNA double helix (affectionately known as the W-C), a short note appeared in that journal entitled "Terminology of Enzyme Formation." Signed by Monod and several distinguished collaborators (Washington University immunologist Melvin Cohn, British microbiologist Martin R. Pollock, microbiologist Sol Spiegelman from the University of Illinois, and the Berkeley microbiologist Roger Y. Stanier), the note was a kind of scientific manifesto, mandating a terminological change that reflected the recent paradigm shift in the field of microbial physiology from enzyme adaptation to enzyme induction. "We propose the following terms and designations; previously used terms are placed in parenthesis," they proclaimed about their enzyme-forming system.

> A relative increase in the rate of synthesis of a specific apo-enzyme resulting from exposure to a chemical substance is an "enzyme induction" (enzyme adaptation). Any substance thus inducing enzyme synthesis is an enzyme "inducer." An enzyme-forming system which can be so activated by an exogenous inducer is "inducible," and the enzyme so formed is "induced" (adaptive). . . . Many enzymes are formed in considerable amounts in the absence of an exogenous inducer. Such enzyme formation is "constitutive." . . . Thus "constitutivity" and "inducibility" are properties of enzyme-forming systems, not of enzymes *per se*, and can be used as significant expressions *only in a biological frame of reference, not in a chemical one* [emphasis added].

With the delineation of the chemical from the biological, a distinction between brute matter from organized systems, the authors formalized a critical reorientation in the studies of enzyme adaptation, a problem which for decades had formed a major research challenge for microbial physiology.[6]

It was common knowledge early in the century that enzymatic properties of microbes depended on the medium on which they had grown, that microbes could be "trained" for, or adapted to, growth in a various environments. In the mid-1930s, when Monod began his studies of bacterial growth, bacterial enzymes were divided into two classes: adaptive enzymes, formed only in the presence of their substrate in the medium (such as carbohydrate-splitting enzymes in the presence of sugars); and constitutive enzymes, whose formation proceeded regardless of the nature of the medium. By the early 1940s researchers knew that the appearance of new enzyme activity in multiplying bacteria resulted either from a chemical stimulus from the medium, or from gradual selection of spontaneous genetic variants. Max Delbrück and Salvador Luria's landmark paper on the "fluctuation test experiment" (1943), which seemed to have proved statistically the spontaneous character (random rather than directed) of bacterial mutations, was seen as a severe challenge to Lamarckian (and "adaptational") thinking still dominating French biology.[7] This knowledge informed Monod's work (collaborating with Alice Audureau in the mid-1940s) on enzymatic adaptation of β-galactosidase (lactose-digesting enzyme) in E. coli, conducted in wild-type and mutant bacteria. The apparent duality of the control mechanisms of enzyme biosynthesis—genetic and chemical—became a key problem in bacterial physiology by the late 1940s, with Joshua (and Esther) Lederberg's pioneering work on bacterial sexuality and genetic recombination contributing novel modes of reasoning and powerful techniques for analyses in the field.[8]

But as Monod's prodigious and often cited 1947 review, "The Phenomenon of Enzyme Adaptation and Its Bearings on Problems of Genetic and Cellular Differentiation," reveals, the larger significance of this phenomenon derived from its centrality in the long-standing debates over cytoplasmic inheritance and the quest of the secrets of biological specificity. In the late 1940s specificity was the unifying theme in the study of life—spanning the spectrum from molecules and organisms to species—and was generally regarded as the key to understanding ontogeny, differentiation, development, and speciation. As such, specificity was a key element within the discourse of organization. That older world-picture life was complex, fluid, and contingent. As Monod explained, "The cell is viewed as a complex population of specific molecules and molecular groups, cellular organizations resulting from the interactions, competitions and regrouping of elementary units": the information discourse had not yet come into being.[9] Within the protein paradigm of life, which framed Monod's article, enzyme adaptation supplied experimental tools for probing the mechanisms of molecular specificity inherent in the patterns governing gene action and antibody formation.[10] As Jean-Paul Gaudillière has shown, the protein paradigm defined the disciplinary combination of metabolic biochemistry and immunology, producing research programs that colored the local scientific culture at the Pasteur Institute. Before the 1953 paradigm shift to DNA, the immunological methods honed by Mel Cohn (collaborating with Monod at the Pasteur Institute from 1948 to 1953) were applied to studies of β-galactosidase (in collaboration with Annamaria Torriani) in an effort to trace lines of descent and kinship between protein molecules between genes, enzymes, and antibodies.[11]

Predicated on extremely stable and finely controlled conditions for bacterial growth, these experiments were enabled by Monod's technical innovation of continuous bacterial culture—the bactogène (1950)—where continuous dilution at constant volume maintained a physiologically stable state. Curiously, physicists Leo Szilard and Aaron Novick at the University of Chicago, both recent converts to biology studying the kinetic of bacterial growth, "independently" developed in 1950 a very similar technology of continuous culture, the chemostat. Szilard shared his satisfaction with Erwin Schrödinger (whose *What Is Life?* had brought Szilard "much pleasure").[12] (See Fig. 22.) Szilard had developed the chemostat partly in response to Monod's investigations on bacterial growth. In turn, the new technology led Novick and Szilard in 1953 to the observation of feedback control by end-product inhibition in E. coli (or negative feedback): they showed that the synthesis of

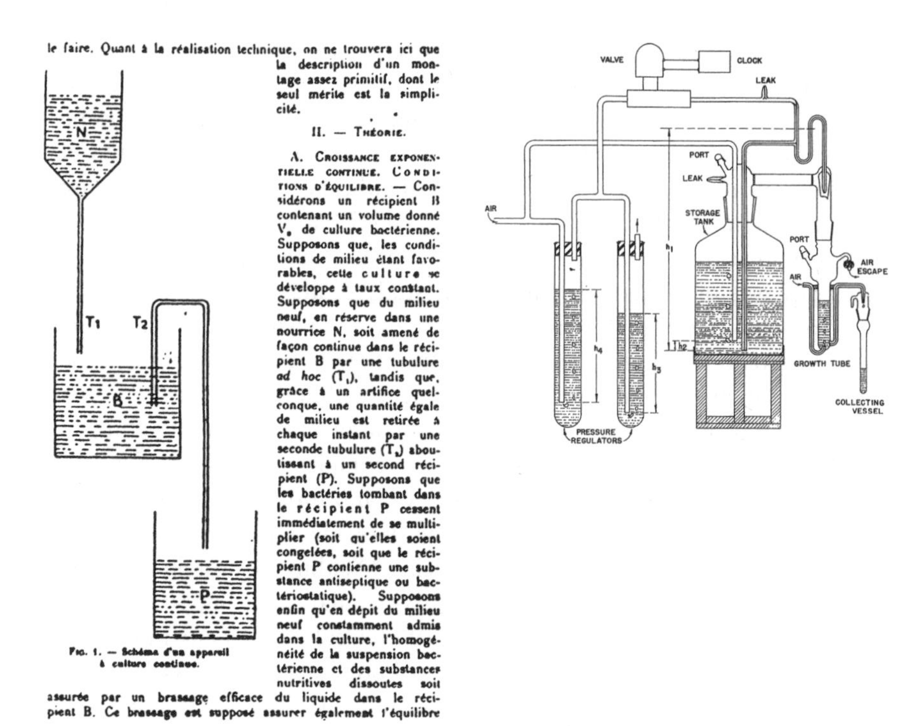

le faire. Quant à la réalisation technique, on ne trouvera ici que la description d'un montage assez primitif, dont le seul mérite est la simplicité.

II. — Théorie.

A. Croissance exponentielle continue. Conditions d'équilibre. — Considérons un récipient B contenant un volume donné V_0 de culture bactérienne. Supposons que, les conditions de milieu étant favorables, cette culture se développe à taux constant. Supposons que du milieu neuf, en réserve dans une nourrice N, soit amené de façon continue dans le récipient B par une tubulure *ad hoc* (T_1), tandis que, grâce à un artifice quelconque, une quantité égale de milieu est retirée à chaque instant par une seconde tubulure (T_2) aboutissant à un second récipient (P). Supposons que les bactéries tombant dans le récipient P cessent immédiatement de se multiplier (soit qu'elles soient congelées, soit que le récipient P contienne une substance antiseptique ou bactériostatique). Supposons enfin qu'en dépit du milieu neuf constamment admis dans la culture, l'homogénéité de la suspension bactérienne et des substances nutritives dissoutes soit assurée par un brassage efficace du liquide dans le récipient B. Ce brassage est supposé assurer également l'équilibre

FIGURE 22. *Left*, Jacques Monod, *Annales de l'Institute Pasteur* 79 (1950): 390. The bactogène apparatus for continuous bacterial culture. *Right*, A. Novick and L. Szilard, *Cold Spring Harbor Symp. Quant. Biol.* 16 (1951): 338. The chemostat device for continuous bacterial culture. Reprinted by permission.

the precursor for the amino acid tryptophan was inhibited by tryptophan itself (though they did not demonstrate the mechanism on the enzymatic level). Corroborated by others (notably Arthur Pardee and Edwin H. Umbarger), negative feedback would soon be everywhere and assume major significance in Jacob and Monod's construction of enzymatic cybernetics. Monod, Cohn, Szilard, Novick, the Pasteur Institute, and the Centre National de la Recherche Scientifique (CNRS) shared the proceeds from the U.S. patent rights (1958) on the new biotechnology. This coincidence arose from the commitment to a shared paradigm and the technoepistemic imperatives of the experimental system.[13]

This technoepistemic overlap reflected the frequent contact, deep intellectual bond, and affection between Monod and Szilard, which lasted until Szilard's death in 1964. For Monod, the eccentric man he had first seen in 1946 snoozing at the front row of the Cold Spring Harbor symposium, "with his round face and fat belly, looked like a petty Italian fruit-merchant, dozing in front of his shop," became his close colleague, scientific guru, and political soul mate. In turn, Szilard's restless intellectual temperament and peripatetic science depended on constant stimulation and experimental novelties generated by young brilliant researchers like Monod. Szilard fascinated Monod with the mysteries of "Maxwell demons" and their potential solution through processes of cellular memory, operating via genetic control of enzymes. Later, through his ongoing communications with members of the phage group and the RNA Tie Club, Szilard would keep the Pasteur group abreast of the recent efforts in genetic decoding.[14]

By 1953, when the Monod-Cohn collaboration ended and Cohn returned to St. Louis (Arthur Kornberg's microbiology department), the two had prepared a comprehensive review outlining the salient features of enzymatic synthesis in E. coli. The buffering inhibitory effect of lactose on β-galactosidase was attributed to the displacement of some endogenous inducer responsible for constitutive synthesis. Namely, they then conceptualized these processes in terms of *positive* control; negative feedback had not yet entered the picture. They also established that, contrary to prevalent explanations, the inductive power was not related to the enzyme's action on the inducer (such as β-galactosidase action on lactose), or even to their chemical affinity. They reasoned, instead, that it had to act at the level of another specific molecular constituent. Curiously, (constitutive) bacteria growing on a medium containing a sugar different but closely related to lactose, which they could not metabolize, nevertheless produced the enzyme β-galactosidase, a "theater of the absurd," as Cohn put it. Why would bacteria produce a useless enzyme in response to a substance they could not metabolize? "Perhaps this can be most readily explained in terms of an all-or-none response of individual cells. Perhaps the terminology of neurophysiology would be as appropriate for enzyme studies as for embryology," suggested Joshua Lederberg, intrigued by cybernetics' potential application to molecular biology. For Lederberg, Cohn and Monod's review "for the first time seems to discard speculative superstitions about enzymatic 'adaptation.'"[15]

For Monod, these findings represented another blow to French neo-Lamarckianism with its emphasis on progressive evolution and a challenge to the widespread teleological explanations of life. French biology was still gripped by accounts of vital forces and final causes, guiding the course of progressive orthogenetic development and the harmonious functioning of complex organs and structures; Teilhard De Chardin was a strong presence among French intellectuals in the 1950s. These traditional commitments were further reinforced by what Jan Sapp has called "the cold war in genetics" and the appeal of Lysenkoism with its emphasis on organismic adaptation. As late as the 1940s, Lucien Cuénot, the "grand old man of French science" and one of the few French geneticists, envisaged the germ cells to possess a "teleological power of invention."[16] While Monod's 1947 essay on enzyme adaptation contained the discursive vestiges of that world picture and its discourse of organization, by 1953 he pronounced these trends as the enemy of scientific progress. Among the many philosophical musings scattered in Monod's personal notebooks, one finds in 1953 tirades against the Aristotelian doctrine of final causes in biology, against the futility of vitalism, the threat of teleology to science, the dangers of Lysenkoism, criticism of Cuénot's "morphological networks" (referring to the organism as a whole), which are "much less rich and complex than enzyme networks," and the championing of chance as central to biological accounts and general epistemology.[17]

Thus, beyond terminological correctives, the 1953 announcement in *Nature* constituted an ideologically charged scientific encyclical, an epistemological turn, and a rhetorical move adopted by "the Adaptive Enzyme's College of Cardinals," as Cohn put it. Clearly, on one level it was just an explicit response to the specific paradox of useless bacterial adaptation; but on another level, it was an implicit confrontation, a challenge to the long-standing use of teleology in biology and to the contemporary ravages of Lysenkoism. The term *adaptation*, which in Darwinian parlance denoted modifications that increased an organism's biological (genetic) fitness, seemed utterly misleading, the authors argued. For it had been used to explain mechanisms in microorganisms possessing both constant and changing genetic constitution and mechanisms that also seemed to be functionally nonadaptive and therefore biologically useless.[18]

The technical precision and theoretical clarity gained from the linguistic substitution of "enzymatic adaptation" by "enzymatic induction" thus entailed a profound gestalt switch in biological thought. It constituted the first step in a sequence of interrogations that, within the next five years, would culminate in powerful representations of organisms as reservoirs of genetic potentialities, as informational programs impervious to the environment, regulated purely internally and executed through a series of transformations. "The programme does not learn from experience," as Jacob would later put it.[19] Ironically, these new representations would necessitate the reembracing of teleology; not through Aristotelian or Lamarckian atavism, but through the reinvention of teleology as teleonomy, the manifest property of cybernetic machines, computer programs, and now biological systems.

The year 1953 was not merely scientifically eventful for Monod. That year he was assigned to create and direct the Department of Cellular Biochemistry (Service de Biochemie Cellulaire) at the Pasteur Institute. While obviously organized around Monod's own research program, on the disciplinary level this event reflected his commitment to revitalize French biology (generally still parceled along the traditional disciplinary boundaries of botany, zoology, physiology, and embryology; genetics was just beginning to coalesce as a viable discipline) by emphasizing cellular and molecular mechanisms in microorganisms. It also signaled the modernization of Pasteurian biochemistry, away from antibody production and metabolic economy and toward processes of biosynthesis informed by the perspectives and materials of the young specialty of microbial genetics. Although in the early 1950s Monod's group had not yet utilized genetic techniques, Monod had a deep understanding of genetic theories and practices. These were interests he had developed through close ties with his Sorbonne mentor Boris Ephrussi, a postdoctoral Rockefeller Fellowship at Caltech's biology division (1936), constant interactions with Lwoff's group, and frequent meetings with British and American molecular geneticists.[20]

The modern, and in some sense American, style of Monod's science was also reflected in his fund-raising activities. His spacious and well-equipped new department was supported by money raised from CNRS, Rothschild family donations, the Rockefeller Foundation, the National Science Foundation, and the National Institutes of Health. And, like Caltech and Cambridge (England), Monod's laboratory became an international hothouse for the consolidation of the bold new field of molecular biology. Its prominence also buttressed Monod's campaign against French scientific "sclerosis." From the mid-1950s on he promoted modernization policies (with American science as a model) of boosting government investments in basic research, notably the new biological fields. "I have immediately taken up my new duties, starting with a detailed study of the present situation in the department (facilities, equipment and personnel), and planning a complete reorganization," Monod reported to the Rockefeller Foundation requesting their support. He envisioned a department of about twenty-five or thirty people, including both researchers and technical support. He wrote to the foundation:

> Now, as regards the scientific orientation which I wish to give to the department, it could be defined very broadly, as the study of elementary processes of cellular growth. This in my mind would involve parallel studies on the more chemical, and on the more physiological aspects of growth, i.e. of biosynthesis, and I am envisaging the formation of three or perhaps four research groups, working respectively on: 1. the biosynthesis of proteins, particularly enzymes. 2. the biosynthesis of nucleic acids. 3. the biosynthesis of amino acids. 4. the biosynthesis of puric and pyrimidic bases and nucleotides. . . . This, of course, is an exceedingly ambitious and exceedingly broad program.

Although he admitted his program's ambitiousness, he also believed that since it concentrated on the same microbiological material (E. coli) and common methodologies it would be cooperative and quite focused.[21] The foundation appropriated $50,000 for a four-year period. During a visit by the Rockefeller Foundation officer to the new facilities (then still under construction) Monod underscored the special importance of the foundation's support: "Without the grant from the Foundation and the recognition it gave at the international level, it would have not been possible to get this total equivalent to $100,000 [from other sources], and it is a good thing they got this money because it turned out that rearranging the old labs for modern teaching and research purposes was an enormously expensive affair." [22]

As head of a major new department at the Pasteur Institute, Monod became even more publicly visible. And he remained as politically committed as in earlier years, when his Sorbonne laboratory had been a clearinghouse for the political activities of his Resistance group during the Nazi occupation of France and during his brief concomitant membership in the French Communist Party (FCP) (he resigned, partly in protest against the party postwar policies and its affinities with Lysenkoism). After organizing the general strike that led to the liberation of Paris, Monod became an officer in the Free French Forces and a member of Général de Tassigny's general staff. When George Cohen (his future collaborator) first met him in 1944 "he was still wearing the uniform of a French Army Major." [23]

Affiliation with the FCP, which had a strong presence in the early 1950s, was damaging for French scientists supported by American agencies under the aegis of the Marshall Plan, the massive recovery policy ($12.4 billion, or 1.2 percent of the United States gross national product) aimed at strengthening free-market Europe within the cold-war NATO alliance. But such affiliation was far more detrimental in the United States at the height of the McCarthyist witch hunts. When in 1952, following invitations for distinguished lectures by the American Chemical Society and Harvey Society, Monod was denied a U.S. visa, he elected to forego the humiliations of being admitted temporarily, using the occasion instead to publicize his opposition to America's misguided security measures in *Science* magazine. A prepublication copy went to Szilard, who was prodding scientists toward collective action against the excessive security clearance practices of the AEC (Atomic Energy Commission). Szilard was also engaged in a crusade against the H-bomb, sparing no effort to contain the damage inflicted by his friend Edward Teller on J. Robert Oppenheimer for questioning the commitment to build the bomb. (Teller was then involved in breaking the genetic code; see my discussion in Chapter 4.) Monod, who emphasized his love for the United States (his mother was American) and his great admiration for and debt to American science, was convinced that America's security policy was misguided: "It is a plain fact, that such measures represent a rather serious danger to the development of science, and that to that extent at least, they must be contrary to the best interests of the United States itself."[24]

He also staged an extraordinarily moving protest in response to the Rosenberg trial (1953) in the *Bulletin of Atomic Scientists* (again, with a preprint to Szilard). "American scientists and intellectuals," he called out in the conclusion to his long letter about the event that shook Europe. "The execution of the Rosenbergs," he wrote, "is a grave defeat for you, for us, and for the free world. . . . It does testify to your present weakness in your own country. . . . You American scientists and intellectuals bear great responsibilities which you can not escape, and which we can only partly share with you. You must, for civilization's sake, obtain moral leadership and power in your own country."[25] "From here the U.S. seems an unhappy place," Novick reported to Szilard from the Pasteur Institute. "The political events of the last few months in the U.S. frightens everyone here."[26] The advent of the Eisenhower administration (1953–61) further reduced the prospects for science as a civilizing mission. The new intensity of the cold war spelled an even greater commitment to national security, which only fueled Szilard's efforts at arms control.

Neither the State Department nor the Rockefeller Foundation appreciated Monod's political activism and his outspoken manner. When he was granted a visa in 1954 to deliver the Jessup Lectures at Columbia University he was considered somewhat a security risk, though one worth taking, according to the authorities. Considering the State Department's stance (articulated by D.R.), Rockefeller Foundation officer G.R.P. commented:

> Their judgment was, on balance, that if there were any risk in this particular case it was a risk worth assuming. He [Mr. Rudolph, State Department officer] said that the principal anxiety about Monod was really one of indiscretion on political matters—that he might make statements at the wrong place, at the wrong time and in the wrong way which would be disagreeable. He immediately said, however, that we [the Rockefeller Foundation] should have no objections to people having other ideas and expressing them and that we ought not to attach too much importance to this type of criticism.[27]

But for Monod the boundary between the *vita contemplativa* and *vita activa* had always been a permeable one. It would become even more so in subsequent years, as his involvements in restructuring French biology around molecular biology intensified during the politically turbulent regime of Charles de Gaulle.

In the mid-1950s, in tandem with these administrative and political activities, and concurrent with the ongoing research on enzyme induction,

Monod's laboratory focused on a related phenomenon, which soon brought the role of genetic control of enzymes into sharper focus. Following an experimental strand of the Cohn-Monod collaboration, Monod and microbiologist George Cohen (who had just joined the new biochemistry laboratory) turned their attention (1953–57) to the analyses of "cryptics": mutant E. coli (known as Lac−, or lactose-negative mutants) that cannot develop on lactose, even though they possess (or are constitutive for) the enzyme β-galactosidase. Their "crypticity" derived from the observation that intact bacteria did not display galactosidase activity, yet such activity was clearly manifested in the cell extract. Lac− mutants were first isolated in 1948 by Joshua Lederberg at the University of Wisconsin; by the mid-1950s the Lederbergs had isolated and characterized numerous other Lac− mutants, including "cryptics." Through ingenious crossing experiments and complementation tests (modeled after the classic *Neurospora* work of George Beadle and Edward Tatum) these Lac− mutations were located and mapped along the E. coli genome (the corresponding genomic segment became known as the "Lac region"). Thus while Monod was studying the physiology of lactose fermentation, Lederberg was probing its genetic basis. A precision technology, these different mutant strains were central to the material culture of microbial genetics, circulating widely within its international network. The physiological basis of some mutants was simple; others, such as "cryptic," seemed bafflingly complex.[28]

Postulating that "cryptics" had to possess specific permeation factors, Cohen proposed a model that, after considerable tinkering with Monod, soon led to the identification of a functionally specialized enzymelike factor, which concentrates the basic units of lactose in the cell without chemically modifying the cell, a factor that soon after became known as β-galactosidase-permease (at this point permease was only a theoretical construct; it was isolated a decade later). The "cryptic" mutants had lost their permeases. Moreover, Monod and Cohen found that strains constitutive for β-galactosidase were also constitutive for permease, an observation suggesting that both enzymes might be genetically linked.

By 1957 Monod's laboratory classed the mutants into three fundamental types: y, z, and i. Types y+, in the presence of lactose, could synthesize permease; mutants y−, in the presence of lactose, could not synthesize permease; types z+ could synthesize β-galactosidase; types z− could not synthesize β-galactosidase; inducible (wild type) i+ synthesized both permease and β-galactosidase in the presence of lactose; and constitutive types i− synthesized both enzymes even in the absence of lactose. "Thus the role permeases as chemical connecting links between the external world and the intracellular metabolic world appears to be decisive," Cohen and Monod proclaimed in their article. They continued, "Enzymes are the elements of choice, the Maxwell demons which channel metabolites and chemical potential into synthesis, growth and eventually cellular multiplication. Occurring first in this sequence of chemical *decisions*, the permeases assume a unique importance; not only do they control the functioning of intracellular enzymes, but eventually their induced synthesis [emphasis added]."[29]

These were intelligent, decision-making systems. For Monod, "the [significance of] the discovery of Maxwell in order of derivation [was]: Genes, Enzymes, Ideas"; this thought trajectory was probably inspired by Szilard, who had shown (back in the 1920s) that by thinking, Maxwell's demons generated entropy, which accounted for the increased order in systems.[30] It now seemed intuitively clear that the differential enzymatic decisions among these mutants were not only genetically controlled but also their genetic mechanisms were somehow linked. Understanding these control mechanisms and their linkages became the next goal of the Pasteur group, their key to unlocking the mystery of being and becoming. In the following two years, the new technologies of bacteriophage genetics and the negative feedback models of enzyme synthesis would converge to represent these cellular mechanisms as cybernetic and informational systems, a technoepistemic-discursive convergence produced through the (now classic) PaJaMa experiments.

INFORMATION, CYBERNETICS, AND THE REINVENTION OF TELEOLOGY

As evidence for the centrality of genetic controls in the synthesis of permease and β-galactosidase mounted, the need for genetic analyses of enzyme induction assumed a sense of urgency in 1956. Monod now began to collaborate with François Jacob, who had been working in André Lwoff's laboratory, which was a European contact point in Max Delbrück's phage network, or "phage church," as it was sometimes called. Housed in the attic of the Pasteur, the now-legendary *grenier*, where Monod had worked when he first joined the institute in 1945, Lwoff's laboratory was spearheading novel approaches to the study of lysogeny. The lab was at the vanguard of bacterial genetics, defined, to a large extent, by the remarkable studies of Eli Wollman and Jacob in the mid-1950s on bacterial sexuality and genetic organization. Beyond the contributions to bacterial genetics, the new approach now promised to lend powerful tools for the analyses of the problem of enzyme induction.[31]

Jacob had only recently entered biology. The war had interrupted his medical studies at the University of Paris. He joined the Free French Forces in Lon-

FIGURE 23. *Left to right*, François Jacob, Jacques Monod, and Andre Lwoff, 1965. Courtesy of Musée Pasteur.

don as a medical officer and experienced the bloody horrors of the battlefield in Africa and Normandy. (The injuries and vividness of those memories, the sights, sounds, and smells, kept haunting him even in the relative tranquility of postwar reconstruction.) Jacob persevered in making his way into Lwoff's laboratory (having decided to become a biologist soon after completing his medical degree) and by 1950 found himself working on the newly discovered phenomenon of "prophage induction," terms that meant little to him then. Lysogenic bacteria, strains that perpetuate the phages they carry in a noninfectious, or prophage, state may be chemically induced (for example, through exposure to ultraviolet radiation) to become lytic (burst). The mechanisms governing this induced transformation became a "hot" problem in bacterial genetics. While enrolled at the Sorbonne, Jacob studied the properties of lysogenic bacteria and demonstrated their "immunity" by exhibiting the existence of a mechanism that inhibited the activity of the prophage genes; his was a thesis project that led to a doctorate in 1954. By that time he was already deep into the collaboration with Elie Wollman, whose parents, Eugène and Elizabeth Wollman, had pioneered lysogeny studies at the Pasteur Institute before perishing in the Nazi camps. The research of Wollman and Jacob aimed at explaining the relationships between the prophage and genetic material of the bacterium.[32]

Together they developed powerful genetic techniques of bacterial conjugation, or "zygotic technology," as it came to be called. With these new tools, their experiments soon showed that the chromosome of a donor ("male") lysogenic bacterium (carrying a prophage) enters a nonlysogenic receptive bacterium ("female") and stimulates there the expression of phages and a concomitant bacterial lysis. Their hypothesis that, during conjugation, a chromosomal segment was transferred from donor to receptor bacterium, unidirectionally and at a constant rate, was elegantly demonstrated in 1955 in the "interrupted mating experiment," variously also known as the "spaghetti experiment," but more often, the "coitus interruptus" experiment of "erotic induction." Using a Waring blender (from Mrs. Wollman's kitchen) they "shook" the bacterial suspensions at different times during conjugation, which resulted in pulling the pairs apart and breaking their chromosomes, and thus revealed that the "male" (donor) genes enter the "female" (recipient) in a definite order specific to each strain (see Fig. 24). Translating French poet Paul Verlaine, Wollman and Jacob wrote about the sexuality of bacteria, which, they thought, revealed even more about the mechanisms of genetic recombination in man:

> These passions which only they in their
> sport
> Call love: they too are love, tender and
> furious
> And with particularities curious
> Not love of the everyday sort.[33]

Jacob and Wollman used this "coitus interruptus" technique to probe the newly cited phenomenon of "zygotic induction" (or "erotic induction"): how, in conjugations between certain lysogenic, or phage-carrying, "male" bacteria (known as Hfr) and nonlysogenic "female" bacteria, the prophages that enter the "female's" cytoplasm are induced, namely, become infectious. Jacob and Wollman not only showed that, surprisingly, the induction was instantaneous but also that for a period of time the "female" (being the recipient) possessed two copies of chromosome segments. This temporary existence in a diploid, rather than the typical haploid state, created an extraordinary experimental situation, permitting an analysis of genetic dominance (characteristic of higher, diploid organisms) in a greatly simplified form.[34] It was these recent findings of bacterial recombination and prophage induction that spurred the collaboration between Monod and Jacob early in 1957. The "zygotic technology" promised powerful probes for analyzing the mechanisms of genetic control of enzyme induction. It was at this point that Arthur B. Pardee, who had been working on enzyme biosynthesis in bacte-

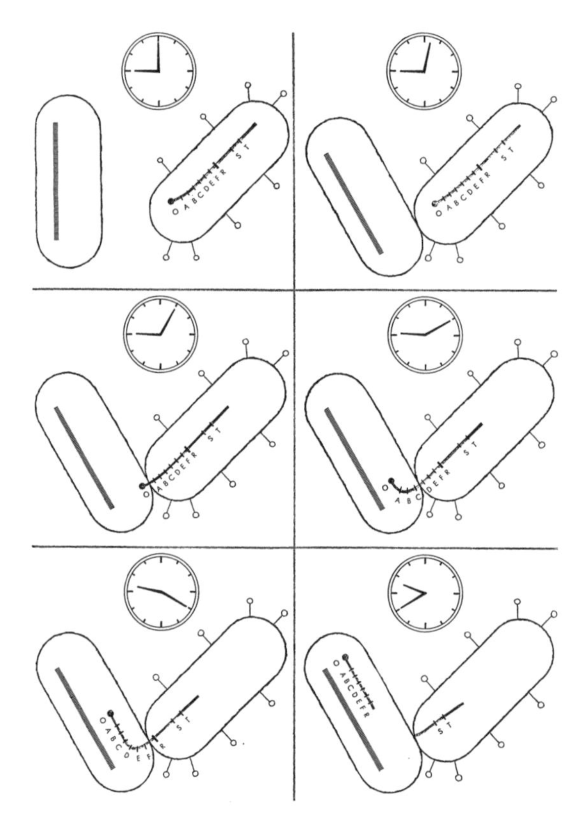

BACTERIAL GENES are transferred from one bacterium to another in the linear order demonstrated in the experiment diagrammed [elsewhere in the article], suggesting that the genes are organized on a chromosome-like structure. The "male" in these diagrams is tagged with virus as in the electron micrograph [elsewhere]. Conjugation begins a few minutes (*at upper right*) after the two strains are mixed. The transfer of genetic material (*at left center*) begins a few minutes later. The letters represent the location along the chromosome of genes for specific, identifiable traits. Experimental interruption of conjugation process at successive stages shows that genes of a given strain always penetrate the "female" cell in the same order.

FIGURE 24. Elie Wollman and François Jacob, "Sexuality in Bacteria," *Scientific American* 195, no. 1 (1956): 109–18. Genetic transfer in bacterial conjugation. Reprinted by permission.

ria at Berkeley's Virus Laboratory, came to spend a year (1957–58) in Monod's laboratory.

Pardee recalled two decades later that during the years 1953–57 "work in my laboratory often was parallel to publications from Monod and Jacob's groups. My interests in enzyme induction arose from several studies I had made: on enzyme changes following viral infection, on regulatory interactions between protein synthesis and nucleic acid synthesis, and from our discovery of feedback inhibition of enzyme activity, and of enzyme synthesis repression." He was thus tacitly arguing for his priority to (or at least simultaneity with) some of the key discoveries credited to Monod's laboratory. Haunted by the swiftness and brilliance of the French competition and their rhetorical skills, Pardee remembered how Gunther Stent, after each yearly pilgrimage to the Pasteur, would predictably react to Pardee's progress reports with, "the Pasteur group had done it already." Pardee mused, "It was a glorious day when I told him about something new that we had done and after a long pause, he responded, 'Well, they haven't done that but they are thinking about it.'"[35]

Pardee did work at the forefront of biosynthesis, though his meticulously detailed articles lacked the power of generalization and presentational flair. His demonstration in 1956 that 5-bromouracil caused mutations in bacteriophage by disturbing the DNA metabolism contributed effective tools to molecular biology. He also independently discovered the phenomenon of negative enzymatic feedback in bacteria (which implied enzymatic repression, though he did not introduce this scientific neologism, using instead the conventional terminology of inhibition), but Edwin Umbarger (at Harvard's microbiology department) beat him to publication of the discovery by a month. Umbarger's sensational report in *Science*, "Evidence for a Negative-Feedback Mechanism in the Biosynthesis of Isoleucine," was bound to strike a resonant chord with a wide scientific audience. The generality of its claims and its analogies to automation and contemporary cybernetic systems were compelling. Umbarger wrote:

> Recent developments in automation have led to the use in industry of machines capable of performing operations that have been compared with certain types of human activity. In the internally regulated machine, as in the living organism, processes are controlled by one or more feedback loops that prevent any one phase of the process from being carried to a catastrophic extreme. The consequences of such feedback control can be observed at all levels of organization in a living animal. . . . A simple, though typical, example is the effect of L-isoleucine on the L-threonine requirement of threonine mutants of Escherichia coli.[36]

Pardee, on the other hand, submitted his comprehensive studies on pyrimidine (single-ring bases of nucleic acids) biosynthesis in E. coli to the *Journal of Biological Chemistry*, reporting matter-of-factly the "existence of a cellular mechanism which links pyrimidine production to the rate of pyrimidine uptake for nucleic acid synthesis," casually adding, "Inhibition by end-product of its own synthesis appears to be a common control mechanism in the cell," with a reference to Umbarger's "feed-back."[37] Nevertheless, in spite of the fact that he wasn't a brilliant promoter, Pardee was a superb experimenter, clearly positioned at the vanguard of biosynthesis research. His interests, which now converged with Monod's, led him to spend his sabbatical leave at the Pasteur Institute.

Despite the differences in technical approaches between Pardee's and Monod's laboratories, Pardee swiftly mastered the intricacies of the mating system of lac+ and lac− bacteria. He even improved on the local biochemical procedures, an example par excellence of scientific trading zones as exchange sites for different theoretical commitments and material cultures. In December 1957, Pardee, Jacob, and Monod embarked on a series of experiments aimed at investigating the genetic control of the lactose system, studies which became known as the "PaJaMa experiment/s." Utilizing the various mutants isolated by Monod (mutants defective in the production of either β-galactosidase or permease, and the constitutive cryptic mutants: the z, y, i trinity) they planned to insert them in different combinations in male or female bacteria. This zygotic technology literally became a device for tracing the location and function of bacterial genes, that is, for generating genomic maps (Marshall Nirenberg even envisioned it as a method for breaking the genetic code; see related discussion in Chapter 7). The experimental system was altered. The conjunctions of the E. coli and phage systems now produced a composite experimental system and hybrid technologies, thus reconfiguring the experimental space, once again, to include new graphemes: representations of enzymatic function in terms of genomic maps.[38]

Within a couple of months, by performing various reciprocal crosses, their experiments established three principal features of the lactose system. First, to their great surprise, β-galactosidase was synthesized at maximal rate within two to three minutes after entry of the gene into the bacterial cell. This synthesis could not be the expression of genetic recombination, which, as Wollman and Jacob had demonstrated, lasted more than an hour; instead, it had to be some kind of a direct chemical signal—a messenger—from the gene to the cytoplasm, a hypothesis which focused attention on RNA as an intermediary and soon after initiated an international hunt for an unstable intermediate template (reported earlier by other researchers) associated with that messenger (see section below). Second, they demonstrated that the i gene, which determined the inducible versus constitutive character of enzyme synthesis, was distinct from the z gene, the gene controlling β-galactosidase synthesis, and also distinct from the y gene, which controlled permease synthesis. Third, against all expectations, they found that inducibility was dominant over constitutivity; it was therefore the zygote's inducible i+ allele, not the constitutive i− one, that was expressed in the cytoplasm as an active product. This finding contradicted Monod and Cohn's earlier hypothesis (1953) that constitutivity resulted from *positive* control, that is, from the synthesis of an endogenous inducer, whose displacement resulted in inhibition.[39]

By January 1958 everything seemed to point toward the presence of a *negative* control mechanism of enzymatic induction. It now appeared the i gene might control the synthesis of some product that inhibits enzyme biosynthesis: Novick and Szilard's identification of endpoint inhibition in tryptophane synthesis; Werner Maas's studies of arginine biosynthesis; Umbarger's cybernetic mechanisms of isoleucine synthesis; Pardee's analysis of inhibition of pyrimidine synthesis; and other prior and subsequent sitings of negative feedback in enzyme biosynthesis. Since the existence and generality of inhibitory systems was now amply demonstrated, why not suppose, then, that induction could be effected through anti-inhibition rather than by inhibition by an anti-induction? Monod recalled several years later:

> Of course I had learned, like any schoolboy, that two negatives are equivalent to a positive statement, and Melvin Cohn and I, without taking it too seriously, debated this logical possibility that we called the "theory of double bluff," recalling the subtle analysis of poker by Edgar Allan Poe. I see today, however, more clearly than ever, how blind I was in not taking this hypothesis seriously sooner.[40]

Szilard's visit to the Pasteur at the end of January 1958 tipped the balance in favor of what became known as the "repressor hypothesis." Szilard had spent a busy winter in 1957 in Europe. After visiting for several days with Crick's group at Cambridge (following up on his aborted coding scheme), he passed through London in December, where he sat on the Pugwash Continuing Committee meeting in an effort to shape the course of future Pugwash and arms control activities. At the same time he was advising the German ministry on restructuring postwar biology around the new trends in molecular biology, and lecturing on enzyme induction and repression in bacteria (for example, to the German Chemical Society in Berlin), ideas which he also presented in Paris. "We all looked forward to Szilard's coming," remembered Cohn, then visiting at the Pasteur Institute. "He was given an office in Monod's laboratory, and we all had to talk to him; everybody lined up for the chance."[41]

Thus, just when Pardee, Jacob, and Monod were struggling with the interpretations of their results, trying to decide on the type of control operating in the lactose system, Szilard stepped in. Arguing for the existence of a "repressor" molecule (shown to be a protein only eight years later), he envisioned it to somehow block the synthesis of both β-galactosidase and permease, probably by binding to a specific site on the i gene (presumably the signaling site from the gene to the cytoplasm); thus in constitutive mutants it was the repression mechanism that was disrupted. Induction was, in fact, derepression; a "double bluff" mechanism after all. Before Pardee's return to Berkeley in July 1958, a short report immediately appeared in *Comptes Rendus de l'Académie des Sciences*, which was soon followed by an expanded article. By the time of that article's appearance in the *Journal of Molecular Biology* (first volume, 1959), their findings had assumed far greater generality and significance, which derived from Jacob's intuition about the analogy between the mechanisms of phage induction in lysogeny and those of bacterial enzyme induction.[42]

It was then, in the late spring of 1958, that Monod (and Jacob) first began to use the idiom and model of information transfer, very specifically, in an attempt to account for the mystery of genetic action-at-distance: for the "transfer of structural information from a specific gene" to the cytoplasm, where its interaction with the inducer occurred. "Information," then, was invoked as a way of visualizing a mechanism that could not be explained solely in material terms, since it involved features such as recognition and communication (qua control).[43]

Up at the *grenier*, Jacob was literally sweating out the mechanisms of phage induction and immunity in E. coli, when their "investigative pathways" (to use F. L. Holmes's term) suddenly converged. It was "late July 1958. Sunday in Paris," Jacob recalled. "A day with no taste for work," despite the pressure of preparing too many lectures, especially the Harvey Lecture, "Genetic Control of Viral Functions," to be delivered in New York in September; a milestone in a biologist's career. But his restless mind was stubbornly at work, even during an escape to the movie theater. He recalled:

> Shadows move on the screen. I close my eyes to heed extraordinary things going on within. I am invaded by a sudden excitement mingled with a vague pleasure. It isolates me from the theater, from my neighbors whose eyes are riveted on the screen. And suddenly a flash. The astonishment of the obvious. How could I not have thought of it sooner? Both experiments—that of conjugation done with Elie on the phage, erotic induction; and that done with Pardee and Monod on the lactose system, the PA JA MA—are the same! Same situation. Same result. Same conclusion. In both cases, a gene governs the formation of a cytoplasmic product, of a repressor blocking the expression of other genes and so preventing either the synthesis of the galactosidase or the multiplication of the virus. In both cases, one induces by inactivating the repressor, either by lactose or by ultraviolet rays. The very mechanism that must be the basis of the regulation.

There was no one with whom to share the excitement; no one to check his wild idea. With both Lwoff and Monod away on summer vacations and Wollman gone too, Jacob pursued the powerful analogy on his own.[44]

He included these new insights in the Harvey Lecture, ideas that were formulated through the information discourse. Implicitly delineating matter from information, Jacob showed how the two-step sequence of phage expression in lysogenic bacteria (corresponding to two "genetic units") was predicated on "information transfer." As he visualized it, the phage's information allowed the phage's genetic material to recognize a specific locus of the host chromosome and to establish itself there as prophage. Drawing on his and Wollman's studies of zygotic induction, he reported that immunity to phage infection in lysogenic bacteria (or inhibition of phage production) was a genetically dominant trait expressed in the cytoplasm. "Strikingly enough, the expression of immunity in zygotes appears to be similar to that found for the expression of the inducible character of an enzyme in reciprocal crosses between constitutive and inducible strains," Jacob boldly announced, citing the PaJaMa experiments. "The analogy between this phenomenon and immunity of lysogenic cells is such that we can hardly escape the assumption that immunity also corresponds to the presence of a repressor in the cytoplasm of lysogenic cells."[45] He expanded these analyses in a concurrent publication, "Transfer and Expression of Genetic Information in Escherichia Coli K12," where the distinction between (extant) genetic material and the transfer (and expression) of genetic information was made more explicit. But the operational identity of the two phenomena—enzyme induction and phage induction—was yet to be elucidated. Jacob now came to envision the processes of genetic and enzymatic regulation as operating not progressively, but discontinuously, like a switch, by an all-or-none, stop-or-go mechanism.[46]

Jacob shared his visions with Monod right after his return from New York in September. Monod was not so quickly convinced. But when he warmed up to the idea, there ensued in the following weeks one of the most intense partnerships in the history of modern science, "the grand collaboration." In the hours-long discussions during the daily revisions of the models they feverishly drew on the blackboard—arrows and diagrams going in every direction—Jacob and Monod deployed analogies and images from contemporary technoculture, including not only those of electronic circuits but also of electronic weapons systems. Jacob later wrote:

> We saw this circuit as made up of two genes: transmitter and receiver of cytoplasmic signal, the repressor. In the absence of the inducer, this circuit blocked the synthesis of galactosidase. Every mutation inactivating one of the genes thus had to result in a constitutive synthesis, much as a transmitter on the ground sends signals to a bomber: "Do not drop the bombs. Do not drop the bombs. If the transmitter or the receiver is broken, the plane drops its bombs. But let there be two transmitters with two bombers, and the situation changes. The destruction of a single transmitter has no effect, for the other will continue to emit. The destruction of the receiver, however, will result in dropping the bombs, but only by the bomber whose receiver is broken.

As Jacob saw it, the bomber (inducer), bomb (enzyme), transmitter (i gene), receiver (z gene), and cytoplasmic signal (repressor) formed a closed communication system. Monod visualized enzymatic regulation as electronically automated ballistic missiles. "Interpretations (or descriptions) are cashed out in experiments," as Evelyn Fox Keller put it. Several new and rediscovered mutants, and a few experiments later, and the phenomenon of simultaneous activation of the two adjacent genes—galactosidase and permease—was confirmed. By the fall of 1958 their representations of induction in terms of "units of activity," modeled as cybernetic systems activated or inactivated as a block via a yes-no switch, no longer seemed to them to be just a daydream.[47]

The timing was fortuitous. Monod was just completing the preparation of the prestigious Dunham Lectures, to be delivered at Harvard in the third week of November 1958. The lecture series was entitled: I. "Properties, Functions, and Interrelations of Galactosidase and Galacotside-Permease in E. coli"; II. "Induction and Repression"; and III. "Genetic Control." He intended to combine the lecture event with a "tour to the far-west," stopping along the way to visit friends (and give seminars), including visits at NIH in Washington, D.C., with Mel Cohn in St. Louis, and with Roger Stanier at Berkeley. At the end of September, Monod concluded that the scope of the three lectures was too vast and his treatment of the subject too limited, so he sent the lecture drafts to Cohn, to serve as the basis for a joint monograph on enzyme cybernetics.[48]

Focusing on the broad phenomenological quandary of "Being and Becoming," and drawing on the explanations which he and Jacob had just constructed, Monod framed his lectures around their cybernetic model of enzyme regulation and information transfer. In the second lecture he observed, "The two mechanisms, induction and repression are almost certainly related, and appear together to constitute one of the major integrating mechanisms in cellular economy. I would like to give an integrated discussion of these integrating mechanisms, but enzyme kibernetics is a very complex science indeed."[49] In a subtle rejection of "enzyme adaptation," he posed the (rhetorical) question: "Does the cell 'learn' something from the inducer or does the inducer only activate, select out, trigger preexisting mechanisms?" Guiding his audience through the intricacies of enzymatic regulation, he moved them toward a representation of the enzyme (β-galactosidase) as possessing very precise (preexisting) structural information. In answer to the more fundamental question, "Where does this information come from?"—thus addressing, more generally, the problem of being and becoming—he skillfully led them to the conclusion that the information was "entirely inborn." The old notion of "cellular memory," that the cell learns from external experience, was now being replaced by computer-informed models of "cellular memory" as internal storage of finite amounts of genetic information. There was no evidence that the inducer communicated any new information to the cell; rather, Monod insisted, it merely released or activated latent genetic potentialities, thus buttressing Crick's newly enunciated Central Dogma. It was an implosion of being and becoming, a genetic collaboration (z+ and i−) achieved through the action of what Monod called "cytoplasmic 'messengers'"; however, he did not specify what the nature and mode of action of such "messengers" could be.[50]

The lectures were apparently a huge success. "I have virtually made up my mind to write, in collaboration with Dr. Melvin Cohn, a monograph based on my Dunham series," Monod informed his Harvard host, Otto Krayer, upon returning to Paris. Comprising seven chapters, two to be written by Cohn, the book, *Enzyme Cybernetics* (never completed), was to be published by Harvard University Press. Though committed to a cybernetic mechanism —its specific role in enzyme synthesis and its generalization to other physiological processes—Monod was nevertheless apologetic about using the fashionable terms of cybernetics (especially popular in France). He wrote in the introduction to the book:

> The subject of the present essays is thus perhaps not limited to enzyme adaptation itself and will be undoubtedly better described under the name of (cellular) enzymatic cybernetics, if only this term which has become fashionable did not invoke for the reader the alarming journalistic resonances. . . . The interest and the difficulty of this enzymatic cybernetics resides precisely in that the theoretical and experimental problems its subject presents are multiple, relevant to several disciplines and may be envisioned from robustly different points of view.[51]

The chapter drafts conveyed an unmistakable sense of excitement about the newly elucidated cellular mechanisms. These were not the clunky machines of prewar years; instead, they were the intelligent, goal-directed, communication systems of the postwar era, which had been reshaping the industrial-military terrain and the contemporary imagination. With this celebration of cybernetics, Monod was obviously moving toward the readmittance of teleological explanations in biology. The apprehension of journalistic sensationalism helps explain why, given the commitment to cybernetics, he and Jacob shied away from its enthusiastic promotion in the scientific community.

Monod's Berkeley lecture, "Induction—Repression—Genes," delivered the week after the Dunham Lectures, was far more narrow and technical; Monod's notes were crisp and clipped. They read:

> Introduction: Main point: a. new light on mechanism of enzyme induction. b. new light on genetic control of enzyme synthesis. Existence of genetic "controlling units" comprising Two Types of Specific Control: information gene—

Génotypes et phénotypes biochimiques du système galactosidase-perméase

Génotype	Phénotype biochimique			
	Avec inducteur		Sans inducteur	
	Galactosidase	Perméase	Galactosidase	Perméase
$z^+y^+i^+$	+	+	—	—
$z^-y^+i^+$	—	+	—	—
$z^+y^-i^+$	+	—	—	—
$z^+y^+i^-$	+	+	+	+
$z^-y^-i^+$ (1)	—	—	—	—
$z^-y^-i^-$ (2)	—	—	—	—
$z^+y^-i^-$	+	—	+	—
$z^-y^+i^-$	—	+	—	+

(1) · Ce génotype, difficile à obtenir par recombinaison, et impossible à sélectionner n'a pas encore été isolé.
(2) · Ce génotype a été observé dans une souche portant une délétion complète de la région »Lac« du chromosome d'*Escherichia coli*.

FIGURE 25. Jacques Monod, "Information, induction, répression dans la biosynthèse d'un enzyme," *Colloquium der Geselschaft für Physiologische Chemie* (9/12 April 1959): 125. Biochemical genotypes and phenotypes of the galactosidase-permease system. Reprinted by permission of Springer Verlag.

(structure); regulatory gene—(release of information). Role of Pardee; β-galactosidase System; Induction; Constitutives; Repression: anti-induction; Mutant Types [the z, y, i trinity]; Locus Structure: z = structural gene = information; i = controlling gene = release of inf. Predictions: how does i+ → i− "liberate information" (three possibilities: activator, inducer, repressor); Zygotic Technology; Speculations: Cancer = breakdown of control.[52]

There is little doubt that the concept of information assumed a precise and operational role in Monod and Jacob's interpretive scheme: it served to explain the difference between structural genes (z and y) and regulatory genes (i); and it accounted for the action-at-a-distance of the abstract "messenger" that mediated "information release."

Fully conscious of the new information idiom, Monod explicated its usage in the lecture, "Information, induction, répression dans la biosynthèse d'un enzyme," prepared for the spring 1959 colloquium of the German Society of Physiological Chemistry in Mosbach-Baden; his colleague at the Pasteur, François Gros, delivered the paper. With a graphic representation of the genetics of the galactosidase-permease system (see Fig. 25 and Fig. 26), Monod proposed a mechanism that explained the nature of the interaction between the z and i mutations. He wrote, "One could suppose, for example, that the locus z contains genetic information relative to the structure of the

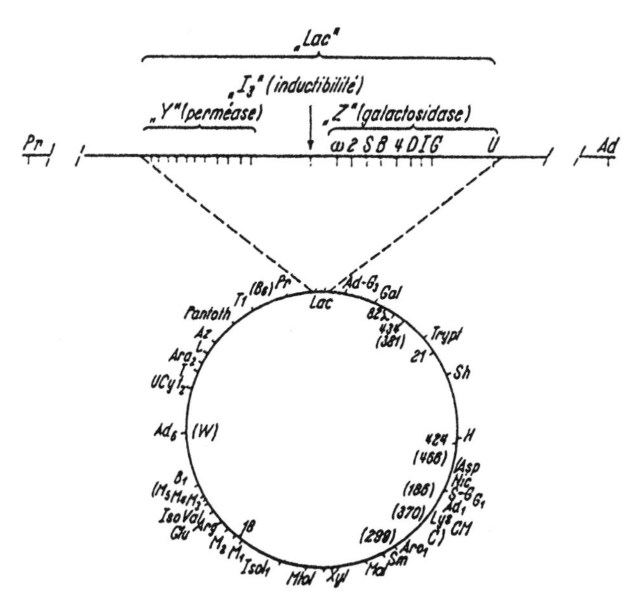

Figure 1. - ***Structure du segment Lac du chromosome d'Escherichia coli.***
A la partie supérieure de la figure: représentation schématique agrandie du segment Lac. Le cercle de la partie inférieure de la figure représente l'ensemble du groupe de liaison d'*Escherichia coli* (cf. JACOB et WOLLMAN, 1958 b) et situe le segment Lac par rapport aux autres marqueurs connus

FIGURE 26. Jacques Monod, "Information, induction, répression dans la biosynthèse d'un enzyme," *Colloquium der Geselschaft für Physiologische Chemie* (9/12 April 1959): 126. Structure of the Lac segment of the E. coli chromosome. Reprinted by permission of Springer Verlag.

protein galactosidase, while the locus i determines under which conditions this information is possibly transferred to the cytoplasm." The transfer was controlled via a "cytoplasmic message emitted by gene i and which determines either the inducible or constitutive synthesis by gene z." There were therefore two kinds of genes: one, *le gène informateur* (e.g., z and y), which contained structural information for making specific protein molecules (e.g., galactosidase and permease); the other, *le gène regulateur*, which determined the synthesis of a specific repressor inhibiting the expression of the "gene informateur."[53]

Monod pointed out that protein synthesis was determined not only by one gene—as the old "one gene-one enzyme" rule had decreed—but also by other genes and, more generally, by genetic and biochemical "*context*," which carried profound implications for physiology. For, he reminded his audience, it was the phenomenon of enzymatic adaptation in microorganisms that would ultimately furnish the explanatory models for embryologists in their studies of differentiation.[54] Returning, at least for the occasion, to questions of development that had engaged him in the mid-1940s, he now envisioned development as genetically determined, defined purely by the "context" of the system, in which the flow of information was regulated through sequential processes of activation and repression, a switching mechanism impervious to external experience. In Monod's Platonic gaze, cellular information was inborn, finite, and eternal.

Unlike many molecular biologists, Monod was not a naive participant in the information discourse. Fully aware of the general features of the mathematical theory of communication—its scope and limits—he knew within that domain information was purely syntactic and devoid of semantic value. His frequent and intense discussions with Szilard had undoubtedly forced him to critically ponder the ideas of Norbert Wiener, Claude Shannon, and Leon Brillouin (Szilard's friend who coined the term *information value*); Szilard, though au courant in communication sciences, and perhaps for that reason, did not seem tempted to cast his own work in cybernetic or informational terms. Given the strictures of information theory, Monod grappled with the problems of applying information theory to biology. "Trying to define the value of information by its 'transmissibility,'" he jotted down in his notebook in May 1959.

> From the point of view of the theory of information, the works of Shakespeare, with the same number of letters and signs aligned at random by a monkey, would have the same value. It is this lack of definition of the value of information that makes it difficult to use in biology. What could be considered as "objective" in the Shakespearean information that would distinguish it from the monkey's information? Essentially the transmissibility. The value of influence, [is] therefore of evolution. The biological information is such that its transmission contributes to its sustainability—the sustainability of a system that transmits it. Therefore, its information that contributes to the reproduction of the system, and consequently, to its own reproduction. Could we find a quantitative definition of this notion?[55]

The answer, at least for now was "no." (See Manfred Eigen's treatment of this issue, below in Chapter 7.) Monod was using "information" qualitatively, namely, the tropes and semiotics of the multivalent information discourse. As such, they were not merely culturally resonant but also scientifically productive. As immunologist Peter Medawar observed, "The ideas and terminology of information theory would not have caught on as they have done unless they were serving some very useful purpose." As he saw it, the qualitative concepts of information theory have been productive in representing biological functions, such as transmission and modification (mutations) of messages across space and time; even if the physical and mathematical analyses of information as a measure of biological organization (e.g., negentropy and Henry Quastler's calculations) have not.[56]

The productivity of metaphors in science is measured by the utility of the models they generate. But the process of "borrowing" is complex. Monod was engaged in ordering not only the elements of his own cellular universe but also the elements of the wired universe of communication and control, transported into his (and Jacob's) experimental space through cybernetic analogies and informational models. In that dialectical process the objects in both universes—bacteria, phages, information, negative feedback, circuits, switches—were reconfigured and transformed. As George Canguilhem, who in 1961 examined the relations between mechanical and biological feedback, put it, "A model only becomes fertile by its own impoverishment. It must lose some of its own specific singularity to enter with the corresponding object into a new generalization." In biology, he observed, analogical models have been more frequent than mathematical models. And they have been more fruitful in the study of function (e.g., genetics) than in investigations of structure and the relation of structure to function (e.g., biochemistry).[57]

The cascade of experimental results and their teleological (cybernetic-informational) interpretations assumed the epic proportions of a religious conversion, which Monod passionately expressed in his personal notebook in May 1959. Having for years demonized the destructive powers of final causes, and having condemned Aristotelian, Lamarckian, and Lysenkoist invocations of teleology in biology, Monod now recanted. Confessing his shortsightedness, he now admitted to the centrality and explanatory power of "finalism" in biology. Under the heading of "The discovery of repression," Monod declared that induction as antirepression had taught him a profound lesson in the form of a decisive reevaluation. "The belief in antifinalism has dominated, guided, and why? It has hidden, misled my work," he confessed. "I fought final causes, stressed the role of chance, and steadfastly rejected explanations of processes governed by an end-point," he recounted. "Life is a chance that becomes necessity. It takes a scientist thirty years to understand and accept that necessity does not exclude chance." His epiphany would soon be formalized through the new evolutionary construct of "teleonomy" (1958): chance mediated by the necessity (goal-directedness) of a computerlike "genetic program."[58]

By that time, the PaJaMa paper, "The Genetic Control and Cytoplasmic Expression of 'Inducibility' in the Synthesis of β-Galactosidase by E. coli," had just come out in the second issue of the *Journal of Molecular Biology*. With John Kendrew, British crystallographer and champion of molecular biology, as its editor-in-chief, the new journal formalized the coalescence of a distinct (if hybrid) disciplinary identity. Deliberately casting their work as "molecular biology," the Pasteur group was consciously participating in that disciplinary consolidation, including its cleavage from medicine and its alliance with the Rockefeller Foundation's policies in molecular biology. The discursive demarcation corresponded to potent institutional strategies. As Jean-Paul Gaudillière has recounted, in 1959 the Pasteur group was involved in discussions with United States Senator Hubert Humphrey to establish a European Molecular Biology Organization (EMBO) within his proposed scientific "Marshall Plan." Though the Humphrey bill never reached Congress, the buildup of European molecular biology intensified. In politics that would set the trend for the 1960s decade, the first De Gaulle government (the Fifth Republic) agreed to boost state funding for scientific research and development. Through the Pasteur influence, the newly established Délégation Général à la Recherche Scientifique et Technique (DGRST)—a cold-war institution modeled after a military mobilization of science—targeted "molecular biology" as the spearhead of future science and biotechnology.[59]

The PaJaMa article quickly became a classic. Its technical and conceptual intricacies, delivered in an elegant expository style, culminated in the authors' decisive conclusions: enzymatic synthesis in E. coli was controlled by three extremely closely linked genes (cistrons) z, i, and y; the i and z factors involved a specific cytoplasmic messenger (i sending the message picked up by z, the region of structural information), although the mode of action of that substance was yet unclear; and the formation of these sequential enzyme systems (analogous to zygotic induction in phage) was repressed by their end product.[60] For many readers outside the Pasteur network these results and interpretations must have been startling, though similar findings were quietly obtained at NIH. But by that spring of 1959, Jacob and Monod had already moved to a higher level of generalization: the operon.

Having established that the transfer of information from structural genes was controlled by specific repressors synthesized by specialized regulator genes, Jacob and Monod now tackled the next problem: the site and mode of the repressor's action. The specificity of the repressor's operation implied it acted by forming a stereospecific combination with some element of the system—somewhere along the i region—possessing the complementary molecular configuration. They reasoned that the flow of information from gene to protein had to be interrupted when that control element, which they called the "operator," was combined with the repressor. The existence of the operator was taken for granted; the problem was to figure out where and how it intervened in the system of information transfer.[61] The regulatory i region became associated with the "O," or operator gene, and the i,z,y segment became the "ozy unit" of information transfer. In 1960 Monod and Jacob

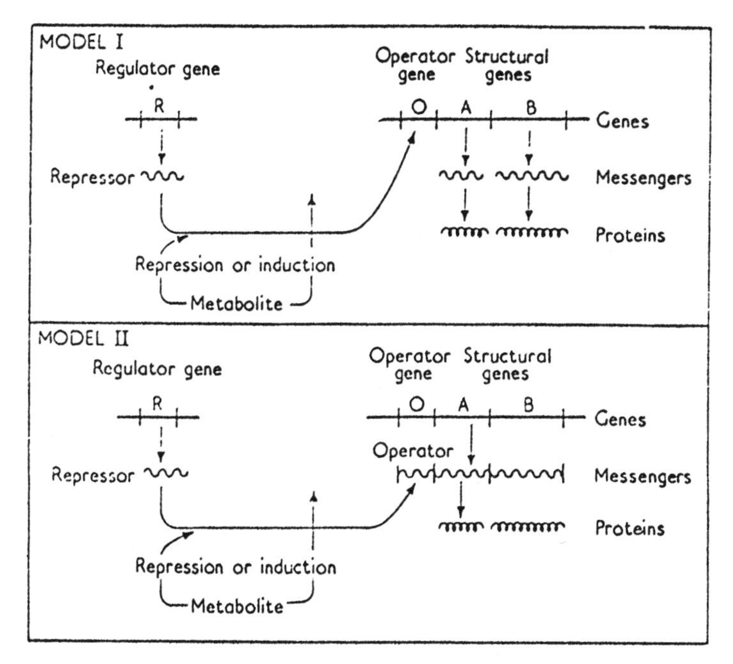

Models of the regulation of protein synthesis.

FIGURE 27. François Jacob and Jacques Monod, "Genetic Regulatory Mechanisms in the Synthesis of Proteins," *Journal Molecular Biology* 3 (1961): 318–59. Reprinted by permission.

christened this genetic "unit of coordinated expression"—an idea informed by Bruce Ames's work at NIH on enzymatic repression—the "operon." Beyond quaint images of genetic blueprints, the operon was visualized as a computerlike "coordinated program of protein synthesis and the means of controlling its execution."[62] (See Fig. 27.)

As the pieces rushed into place, the main remaining snag in the elegant operon model was the cytoplasmic "messenger," that theoretically constructed intermediate, whose function now linked the operon directly to the problem of the genetic code.

TRANSCRIBING THE WORD: THE MESSENGER

The Pasteur group had been following the work on the genetic code through the literature and Szilard's reports. And they first met Francis Crick in the spring of 1955, when he offered to give a seminar at the institute; as Crick put it, his preoccupations with protein synthesis linked up with their work. "For some time now," Crick wrote to Monod, "I have held the view that induction was due to a change in the folding up of the protein *during its synthesis* . . . and I notice that your own idea must be along similar lines, to judge from a reference to antibody synthesis in your last paper."[63] Crick could not manage a talk in French; "my French consists mainly of reproaches to children, and in the second person!" he apologized (his wife was French). He recalled presenting "the quite erroneous [instructional] theory" of an inducer-driven protein folding. But for Jacob, taken with that "formidable intellectual machine," the presentation proved "Francis was not simply Jim's appendage," as the Pasteur group had previously imagined. Crick's subsequent French presentation (his wife's translation) on the genetic code to the Parisian Physiology Club was less spectacular (he belatedly learned, for example, that the French scientific term for "overlapping" (code) was "oh-ver-lap-pang").[64] The interactions of the Pasteur and Cambridge groups became frequent and intense in the late 1950s, especially between Jacob and Sydney Brenner. "Seymour [Benzer] and I have done proflavine on rII and Francis and I have a new code," Brenner informed Jacob in the winter of 1958 (referring to the intricate recombination experiments on the rII region of the phage genome treated with the mutagenic acridine dye proflavine, which deletes DNA bases), "but since they will both be coming to talk to you I shall not steal their thunder."[65]

Often exchanging mutant stocks, Jacob and Brenner followed each other's progress closely. As Brenner and Crick raced to deduce the code from exhaustive sets of genetic changes engineered through phage matings—matings that earned Brenner the attribution "sex maniac"—Jacob was trying to identify the nature of "nonsense" mutations (where a DNA triplet did not specify an amino acid, or protein synthesis was repressed) and of the switching mechanism in the operon; all of which linked to the postulated cytoplasmic "messenger." "I understand you are now racing with Cy Levinthal's team for breaking the code and I'll be very interested to learn the news from your O mutants," Jacob wrote to Brenner in April 1959.

> I am interested by the sequential switch of the phage protein. . . . Now, since nuclear replication as well as protein synthesis is strictly controlled by the cell, there must be a system which can tell to the DNA molecule (and to the RNA molecule): you do your job or you don't do it. Hence the idea that there must exist specialized sites acting as "switches" for the whole molecule (of DNA and/or RNA) which receives the cytoplasmic messenger (inducers or repressors) and transmits it to the molecule under the form of "yes" or "no," "work" or "do not work."

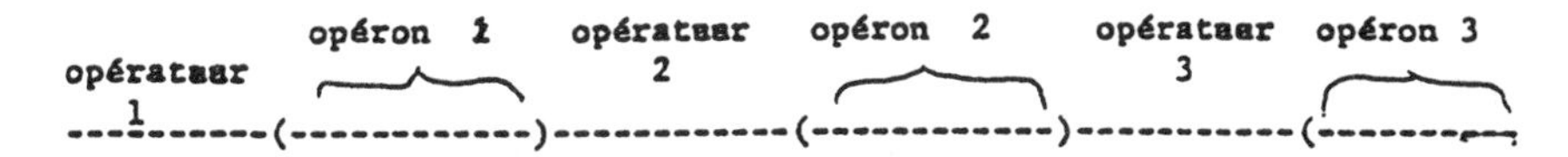

FIGURE 28. Jacques Monod Papers, MS (Mon.MSS.01 #17), "Operon." Monod's model for the transcription of messenger RNA. Courtesy of SAIP, Font Monod; © Institut Pasteur.

This association of the messenger with an inducer or repressor was obviously quite general; the entity was not yet the unstable RNA species that would later become messenger RNA. But this model confirmed for Jacob the selective control operating via linked genes in the biochemical system of both enzyme and phage production,[66] control of genetic information governed by the rules of a code. As Monod explained a couple of months later in his chapter on genetic determinism (Chapter 6) in *Cybernétique Enzymatique*, "A given gene contains, in a form of a particular nucleotide sequence, the information relative to the amino acid sequence in a particular protein. The knowledge from the exceedingly interesting speculations has led to proposing the principle of a cipher or code which would permit the 'translation' of a nucleotide sequence to an amino acid sequence."[67] Crick's (and Delbrück's) elegant theoretical models of commaless codes were then beginning to be replaced by notions of initiation and termination codons as defining the sequential "reading" of the DNA "text." This model, with its textual significations, became central to visualizing the function of the messenger in the operon system. Monod wrote in notes pertaining to the operon model, "The existence of the operon as the main unit of genetic expression *transcribed* by a messenger integral to understanding the information relative to diverse peptide chains poses an important problem which we wish to discuss here: that of the "*punctuation*" of the linear chemical *text* which constitutes the DNA [my emphasis]." Viewing the messenger as mediator in a "system of translation," he reasoned that certain segments of the sequence defining the beginning and ending of the operon corresponded to the points of initiation and termination of the transcription of DNA to an RNA messenger. He thus modified the extant punctuation signs, introducing a "parenthesis" to replace the two commas demarcating an operon (see Fig. 28).[68]

Jacob and Monod estimated in their operon paper, whose publication in the *Journal of Molecular Biology* was delayed despite Jacob's proddings: "If we assume that the message is polyribonucleotide and a coding ratio of 3, the "unit message" corresponding to an operon governing the synthesis of the three proteins of average molecular weight 60,000 would have a molecular weight about 1.8×10^6." By the time the paper appeared, the vague notion of a cytoplasmic messenger had become far more specific. After prolonged resistance, Monod was ready to accept the conclusions of Brachet and the Rouge Cloître group that this intermediary was, in fact, an RNA. Although already in 1958, as Thieffry has shown, Raymond Jeener had argued that it was an RNA with a rapid turnover, it took the Pasteur group about two years to reach a similar conclusion. Decisive experimental evidence pointed to the fleeting nature of the messenger and soon linked it to researchers' earlier work.[69]

The switch to a view of messenger RNA as an unstable intermediate came from several directions. The PaJaMa and subsequent experiments suggested that, once transferred into a bacterium and before the occurrence of genetic recombination, the gene controlling the structure of a protein began to function without delay, producing protein at a maximal rate. These were startling results, inconsistent with the prevailing notions that gene-directed protein synthesis occurred on the enduring templates of ribosomal RNA. To test the new model, Jacob reflected, one could withdraw a gene from a bacterium and search for residues of proteins synthesized off the stable RNA template. But such genetic extraction was technically forbidding. Alternative experimental design entailed the transfer of a gene segment, heavily labeled with radioisotope (^{32}P) and the subsequent destruction of that gene by radioactive decay.[70]

While Jacob and Monod were fine-tuning their operon model, Arthur Pardee and his graduate student Monica Riley performed this technical feat at Berkeley, presented as a four-way transcontinental collaboration. Their results demonstrated unambiguously that capacity to produce the protein does not survive destruction (radioactive decay) of the gene. This finding produced evidence that gene expression did not proceed through formation of stable templates; there had to be a rapid RNA turnover. This was a turning point for the general concept of a messenger to a specific characterization of its fleeting nature. Concurrently, François Gros in Monod's laboratory, a recent convert from older studies of RNA metabolism to the new analyses of RNA synthesis, added decisive evidence. He had developed in 1959 a new centrifugation technique for RNA fractionation and "short-pulse" methods geared toward identifying "transitory" fractions. Now his preliminary experiments showed that addition of the mutagen 5-fluorouracil to a bacterial medium resulted within minutes in the production of abnormal proteins, thus presenting another strong argument against the persistence of any stable templates.[71]

The interpretations of these experimental findings were constituted through discursive practices that represented the transfer of genetically determined structural specificity as an information transfer system. The new, third RNA species (still a theoretical construct), was represented as "information-bearing RNA," or "informational RNA." This designation was not merely a fashionable catchword. Operationally, it was used to distinguish this short-lived RNA species from the soluble RNA fraction (by now being renamed "transfer RNA"), which according to their calculations "appears much too small for it to carry all the information concerning a long polypeptide chain such as that of the monomer of β-galactosidase";[72] even in its qualitative metaphorical version as the circulating currency of structural (immobile) specificity, information served to "size" form. "Information-bearing RNA" was also distinguished from the stable ribosomal RNA fraction, which would soon be conceptualized as a mere translation machine, or the "reading head" for the "magnetic tape" of the genetic "program."[73]

The international race to isolate and characterize messenger RNA, an intricate technical challenge involving top laboratories and leading practitioners of molecular genetics, lasted a little more than a year. It was a fine balancing act of competition and cooperation, local scientific cultures and international networks, biochemical virtuosity and technical genetic savvy. It seemed that almost overnight everyone "remembered" the strange findings of Eliott Volkin and L. Astrachan from the biology division of Oak Ridge National Laboratory. Following up on Alfred Hershey's 1952 "neglected" observation that in (T_2) phage-infected bacteria RNA was synthesized at a rapid rate, in 1956 they reported the "rapid" RNA's base composition and showed it to be different from bacterial RNA and similar to phage DNA.[74] Their works circulated widely but did not readily integrate into any known models of RNA (Jeener drew on it in 1957). Not until the late summer of 1959 did the physical form of "Volkin's RNA," as it was sometimes called, become an object of analysis. In comparing it to the two known RNA species (soluble and ribosomal RNA), Sol Spiegelman and his collaborators at the University of Illinois immediately noted that none of the phage RNA was incorporated into the stable bacterial ribosomes (existing free in low magnesium concentration), a finding simultaneously corroborated by Watson's laboratory.[75]

A picture of phage protein synthesis was now coming into sharper focus. It appeared to take place on genetically nonspecific (bacterial) ribosomes to which the "informational RNA" template attached itself. However, isolation and characterization of the messenger remained ("Volkin's RNA"). To accomplish this technical feat, Jacob and Brenner planned to spend the month of June 1960 at Caltech's biology division, collaborating with their friend Matthew Meselson. His recent investigations of bacterial ribosomes and the precision technology he had developed—combining heavy radioactive labels with centrifugation methods (density-equilibrium gradients)—were crucial to their experimental plan. "This letter is to tell you about the exciting developments here and also to discuss the experiment that we should do in Pasadena. I don't know whether you know of the recent Monod-Jacob work, I will just mention the salient features," Brenner wrote to Meselson early in May 1960. After providing the general outline that led to the equation of messenger RNA with "Volkin's RNA," Brenner suggested a couple of experiments. One that involved determining the sequence of the head protein of phage as a way of correlating it with the RNA and thus breaking the code was already being done in Cambridge, he reported; the second was to be conducted in Pasadena. "It would help us if you had available all the density labels that we might need, heavy bases, heavy amino-acids, etc. You know exactly what we will have to use," Brenner added.[76]

Arguing that all the (T_2) phage messenger RNA should be attached to the "old" bacterial ribosomes (present before phage infection), the three raced to prove their claim within the allotted month in Pasadena. Jacob recounted, "We were to do very long, very arduous experiments. One part was carried out in Weigle's lab [Jean Weigle, Swiss physicist-turned-biologist]; another in the basement where centrifuges and Geiger counters were located. There were interminable dead periods while the centrifuge was spinning and the density gradient was forming." Meselson, haunted by the cold war and by the need to establish better relations with the Soviet Union, would discourse for hours on strategy, tactics, nuclear arms, the Rand Corporation, first strikes, reprisals, and annihilation. They would first phage-infect heavy-labeled bacteria (grown in a medium containing ^{13}C and ^{15}N) in light (^{12}C and ^{14}N) medium, then centrifuge the sample, collect the differentially sedimented RNA fractions, and search for the light radioactive peaks. After repeated failures (due to the wrong magnesium concentration), CsCl equilibrium centrifugation elegantly confirmed their argument—in the nick of time, literally the day before Brenner and Jacob's departure for Europe. Indeed, most of the labeled phage messenger RNA was in the fraction attached to the preexisting ribosomes; soon after, Brenner showed that the nascent viral protein was bound to the old ribosome as well.[77]

"For me it was more painstaking!" Gros recalled. Just when Jacob and Brenner had departed for Caltech, he left for Watson's laboratory at Harvard. "We were equally convinced that similar messenger RNA would be found in uninfected bacteria. Its demonstration then presented greater problems because of the simultaneous synthesis of ribosomal and soluble RNA," Watson recounted in his Nobel Lecture (1962). Using Gros's "pulse-label" technique, they decided to look for labeled messenger molecules in cells briefly exposed to radioactive RNA precursor. If infected cells behaved like uninfected ones, they reasoned, then during any short interval most RNA synthesis would be messenger, broken down as fast as it was made, with no significant accumulation. "The heat was suffocating, the laboratory glassware reduced to nil, the radioactivity counters old, enormous and noisy; many times experiments would trail along late into the night," Gros remembered. But their hypothesis was confirmed. The RNA labeled during pulse exposures was largely attached to the ribosomes, and base ratio analysis revealed that the RNA templates corresponded to the bacterial DNA template, an observation which was immediately elaborated in other laboratories, most notably by Spiegelman's group.[78]

Jacob and Brenner met several times and were in constant communication during the coming months of finalizing and polishing their work. Jacob prodded Brenner early in September:

> I fear that the whole story will be much weaker if we (I should say if you) cannot demonstrate that phage proteins are actually made on old ribosomes. . . . As you probably heard, Francois Gros and Jim have now found the messenger RNA by two methods [^{32}P and ^{14}C pulse-label]. . . . Apparently the whole US biochemists are now working on it. It is therefore worth not to delay too much our publications.[79]

Still trying to isolate nonsense mutants, Jacob now intensified his probings of the genetic code. He also joined forces with Gros to study the relation between messenger RNA and the repressor. With the "genetic transcript" in hand, a messenger translation system was being fashioned by Gros's wife, Françoise. She was working that summer of 1960 with Alfred Tissières at Harvard to develop an in vitro cell-free system for β-galactosidase synthesis. With François Gros as a collaborator and the principle investigator of the messenger work done in a competing laboratory, Jacob and Brenner were engaged in a delicate balancing act. In mid-February 1961 both laboratories were completing the write-up of their experiments (with detailed critical input from Meselson); both papers would be submitted to *Nature* to be printed together in the same issue. "This is by far the best solution and we should have done this before," Jacob informed Brenner.[80]

Both papers appeared in the May 1961 issue of *Nature*: Brenner et al. reported "An Unstable Intermediate Carrying Information from Genes to Ribosomes for Protein Synthesis" in phage-infected bacteria; Gros et al. analyzed the "Unstable Ribonucleic Acid Revealed by Pulse Labeling of Escherichia Coli" in uninfected cells. Both papers interpreted their findings within the framework of the information transfer system and designated the messenger as "information carrier" to distinguish it from noninformational RNA, notably the ribosomes. Completing the unidirectional model of information transfer enunciated by the Central Dogma and elaborated in the operon model, Brenner et al. concluded, "Although the details of the mechanism of information transfer by messenger are not clear, the experiments with phage-infected cells show unequivocally that information for protein synthesis cannot be encoded in the chemical sequence of the ribosomal RNA. Ribosomes are non-specialized structures which synthesize, at a given time, the protein dictated by the messenger they happen to contain."[81] Gros et al. confirmed, "Our working hypothesis is that no fundamental difference exists between protein synthesis in phage-infected and uninfected bacteria. In both cases typical ribosomal RNA does not carry genetic information, but has another function, perhaps to provide a stable surface on which transfer RNA's can bring their specific amino-acids to the messenger RNA template." Interesting evidence was also unearthed from an unexpected source: Martynas Yčas, pursuing the function of RNA in yeast, had identified a similar RNA species, which Brenner viewed as significant support for the universality of the code.[82]

Information, a privilege formerly reserved for the specificity of DNA and RNA, was now bestowed only on select types of nucleic acids: DNA and messenger RNA. Though Spiegelman, who had been conceptualizing his work on enzyme and RNA synthesis through cybernetic analogies, computer models, and tropes of information transfer for more than a decade, offered a more nuanced, if confusing, distinction. In elucidating the "relation of informational RNA to DNA," he offered a terminological primer. The terms "complementary" and "information" had "well-defined operational definitions," he insisted:

> A given RNA molecule is defined as falling within the informational class if its base ratio is homologous and its sequence is complementary to a specific DNA molecule. . . . Every "complementary" RNA is "informational" in at least one sense. Even if it is a complementary copy of a nonsense DNA sequence, it still contains the information necessary to specify the order of the bases. . . . It is important to emphasize that the word "informational" is not proposed as a substitute for the term "messenger" introduced in the elegant experimentations and theorizations of Jacob and Monod (1961). It seems that both terms will be useful. Thus, a given messenger RNA is presumed to constitute the structural *program* for the synthesis of a particular protein. It obviously must, therefore, be informational. However, not all informational RNA need serve a messenger function. It is conceivable, as is indeed implicit in the operon theory of

> Jacob and Monod (1961), that informational RNA molecules will be found which serve regulatory rather than programming functions [my emphasis].[83]

A noncorporeal quality inherent in the intelligence of the programmed system, biological information served to demarcate the RNA transcribed off the gene from mere brute matter.

With their messenger work completed and written up, Jacob began to use novel strategies for breaking the code. Indeed the existence of messenger RNA completely revolutionized the "decoding" approach. Instead of the indirect black-box approach of matching nucleotides with amino acids (including the laborious amino acid substitution method in TMV) one could use a direct biochemical approach: the messenger, as the DNA transcript, could be used in the E. coli cell-free system to produce a direct translation into proteins; it could be the key to the code. There was excitement in the air. A new experimental approach to decoding had just opened up at MIT and the Pasteur Institute by the sitings of genetic transfers between E. coli and Serratia bacteria (a well-researched microorganism at the Pasteur). Preliminary studies by both groups suggested that the E. coli genes were transcribed correctly in Serratia. This seemed to indicate that the 20 percent difference in the base ratio G+C/A+T between the two genera was not due to different codes. Whatever the form of the code, this evidence suggested it was universal. Jacob reported to Brenner in March 1961:

> It is already a few months that I am fed up with messenger. What I am mostly interested in now is to try to show that it is actually the synthesis of the messenger which is actually switched on or off by the inducers and repressors. . . . The other way we want to use is with Serratia strains and we hope to be able to distinguish between coli and Serratia messengers. The Serratia system may be very interesting.

The genetic mechanisms of interbacterial transfer "would favor a degenerate binary code," he thought, in support of Sinsheimer's informationally efficient two-letter code.[84] Soon they were coordinating travel plans for the Cold Spring Harbor summer symposium, where the Pasteur contributions would occupy center stage.

The symposium "Cellular Regulatory Mechanisms" (sponsored by NIH, U.S. Public Health Service, National Science Foundation, Rockefeller Foundation, AEC, U.S. Air Force, and monitored by the Air Force Office of Scientific Research of the Air Research and Development Command) was a memorable event, with nearly two hundred researchers attending and forty presentations delivered. It was a milestone in the history of molecular biol-

FIGURE 29. Leo Szilard (*far left*), Jacques Monod (*left, facing forward*), and François Jacob (*right, facing forward*) at the 1961 Cold Spring Harbor Symposium. SAIP, Fond Monod; courtesy of the archive of the Institut Pasteur.

ogy, enshrining the phenomenological trinity of protein synthesis as a scriptural operation: DNA replication; RNA transcription; and protein translation. The opening address by Harvard microbiologist Bernard Davis, entitled "The Teleonomic Significance of Biosynthetic Control Mechanisms," left no doubt that the cellular machinery had been recast as a cybernetic communication system. Davis began his introduction:

> But now that many biosynthetic pathways have become more or less completely known, it has become possible not only to describe flow rates but to analyse in detail the mechanisms that control them. And as is reflected in the widespread use of the term "feedback," such studies of cellular regulatory mechanisms have been influenced to some extent by concepts that have developed in communication engineering.

He then proceeded to paint a masterful canvass of the topics to be covered in the next few days.[85] News from the decoding front was presented in the session on the "Role of DNA in Protein Synthesis." There were two small sessions on the "Control of Nucleic Acid Synthesis" and on the "Role of DNA in RNA Synthesis," where Spiegelman presented his new findings on "informational RNA." Brenner and Gros announced the birth of the messenger in the session on the "Role of RNA in Protein Synthesis." The enormous session on "Regulation of Enzyme Synthesis" paraded the recent findings on repression and derepression, with Jacob and Monod's paper as a centerpiece, followed by the sessions "Control of Enzyme Activity: Competition Between Enzymes," "Control of Enzyme Activity: Feedback Control," and "Control of Enzyme Formation and Activity in Animal Systems." Jacob and Monod presented the symposium's grand finale, entitled "Teleonomic Mechanisms in Cellular Metabolism, Growth, and Differentiation."

That neither they nor Bernard Davis had stopped to define or explain the notion of teleonomy—the goal-directed process of ontogeny determined by a genetic program—suggests the concept of teleonomy had already been assimilated in this scientific community. Monod and Jacob's subsequent representations were more graphic, visualizing the "cell as a society of macromolecules, bound together by a complex system of communications regulating both their synthesis and activity"; the basic regulatory elements of the system were "like basic elements of electronic engineering [which] can be organized into a variety of circuits fulfilling a variety of purposes": they thus depicted a finely tuned system of information and control.[86] Although only a couple of decades separated these representations of life as communication from the organizational metaphors of Paul Weiss, Walter Cannon, and even the young Monod, they belonged to a new world picture, to a postwar technoculture marked by the discourse of information and the nascent computer age.

Mindful of the dominance of research on microorganisms and of larger biological questions, they raised "A constantly recurring question":

> To what extent are the mechanisms found to operate in bacteria also present in tissues of higher organisms; what functions may such mechanisms perform in this different context; and may the new concepts and experimental approaches derived from the study of microorganisms be transferred to the analysis and interpretation of the more complex controls involved in the functioning and differentiation of tissue cells?

Following a telescopic display of experimental strands and pithy theoretical generalization of the teleonomic cellular mechanisms, they returned to that question. The answer hinged on the nature of the genetic code governing the information transfer from DNA to enzyme synthesizing systems. The unity of cellular mechanisms across species divides would hold if the code were shown to be universal, they argued.[87]

Proceeding to outline up-to-the-minute advances in the decoding competition, they cited the colinearity matches, chemical mutagenesis, reverse mutations, and "direct chemical attack, involving the determination of partial sequences in both a protein and the corresponding messenger RNA, [which] may become possible, assuming the mRNA theory to be correct, if and when methods of isolating a specific message will be available." Whether in the realm of fine-tuning cell-free translation systems for messenger RNA, or in the intricacies of phage-recombination analyses, the code work was moving forward surely but very slowly. They forecasted in the notorious statement, which has since become emblematic for the disciplinary monopoly of molecular biology over the "secret of life": "If the codes in Serratia and Escherichia and perhaps a few other bacterial genera turn out to be the same [referring to the recent work at MIT and Pasteur], the microbial-chemical-geneticists will be satisfied that it is indeed universal, by virtue of the well-known axiom that anything found to be true of E. coli must also be true of Elephants."[88] The usually effusive biochemist Gordon N. Tomkins from the NIH sat in tortuous silence through the presentation. Just three weeks earlier Marshall Nirenberg and Heinrich Matthaei in his department had "cracked the code." However, not a member of the inner circle gathered at Cold Spring Harbor, Nirenberg had been denied admission to the meeting. The news would spread like brushfire following Nirenberg's presentation of his findings at the Fifth International Congress in Moscow in August 1961, thereby ushering in the second, mostly biochemical, phase of genetic decoding.

CHAPTER SIX

Matter of Information: Writing Genetic Codes in the 1960s

IN SEARCH OF GENETIC INFORMATION IN PROTEIN SYNTHESIZING SYSTEMS

When Marshall W. Nirenberg (b. 1927) arrived at the National Institutes of Health (NIH) in Bethesda in the summer of 1957 as a postdoctoral fellow of the American Cancer Society, NIH, the principal medical research arm of the Public Health Service (PHS), was emerging as a major player in the biomedical sciences. Since the PHS's reorganization and expansion (initiated by the Public Health Service Act of 1944) its annual appropriations quadrupled to $840 million by the end of the Eisenhower administration. Its mandates reflected the changing political climates, from the social activism of the New Deal and pragmatism of war, to the political conservatism of the McCarthy era and the Korean War. National health insurance, a former PHS commitment, was dead, in part because the American Medical Association (AMA) intensified its campaign against "socialized medicine," and McCarthyism swept the PHS with loyalty investigations, dismissals, and blockage of extramural grants. In this political environment the emphasis shifted from medical care to the politically more neutral pursuit of medical research. When in 1953 Eisenhower elevated the Federal Security Agency to cabinet status as the Department of Health, Education, and Welfare (HEW), the PHS became part of HEW, and its research flourished within the expanding new institutes of NIH.[1]

As a major patron of biomedical research, NIH was increasingly displacing the dominance of military support. In 1955, out of a total of $2,744.7 million in government spending for scientific research and development, the Department of Defense (DOD) had obligated $2,084.2 million and the Atomic Energy Commission (AEC) another $372.9 million, while HEW (including NIH) squeezed its science out of a budget of $70.9 million and the National Science Foundation (NSF) only $10.3 million. Approximately 45 percent of government support in 1955 for the biomedical sciences came from the DOD and AEC (down from 60 percent at the beginning of the decade). By

1957, even though direct military support for the biomedical sciences streamlined at 19 percent, NIH had entered a period of unprecedented growth, driven by the general boost to science and technology generated by the space program, which had been itself expanded preeminently by the traumatized response to the Soviet's launching of *Sputnik I* on October 4, 1957. Each year from 1957 to 1963 the NIH budget increased by an average of 40 percent annually; appropriations grew from $98 million in 1956 to $930 million in 1963, with a twelvefold expansion in grants for extramural research. From a handful of buildings in the 1940s, by the early 1960s NIH sprawled to fifty buildings housing 13,000 people.[2]

Biochemistry—notably protein research, enzymology, and nucleic acids metabolism and synthesis studies—occupied a privileged place at NIH in the 1950s, increasingly featuring young talent, such as Arthur Kornberg, Alexander Rich, Bruce Ames, Maxine Singer, Marshall Nirenberg, and Philip Leder. From several testimonies, NIH appears as a crucible of postwar biological sciences, from which numerous luminous careers in molecular biology were forged. For Lewis Thomas, "All by itself, this magnificent institution stands as the most brilliant social invention of the twentieth century, anywhere." Nobel laureate Kornberg said, "More than any university, NIH is my alma mater." Like other NIH alumni, Kornberg praised the "untrammeled atmosphere of well-equipped, well-managed laboratories, [where] young MDs and PhDs were introduced to professional science." Many remained at NIH, but more than 25,000 researchers went on to leadership positions in basic and clinical biomedicine in the United States and abroad. Nirenberg was among those who stayed on, despite the offers he received from leading institutions.[3]

With cultlike reverence reminiscent of that professed by pilgrims to the Pasteur Institute, Cambridge (England), or Caltech, and perhaps in reaction to their deifications, NIH biochemists celebrated their own scientific culture and heroes. Just around the corner from Nirenberg's laboratory, Leon Heppel—"intense, excitable, birdlike"—pioneered world-class approaches to polynucleotide synthesis. Only his enthusiasm for life exceeded his delight in research. Biochemist Robert G. Martin (Heppel's protégé and participant in the code work) recalls:

> Leon set the tone for the lab: careful, meticulous, childish, and screwy, but always calculatedly so. A member of the NIH staff since 1942, Leon had done much to clarify the structure and metabolism of nucleic acids—work that was fundamental to Arthur Kornberg's discovery of an enzyme that synthesizes DNA [DNA polymerase]. It was characteristic of him to keep a freezer full of polynucleotides he had synthesized over the years [polynucleotides that would "break the code"]—a freezer he never failed to inspect daily.[4]

The laboratories of Heppel at NIH and Severo Ochoa, an invited collaborator at New York University, became principal centers for polynucleotide synthesis in the late 1950s. Biochemist Gordon Tomkins was equally influential. Actually, "Gordon was an MD, not a biochemist, but he learned fast," Nirenberg recalled. After serving as section chief in Heppel's laboratory (succeeding Herman Kalckar), he had been appointed a laboratory chief in the Arthritis Institute just prior to Nirenberg's arrival at NIH. "Tomkins was something special," broadly educated, scientifically versatile, an accomplished jazz musician; he was an intellectual with infectious enthusiasm. As with Max Delbrück's phage cult, "A whole generation of scientists has adopted his [Tomkins's] speech and mannerism."[5]

Like most biochemists, NIH biochemists too harbored mistrust for the fast tricks of molecular biology, its glamour, and its high priests, Jacques Monod, François Jacob, Francis Crick, James Watson, and Sydney Brenner. "Not that there is anything inherently wrong with the techniques of molecular biology. Rather, the impetuous and creative new breed of scientists often went off half-cocked on inadequate data."[6] In a tone reminiscent of Erwin Chargaff's admonitions of Watson and Crick, Seymour Cohen's resentments of the phage group, and of Arthur Pardee's bitterness toward Monod, NIH biochemists had their own gripes. They remembered Monod's 1958 visit to NIH (part of his grand tour of the United States), when Bruce Ames, who had recently perceived histidine biosynthesis as the activation (derepression) of coordinated genes, explained his ideas to Monod, thus both preempting and inspiring the operon theory. Some also recalled Monod's vehement insistence on generalizing his findings to all enzymatic and bacterial systems as well as his abrasive dismissals of dissenters from the dogma of negative feedback—those who were not members of the inner sanctum but whose work was backed by sound evidence. Martin even suggests that the remarkable achievements of the Pasteur group derived to a large extent from young American talent. "And with all due respect to the French scientific ambience," he wrote, "it is hard to imagine failure with the talents of Ames, Beckwith, Cohen, Hogness, Pardee, Miller, Stent, Tomkins, Yarmolinsky, and many others."[7] His was another perspective on the French origins of molecular biology.

Yet these local and global tensions between molecular biology and biochemistry are better understood as a reflection of the long-standing alienation of biochemistry from classical and molecular genetics, younger fields which generally did not share biochemistry's medical lineage, its disciplinary and material culture, and its discursive practices. Indeed, like the Rockefeller Institute, which had provided inspiration and a model, the biomedical sciences at NIH had not included any genetics research—no *Drosophila* studies, no microbial genetics, no phage research. As Martin admitted, "The classical biochemists had an ingrained distrust of the data that the geneticists insisted were compelling. It was hard for them to accept proofs that involved the abstractions of microbial genetics." It is therefore all the more remarkable that, in that environment, two of the memorable courses of the "night school"—a "continuing education" program by and for NIH researchers—were Robert DeMars's course on phage genetics and Bruce Ames's course in biochemical genetics. DeMars's phage course left a deep impression on Nirenberg. It opened up new ways of thinking about cellular regulation and exposed him to genetic modes of reasoning and of experimenting with enzyme systems. It also led him to conceptualize the genetic material as matter of information, which was an increasingly visible trend among biochemists in the late 1950s.[8]

This genetic sensibility and informational thinking did not originate from his academic training—undergraduate studies in zoology and master's work on the ecology of caddis flies, both at the University of Florida. Nirenberg's graduate work in biochemistry at the University of Michigan, on the enzyme permease for the transport of the sugar hexose in ascites tumor cells, placed him squarely within the metabolic biochemistry of cancer research far removed from molecular genetics. By choosing his postdoctoral training at NIH under De Witt Stetten, Jr. (intermural director of the National Institute of Arthritis and Metabolic Diseases) and biochemist William Jakoby, Nirenberg was deepening that traditional disciplinary orientation, which was further extended through a subsequent PHS fellowship.[9]

His joint papers with Jakoby (1959–60) were meticulous displays of classical enzymology, while as is generally the case with published works, the author's scientific musings, personal speculations, and meandering thought tracks have been erased. In contrast to the published accounts, Nirenberg's "behind the scenes" preoccupations convey a fascination with the role of genetic control and notions of "information flow" in biochemical reactions and cellular functions. His professional diaries are filled with analyses informed by molecular genetics. One can catch a glimpse of this in Nirenberg's short but remarkable published communication, "The Induction of Two Similar Enzymes by One Inducer: A Test Case for Shared Genetic Information," in which—like Monod's group—he represents enzyme systems in a new light, reconfiguring them through the technoepistemic reasoning of genetic regulation and the discourse of information. Although Nirenberg was not particularly self-aware of his usage of the "information" idiom (only of its trendiness), "there are no innocent names," as Hans-Jörg Rheinberger has shown; names do matter. As in other contemporaneous instances in biochemistry (e.g., Heinz Fraenkel-Conrat's studies of tobacco mosaic virus, TMV; Severo

Ochoa's investigations of enzymatic mechanisms; and Paul Zamecnik's work on transfer RNA) information functioned as the Derridian *supplement*, a tag-along term which, by smuggling in genetic representations—modes of thinking and modes of doing—eventually necessitated the reconfiguration of the entire representational space and discourse of enzyme regulation and protein synthesis.[10]

In a biomedical culture that valorized team science Nirenberg worked largely alone, quietly and intensely. In the 1968 Nobel Prize profile in the *New York Times* (a prize Nirenberg shared with Robert W. Holley and H. Gobind Khorana) Nirenberg was described as a "genius because he does one thing superlatively well, but he has trouble driving cars and he has been known to trip over his feet. . . . Works 12 hours a day 7 days a week and has no outside hobbies."[11] He has also been perceived as a straight and narrow biochemist devoid of biological sensibilities and theoretical insights, his accomplishments thereby reduced to a matter of cookbookery and luck. But his work diaries tell otherwise. From detailed entries—"things to read" and "things to do"—written nearly daily over the three-year period preceding the sensational experiments that "broke the code," Nirenberg emerges as widely curious and intensely driven; pages are peppered with self-prodding admonitions to work harder, know more, and think faster. His distant fascination with the "code of life" belonged to a relentless search canvassing some of the key problems in cellular and molecular biology.[12]

While in the first few months at NIH his conceptualization of enzymatic activation of glycogen phosphorylase was indeed purely biochemical, by the end of 1957, he had begun to raise questions in his diaries about the transforming role of DNA in ascites tumor cells. He wrote, adding to his intrigue with DNA, "Areas I should learn something about," and included brain chemistry (a field he turned to immediately after solving the code problem), biochemical genetics, biochemical embryology, immunology, differentiation, dedifferentiation, and tissue recognition. There were not enough hours in the day to meet even a fraction of these goals. Below, under a heading "Overall Philosophy," Nirenberg conveyed a concern over the disparity between visions and realities. Working alone, he felt, retarded the realization of his potential for tackling the big questions in biology. He mused:

> I am primarily interested in looking for problems—devising questions and experimental approaches to answers—than turning out volume research. Would like to spend my time almost entirely on *nonproductive* nonpublishable work. After a system is going well, turn it over to someone else. . . . Personal ambition should not mean anything to me. In fact—it does mean something to me, for I feel that I can do *more* work if I have a job where 1, 2 or 3 people will work with me. I cannot approach my potential by working alone.[13]

By the fall of 1958, though still working alone on glycogen phosphorylase, his involvement with phage and biochemical genetics intensified, undoubtedly nurtured by the presence of DeMars and Ames. His notes convey a close familiarity (even before the 1959 publication of the PaJaMa experiment) of the relation between phage genetics and inducible enzymes. Being a great admirer of the work at the Pasteur Institute, Nirenberg hoped, among his other career goals, to spend a year in Jacob's laboratory.[14]

His transition from biochemistry to molecular biology accelerated at NIH. A note, "Get Chemical Basis of Heredity" (referring to *The Chemical Basis of Heredity*, edited by William D. McElroy and Bentley Glass in 1957), signaled that transformation. McElroy and Glass's book, according to Robert Martin, had transcendent qualities: it became "the second generation's [of molecular biologists] New Testament." It included classic essays on cellular units of heredity, by George Beadle, Seymour Benzer, and David Nanney; on the role of nucleic acids in protein synthesis, by Sol Spiegelman and Henry Vogel; on viruses and genetics, by Roger Herriott, François Jacob and Elie Wollman, and Fraenkel-Conrat; on the structure of nucleic acids by Chargaff, Crick, Watson, and Alexander Rich; nucleotide synthesis by Kornberg, Ochoa and Heppel, Seymour Cohen, and Elliot Volkin and Lazarus Astrachan; on DNA replication and recombination by Max Delbrück and Gunther Stent. These scientists defined the vanguard of their respective research areas.[15] Like Monod, Nirenberg's intrigue with the genetic control of enzyme induction was closely linked to an avid interest in embryology and development and to generating ideas for experiments on fertilization, sexual development, and differentiation in different model organisms, from sea urchins to slime molds and frogs. Nevertheless, Nirenberg readily admitted that in casting about for an interesting problem, he "seemed to be wildly going around in circles."[16] By the end of 1958, inspired by Jacob's work, he speculated that their zygotic (bacteriophage) technology could generate a biological method for obtaining the specific nucleotide sequences of biologically specific genes (for example, the transforming gene for cancer) as a way of breaking the genetic code. In a crisp note below, in which he captured the technoscientific imaginary, he surmised, ". . . would not have to get polynucleotide synthesis very far to break the coding problem. Probably 30 nucleotides & equal number of AA [amino acids] would do it. *Could crack life's code!*"[17]

Yet the intensity of this insight did not lead to a single-minded pursuit; "the code" had only just begun to constitute itself as a concrete object within the material and semiotic space of biochemistry. For most biochemists "the code" still served primarily as shorthand for "protein synthesis." Like most biochemists, Nirenberg was then only vaguely aware of the RNA Tie Club and its theoretical and mathematical strategies of decoding. Rather, the code fermented in the back of his mind as he probed the relations between genes, nucleic acids, and proteins, formulating his queries within the field which he knew best: enzymology, though not the metabolic enzymology of yesteryear, but the new paradigm of enzyme regulation as a sequential, gene-controlled process of induction and repression. In fact, being acutely aware of the fierce competition by leading laboratories to solve the problem of protein synthesis, Nirenberg felt he stood no chance. He instructed himself in the spring of 1959, "Protein syn. will be over in 2–5 years. Prob. 2 years. After the correlations ['the code'] have been worked out one can ask the following questions . . . [genetic activation, enzyme induction, and lysogeny]. Do not compete on protein synthesis. . . . Use the time to plan expts to do when protein syn. breaks & be all set up physically & mentally to jump on it hard." He reiterated a couple of days later (as if to restrain himself), "My main aim is not to crack protein synthesis but to have everything ready to study enzyme induction."[18]

It was around that time, before attending a summer course on bacterial genetics at Cold Spring Harbor, that Nirenberg embarked on studying the cell-free synthesis of the enzyme penicillinase in B. cereus bacteria, an enzyme system the prominent British microbiologist Martin Pollock had studied since his stay at Monod's laboratory in 1953. It was an interesting system since, like enzymatic induction and repression of β-galactosidase in E. coli, inducible B. cereus strains produced penicillinase only in the presence of penicillin. But unlike the β-galactosidase system, it seemed improbable that constitutive mutants would produce penicillinase even in the absence of the (penicillin) inducer, since penicillin is a drug, a foreign substance that is not part of the metabolic repertoire of the bacteria (Pollock too disagreed with Monod's generalization from E. coli to all other bacteria). Pollock and his colleagues had published widely on the regulation of penicillinase synthesis in vivo, showing that it was a small enzyme (low molecular weight) lacking the amino acid cysteine, which proved to be a useful clue.[19]

Nirenberg therefore reasoned that it seemed likely he could selectively inhibit the synthesis of proteins requiring cysteine and simultaneously stimulate penicillinase synthesis in vitro by adding nucleic acid templates to cell extracts and then assaying for penicillinase activity (measuring the degradation of penicillin). Since one could measure biologically minute amounts of penicillin, the assay promised to be exceptionally sensitive. For nearly two years he studied the properties of the penicillinase system, determining the effects of nucleic acids and numerous other factors on the rate of cell-free protein synthesis and devising a highly sensitive assay for penicillinase. It was Nirenberg's "general philosophy in protein synthesis" not to leave a stone unturned; he resolved to "supplement with everything including the kitchen sink & try to get *new* protein syn. [synthesis]. Then and only then, run many controls. a) Also throw in Ochoa's enzyme. etc. [polyribonucleotide phosphorylase, isolated by Marianne Grunberg-Manago and Severo Ochoa in 1955; see below]." He followed closely the mounting evidence for enzymatic feedback inhibition in microorganisms, as indicated by the fact that the genetic approach to enzyme regulation of the Pasteur group was consistently present in his interpretive framework of the penillinase system. By the summer of 1959, traces of the recently published PaJaMa paper (postulating the existence of a "cytoplasmic messenger") were evident in his work.[20]

Racing against the clock, Nirenberg drove himself hard. "Life is so short —make every minute count in thought and deed. . . . If your brain works 2x as fast it is almost the same as leading a life 2x as long." Though he softened his single-mindedness with a reminder, "Give of yourself to others. Give people love & understanding, kindness and cheerfulness. . . . Give without the expectations of anything in return. Don't be so wrapped up in yourself." All the while he continued his musings about the relation between nucleic acids and protein synthesis through perspectives of phage technology, the logic of (the then unproven) gene-protein colinearity, the biochemical meaning of the position effect (relative changes in gene order), and cellular differentiation as the biochemical link between genotype and phenotype. He speculated, "When protein synthesis is cracked, could follow syn. of particular protein before & after diff. [differentiation]. Could test whether particular cystron [cistron, Benzer's smallest unit of genetic function] is present all the time. Might be tough to do if cystron [gene] is repressed."[21] It was around this time that he wrote his short communication to *Federation Proceedings*, conceptualizing enzyme systems in terms of shared genetic information.

Nirenberg's penicillinase experiments yielded only minute amounts of *de novo* enzyme synthesis, which required a more sensitive assay. He generated voluminous information on the system but, unlike Pollock (and most junior researchers), he never bothered to publish the inconclusive results. The precious time it would have taken to prepare manuscripts for publication was better spent on forging ahead, he felt, exhibiting a rather unorthodox strategy even by the standards of the time. By fall of 1959, with his PHS fellowship soon to expire, Nirenberg considered spending a year honing his molecular genetics techniques. Of several possible training grounds (the laboratories of Joshua Lederberg, Edward Tatum, Salvador Luria, and Jacob) he chose to apply to the Pasteur Institute but was turned down (as was his attempt to join Kornberg's laboratory). It was to Tomkins's credit and a testimony to NIH's enlightened approach to research (which Nirenberg has repeatedly acknowledged) that in 1960, despite the penicillinase publication blank, Nirenberg was offered a position as a research biochemist in Tomkins's Section of Metabolic Enzymes. There, Nirenberg gradually eased out the B. cereus system. His revised search for a protein synthesizing system designed to amplify genetic signals soon led him to E. coli, the principal experimental system in molecular biology. E. coli's cell-free extracts had less penicillinase activity than uninduced B. cereus extracts, which improved the sensitivity of the assay.[22]

As Nirenberg was all too painfully aware, cell-free protein synthesis systems had become an "industry" by the late 1950s. New findings were being reported monthly by various groups, notably by Paul Zamecnik's laboratory at the Collis P. Huntington Memorial Hospital of Harvard. As Hans-Jörg Rheinberger has detailed, Zamecnik's pursuit of a cell-free system grew out of cancer research, from his goal of elucidating the mechanism of protein synthesis in malignant cells. Using rat liver cells as the experimental system, those investigations had established by 1958 the presence and necessity of small RNA molecules in protein synthesis. Classified operationally as soluble RNA (sRNA), it was the fraction remaining in solution in the supernatant at pH 5 after the *microsomes*, the term for ribosomes with fragments of endoplasmic reticulum and related material attached, were centrifuged down for two hours at 100,000 g. These studies also demonstrated that there were many different sRNAs and that they combined enzymatically with activated amino acids—each amino acid had its own sRNA carrier—transporting them to the ribosomes, where they were strung together into polypeptide chains on the ribosomes' surface. Within a couple of years sRNA was resignified as transfer RNA, or tRNA.[23]

Much has been made of the hindsight reconstruction that these small sRNA molecules were quickly "recognized" as the "adaptors" postulated by Crick in his 1955 note to the RNA Tie Club in order to explain the process by which amino acids were lined along the nucleotide template (as I discussed in Chapter 3).[24] But such ex post facto reconstructions seem facile. Crick's devices were simplistically tiny, direct triplet adaptors for the genetic template; the complex and larger clover leaf–shaped sRNA, loaded with the specific amino acid, attached itself to nontemplate ribosomes; its anticodon complementary to the messenger's codon. More importantly, Crick's "adaptors" were conceptualized within a different representational space, that of the coding problem. Crick's was a representation of protein synthesis constituted through the discourse of information and its scriptural significations. As Rheinberger has shown, the transition from Zamecnik's carriers to Crick's adaptors, from the operationally defined sRNA to the functionally conceptualized tRNA, was neither an innocent name change nor a frictionless "realization." Acting as the Derridian *supplement*, the information idiom signaled a reorientation of the framework of protein synthesis. In that reconfiguration from purely material and chemical to genetic and informational representations, epistemic and disciplinary commitments had to be negotiated. As in the contemporaneous case of the TMV studies at Berkeley and in the interpretation of the PaJaMa experiments in Paris, Zamecnik and his colleagues now began to conceptualize protein synthesis as gene-based information transfer. Suddenly linguistic and textual icons, absent from the earlier semiotic repertoire of biochemistry, began to merge with molecular biology, recasting protein synthesis as a communication system. In 1958 tRNA was envisioned as the Rosetta stone that would decipher the language of the gene.[25] (tRNA did form a key element in the work on the genetic code, especially with Robert Holley's feat at Cornell of sequencing yeast tRNA in 1965.)

Despite his increased interactions with Crick, Zamecnik did not jump on the bandwagon to crack the code (observing that the code already commanded more than sufficient attention). But his laboratory soon made a pivotal contribution to that quest by developing a protein synthesizing system from E. coli, the primary experimental organism for decoding. Modeled closely after the rat liver cell-free preparations, the Lamborg and Zamecnik bacterial system too required the presence of ribosomes, as well as the two-hour, 100,000 g supernatant fraction containing nucleic acids and enzymes, energy-generating molecular fuel ATP and GTP (adenosine and guanine triphosphate), and a supply of amino acids (one radioactively labeled for tracking) to be strung together into a protein. After considerable tinkering, the new E. coli system outperformed the older rat liver one: the rate of incorporation of amino acids into proteins (with radioactive traces as indicators) was severalfold higher.[26] That system, however, was mute with respect to the genetic "information" since the DNA nucleotide sequence ("the code") specifying the protein was not known and its "transcript" messenger RNA did not yet exist.

The urgency to recover that genetic signal—genetically and biochemically—intensified with the international hunt for the RNA messenger that transported the DNA "message" to the cytoplasm, which was also an "informational RNA." While in the summer of 1960 François Gros from the Pasteur Institute and his collaborators at Watson's laboratory were racing against Jacob, Brenner, and Meselson at Caltech (and other laboratories) to identify that unstable RNA intermediate, Françoise, Gros's wife, was collaborating with Swiss biochemist Alfred Tissières at Harvard in an effort to fine-tune their E. coli cell-free system. Messengers could potentially serve as probes into protein synthesis. As Jacob envisioned it, such a system could be used to test the expression of the messenger as a polypeptide. Tissières's findings were similar to those of Lamborg and Zamecnik (with whom he shared a cordial cooperative competition), but they added a critical insight: they established that DNAase (a DNA-chewing enzyme) inhibited protein synthesis. These observations, reported also by other researchers, demonstrated conclusively that cell-free protein synthesis was dependent on the DNA template that specified the RNA messenger. It also suggested that DNAase treatment could be used to destroy the system's internal signal (and thus also the complementary endogenous RNA) and then test for the incorporation of proteins built according to the specification of an external, well-characterized RNA signal from well-established sources—a dialectic of representing and intervening.[27]

By the summer of 1960 Nirenberg had reached similar conclusions with his new E. coli system. Certainly, Nirenberg had access to key sources, notably the leading authorities on nucleotide synthesis at NIH, the expertise of Richard B. Roberts's group at the Carnegie Institution of Washington in ribosomes' function and protein synthesis, and consultations with biochemist Roger Herriott, who specialized in phage expression in E. coli at Johns Hopkins University. Furthermore, working alone and racing against several large laboratories, he resolved (once again) to "work nights as well as days," to test various aspects and components of the system. Throughout that spring and summer his diaries record queries and procedures for handling DNA-RNA complexes, numerous experiments (actual and planned) on preparation of ribosomes, and analyses of enzymatically activated sRNA–amino acid complexes. A new method for determining the sedimentation of enzymes with direct application to protein mixtures (developed that summer by Martin and Ames) improved Nirenberg's techniques, which he described as "the most exciting thing I have seen in months." His notes are flooded with references to recent updates on protein synthesis in E. coli systems (by Spiegelman, Novelli, Nisman and Fukuhara, and Tissières), repetitions and modifications, and questions about interpretations of observations.[28]

When Nirenberg eventually zeroed in on his problem, scientifically speaking, he did "one thing superlatively well," as the *New York Times* put it. The meandering fascinations with the big questions in biology, the intrigue of embryology, and even the preoccupations with phage genetics, which had informed his conceptualization of cellular regulation, had now given way to a sharply focused mandate: to solve protein synthesis (which was to "break the code"). "Hurry up exps. Shouldn't take 1 week to know whether system will work. *Work -Work -Work*," he prodded himself early in August 1960.[29] Soon after, however, a fortuitous event occurred that would substantially accelerate his progress. Unannounced, plant physiologist Heinrich Matthaei, a

German postdoctoral fellow from Cornell University, literally presented himself at Nirenberg's doorstep inquiring whether he could work with him on protein synthesizing systems. They decided Matthaei would join Nirenberg in November. Nirenberg's diary notes on tasks to be performed "when Heinrich comes" signaled a new phase in the project.[30]

BREAKING THE "CODE OF LIFE"

For Heinrich Matthaei, NIH represented an urgent chance for realizing his scientific goals after some frustrating experiences. He had come to the United States from Bonn in June 1960 on a NATO Fellowship (fellowships established to stimulate the growth of science as a springboard for the technological and military competitiveness of the Western alliance, carrying additional significance for the reintegration of Germany, the linchpin in American policy in Europe). Since his 1956 doctoral thesis on the protein balance in growing plant tissues, he had been following the general outline of the research on cell-free protein synthesis, and as his publications reveal, he was aware of RNA's role in protein synthesis from the writings of Jean Brachet and Torbjorn Caspersson. But his research goals were retarded at the University of Bonn, where there were no facilities for radioactive work (needed for tracking protein incorporation) and few colleagues to consult about cell-free systems.[31]

His NATO Fellowship project aimed to accomplish cell-free protein synthesis, using radioactive amino acids, in order to answer central questions in cell physiology: protein synthesis, enzymatic regulation of cell metabolism, and cellular development. Cornell University's botany department (leading research center in plant physiology) and the laboratory of Frederick C. Stewart, a specialist in protein synthesis in carrot tissue, seemed like an excellent match. But Matthaei quickly found out that in addition to equipment problems, Stewart was not enthusiastic about Matthaei's independent project. Matthaei would have to look elsewhere, a complicated process requiring the permission of the German Academic Exchange Service and the relocation of his family. California was too far, Zamecnik was out of the country, and Fritz Lipmann (Nobel laureate at the Rockefeller Institute who had mapped the energy flow of cell metabolism) could not accommodate him until the following year, but he did suggest NIH. "Fortunately, I have found a very good workplace at NIH," Matthaei announced in the transfer requisition, detailing his earlier frustrations. Indeed, Nirenberg's approach most closely matched his own.[32]

When Matthaei arrived in November, Nirenberg was shuttling back and forth between the B. cereus system (still attempting cell-free synthesis of penicillinase) and the E. coli system (probing the mechanisms of amino acids incorporation). A virtual bricolage, the suturing of some key ideas for the subsequent experiments with Matthaei was formulated within the B. cereus system, then grafted onto the E. coli. Nirenberg noted:

> [Saturday night in mid-November] In absence of cysteine [the amino acid absent in penicillinase], formation of *messenger RNA* might slow down. Would have to separate mess. RNA & sRNA. . . . [and a week later] Get AA [amino acid] incorp. [incorporation] system dependent upon DNA. B. cereus best, but use E. coli if necessary. . . . [a list of items to be tested in the system] Get oX-174 DNA [circular DNA from a tiny phage]. Will single stranded DNA work? +/− Nitrous acid [RNA mutagen]. Heterologous DNA, *Synthetic RNA polymers* [the macromolecules that a few months later would act as the key to the "code"], RNA transforming factor. . . . Can you swamp system with *messenger RNA*? Or is DNA necessary. Could use AA [to] analyze that one incorp into protein. i.e., C^{14}-phenylalanine [exceptionally easy to recover from the reaction mixture].

These notes underscore two critical points: the notion and discourse of messenger RNA was by then widespread (even though NIH researchers claimed to have been unaware of the "messenger hypothesis"); and synthetic RNA polymers were intended to be tested for their template function in protein synthesis (as was TMV RNA soon after). The notes also reflect Nirenberg's frequent consultations with Tomkins and with Maxine Singer, Heppel's postdoctoral fellow, who provided materials and expertise in polynucleotide synthesis. In fact Nirenberg invited Singer to collaborate with him, but despite Heppel's persuasion, she declined. She would help Nirenberg in all ways possible but forego a formal collaboration, insisting on making a name for herself based on her own project. "There goes your Nobel Prize," Heppel allegedly told her.[33]

By December Nirenberg and Matthaei had concentrated their energies solely on the E. coli system. Sharing a small bench in Tomkins's laboratory, they worked very closely, literally and figuratively (conveniently, both preferred to work in the solitude of night). Their efforts to fine-tune the system consisted not only of accomplishing the incorporation of amino acids in the DNAase supplemented system (as Tissières had just reported) but also of establishing definitively that the system was dependent upon an RNA template (as Tissières's work implied). An elaborate diary entry in mid-January 1961 makes evident that their goal was a part of a strategy for breaking the code with a polynucleotide "key," and that the prevalent model of a triplet code guided their thinking:

> *Idea. Approach to code.* 1. Charge sRNA mixture with C^{14} protein hydrolysate so that all AA would be charged. Add sRNA-AA to poly A, U, C, G & do density gradient cent. [centrifugation]. . . . 2. Use either mRNA or TMV RNA & show that all AA-sRNA interact. Will they all interact? 3. Use something like AU polymer [from Heppel's freezer]. Have many diff. permutations. Might find that some interact, others no. Could specify limits. Compare AG and UC, purine vs. pyrimidine. Also AU, AC, GU GC. Might be able to get enough info. to establish limits of a code. . . . If you need all 4 bases [presumable for the detection of a particular amino acid], could not be triplet code. If three bases do not work.[34]

The "code of life" as a semiotic for gene-directed protein synthesis had been fermenting in the back of Nirenberg's mind. But now the protein synthesis project was beginning to be formulated within the representational and discursive space of the coding problem. (For Matthaei the reorientation from plant biochemistry to molecular biology would be facilitated through a summer course in biochemical genetics at Cold Spring Harbor.)

By the end of January they had attained their preliminary goal of establishing a remarkably sensitive and stable cell-free system that incorporated C^{14}-valine into protein at very rapid rate. Craftsmanship and nuances of material practice were essential to that feat. Matthaei, a meticulous and dexterous experimenter, markedly refined and stabilized the system. Notably, he fine-tuned the system's components (e.g., ATP and the critical magnesium concentration), helped Nirenberg in devising a rapid protein assay that reduced experimental time by four, and developed a procedure for storing the system in the freezer for months, thereby raising the experimental efficiency and the uniformity of parameters across many experiments. "I could not have done this without him," Nirenberg reiterated. They were preparing a short communication of their results in time for the mid-February Federations Meetings. Beyond the description of "Some Characteristics of a Cell-Free DNAase Sensitive System Incorporating Amino Acids into Protein," they reported that amino acid incorporation shut down in the presence of DNAase and thus could well be DNA dependent.[35] They received little notice.

On March 22, they sent a somewhat longer and more definitive version to *Biochemical and Biophysical Research Communication*, a newly established journal designed for rapid bulletins. The cautious title, "The Dependence of Cell-Free Protein Synthesis in *E. coli* upon RNA Prepared from Ribosomes," seemed somewhat misleading, for they were, in fact, referring to protein synthesis via messenger RNA. They were reporting "a novel characteristic of the system; that is, a requirement for 'high molecular weight ribosomal RNA,' (a third species of RNA) needed even in the presence of soluable RNA and ribosomes"; the RNA requirement was evidenced by the fact that RNAase (an RNA-chewing enzyme) ground protein synthesis to a halt. But what they meant—as Nirenberg's diary notes show—was the requirement of an RNA messenger attached to the ribosomes. (This feature of messenger RNA was demonstrated in *Nature* by Brenner, Jacob, and Meselson only a couple of months later, though Nirenberg and Matthaei seemed unaware of that work.) Only at the end of the paper did Nirenberg and Matthaei suggest, "It is possible that part or all of the ribosomal RNA used in our study corresponds to template or messenger RNA."[36] As the diaries indicate, for months already, messenger RNA had been central to conceptualizing the experimental design. Significantly, the discursive interchangeability of "template" and "messenger" RNA captures a telling moment in the transition from the older representational framework of biochemistry to the new one of molecular biology: from matter-bound chemical specificity to a recognition and communication apparatus of information transfer. This report too received scant attention.

Curiously (and ironically) poly-A, a synthetic polymer of polyadenylic acid from Heppel's freezer, was added to the reaction mixture, not as messenger but as a polyanion, as a protection of the endogenous messenger RNA. Matthaei explained:

> We wanted to find out whether or not there might be at least some unusual sorts of RNA with *no template activity* at all that would have helped to rule out the possibility that the addition of RNA would just mean engaging some ribonucleoease and preventing it therefore from degrading the endogenous RNA. . . . The hypothesis at the time was that there are some nucleotide sequences that have *no coding activity* and in that case if they appeared to stimulate amino acid incorporation—indirectly by engaging the ribonuclease—it would have meant that perhaps our messenger assay is not very good. . . . Of course we were aware even in February 1961 that *poly-A might have contained a codon* [nucleotide triplet] for one of the 20 amino acids which we just did not happen to test on this occasion; we left this for later [my emphasis].[37]

Contradictory reasoning and a strange turn of events. According to the then dominant model of Crick's commaless code, the nucleotide triplet AAA represented a "nonsense" word (not specifying an amino acid). Had Nirenberg and Matthaei been following that model, then poly-A would have indeed served merely to protect the endogenous RNA. If they were unfamiliar with the strictures of the commaless code, then they surely had no reason to rule out poly-A's template activity. But in groping in the grayness between "sense" and "nonsense" and being only vaguely aware of the various codes and their features, they did not adhere to any particular version, admitting all possibilities. Poly-A did in fact possess template activity, as would be shown a year or so later. It gave rise to the monotonous polypeptide poly-lysine (namely, AAA specified the amino acid lysine); but, being soluble in the reaction mixture (with precipitating reagent, 10% trichloroacetic acid) poly-lysine was technically difficult to recover, even if their experiment had been set up to test its incorporation. Experimental systems are not black boxes that automatically translate input into output; instead, they are sensitive differential generators that have to be adjusted for the detection of specific signals.

While Nirenberg and Matthaei demonstrated that endogenous messenger RNA did stimulate protein synthesis, the effect was relatively small, just above background, in the hundred-count range. They desperately needed a better, less contaminated RNA preparation. In the following weeks Nirenberg and Matthaei checked various features of the system in preparations for testing the effects of different types of synthetic and natural RNA samples: "David's RNA," "Crestfield-RNA," ascites RNA, yeast RNA, and finally TMV RNA. Nirenberg wondered in early May, "*Idea:* Is viral RNA acting as template? Use TMV RNA & another RNA of different base composition, preferably Turnip Yellow Mosaic Virus RNA [TYMV]. Assay ones which could be obtained. Use same enzyme function. Test about 10 diff AA. Are ratios of AA incorp. into protein in presence of TMV RNA same as those incorp. in presence of TYMV RNA?"[38] He thus unknowingly echoed Gamow, Yčas, and Rich's queries of the 1950s. His experiments supported the hunch that TMV might be a viable template, because the protein incorporation was relatively high. These preliminary findings were exciting not only for their validation of the new cell-free system but also for their enormous potential for breaking the genetic code. As such, Nirenberg and Matthaei's work converged with the recent efforts of Heinz Fraenkel-Conrat at the Berkeley Virus Laboratory.

Fraenkel-Conrat's laboratory was then leading in the race to decipher the code. Having just a year earlier (1960) published the complete amino acid sequence of TMV, they were now deploying RNA alteration techniques (nitrous acid mutagenesis introduced by their German competitors in 1958) to correlate the nucleotide changes in the viral core with amino acid replacement in its protein coat; "a Rosetta stone for the language of life," as scientists and the media put it. Nobel laureate Wendell Stanley, the powerful department head, even nominated Fraenkel-Conrat in 1960 for the Nobel Prize.[39] Nirenberg greatly admired Fraenkel-Conrat. He also felt that the

Virus Lab was then the most exciting place, in part because of the potentiality of the lab's large mutant stocks and amino acid correlation data, which could be utilized for studying the viral proteins produced with a cell-free E. coli system. Nirenberg thought a month at Fraenkel-Conrat's laboratory would improve his technical facility with TMV and answer many of the questions about the nature of the TMV RNA's template activity; it could even yield substantial clues into the genetic code. Nirenberg planned numerous experiments and a list of two dozen items, namely, notebooks, manuscripts, laboratory accessories, and various samples, for his Berkeley trip in mid-May. He also detailed protocols for polynucleotide experiments to be performed by Matthaei.[40]

Matthaei resented this scientific paternalism. An accomplished experimenter and only a year younger than Nirenberg, he was clearly in a position to design his own experiments. The resentment quietly but steadily bubbled up to the surface. In the meantime Matthaei executed Nirenberg's protocols efficiently and meticulously. He began the experiments on May 15, testing poly-A—this time as messenger—as well as poly-U (poly-uridylic acid), and poly-(2A)U and poly-(4A)U (polymers consisting of different ratios of uridylic and adenylic acid) for amino acid incorporation. On May 22 he noted significant activity for poly-A. In preparation for the follow-up experiments with poly-U, a full set of radioactive, C^{14}-labeled amino acids were to be added to the cell-free system. Assembling and testing all the required materials took a few days. He recalled:

> Dr. Heppel could only give me 1 mg of poly-U at that time and into each of the two reaction tubes that I set up I put 200 micrograms, so 400 of the 1000 gammas [one milligram] were spent already, but it was still enough to do many experiments. . . . It was enough to find out the specific amino acid and to get enough product to characterize it. . . . So I got the 18 amino acids "borrowed together," we always called it borrowing, we never gave anything back I guess.[41]

The logic of the experiments consisted of testing each synthetic polynucleotide in the presence of nineteen unlabeled (cold) and one labeled (hot) amino acid, and through systematic variations, determining which radioactive amino acid was incorporated into a polypeptide in response to a specific synthetic messenger RNA. Matthaei performed experiment 27Q at three o'clock in the morning of Saturday, May 27, placing ten micrograms of poly-U and nineteen cold amino acids plus hot phenylalanine into the cell-free system. After an hour of incubation, the Geiger counter registered more than thirty-eight thousand counts per milligram of protein, while the control samples measured around seventy counts, just slightly above background level (see Figs. 30 and 31). It was unambiguous: poly-U specified the assem-

(see M1, p. 107) 9/

27-Q incub. 5-27-61, 3 a.m. for 60' at 36°, 10% TCA for 60', shot sick.

#	System	Special treatment	H_2O	min for 10,240 cts.	cpm (+bckgd.)	cpm −bckgd.
1	Complete		640	50.53	202 >206	167 >156
2			640	48.69	210	144
3		+ 10γ Poly-U	550	2.69	3810	3748 26x (23x)
4		+ 100γ RNAase at	540	261.73	39.2 >?	
5		0-t.	540	164.12	62.4	
6	Complete, like 1/2, but 20AA−tyr, 25λ C14-tyr		640	113.29	90 >92	36
7			640	108.39	94	
8		+ 10γ Poly-U	550	94.81	108 108	52 1.45X
9		+ 100γ RNAase	540	181.87	56 56	0
10		at 0-t.	540	184.48	55.5	

FIGURE 30. Heinrich Matthaei's Notebook M_1. Experiment 27Q. Courtesy of the author.

bly of polyphenyalanine; it had to be the first "word" to break the genetic code. Matthaei later recalled, "And when Gordon Tomkins came in, well I guess maybe nine, eight or nine in the morning, I already told him I now know it's only *this* one which is coded."[42]

The news was kept quiet; not a word was mentioned when Sydney Brenner stopped to give a talk at NIH on the messenger work, just a week before the 1961 Cold Spring Harbor symposium.[43] Nirenberg and Matthaei, not members of molecular biology's inner circle, had been denied participation in the meeting, and Tomkins remained silent throughout. Matthaei phoned the poly-U results to Nirenberg at Berkeley, but Nirenberg kept the news to himself. He soon returned from Berkeley, having completed the experiments which, he thought, demonstrated conclusively the messenger activity of TMV RNA. By Sunday, June 11, Nirenberg was back in the laboratory, repeating the poly-U experiment and extending the procedure to poly-A, poly-C, and poly-I (a variant of poly-G, with inosinic replacing guanylic acid). Nirenberg's "takeover" only deepened Matthaei's resentment; on the other hand, Nirenberg did not feel he excluded Matthaei from the polynucleotide work, only that he tried to divide up the work to avoid duplication of effort. "There was so much to do and there were only the two of us doing it," he remembered. They (and their NIH colleagues) continued to keep the sensational news under their hats, presumably to stave off the enormous competition as they conducted follow-up studies. There were papers to be prepared for publication and a short talk to be delivered by Nirenberg in mid-August at the Fifth International Congress of Biochemistry in Moscow.[44]

Their two papers reached the *Proceedings of the National Academy of*

FIGURE 31. *Medical World News*, 5 January 1962, cover. J. Heinrich Matthaei and Marshall W. Nirenberg, ca. 1961.

Sciences on August 3, 1961: the first, "Characteristics and Stabilization of DNAase-Sensitive Protein Synthesis in E. Coli Extracts," by Heinrich Matthaei and Marshall W. Nirenberg; the second, "The Dependence of Cell-Free Protein Synthesis in E. Coli upon Naturally Occurring or Synthetic Polyribonucleotides," by Marshall W. Nirenberg and Heinrich Matthaei. The understated title gave no clue that they had broken the code. This alternating order of authorship carried significance in conveying the credit differential. The first paper reported the system's technical features, its storage capabilities, active amino acid incorporation, and its relationship to "messenger RNA"; the second reported the remarkable finding that "the synthetic polynucleotide appears to contain the code for the synthesis of a 'protein' containing only one amino acid." TMV RNA was twenty times as active as endogenous RNA, poly-U nearly nine hundredfold. Their conclusions were broadly cast.

> The results indicate that polyurididylic acid contains the information for the synthesis of a protein having many of the characteristics of poly-L-phenylalanine. . . . One or more uridylic acid residues therefore appear to be the code for phenylalanine. Whether the code is of the singlet, triplet, etc., type has not yet been determined. Polyuridylic acid seemingly functions as a synthetic template or messenger RNA, and this stable, cell-free E. coli system may well synthesize any protein corresponding to *meaningful information* contained in added RNA [my emphasis].[45]

"Information" signified the genetic specificity of poly-U; not any information, but the meaningful kind, corresponding to "sense," to "words." From his earlier studies of enzymatic repression Nirenberg knew that not all genetic information translated into proteins.

By the time of the papers' appearance in October 1961, the news had spread like brushfire among molecular biologists, who had first gotten wind of it in August at Nirenberg's presentation at the International Congress of Biochemistry in Moscow. Only a handful of people attended Nirenberg's fifteen-minute presentation, among them Walter Gilbert, Tissières, and Meselson. Nirenberg's abstract, sent before the poly-U experiment, offered no clue of things to come. Meselson immediately alerted Crick to the startling news, whereupon arrangements were made for Nirenberg to give his talk again in front of the wide audience that flocked to the session chaired by Crick. Understandably, Nirenberg's announcement was a major upset for Crick, who had been the spokesman for the genetic code and whose group was then blazing through phage recombination studies aimed at breaking the code. As he had informed Max Delbrück enthusiastically only a few weeks earlier, "I have been hard at work doing phage genetics; in particular this question of suppressors. I believe Dick Feynman has done something similar, but I have only heard rumours. . . . We have an ingenious theory for our results which, if true, would be very important for decoding, but it needs much more work to establish it. If only we had a protein!"[46] Nirenberg had the protein (phenylalanine). Indeed, the breaking of the code by Nirenberg and Matthaei was one of the most stunning events in the history of modern science. It represented a victory of material ingenuity over Pythagorean ideals and is a David versus Goliath tale of an obscure young scientist defeating the eminent gray matter of physicists, mathematicians, biochemists, and geneticists, some of them Nobel laureates.

Many begrudged the success. "We had no way to decode it mathematically and Nirenberg started it. And Nirenberg didn't even quote me in his paper," George Gamow protested (Nirenberg did not know of Gamow's work until later). In retrospect, some thought the whole thing was simplistic, too obvious. Jacob, who together with Françoise Gros had tried—and failed—to isolate messenger RNA for β-galactosidase in a cell-free system of induced E. coli, remembered, "We used to talk as a joke of putting poly-A or poly-U [into the system]." Tissières too had a good cell-free system and access to Paul Doty's poly-U but considered it an "idiotic thing" to try, a view that made perfect sense according to Crick's commaless code. Alexander Rich conceded that "Nirenberg has made an enormous contribution to the problem. What puzzles me now," he wrote to Crick, "is why it took the last year or two for anybody to try the experiment, since it was reasonably obvious."[47] Some viewed it as a lucky strike. Gunther Stent subtly implied it was an unintentional move: "One day, an artificially synthesized polyuridylic acid was added to this reaction mixture instead of natural mRNA and a most surprising result was obtained," he explained in his widely read textbook.[48] Zamecnik too believed that Nirenberg was very lucky, since with just a minute decrease of the magnesium concentration, protein synthesis would have shut down (Nirenberg and Matthaei had tested a wide range of magnesium concentrations).[49] Others, even at NIH, believed that poly-U (like poly-A earlier) was added as a negative control. Tomkins, after supposedly having suggested using synthetic polynucleotides as negative controls, "eventually came to believe that he had recommended them to break the code."[50] Both Nirenberg and Matthaei have strongly disagreed with these perceptions of skeptics and critics, insisting that the experiments worked according to plan. Their stand is strongly supported by the abundant ideas, intentions, strategies, and the design of experiments sketched out in Nirenberg's diaries.

With the remarkable feat of Nirenberg and Matthaei, studies of the genetic code—the correlations of nucleotides and amino acids—entered a second phase: a predominantly biochemical phase supplemented with genetic analyses. With some exceptions (notably Carl R. Woese from the University of Illinois), the inferential constructions derived from theoretical and mathematical approaches initiated by Gamow and the RNA Tie Club were now eclipsed by direct experimentation. The genetic code could be now completed by systematically studying the effects of synthetic RNA messengers placed into the protein synthesizing system. It was expected that the entire code would be solved within a year or two, although it finally took six, if one counts the termination signal triplets, and it entailed additional feats of technical and analytical ingenuity. Given the impressive directness, reliability, and efficiency of the new biochemical methods, the earlier indirect approach was devalued. Many researchers came to downgrade the "decoding" work conducted in the first, mathematical and theoretical phase, viewing it as naively optimistic at best, misguided and unproductive at worst. But as Carl Woese, who persisted productively in his theoretical analyses of the code (see below), observed, "What has not been generally appreciated is that the subsequent spectacular advances in the field, occurring in the second period [1961–67], were interpreted and assimilated with ease, their value appreciated and new experiments readily designed precisely because of the conceptual framework that had already been laid."[51]

Indeed, as numerous researchers joined the race, they adopted the scriptural representations and the informational discursive practices of the coding problem, as formulated by the RNA Tie Club in the 1950s, in structuring their approaches, materials, and methods. By 1962 biochemists had come to perceive nucleic acids as "informational macromolecules" (namely, bearers of genetic specificity). Both nucleic acids and proteins acquired linguistic attributes, which reconfigured protein synthesis as a communication system (as in the case of the protein regulation work at the Pasteur Institute and Fraenkel-Conrat's TMV analyses at Berkeley). The cell-free protein synthesizing system was soon recast as a "translation system" for the messenger's "transcriptions"; nucleotides were elevated to "letters" ruled by "punctuations"; amino acids to "words"; and the full set of code words reemerged as the "dictionary." This trend intensified even if biochemist Erwin Chargaff who had embraced models of cybernetics and communication theory in the 1950s had by then long retracted them, even mocking these informational representations as the pretentious technobabble of molecular biology. But these were not merely scientific popularizations, fashionable parlance, or rhetorical and disciplinary strategies (although they did serve such purposes). As discursive practices—modes of articulating, representing, and intervening—informational tropes and models guided the conceptualizations, interpretations, and material practices of the subsequent experiments to determine the other code "words."

"INFORMATIONAL MACROMOLECULES": LETTERS, WORDS, NONSENSE

With the decoding impasse cleared, the race to complete the code reached new levels of urgency and fierceness, meaning that suddenly Nirenberg faced the overt challenge of competition with at least half a dozen other groups. Robert Martin remembered walking into the laboratory on a Saturday afternoon of the Indian summer of 1961; he found almost no one around and Nirenberg sitting alone at a table with his head bowed and his eyes glassy, obviously upset and depressed. He had just given a paper at the September meeting of the New York Academy of Medicine, where Severo Ochoa announced that his laboratory, staffed by about twenty researchers, had moved full swing into the coding problem.[52] Nirenberg did not welcome Ochoa's entry into this race. Having only two years earlier received the Nobel Prize for his work on the enzymatic synthesis of ribonucleic acid (which he shared with his former postdoctoral fellow, Arthur Kornberg, for the discovery of the DNA synthesizing enzyme, DNA polymerase), Ochoa did not need to compete, especially not with the NIH group with whom he had collaborated by invitation in the last five years. Nirenberg's NIH colleagues shared his sentiment. As Martin recounted, "We at NIH were terribly angry with Ochoa and his colleagues for jumping in on Marshall and Heinrich's discovery. But of course, they too were working on protein synthesis at the time. Peter Lengyel, Ochoa's director of operations, says they had planned those experiments before hearing of Marshall's work. I am sure he's right."[53] Matthaei too had suspected this development, remembering that when he was at Cold Spring Harbor in July 1961, he noticed, "Ochoa telephoned my friend Jo Speyer, and told him to cancel his vacation and come back to the lab because they must resume their work on polynucleotide coding. They must have tried it some time early in 1961, actually before us, but they must have missed the poly-U effect."[54]

Undoubtedly, missing the poly-U effect was an uncomfortable experience for Ochoa the Nobel laureate. Like most biochemists, until that time (ca. 1960), Ochoa had worked strictly within the traditional framework of metabolic biochemistry (focusing on chemical composition and energy exchange). A native of Spain, he had completed his medical degree in 1929 at the University of Madrid and his postdoctoral training in biochemistry with Otto Meyerhof at the Kaiser-Wilhelm Institute in Heidelberg, subsequently becoming a lecturer in physiology and biochemistry at the University of Madrid. He did not spend much time in Madrid; with the onset of the Spanish Civil War in 1936 he departed for Germany. From there he moved through several laboratories in England and finally landed in the United States. (These wartime "wander years" also marked the careers of other European biochemists, for example, Heinz Fraenkel-Conrat.) After a year as an instructor at the legendary laboratory of Carl and Gerty Cori (who shared the 1947 Nobel Prize for their studies of carbohydrate metabolism) at Washington University in St. Louis, Ochoa moved to New York University School of Medicine in 1942. "A courtly, charming, El Greco figure excited about his

FIGURE 32. Severo Ochoa, ca. 1960. Courtesy of the Elmer Holmes Bobst Library, New York University Archives.

work" (as Kornberg described him), Ochoa had risen from a research associate, constrained by meager resources, to full professorships in biochemistry and pharmacology in 1954, and then to the chairmanship of a rapidly expanding department. His research had dealt mainly with enzymatic processes in biological oxidation, synthesis and energy transfer in the metabolism of carbohydrates and fatty acids, and the utilization of carbon dioxide.

More recently he had concentrated on the biosynthesis of nucleic acids (their metabolic aspects) and on protein synthesis in cell-free bacterial systems.[55] Genetic concepts and models of information transfer played no role in these investigations.

In 1954, biochemist Marianne Grunberg-Manago from Paris, a postdoctoral fellow studying oxidative phosphorylation in Ochoa's laboratory, made an exciting discovery. She identified a bacterial enzyme that gave rise to an RNA-like product containing only one of the four nucleotide bases. This was the first time an RNA-like polynucleotide had been synthesized outside the cell. Christened as polynucleotide phosphorylase, the enzyme was first thought to catalyze RNA synthesis in vivo. As such it became a focus of vigorous research (and the subject of many publications), despite ensuing criticisms.[56] It soon turned out that the enzyme played no role in RNA synthesis in vivo (the discovery of RNA synthesizing enzymes in 1959 finally dispelled Ochoa's perceptions). But for several laboratories working on the biochemical and structural properties of RNA (notably Alexander Rich, who had moved from NIH to MIT in 1957), synthetic RNA polymers provided handy models, "stencil RNA" as Ochoa visualized it. Polynucleotide phosphorylase was a critical tool for their biochemical craftsmanship. In the late 1950s, a major collaboration between Ochoa's and Heppel's laboratories (with significant contributions by Maxine Singer) had established that polynucleotide phosphorylase made RNA-like polymers from a mixture of the four nucleotide bases, though the resulting base sequence was randomly ordered (the ratio of nucleotides, say, 2U 1G, or 2C 2U, could be determined, but not the sequence).[57]

At least in retrospect, Ochoa regarded polynucleotide phosphorylase as "the Rosetta stone of the genetic code." As he remembered it, by 1960 the new concept of messenger RNA suggested the use of synthetic polynucleotides as messengers in cell-free systems for deciphering the genetic code.

> Peter Lengyel [then working on his Ph.D. project] and Joe Speyer firmly believed that this approach would open the way for deciphering the code. In early 1961 they started to work with cell-free protein synthesis systems. The expectation was that systems depending on the addition of mRNA for incorporation of amino acids into protein might respond to the addition of synthetic polynucleotides and incorporate different amino acids depending upon their base composition. When we were beginning our work Nirenberg reported that a system from E. coli translated poly(U) into polyphenyalanine.[58]

Miffed by Nirenberg's priority, Ochoa's group now concentrated all their energies on completing the code, exploiting their synthesizing machinery and stockpiles of synthetic RNA polymers. In the next two years, Nirenberg's

AMINO ACID INCORPORATION IN *E. coli* SYSTEM WITH VARIOUS POLYNUCLEOTIDES*

	Polynucleotide					
Amino acid	None	Poly U	Poly C	Poly UC	Poly UA	Poly CU
Phenylalanine	0.03	**13**		7	3	0.02
Serine	0.02	0.02	0.01	1.6	0.01	
Tyrosine	0.02			0.02	**0.75**	
Leucine	0.02	0.3		1.5	0.46	0.03
Isoleucine	0.01	0.09		0.32	**0.62**	0.007
Proline	0.02	0.02	0.06	**0.6**	0.03	0.14

* mμmoles/mg of ribosomal protein. 19 amino acids were tested individually in all cases, but the ones giving negative results have been omitted from the table. All values (except those for poly CU) are averages of at least two separate experiments.

FIGURE 33. After P. Lengyel, J. Speyer, and S. Ochoa, "Synthetic Polynucleotides and the Amino Acid Code," *PNAS* 47 (1961): 1936–42.

and Ochoa's laboratories would run neck-to-neck; their competition and redundancy of effort mediated somewhat by frequent communications and exchange of manuscripts prior to publication.

On October 25, 1961, Lengyel, Speyer, and Ochoa had submitted their paper, "Synthetic Polynucleotides and the Amino Acid Code," to the *Proceedings of the National Academy of Sciences* (hereafter *PNAS*), the first of a nine-part series. In contrast to the cumbersome title of Nirenberg and Matthaei's paper ("The Dependence of Cell-Free Protein Synthesis in E. coli upon Naturally Occurring or Synthetic Polynucleotides"), Ochoa's crisp heading left no doubt about his group's claims. They confirmed the NIH poly-U studies and established other features of protein synthesis. Addition of E. coli transfer RNA (sRNA) caused a pronounced increase of the incorporation of phenylalanine, indicating that poly-U affected the transport of activated amino acids from transfer RNA to the ribosomes, and thus acted as a messenger. This point was crucial because it provided an experimental warrant for the conceptual interchange between synthetic polynucleotides and endogenous messenger RNA. But the warrant also produced a conflation of the messenger as tool and messenger as object of study: the tool and object informed each other as a dialectic of epistemic and technical things (to use Rheinberger's analytics), an interlocking of representing and intervening. Ochoa's group also determined that, whereas poly-U promoted incorporation of only phenylalanine, mixed polynucleotides like poly-UC promoted the incorporation of phenylalanine, serine, and leucine; and poly-UA that of phenylalanine and tyrosine (see Fig. 33).[59]

Ochoa later remembered, "Lengyel, Speyer, and I were watching the counter and were thrilled. This results, obtained for the first time anywhere, showed that incubation of E. coli extracts with copolynucleotides containing C or A besides U residues promoted the synthesis of polypeptides containing serine, leucine, and tyrosine, along with phenylalanine. I remember this as one of the most exciting moments of my life."[60] Ochoa and his group

TRIPLET CODE LETTERS FOR AMINO ACIDS*

Amino acid	Code letter†
Cysteine	2U 1G
Histidine	1U 1A 1C
Isoleucine	2U 1A
Leucine	2U 1C
Lysine	1U 2A
Phenylalanine	UUU
Proline	1U 2C
Serine	2U 1C
Threonine	1U 2C
Tyrosine	2U 1A
Valine	2U 1G

* From data for *E. coli* system from this and the previous[1] paper.
† Sequence unknown except for phenylalanine.

FIGURE 34. After J. Speyer, P. Lengyel, C. Basilio, and S. Ochoa, "Synthetic Polynucleotides and the Amino Acid Code, II," *PNAS* 48 (1962): 63–68.

concluded their paper with a tone of priority, "These and other results reported in this paper would appear to open up an experimental approach to the study of the coding problem in protein biosynthesis." A note added in proof announced that by using poly-UG and poly-UAC they had extended the amino acid list to eleven: cysteine, histidine, isoleucine, leucine, lysine, phenylalanine, proline, serine, threonine, tryrosine, and valine. Further results would soon appear in *PNAS*.[61] Their protein synthesis machine was officially set in high gear.

About three weeks later, their paper, "Synthetic Polynucleotides and the Amino Acid Code, II," reported that poly-UC stimulated the incorporation of phenylalanine, serine, leucine, proline, and threonine; poly-UA that of phenylalanine, tyrosine, isoleucine, and lysine; poly-UG that of phenylalanine, valine, and cysteine; poly-UAC that of phenylalanine, serine, leucine, tyrosine, isoleucine, proline, threonine, lysine, and histidine; poly-UCG that of phenylalanine, serine, leucine, proline, valine, and cysteine; and poly-UAG that of phenylalanine, tyrosine, isoleucine, lysine, valine, and cysteine. The nucleotide sequence of these mixed polymers was, of course, not known (only their composition), thus assignments of amino acids were still obtained indirectly, through inferential estimates. Assuming a code of nucleotide triplets, and based on a comparisons with the length of poly-U and its frequency ratio of UUU triplets, they could assign triplet code letters to eleven amino acids (see Fig. 34). Their proposed code letters, they verified, were in excellent agreement with amino acid replacement data in nitrous acid mutants of TMV. A note added in proof announced, "The assignment of the following additional code letters [which] will be shown in a subsequent article in these *Proceedings*: arginine 1U 1C 1G, glycine 1U 2U 2G, and tryptophan, 1U 2G."[62]

With their earlier findings that poly-U played a role in the transport of the activated amino acid from transfer RNA to the ribosomes, Ochoa's group beat Nirenberg's group by three days, though Nirenberg's demonstration was far more rigorous. On November 24, 1962, Nirenberg's group submitted to *PNAS* their paper, which demonstrated that phenylalanine-transfer RNA was an obligatory intermediate in poly-U dependent polyphenylalanine synthesis.[63] Ochoa's reports of the amino acids list preceded Nirenberg's by six weeks. To match the competition's prowess, members of Tomkins's and Heppel's groups teamed up with Nirenberg in a remarkable esprit de corps and round-the-clock collaboration. Singer provided enzymes and expertise, Martin synthesized polynucleotides from three in the afternoon until one in the morning, and

> Heinrich, who had a penchant for night work (when the radioactivity counters were likely to be available), took the lobster shift from midnight to noon. He tested polynucleotides for protein synthesis. Marshall, working from about 9 to 6 in the lab and the rest of the evening at home, analyzed the data. . . . Stetten and Tomkins were supportive, supplying Marshall with space and postdoctoral fellows.[64]

Nirenberg has never failed to acknowledge great gratitude for this support. And he has observed that he "discovered to his horror that he liked to compete."[65]

For rapid publication, on December 4 Nirenberg's group submitted their paper, "Ribonucleotide Composition of the Genetic Code," to *Biochemical and Biophysical Research Communications*, reporting "the genetic code" for fifteen amino acids (partly confirming Ochoa's results).[66] (See Fig. 35.) And like others, they got tangled in the linguistic slippages inherent in the code idiom: Was it the amino acid "code" of proteins, the nucleotide base "code" of DNA (or RNA), or the "code" as the correlation between the two?

Beyond expanding on the correlations of nucleotides with amino acids (they now had accounted for fifteen amino acids) their paper was remarkable for its discursive and epistemic turn, signifying a move from biochemical representations to the scriptural significations of the coding problem. It is clear that they were now also communicating with molecular biologists and addressing their preoccupations with the structure and formalisms of the code. The authors redefine nucleotides as "letters of the genetic code" and amino acids as "words of the code," slipping between two semiotic versions of the code: code letters as single nucleotide bases and triplets as words; and code words as amino acids and proteins as sentences, or text. They also began to broach the still-unresolved question of whether or not the code consisted of the theoretically postulated triplets.

Genetic Code for Fifteen Amino Acids

Amino Acid	*Nucleotide Composition of Coding Unit**
Phenylalanine	UUU . . .
Valine	UG (U > G)
Leucine	UG, UC (U > G) (U > C)
Cysteine	UG
Tryptophan	UG (U ≤ G)
Glutamic Acid	UGC
Methionine	UG
Glycine	UG
Arginine	UGC
Alanine	UGC
Serine	UC, UGC (U > C) (U > G or C)
Proline	UC (U < C)
Tyrosine	UA (U > A)
Isoleucine	UA
Lysine	UA (U < A)

* The order of the nucleotides in a coding unit is not specified.

FIGURE 35. After R. G. Martin, J. H. Matthaei, O. W. Jones, and M. Nirenberg, "Ribonucleotide Composition of the Genetic Code," *Biochemical and Biophysical Research Communications* 6, no. 6 (1961/62): 410–14.

Informed by the scriptural representations of the coding problem and speaking to audiences beyond biochemistry, they used their data to "rule out the possibility of singlet and doublet codes. The *minimum* coding ratio must be three; *very possibly larger*." They also noticed, "If a polynucleotide containing two bases stimulated the incorporation of an amino acid, inclusion of third base in the polynucleotide did not prevent this stimulation," indicating that not only the code but also the "coding units" (triplets) themselves were partly "degenerate." (This observation would soon lead Nirenberg to ingenious experiments designed to determine, once and for all, the validity of the triplet model; see discussion below.) All of this conceptual terminology originates not in biochemistry but in mathematical and physical concepts that had been transported into molecular biology in the 1950s by Henry Quastler and the RNA Tie Club. Unfortunately, as later shown, the code would also turn out to be ambiguous (more than one amino acid, e.g., leucine or serine, was specified by one triplet), though they did not point up that feature. They sent a more detailed paper to *PNAS* two months later, and there for the first time did they cite Gamow's 1954 paper in *Nature*.[67]

Clearly, during the fall of 1961 Nirenberg had read some of the literature on theoretical approaches to the code, familiarizing himself with key concepts and terminology. In support of the experimental results, he cited the amino acid replacement data from mutant TMV, significantly, drawing on Martynas Yčas's 1958 article, "The Protein Text," in *Symposium on Information Theory in Biology* (organized by Henry Quastler). In that comprehensive review Yčas had examined the problem of "storage, transfer, and replication of the information contained in the protein molecule," signifying the nucleotide sequence as "a text, written in a four-symbol alphabet, which encodes another text, the protein, written with about twenty symbols." He also analyzed statistically the correlations between adjacent amino acids and their frequency of occurrence (recall that the amino acid frequency distribution contradicted the analogy of proteins with linguistic texts). And he had sketched out "the coding problem." There Yčas introduced his and Gamow's clumsy "combination code," where only the combination of triplets but not their order mattered, and Crick's elegant commaless code. Though Crick's code had been designed to circumvent the "punctuation mark problem," Yčas underscored that it "could, of course, also be solved if amino acids were selected in a sequential manner starting from one end of the template."[68] (This, in fact, became the salient feature in the Crick group's solution to the code's structure, which Crick published in December 1961.)

That these scriptural representations did not serve for Nirenberg as rhetorical veneer but formed conceptual structures shaping the experimental practice is abundantly evident from the entries in Nirenberg's work diaries in the fall of 1961. "*Discussion*: Discussion of the coding problem as modified specifically by our findings. . . . *Paper* 2 [soon to be submitted to *PNAS*]: Polarity problem [directionality of sequence translation]," he jotted down in mid-September 1961. A few weeks later experimental plans included "other letters of the code; coding ratio." In light of the new insights into the code and the role of messenger RNA, Nirenberg also outlined a possible "Theory of Repression" (a problem that had been foremost in his—and Jacob's—mind), speculating, "One strand of DNA makes template RNA, other strand makes repressor RNA . . . a repressor makes nothing for its nonsense. Complement of code is repression. No two letters of code will be complement of one another. Will have 20 letters, 20 complementary letters (repressor) & 2[?] nonsense if it is triplet nondegenerate code."[69] (Note that here "code" denotes the base sequence, not a correlation.) Indeed, it was an epistemic conundrum to devise a method for measuring the effect of a biochemical "nothing," or to detect a genetic silence, a problem that phage recombinations techniques were designed to solve. By early December plans for "2nd Paper: Characteristics of the Genetic Code" had concretized as: "1. Commaless code; 2. Need at least 3 nucleotides (arg. & hist.); 3. Nonsense codes for nothing; 4. Degenerate (?); 5. Single-Strandedness." These were followed by ideas for a "Degeneracy Test"; more on the status of nonsense words; constant preoccupations with "commas"; and queries whether the code was, after all, overlapping.[70]

By December 20, Ochoa's group had almost caught up with Nirenberg's. In submitting the assignment for fourteen amino acids, only six—alanine, aspartic acid, asparagine, glutamic acid, glutamine, and methionine—were still unaccounted for; curiously, all base triplets contained uracil, as did Nirenberg's assignments. (In fact, the "Nirenberg-Ochoa U rich codes" as they were called, were soon challenged.) A note in proof paved the way for the subsequent publication in *PNAS* of an additional three (alanine, asparagine, methoionine) or possibly five (aspartic and glutamic) amino acids.[71] Perhaps coincidently, on the same day as Ochoa's group published this note, an announcement reached the *New York Times*, which reported the news (apparently for the first time): "'Genetic Code' Partly Broken by U.S. Researchers."

> Government scientists said today they had succeeded in cracking partly the "genetic code." . . . The "genetic code," it was explained in a report by the scientists of the National Institute of Arthritis and Metabolic Diseases, is a system of messages between two chemicals [DNA and RNA]. . . . Together, these two nucleic acids work to bring about the manufacture of specific proteins. . . . The theory has been that the DNA acts as a general storehouse of genetic information. . . . RNA serves as a kind of chemical messenger to transmit the information. . . . In other words, a "code" is in operation, with four basic chemicals of DNA and RNA directing the selection of twenty amino acids in much the same way that the sequence of the two alternates of Morse telegraph code—dots and dashes—spell out meaningful words from the twenty-six letters of the alphabet. But Dr. Marshall W. Nirenberg and Dr. J. Heinrich Matthaei reported, up to now there has been no experimental evidence that permits direct translation of such a code. . . . Scientists all over the world have been working to break the "code of life". . . . The research has been largely independent at each laboratory. . . . Recently, scientists at New York University School of Medicine, led by Nobel Prize winner, Dr. Severo Ochoa have also made public important results in the same field. Their disclosures appear to cover much of the same territory of that of the scientists at the institute, even going beyond them in some respects. In both cases, that accomplishment represents only a beginning, though an important one, in revealing the details of the code.[72]

Though earlier media accounts had sporadically called attention to the hunt for the code's "Rosetta stone," this announcement initiated a steady stream of reports that soon turned "the code" into a public icon of the atomic age. Meanwhile Nirenberg, like Ochoa only a few years earlier, became a scientific celebrity, manager of an impressive research team, and in 1962, the head of the new Section of Biochemical Genetics (a more precise term for molecular biology).

Ten days later, on December 30, a major article appeared in 1961's last issue of *Nature*: "General Nature of the Genetic Code for Proteins," by F. H. C. Crick, Leslie Barnett, S. Brenner, and R. J. Watts-Tobin. As Crick had informed Delbrück a few weeks prior to the Moscow congress, they had expanded their earlier genetic studies that probed the B gene of the rII region of the bacteriophage system T_4, "so brilliantly exploited by [Seymour] Benzer."[73] (The rII region refers to the location of specific mutations on the phage genome map, identified by their peculiar plaques on a petri dish of E. coli K12; see my related discussion in Chapter 4.) Nirenberg's Moscow announcement and the subsequent developments in biochemical decoding only intensified Crick's efforts in a heroic attempt to catch up. Just two weeks before Crick's publication, Jacob told Brenner, "We had the visit of Francis who gave a remarkable seminar. This story is really astonishing. I tried to persuade Francis that the coding problem has no more interest. I doubt that I succeeded to convince him!"[74]

These genetic experiments, along with studies of others, pointed to four principal features of the genetic code. The Crick et al. article proclaimed:

> (a) A group of three bases (or, less likely, a multiple of three bases) codes for one amino acid. (b) The code is not of the overlapping type (see Fig. 7) [see Fig. 36 in this volume]. (c) The sequence of the bases is read from a fixed starting point. This determines how the long sequences of bases are to be correctly read off as triplets. There are no special "commas" to show how to select the right triplets. If the starting point is displaced by one base, then the reading into triplets is displaced, and thus becomes incorrect. (d) The code is probably "degenerate"; that is in general, one particular amino-acid can be coded by one of several triplets of bases.[75]

In the process they also explicated the mechanisms of nonsense mutations, cleverly detecting the silences of a biochemical "nothing" by genetic means. (See Fig. 36.)

The evidence that the code was not overlapping did not originate from their work, they readily conceded, but from the studies of H. G. Wittman (at the Max-Planck Institute in Tübingen) and of A. Tsugita and Fraenkel-Conrat on the nitrous acid mutants of TMV. In an overlapping triplet code, an alteration to one base would usually change three adjacent amino acids in the polypeptide chain. But amino acid replacement data showed that, as a rule, only one amino acid at a time was changed in response to nitrous acid treatment. If the code was not overlapping, as Crick and Yčas had underscored back in the mid-1950s, then there had to be an arrangement for se-

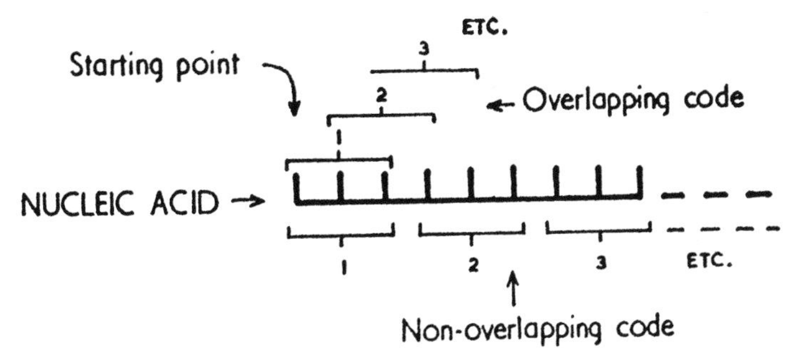

FIGURE 36. F. H. C. Crick, L. Barnett, S. Brenner, and R. J. Watts-Tobin, "General Nature of the Genetic Code for Proteins," *Nature* 92 (1961): 1227–32. "Reading frame" of nonoverlapping triplet code. © 1961 Macmillan Magazines Ltd.

lecting the correct triplets along the base sequence. Crick's elegant comma-free code did provide such an arrangement, by having certain triplets make "sense" and others "nonsense." But in light of the new genetic results, he had come around to Yčas's old solution: a correct selection could be made by starting at a fixed point and moving along the base sequence three at a time.[76]

They had arrived at their conclusions by ingenious manipulations of the bactiophage experimental system, sophisticated genetic analyses, and intricate deductive reasoning: they utilized Benzer's fine-structure map and the voluminous data generated by the precision technology of acridine mutants (phage treated with acridine dyes, e.g., proflavin), in which a nucleotide base is either added or deleted. These "FC O" mutants fail to grow on E. coli K (they grow only on E. coli B) but can be restored to normal function ("wild type," which grows on both E. coli K and B) with a second, "suppressor" mutation a bit further along the gene. Like the FC O mutant, "suppressor" mutants also fail to grow on E. coli K. Thus if an FC O mutation consists of a base deletion (−), then the suppressor restores genetic function by base addition (+) (or vice versa, an addition is neutralized by deletion). However, mutants of two pluses or two minuses would not plate on K.[77]

Crick's group used about eighty independent mutants within that limited region of the gene, all suppressors of FC O, or suppressors of suppressors, or suppressors of suppressors of suppressors. By intricate methods of genetic recombination, they inserted three mutants of the same type into one gene, namely, (+ with + with +) or (− with − with −), showing that this triple alteration resulted in an active gene, while (+ with +) and (− with −), and

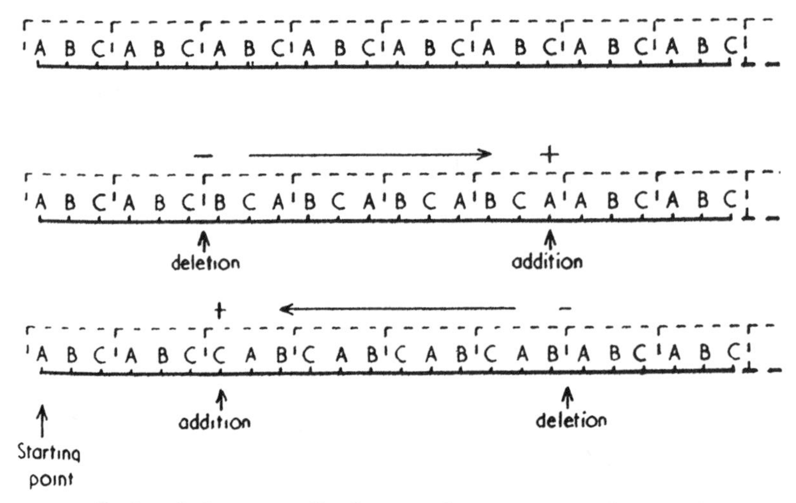

To show that our convention for arrows is consistent. The letters *A*, *B* and *C* each represent a different base of the nucleic acid. For simplicity a repeating sequence of bases, *ABC*, is shown. (This would code for a polypeptide for which every amino-acid was the same.) A triplet code is assumed. The dotted lines represent the imaginary 'reading frame' implying that the sequence is read in sets of three starting on the left

FIGURE 37. F. H. C. Crick, L. Barnett, S. Brenner, and R. J. Watts-Tobin, "General Nature of the Genetic Code for Proteins," *Nature* 92 (1961): 1227–32. Model of frame-shift mutations. © 1961 Macmillan Magazines Ltd.

just (+) and (−) completely inactivated it. From these manipulations they deduced that a base addition or deletion produced a "shift of the reading frame." A shift (either to the right or left) by one or two bases produced triplets, the reading for which was "nonsense," or chain termination, or perhaps other defects in protein structure. A shift of three bases, on the other hand, restored gene activity or "the correct reading" (see Fig. 37). They concluded therefore that the coding ratio was three (though multiples of three could not be ruled out), and thus the code was degenerate; "however how many triplets code amino-acids and how many have other functions we are unable to say."[78]

Apart from the pristine logic buttressed with experimental ingenuity, the other remarkable aspect of the article was its near-oblivion to Nirenberg and Matthaei's discovery and the subsequent elaborations by Nirenberg's and Ochoa's groups. Only at the very end did the authors mention:

> At the recent Biochemical Congress at Moscow, the audience of Symposium I was startled by the announcement of Nirenberg that he and Matthaei had produced polyphenylalanine . . . by adding polyuridylic acid . . . to a cell-free system which can synthesize protein. This implies that a sequence of uracils codes for phenylalanine, and our work suggests that it is probably a triplet of uracils.

> It is possible by various devices, either chemical or enzymatic, to synthesize polyribonucleotides with defined or partly defined sequences. If these too will produce specific polypeptides, the coding problem is wide open for experimental attack, and in fact many laboratories, including our own, are already working on the problem. If the coding ratio is indeed 3, as our results suggest, and if the code is the same throughout Nature, then the genetic code may well be solved within a year.[79]

The Cambridge group, and molecular geneticists in general, chafed at the biochemical takeover of the coding problem, even if their retrospective accounts positioned biochemist Alexander Dounce next to Erwin Schrödinger in the "Genetic Code Hall of Fame." They had hoped to break the code strictly with the power of deductive reasoning, without touching the biochemistry, or opening the black box. As Brenner later put it, "There was a culture—well a *cult*, almost—that became typical of molecular biology. What became prized was ingenuity. You know? . . . I mean it doesn't need all the bloody tubes and counters and so on. And I think that the cult got founded around these ideas of how to solve the code without ever opening the black box. Okay?"[80] A cult initiated with Max Delbrück's phage group in the 1940s. Not only Brenner and Crick but also many molecular biologists conveyed little appreciation for material intricacies, craftsmanship, or the ingenuities and deductive reasoning in the experimental design and interpretations of biochemical studies. They also had no way of knowing the pivotal role that genetic savvy had played in Nirenberg's conceptualization of protein synthesis.

While Nirenberg's and Ochoa's groups continued to chip away at the code, their exciting findings stimulated a flurry of theoretical papers. Analyses from several sources, old and new, got heaped on the code bandwagon, "most of which are best forgotten," Crick thought. Richard V. Eck from NIH, having a year earlier used information theoretical calculations of amino acid data to argue that the code might be overlapping, now offered a "paper experiment" for sequence analysis of large proteins, which "will eventually permit the solution of the nucleic acid-protein cryptogram." All the proposed codes had required an additional unknown source of information such as "commas," "spacers," "stepping by three," "forbidden combinations," and so forth, he pointed out and then insisted, "As long as the nature of that additional mechanism is undiscovered the possibility remains that it could contain enough information to supplement an overlapping code." (His ideas garnered him coverage by the *New York Times*.)[81] Nirenberg's group, in fact, made use of Eck's insights in determining the structure of the code. Robert Wall, from Harvard's Computing Laboratory, drew on Eck's and Woese's studies to offer mathematical arguments in support of the possibility of overlapping codes.[82] On the other hand, R. T. Hersh, from the biochemistry department of the University of Kansas, published theoretical analyses of TMV peptide sequences to confirm the nonoverlapping code and the conclusion that "the message is read from a fixed starting point."[83]

Richard Roberts, the ribosome expert with whom Nirenberg had been consulting in the previous two years, took a radical stand. "The triplet code can be converted to a doublet by discarding U common to all of the code words." His model's price tag was enormous, implying that many of the synthetic RNA polymers were not templates. He pointed out that a doublet code, on the other hand, was compatible with the possibility that ribosomal RNA was a template and with amino acid replacement data in TMV. "The doublet code has the possible theoretical advantages that it takes fewer letters, contains no 'nonsense,' and agrees with the correlations found by Sueoka"; a year earlier, Noboru Sueoka (at the microbiology department of the University of Illinois) had examined the correlations between the compositions of the RNA bases and amino acids in several bacterial species, also suggesting the possibility of quadruplet codes. One of the best proposals for a predominantly doublet code came from a little-known paper by Czechoslovak researchers I. Rychlik and F. Šorm, whose extensive use of amino acid replacement data from TMV led to a close approximation to the code's doublets (namely, matching two out of three bases assigned to an amino acid).[84]

Carl Woese (then still at General Electric) felt that it was "certainly now possible to develop a theory which encompasses all the existing facts regarding the code." He outlined a degenerate coding scheme derived from an "information point of view," which fit closely with the experimentally observed triplet assignment, accorded with amino acid replacement data, and reconciled the experimentally determined composition of code triplets with the enormous variations in the G+C content of DNA of different organisms. He later continued to elaborate this scheme with rather remarkable accuracy.[85] Zhores Alexandrovich Medvedev, senior biochemist at the Agricultural Academy in Moscow and participant in the Fifth International Congress of Biochemistry (who also wrote a book about his superior Trofim Lysenko), favored the idea of an overlapping code, "not because it seems more economical, but because it can better explain the parallelism of distances between nucleotides and amino acid residues and because the deletion or addition of a single nucleotide means in this case a point mutation only, but not the wrong reading of all the genetic information and the production of completely nonsense protein molecules."[86] And Mario Ageno (Instituto Superiore di Sanita, Laboratori di Fisica, Rome) suggested a scheme, where complementary (hydrogen bonding) triplets (such as TAC and ATG) corresponded to the same amino acid, thereby avoiding the problem of determining which of the two DNA strands had to be read.[87] Other schemes abounded.

Ochoa's group submitted five papers to *PNAS* with additional assignments of nucleotide triplets to amino acids—all containing uracil[88]—while Nirenberg confronted the challenges raised by Roberts (and others) about the U-richness of the code and the possibility of it consisting of doublets. Viral RNA did not contain such a preponderance of U, Nirenberg admitted, and their earlier findings that poly-AC directed small amounts (previously neglected) of proline and threonine into proteins did suggest the existence of non-U code words, results just then supported by the meticulous study of M. S. Bretcher in Cambridge, England, and M. Grunberg-Manago in Paris. To resolve the dilemma, Nirenberg used synthetic polynucleotides containing four, three, or two bases, all of which directed amino acids into protein with high efficiency and specificity. Many additional RNA code words not containing U were now found. Significantly, almost all amino acid could be coded by polynucleotides containing only two bases, Nirenberg's group reported, thus lending support to Robert's arguments.[89] Notably, Matthaei's name was not on the paper; he had returned to Germany and joined the Max-Planck Institute for Biology in Tübingen (then the principal German center for molecular biology). Soon after, he settled down at the Max-Planck Institute for Experimental Medicine in Göttingen.

By the end of 1962, when Crick received the Nobel Prize (shared with James Watson and Maurice Wilkins for elucidating DNA's structure), the uncertain status of the code—broken yet unsolved—detracted from his satisfaction. Responding to Delbrück's congratulations, Crick wrote, "As Jim always thinks of you as his scientific father I always regard you as my scientific uncle. The phage field which you created has been the ground in which molecular biology grew, and these years it really seems to have arrived! All the same it would be even nicer if we could solve the code."[90] (Delbrück, by the way, was awarded the Nobel Prize in 1969 for his phage studies.) Nearly a decade after Watson and Crick signified the DNA sequence as "the code that carried genetical information," and despite Crick's recent predictions that the code (the correlation of bases and amino acids) would be solved within a year, many unresolved questions about its structure and content remained: the *sequence* of a coding unit; its size (triplet or doublet); mode of "reading" the code; and its universality. These were the problems Crick chose to discuss in his Nobel address, entitled "On the Genetic Code." He continued to preside over the code problem through contemporary articles in *Scientific American* (Nirenberg's sequel to Crick's article on the genetic code appeared only months later), and in *Discovery* (*Scientific American*'s British counterpart). At the end of his Nobel lecture, Crick concluded ceremoniously, "We are coming to the end of an era in molecular biology. If the DNA structure was the end of the beginning the discovery of Nirenberg and Matthaei is the beginning of the end."[91]

His coding presidency was extended in his comprehensive review, "The

Recent Excitement in the Coding Problem" (presumably a sequel to his pessimistic 1959 resume, "The Present Position of the Coding Problem").[92] With his sequence hypothesis (or Central Dogma) as a point of departure, Crick surveyed the mechanism of protein synthesis and offered a glossary of terms, christening the coding unit as the *codon* (apparently the term was actually coined by Brenner). Most of the evidence rendered unlikely either fully or partially overlapping codes, he stated. Though with ingenuity, it was possible to construct overlapping schemes that fit the data (referring to Wall's work). "Is the message read from one end?" His genetic evidence supported that proposition and was compatible with recent biochemical findings, he reported. It also indicated the codon size to be three bases, though a doublet could not be ruled out. Is the code universal? With few exceptions, comparison of results from different bacterial species, TMV, and mammalian hemoglobin suggest universality.

Admitting that Nirenberg and Matthaei's "discovery completely revolutionized the biochemical approach to the coding problems," Crick was nevertheless critical of Nirenberg's and Ochoa's work. "There are so many criticisms to be brought against this type of experiment that one hardly knows where to begin," he said, referring mainly to the unknown base composition of the polynucleotides, but also to the assumptions and interpretations of the researchers.[93] He did acknowledge, however, that their work established significant aspects about protein synthesis and definitely disproved Sinsheimer's clever two-letter code. Comparing the Nirenberg-Ochoa results with those derived from amino acid replacement data from TMV and human hemoglobin, Crick pronounced, "The cell-free system is not a complete artifact, and is related to genuine protein synthesis."[94] He conveyed appreciation for Woese's theoretical attempts to deduce the whole structure of the code from only part of it, accomplishing his codon assignment with remarkable fidelity. Crick firmly believed that good theoretical work—both interpretations of experimental data and pure (paper) theorizing—were essential to the code project. But, he lamented, there was too much bad theorizing around.

> In the long run we do not want to *guess* the genetic code, we want to *know* what it is. It is after all, one of the fundamental problems of biology. The time is rapidly approaching when the serious problem will be not whether, say, UUC is *likely* to stand for serine, but what evidence can we accept that establishes this beyond reasonable doubt. What in short, constitutes proof of a codon?[95]

Crick concluded his comprehensive survey by underscoring the uncertainties behind the recent excitement.

Although unsolved, "the coding problem" with its scriptural and informational representations had by then permanently reoriented the conceptual framework and material practice of protein synthesis research. The semi-permeable boundary between biochemistry and molecular biology had become nearly porous, as practitioners recast both fields as information sciences. In a symposium in honor of biochemist and Nobel laureate, Albert Szent Gyorgyi, Ochoa, who only three years earlier conducted his studies of RNA synthesis within the paradigm of metabolic chemistry, reframed them in informational terms. This was certainly not a popularization for a lay audience or a mere expository device, but a re-representation of RNA synthesis from a genetic standpoint and around Crick's Central Dogma of information transfer. Ochoa explained in his paper, entitled "Enzymatic Mechanisms in the Transmission of Genetic Information," "The translation of the DNA code into the corresponding code of the RNA messenger becomes therefore the focal point in the transmission of genetic information."[96] Note that he used the code idiom not to denote a correlation between codons and amino acids, but to signify the nucleotide sequences (even adding an RNA code). There were now four meanings to the genetic code.

The subsequent presentation in the Szent Gyorgyi symposium, "A Book Model of Genetic Information—Transfer in Cells and Tissues," by Chicago biophysicist John R. Platt (Leo Szilard's friend then visiting at MIT), exploited the scriptural analogies even further. "The expression of genetic information in cells and whole organisms is like the reading out of a complex instruction manual, but the analogy extends to more details than is generally realized," he stated in the abstract.

> The information is linearly arranged in "words" that are "read out" sequentially in time. There is one copying mechanism (DNA polymerase) for reprinting the whole book, and another (RNA polymerase) for selecting readout into cell chemistry. The read-out is by "paragraphs" (genes) and by pages (operons) that can either be "closed" (repressed) or "opened" (induced), according to contingent "instructions" (repressor-corepressor complexes) from "references" (regulator genes) on earlier pages or in the "books" of adjacent tissues. . . . Between "readings," the "books" can be "locked away" in compact "storage" forms (phage heads, chromosomes).[97]

This Book of Life metaphor had multiple allusions. It was simultaneously envisioned as a biochemical manuscript, a printed manual, and an electronic text (recall that the operon was visualized as a magnetic tape of a computer program and ribosomes as "reading heads," as discussed in Chapter 5). As in previous instances since the 1950s, Platt's scriptural representations of the code governing that cellular "reading" revealed the ambiguity inherent in the triple sense of the code idiom: the DNA code, the amino acid code, and the code as the correlation between the two. Like others, he finessed his way out of the semiotic conundrum of intertextuality and its deconstruction as a code of codes: "If we are to have two different kinds of codes related to each other in cellular biochemistry—the base-sequence codes of the nucleic-acid chain, and the amino-acid sequence code of the protein chain—we must have, at every point of interaction between them, 'translator molecules' containing *both* codes and able to 'speak' both languages."[98]

The biochemists' shift to the information discourse was equally striking at the "Symposium on Informational Macromolecules" in the fall of 1962 (where Nirenberg presented his latest revisions of the code), an enormous gathering of two hundred and twenty-five life scientists at the Institute of Microbiology of Rutgers University. The symposium's mandate formally blurred the boundaries between biochemistry and molecular biology through the information discourse and its scriptural representations.

> In the area of molecular biology, efforts converging from various disciplines have recently led to breakthroughs in our understanding of the molecular bases for the storage, transmission, and expression of genetic information. These efforts largely concern the synthesis and functioning of what may be viewed as the two fundamental classes of biological polymers, the nucleic acids and proteins. The bodies of knowledge regarding these two kinds of macromolecules are now being unified through the genetic code, a concept of broad scope and high precision, which furnishes the key to the translation of the languages of nucleotides into the language of amino acids. In such translation, particular interest attaches to the informational macromolecules which carry instructions for the amino acid sequence in proteins.[99]

Chemical specificity, formerly the overarching theme in biochemistry and other life sciences, had been recast as information transfer: carriers of specificity became instruction carriers. What is in this name change? The information discourse paired down the multitudes and complexities of biochemical pathways involving myriad different molecules to the primordial binaries of life: nucleic acids and proteins.

Joseph Fruton, biochemist and historian of biochemistry, has been skeptical all along about these metaphorical constructions borrowed from information theory and has questioned the significance of the information discourse in biochemistry, although he granted that its idealized models had stimulated important empirical discoveries.[100] But the information discourse did a great deal more: it reconfigured biochemistry around the precepts of molecular biology toward gene-based conceptualizations of animate matter, as evidenced by the structure of the symposium. The sessions on polynucleotide properties and synthesis, protein structure and synthesis, and the genetic code—with few exceptions, presented by biochemists—were obviously conceptualized around the Central Dogma of unidirectional information transfer from nucleic acids to proteins. Protein synthesis had become a programmed communication system.

An equally prominent discursive turn was visible at the International Colloquium on Information in Biology held at the end of 1962 at Abbey of Royaumont, near Paris. "The aim of the organizing committee was to bring together workers interested in various fields of biology in which problems of information are faced—genetics, biochemistry, immunology, embryology, neurophysiology." Max Delbrück chaired a seminar on genetic information, which occupied the first half of the colloquium, where Nirenberg, Lengyel, Bretcher, Wittmann, and Woese each reported recent interpretations of the coding problem. Benzer and Alan Garen from Yale presented updates on mutagenesis. Jacob outlined the problem of "transfer of information from the gene to the machine which makes proteins," graciously spotlighting the support his work had received from Nirenberg's work (whom he had twice rejected: once from the Pasteur Institute and once from Cold Spring Harbor). François Gros had accounted for various RNA fractions in Nirenberg's system and Sol Spiegelman reported on phage RNA. A series of lectures on "information in protein synthesis" was followed by Melvin Cohn's presentations on immunology, C. H. Waddington on embryonic differentiation, and several lectures on feedback mechanisms in enzymatic regulation, concluding with the problem of memory as an information storage process.[101] Obviously far removed from the mathematical theory of communication, in its generic and multivalent form the information discourse was nevertheless remarkably productive. While functioning operationally (albeit not always consistently) as a model-generating metaphor for the transport of genetic specificity within molecular biology, it also served to link that community, through a dialectical process, to informational discourses of other life and social sciences and to the wider culture of the atomic age.

The *New York Times* assumed an important function in this cultural linkage by closely following and interpreting the relevant advances. Continuously touting the enormous potential impact of the genetic code as the agent of life's information, the newspaper portrayed the code as the spearhead of an impending scientific revolution and as a problem to be solved within a year. The paper announced in a January 1962 article, entitled "Structure of Life," "In its quest for understanding the chemistry of life and the basic mechanism of heredity, whereby all things living reproduce themselves in their own image, the science of biology has reached a new frontier said to be leading to "a revolution far greater in its potential significance than the atomic and hydrogen bomb." The journalist concluded by adding a warning by the Swedish biochemist and Nobel laureate Arne Tiselius that "these new discoveries if misused will lead to methods of tempering with life, of creating new diseases, of controlling minds, of influencing heredity, even perhaps in certain desired directions." In a later *New York Times* article biochemist Erwin Chargaff was quoted as issuing a "clarion call for prudence," pointing out that some of the findings were inconsistent with well-established experimental data. Chargaff and his colleagues gathered at a meeting at Columbia University concluded, "It is clear that not much is clear."[102]

Reporting in yet another article, entitled "New Gains Cited on Genetic Code," from a meeting held at Indiana University, the *New York Times* described the field as "moving at a pace so fast that even those scientists who are most deeply involved in it are dazzled."[103] And an impressively long *Times* article, "Biologists Hopeful of Solving Secrets of Heredity This Year," forecasted:

> Biology is undergoing a revolution whose meaning and magnitude have become apparent only in recent weeks. The pace has been so swift, in fact, that many scientists do not fully appreciate the potential of the powder keg they are sitting on. . . . That assessment is based upon the great likelihood that the chemical code of inheritance . . . will be cracked before the year is out. . . . Indeed, it is safe to say that some of the biological "bombs" that are likely to explode before long as a result of that achievement will rival even the atomic variety in their meaning for man.

The newspaper even published the partially solved "Amino Acid Code" as it then stood (the Nirenberg-Ochoa U-rich codes). As Jean Baudrillard later observed, these codes were icons of the postwar culture of simulation: "Between the two, caught between the nuclear and the genetic, in the simultaneous assumption of the two fundamental codes of deterrence, every principle of meaning is absorbed, every deployment of the real is impossible."[104] The *New York Times* also covered the enormous "Symposium on Informational Macromolecules," quoting Nirenberg's and Ochoa's belief that the code will be soon fully known.[105] And about a month later, at the dedication of the new science center at Case Western Reserve in Cleveland, the newspaper announced, based on "mounting evidence" (mainly from viruses and bacteria but also human hemoglobin) that the code was probably the same for all life forms. "The Genetic Code Is Held Universal," it proclaimed.[106]

Universality was of course a highly prized feature. If true, then on the phenomenological level it would elevate the genetic code to the pedestal of universal laws of nature, a privilege generally reserved for the Olympian reaches of physics. On the technological and social level universality would open the door for genetic and biomedical engineering. Joshua Lederberg predicted that in "no more than a decade" the molecular knowledge of microbes would be applied to the human genome.[107] Pragmatically, universality meant economy. The ubiquitous studies conducted in E. coli and phage would be then valid for all organisms—from bacteria to elephants, as Monod put it—and that well-tested experimental system could be further interrogated in order to determine the assignments of codons to amino acids and to ascertain *biochemically* (not just by genetic inference) that the coding unit was indeed a nucleotide triplet. In the following four years (1963–67), after a two-year lull, new precision techniques for synthesizing trinucleotides of known sequences, developed by Har Gobind Khorana, and an ingenious technique for binding trinucleotide messengers to ribosomes, devised by Leder and Nirenberg, would establish biochemically the triplet nature of the code and fix the "dictionary" of codons and amino acids. The deployment of bacteriophage technologies and genetic analyses by Brenner, Garen, and their coworkers led to the identification of the function of the nonsense codons for chain termination. Soon after, Nirenberg would present a compelling demonstration of the code's universality across phylogenetic divides from bacteria to mammals; the implications for genetic engineering were staggering.

FIXING A (UNIVERSAL?) DICTIONARY

The year 1963 was an intellectual, institutional, and social turning point for molecular biology. While according to Crick the previous few years' events had marked the "beginning of the end," Stent viewed the changes as ushering in the "Academic Phase" of molecular biology (the third and last in the progression from the "Romantic Phase," 1938–53, to the "Dogmatic Phase," 1953–63), identified by so-called mopping up operations of "normal science."[108] Institutional and funding patterns changed as well. The assassination of John F. Kennedy in November 1963 triggered a chain of events with profound effects on the PHS, beginning with Lyndon Johnson's journey of legislative activism toward "The Great Society." In the spirit of the early days of the New Deal, Congress enacted laws aimed to combat the "War on Poverty," Medicare and Medicaid, Model Cities, the Voting Rights Act, as well as an avalanche of health-related legislation. Conversely, the exponential growth of NIH budgets (at 40 percent annual increase) dropped to a steady state of about 6 percent annual increase. But these cuts, in fact, represented the price of international success. NIH grants flowed to many foreign institutions (NIH had offices in Paris, Tokyo, and Rio de Janeiro), which together with grants and contracts from the United States Air Force, Army, and Navy as well as the support of the Rockefeller Foundation had shaped the molecular life sciences in the postwar era. By 1963 the United States was preeminent in biomedical research, and molecular biology became the model for biology in the developed world.[109]

These shifts in fiscal patterns reflected policy decisions in Congress and private foundations to wean European science off American dependency and step up assistance to underdeveloped countries. Cumulatively, they had an immediate and lasting impact on the international community of molecular biology. In response to the cuts, intended to stem the brain drain and restore the Old World as the center of science, European countries began to invest heavily in research infrastructures based on the American paradigm, to which so many European postdoctoral fellows (like Matthaei) had been exposed: "well-financed, gossipy, and democratically organized."[110]

In 1963 the British Department of Scientific and Industrial Budget jumped to $2.65 million (from $560,000 in 1959), and the Medical Research Council's budget had doubled to about $20 million (the principal source of support for British molecular biology). In Sweden the budget rose from $1.7 million in 1963 to $2.4 million in 1964. The total income of the Max Planck Society climbed from $1 million in 1961 to more than $30 million in 1963 (then about one thousand scientists and forty-one institutions); and about $50 million was earmarked for university construction in the coming decade, with special attention to molecular biology. In the French Fifth Republic $8.2 million for a five-year period, beginning in 1961, was allotted for molecular biology (due mainly to the campaigns of the Pasteur group). The budgets of the National Institute of Hygiene, CNRS, and the Pasteur Institute increased severalfold, resulting in vastly improved research facilities. According to Victor McElheny, European correspondent for *Science* magazine:

> Professor Marianne Grunberg-Manago of the Institute de Biologie Physico-Chimique in Paris, one of the beneficiaries of the "action concertée" in molecular biology, takes visitors to look at remodeled laboratories crammed with new equipment. She says that there has been a "dramatic improvement" in the last 5 years. George Cohen of the CNRS biology laboratories in Gif-sur-Yvette [who nearly accepted an offer from NIH after years in Monod's laboratory] says that much of the equipment shortage has been solved. "Now if something is wrong," he remarks, "we know it is with *us*."

The French investments intensified after Monod and Jacob used their 1965 Nobel Prize (shared with Andre Lwoff) as a platform for promoting molecular biology and scientific reforms. Beyond these national initiatives, the plans for the European Molecular Biology Organization (EMBO) were now receiving institutional and financial commitments (the vision of EMBO was promoted by Leo Szilard to create a union of researchers in Europe and Israel and an international laboratory akin to CERN). The molecularization of biology was produced through international authorship and through a global process tied to forecasts of immense technological, economic, and social potentials.[111]

Indeed, 1963 also marked the beginning of the American debates on the future implications of biological engineering (enabled by the genetic code) and on the social responsibility of scientists. At the symposium "Man and His Future," sponsored by the Ciba Foundation, two-score distinguished scholars, mainly biologists (among them Hermann J. Muller, Joshua Lederberg, and Crick), discussed the scope and limits of a new eugenics. Researchers at the exceptionally large Eleventh International Congress of Genetics (convened at The Hague) speculated about controlling and creating life. After its lengthy coverage of this congress, a *New York Times* editorial issued warnings on the new biopower emanating from the code.

> Is mankind ready for such powers? The moral, economic and political implications of these possibilities are staggering, yet they have received little organized public consideration. The danger exists that scientists will make at least some of these God-like powers available to us in the next few years, well before society—on present evidence—is likely to be even remotely prepared for the ethical and other dilemmas with which we shall be faced.[112]

The response of Basil O'Connor, president of the National Foundation (a significant patron of molecular biology), conveyed lesser concern about the consequences of the new knowledge than about the "rash of premature fears, superstitions and resistance." He wrote, "To all these the scientist has one answer: his job is the pursuit of knowledge, wherever it may take him, however lovely or unlovely, safe or dangerous—and however prepared or unprepared the world may be to cope with the truth he sets before it. Where there is danger of knowledge, the only safeguard is more knowledge." This was a troubling statement for one who had witnessed the technological sweetness and destructive powers of the atomic bomb.[113]

Similarly, at a conference on the "control of human heredity and evolution," prominent phage geneticist Salvador Luria expressed his position on the issue, "Our task was to discuss only the technicalities of the work that could conceivably be done. . . . I expect that the ethical and moral issues will be the topics of much discussion later."[114] The National Academy of Sciences, however, took a more thoughtful stand. In a prelude to the celebration of its centennial, its president, Frederick Seitz, expressed both excitement and caution toward the new biological findings (though his greatest concern was the recent congressional cuts in the appropriations to science). A *New York Times* article reported, "Dr. Frederick Seitz warned that although advances toward completely cracking the genetic code could conceivably lead to great benefits, there might be unforseen adverse effects of applying such knowledge directly to man. Experiments in these fields should be closely watched 'to assure they don't get out of hand,' he said."[115] At stake, in all these debates, were not only adjudications between scientific truths and ethical concerns but also the perceived imperatives of economic competitiveness and national security.

These dimensions of the promise and peril of the new biology soon became embedded in the larger debates defining the academic political activism of the Vietnam War era: resistance to the war, the industrial-military-academic complex, and biological warfare; challenges to the constitutionality of the loyalty oath and the House Committee on Un-American Activities (HUAC); and renewed commitments to the social responsibility of science.[116] Nirenberg eventually voiced concern about the social implications of the genetic code and wondered about society's preparedness and wisdom to manage the impending sweeps of biological engineering.[117] However, in 1963 Nirenberg, who was by then a member of the National Academy of Sciences, winner of the Academy's Molecular Biology Award, and a recipient of numerous offers from leading institutions, including the Virus Laboratory at Berkeley, had only one goal—to complete the genetic code—a task soon facilitated by Khorana's technical feats of preparing defined RNA messengers.

Like Ochoa and Kornberg, Har Gobind Khorana's interests in nucleic acids dated back to the early 1950s; not as a response to Watson and Crick's elucidation of DNA's structure, but as an outgrowth of his interest in energy exchange in metabolic processes. From humble beginnings in the only literate family in a Punjabi village of one hundred people and educated through monthly visits by an itinerant teacher, Khorana advanced to a master's degree in science at the Punjab University in Lahore. A Government of India Fellowship in 1945 enabled him to go to England for doctoral studies in organic chemistry at the University of Liverpool. After a succession of fellowships in laboratories in Zurich and Cambridge, England, and exposure to analyses of proteins and nucleic acids (especially to Alexander Todd's Nobel-winning studies on internucleotide linkages in nucleic acids), Khorana landed a job at the University of British Columbia in 1952. This position offered meager facilities but "all the freedom in the world" to pursue his research. Together with a few colleagues he began work "in the field of biologically interesting phosphate esters and nucleoic acids." That research greatly expanded after his move in 1960 to the well-endowed Institute of Enzyme Research at the University of Wisconsin (renowned for its biochemistry department).[118]

Up until that time Khorana had little contact with molecular biology or the coding problem (while eventually he came to analyze his results in terms of the then-dominant scriptural representations, his structural investigations could benefit little from informational models of macromolecular function). Kornberg remembered that once when he, Crick, and Khorana happened to be together, Crick asked what prompted their work on DNA. "Gobind answered that his success in the chemical synthesis of ATP [the energy currency in metabolic processes] led him to the more complex coenzyme A, which in turn led him to more and more difficult condensations of chains of nucleotides."[119] That was back in the 1950s. In 1961 Khorana, a leading authority on polynucleotide synthesis ("someone like a God," according to Nirenberg),[120] became strongly motivated by recent developments. The field of polynucleotide synthesis was burgeoning around the cascade of findings: the isolation of DNA polymerase and DNA-dependent RNA polymerase, the production of synthetic polynucleotides, and means of tracking of tRNA as it bound to ribosomes. In a constant dialectic, these epistemic things quickly turned into technical things, that is, into molecular tools for interrogating other molecular processes and entities. Khorana's research interests were now shaped within the representational space reconfigured by the convergence of biochemistry and molecular biology, nucleotide and protein synthesis, and the genetic code, through the merging of matter and information. Nirenberg and Matthaei's breakthrough, its subsequent elaborations by Nirenberg's and Ochoa's groups, and the unresolved problem of codon assignment captured his interest. There was no way then to ascertain a codon's length or sequence from merely knowing the synthetic RNA's composition. To solve the code, to determine which codon of a given empirical formula designated an amino acid (for example, which of the three permutations of U_2G—UUG, UGU, or GUU—represented valine, or cysteine, or leucine) the precise base order in the messenger had to be known. "The hope in my laboratory was to prepare ribopolynucleotide messengers of completely defined nucleotide sequences," Khorana explained.[121]

That task proved to be far more challenging technically than anticipated. The chemical technology for RNA synthesis lagged behind the virtuosity of DNA synthesis; in contrast to DNA chains of ten to fifteen bases (prepared with Kornberg's enzyme) only short RNA segments containing a handful of bases could be crafted. To circumvent this limitation, Khorana used the recently isolated RNA polymerase, with it hoping to synthesize ("transcribe") an RNA messenger off a known DNA sequence and match it in length and complementary bases. But the elegant scheme quickly got snagged on technical incongruities. Rather than producing complementary copies equal in length to the DNA segment, the RNA product was always much longer—more than a hundred nucleotides. After initial discouragement, that feature was turned into a useful tool. The reiterative synthesis (or "copying") would become a device for amplifying the "messages" in the short synthetic polynucleotide.[122]

Some months later (early 1963) Khorana began experiments with DNA polymerase in Kornberg's laboratory at Washington University ("one of the many pilgrimages that I have made to this great laboratory," he wrote). There he succeeded in obtaining a DNA polymer comprising alternating A and T bases (from which complementary RNA messenger with alternating U and A bases could be synthesized). "Everything from this point on [spring of 1963] went remarkably well," Khorana recounted. His paper (coauthored with Arturo Falaschi and Julius Adler), entitled "Chemically Synthesized Deoxynucleotides as Templates for Ribonucleic Acid Polymerase," was submitted for publication in April 1963. Meanwhile, the intricate syntheses of precisely defined di- and tri-nucleotides proceeded smoothly.[123] But less than a year later (well before Khorana's custom-made messengers became available), Nirenberg and the remarkably talented postdoctoral fellow Philip Leder devised an ingenious procedure for directly determining the size and sequence of codons.

A native of Washington, D.C., graduate of Harvard College and Medical School, and veteran of summer jobs in NIH labs, Philip Leder was awarded an NIH Research Fellowship in 1962. Apparently his consultations with Robert Martin (a Harvard classmate) reinforced his decision to work with Nirenberg; he joined the laboratory in 1963. Testing Khorana's scheme of using a short synthetic DNA segment comprising Ts only (oligo dT) together with RNA polymerase in their cell-free system, they demonstrated within a couple of months that the resulting messenger poly-A indeed directed the synthesis of amino acid polylysine (well established from the amino acid assignments of Ochoa's group). They presented their findings at the Cold Spring Harbor symposium "Synthesis and Structure of Macromolecules," which devoted two sessions to the "Amino Acid Code," including not only biochemical approaches to the code but also amino acid replacement studies in E. coli (Yanofsky) and TMV (Wittman). (The year 1963 also happened to be a turning point in the history of Cold Spring Harbor: the withdrawal of the Carnegie Institution of Washington—its patron since Cold Spring Harbor's founding in 1904—initiated a complete reorganization of the laboratories and their administration.)[124]

Nirenberg's Cold Spring Harbor paper, "On the Coding of Genetic Information," was instructive also for its discursive practices and linguistic slippages.[125] Clearly the movement between the definitional differences kept destabilizing the meanings assigned to these molecular entities. In presenting (graphically) their correlations between "observed frequency of amino acids" and "theoretical frequency of RNA code words," the authors worked at constructing their analyses (now probably knowingly) within the scriptural framework of Gamow, Yčas, and the RNA Tie Club. But as was established already in the 1950s, the random (Poisson) distribution of amino acids in proteins (analogized with letters, not words) contradicted the non-random (neighbor-restricted) letter distribution in any known language. And like other code researchers, Nirenberg got tangled in the inconsistencies between analogical definitions of letters (nucleotide bases or amino acids?) and words (codons or amino acids?); in fact he had (unconsciously) replaced his earlier definition of amino acids as words and RNA bases as letters.

The shifting uses of the code idiom revealed even more: in a self-erasing manner, the table of correlations of codons and amino acid was now termed the "code word dictionary." Was it a code? Or was it a dictionary? (A dictionary, by definition, contains words; analogically, in coding theory too the dictionary was defined as the set of k-letter words. See my related discussion in Chapter 4.) These analogical incongruities inhered in the linguistic significations of the *two* entities: nucleic acids and amino acids. For the analogy to be consistent, there had to be two related "*texts*" (as Platt pointed out)—the DNA message, and its protein "translation" (via RNA "transcription")—written in *two* kinds of codes. Their correlation was a problem of intertextuality, or a *code of codes*. Indeed, with the dictionary metaphor, the code metaphor, though still guiding the scientific imagination, had in effect become logically redundant. It would become even more so with the mounting observations of the complexities and contingencies of its functions. Most scientists, however, gave little thought to such semantics.

While Khorana crafted the defined oligonucleotides, a technically demanding and labor-intensive process, Nirenberg and Leder announced a major breakthrough at the Sixth International Congress of Biochemistry in New York (July 26 to August 1, 1964). To establish the sequence of RNA code words, they devised a rapid and sensitive method for measuring the direct effect of trinucleotides on the binding of transfer RNA to ribosomes. In fact, they had already submitted their paper to *PNAS*, using the new technique to determine amino acid assignments. The method was based on technically intricate findings emerging simultaneously from several laboratories. Those studies showed that an aminoacyl-tRNA (the complex comprising an activated amino acid attached to tRNA) could bind to ribosomes in the presence of a synthetic RNA messenger containing the corresponding triplet (for example, phenylalanine-tRNA bound to ribosomes only in the presence of poly-U).[126]

Through careful manipulations, Leder and Nirenberg discovered that, while aminoacyl-tRNA itself did not stick to disks of nitrocellulose filter paper, it did stick when bound to ribosomes, which also adhered to the disks. They quickly learned that even an isolated triplet effected the bind-

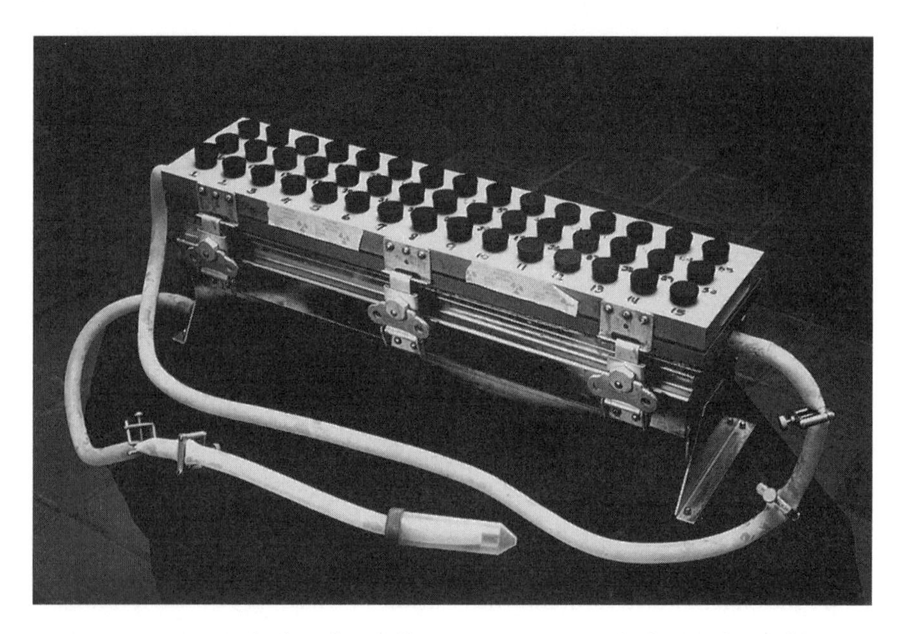

FIGURE 38. Multiple Millipore filtration apparatus ("multiplator") designed by Philip Leder for determining the sequence of bases in triplet codons. Courtesy of the DeWitt Stetten, Jr., Museum of Medical Research, National Institutes of Health, Bethesda, Md.

ing of the aminacyl-tRNA. Testing the procedure on the undisputed assignments of amino acids, they confirmed that pUpUpU directed the binding of phenylalanine-tRNA (p to the left of a base indicates 5′-terminal phosphate; to the right a 3′-terminal phosphate, thus defining the direction, or polarity, of "reading" the message); pApApA of lysine-tRNA; and pCpCpC of proline-tRNA. A doublet had no binding effect; hence the minimum size of the coding unit was three (though the mechanism turned out to be far more subtle, as I will explain below). The coding impasse was cleared. They had established a quick and reliable experimental approach for determining base sequences of the sixty-four codons, one at a time. "Again the race on," this time against hundreds of researchers including senior scientists, foreign visitors, postdoctoral fellows working in American laboratories, and laboratory technicians (many of them women). "And again NIH came to Marshall's aid," with collaborations and an army of postdoctoral fellows.[127]

"Breaking the Code. New Approaches Are Suggested in Understanding Basis of Life," the *New York Times* announced in its coverage of the International Congress. Recounting the sensational story of the 1961 Moscow Congress and the "triumph . . . hailed by scientists throughout the world," they reminded readers of Crick's prediction that the entire coding problem would be solved within a year. "Well, it was not and still is not solved." However, it continued:

> Last week . . . Dr. H. Gobind Khorana of the University of Wisconsin and Dr. Nirenberg revealed new approaches to the coding problem that promise to overcome most of those difficulties if not actually solve the problem. . . . Dr. Nirenberg's approach is simply to make or isolate single triplets, get them to match up with single amino acids and then "look" at the combination chemically, thereby reading the code word for word, as it were. Dr. Khorana's technique is by straight chemical means, to put DNA nucleotides together one by one in six-member chains or to hook three or four three-member chains together, all of known sequences. With such synthetic DNA fragments, he then makes synthetic RNA's of known sequences which make synthetic protein whose amino acid sequence can be analyzed and finally decoded.[128]

It sounded rather straightforward. Indeed, Nirenberg and Leder's fractionation of poly-(U,G) digests quickly yielded three trinucleotides, GUU, UGU, and UUG, shown to specify valine, cysteine, and leucine, respectively. Trinucleotide synthesis, however, proved to be a major technical obstacle. Heppel offered guidance, materials, methods, and cooperation, yielding by 1965 many trinucleotides. Soon after, the correlation of forty-five trinucleotides with amino acids resulted in a cascade of papers from Nirenberg's laboratory.[129]

The lab staff's 1965 paper, "RNA Codewords and Protein Synthesis, VII. The General Nature of the RNA Code," was particularly instructive. Unifying findings from disparate sources, it spotlighted many of the contingencies in the code's features, the subtleties of its recognition apparatus, and its convoluted patterns. With nearly 70 percent of the code determined and additional genetic data accumulated, it was now possible to reassess the question of nonsense codons. Through intricate recombinations of phage mutants ("amber"), Brenner's group had just demonstrated the (previously assumed) colinearity of the gene with the polypeptide chain (shown simultaneously by Charles Yanofsky with his E. coli tryptophan synthetase system). Brenner's results, demonstrating that UAG or UAA might specify the termination of a polypeptide chain in E. coli, were also obtained independently in Alan Garen's laboratory at Yale. (Somewhat later updates from Cambridge, England, suggested that UGA too was a chain terminator.) From biochemists' standpoint, however, these findings constituted indirect evidence. And Nirenberg's laboratory confirmed biochemically that these two codons belonged to a group that did not specify any amino acids.[130]

Furthermore, Nirenberg could now attempt to resolve the doublet versus triplet coding conundrum. For while Leder and Nirenberg had demonstrated that the minimum codon size was three nucleotides, previous studies with randomly ordered polynucleotides showed that "synonym" codons (coding for the same amino acid) often differed in composition by only *one* base. This suggested that bases common to synonym codons occupied identical positions in the sequence and that either two out of three bases in a triplet might be sometimes recognized, or a base might be recognized correctly in two or more ways (Woese's and Eck's theoretical analyses offered insights in explaining these incongruities). "Several generalizations can be made concerning the code," Nirenberg's group concluded: structurally and metabolically related amino acids often have similar codons; many codons may be recognized partially; recognition at the 3′ end of the codon was most variable; and in most cases the apparent template activity of one member of a set of synonyms differed from that of another set. These complex and contingent patterns—hardly a predictive code—seemed to define the general recognition mechanisms, notably the interactions of tRNA molecules with the triplets in messenger RNA. "The nucleotide sequence found for an alanine codon (GCU found, GCC, GCA and GCG predicted), together with the nucleotide sequence of an alanine sRNA [tRNA] isolated from yeast, reported by Holley et al., may provide clues to the recognition process," they stated in their conclusion, referring to the work of Robert Holley at Cornell, who had just elucidated the salient features of that recognition mechanism, by all accounts a stupendous accomplishment.[131]

A modest and reserved Midwesterner from Urbana, Illinois, Robert W. Holley received his graduate training in organic chemistry at Cornell in 1947, becoming (after a completing a postdoctoral fellowship at Washington State University) an assistant professor of organic chemistry at Cornell's Geneva Experiment Station and later a research chemist at the United States Department of Agriculture Laboratory on the Cornell campus. By 1964 he was a professor of biochemistry and molecular biology at Cornell. His long-standing interest in the chemistry of natural products eventually led him toward more complex biological problems, to amino acids and peptides, and finally to the central problem of protein synthesis. By the time Nirenberg had demonstrated that tRNA was an obligatory intermediate in the formation of phenylalanine (1962), Holley had already been working on tRNA for five years.[132]

He had pursued Zamecnik's (and others') research path, which had shown that the cell contained many kinds of tRNA molecules, each specific for a particular amino acid. By 1961, it was known that a series of enzymes, each specific for one amino acid and a particular tRNA, catalyzed the formation of a covalent linkage between the amino acid and the tRNA (forming the aminoacyl-tRNA complex). Even before Nirenberg's experiments, researchers strongly suspected that tRNA transported amino acids during protein synthesis. Early work on the mixture of all these tRNAs showed them to be short. At less than a hundred nucleotides per chain, they were the smallest biologically active nucleic acids known—similar, if not identical, in the nucleotide composition at one end of the chain. Also, they contained small amounts of unusual nucleotide bases (other than A, G, T, and C). Mapping the nucleotide sequence of a tRNA was a formidable task, requiring major technical innovations: methods for isolating a particular tRNA from a large mixture of similar molecules and procedures for sequencing its bases. After five years of work, Holley's group announced early in 1965 that they had mapped the entire sequence of tRNA specific for alanine (in yeast).[133]

As Holley pointed out, in reference to Nirenberg's codon assignment for alanine, a significant feature of their structure was that several trinucleotide sequences emerged as candidates for the coding triplet, or "anticodon," for the transfer of alanine. The term *anticodon* had just been coined to denote the tRNA triplet complementary to the codon in the RNA messenger. Thus Holley's structure and its base sequences explained not only tRNA's function but also could be used as a probe into the code. Studies of anticodons on various tRNA could now become the "Rosetta stone" envisioned back in 1958. Holley's findings inspired Crick to propose a model for adjusting his adaptor hypothesis and for explaining some of the contingencies of coding, a scheme for codon-anticodon pairing aimed at resolving the apparent contradiction between doublet and triplet codons and a molecular mechanism for explaining degeneracy. Crick's resulting "wobble hypothesis" allowed some leeway in the binding of codons at the rounded extremity of the pinlike tRNA structure (see Fig. 39). "While the standard base pairs may be used rather strictly in the first two positions of the triplet, there may be some wobble in the pairing of the third base. . . . Such a wobble could explain the general nature of the degeneracy of the genetic code," Crick suggested, theoretically validating Holley's tRNA structure. Given the formidable challenge and the scope of Holley's findings, biochemists' and molecular biologists' admiration for Holley's work was understandable.[134]

By then (as the *New York Times* reported), Khorana had perfected his technique for synthesizing long RNA chains of completely defined sequences, obtained through an elegant combination of chemical and enzymatic methods. In response to Nirenberg and Leder's efficient technique for binding trinucleotides, he had also prepared and characterized all sixty-four trinucleotides (all permutations of A, C, U, and G).[135] With Khorana's supply of precision-made trinucleotides and Nirenberg and Leder's binding method, it was possible to work out the entire code within the next year (namely, all amino acids were correlated with codons). About a dozen codons remained unaccounted for because their binding tests had been either negative or ambiguous. The meaning of these triplets, including nonsense codons, still had to be determined, which became the subject of subsequent studies utilizing other

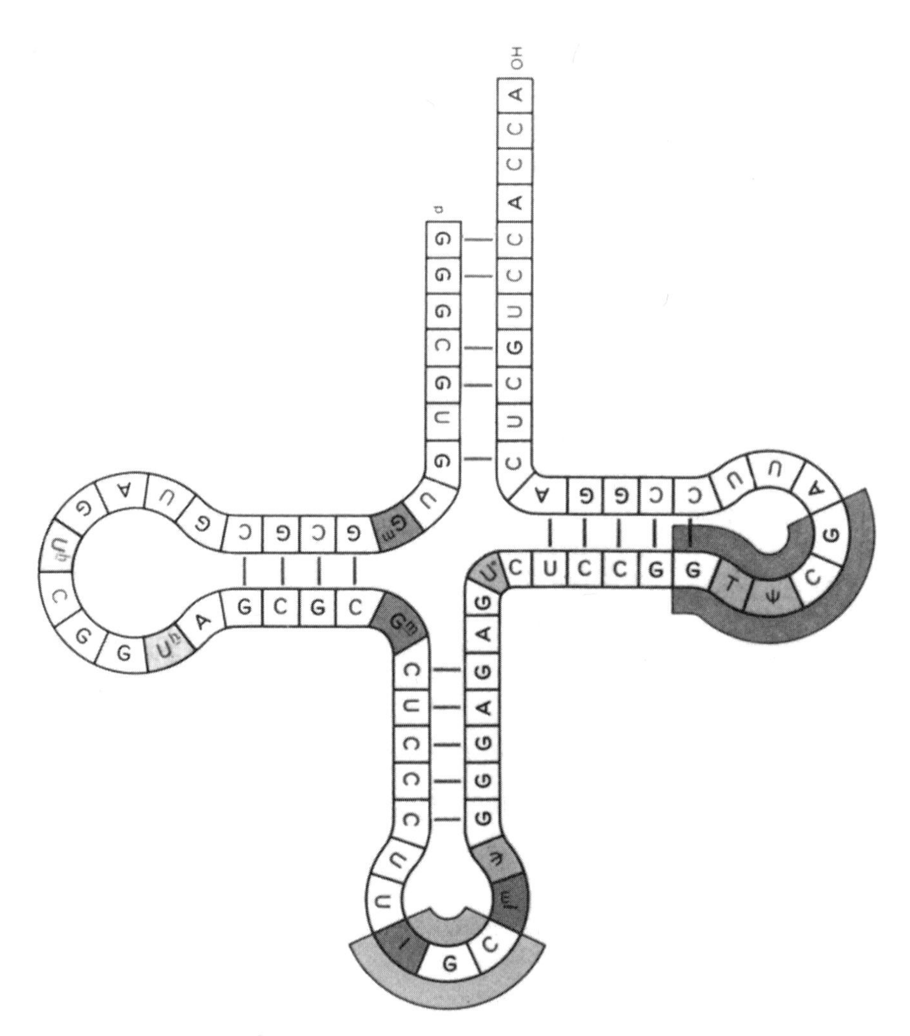

FIGURE 39. Robert Holley, "The Nucleotide Sequence of a Nucleic Acid," *Scientific American* 214 (1966): 31. Reprinted by permission. The caption read: Hypothetical models of alanine transfer ribonucleic acid (RNA) show three of the many ways in which the molecule's linear chain might be folded. The various letters represent nucleotide subunits. . . . In these models it is assumed that certain nucleotides, such as C—G and A—U, will pair off and tend to form short double-strand regions. Such "base-pairing" is a characteristic feature of nucleic acids.

"decoding" procedures.[136] But essentially, the (universal?) "dictionary" was completed by 1966.

The 1966 Cold Spring Harbor symposium was devoted entirely to the genetic code, and its program was prepared by Nirenberg, Speyer, Watson, and Crick. "In 1961, the genetic code was shown to be made up of three-letter words and the word UUU was found to code for phenylalanine. Five years have passed, and the code is now known, to all intents in its entirety," the new director at Cold Spring Harbor, molecular geneticist John Cairns, stated in his opening remarks (Milislav Demerec, Cold Spring Harbor's revered director of twenty-five years, had recently passed away). "The effort that has gone into this decipherment, the strange sense of urgency, and the remarkable variety of approaches that have together led to the solution must be without parallel in the history of biology. It seemed therefore to be the right moment to hold a meeting and draw together, once and for all, the many contributions to this solution." Crick delivered the introduction, which he titled "The Genetic Code: Yesterday, Today, and Tomorrow"; it was the first (published) rendition of the code's history and a critical resume of the remaining problems to be addressed in the sessions.[137]

While nonsense codons were being assigned chain termination functions with reasonable certitude, the mechanisms for chain initiation remained obscure, Crick pointed out. At least in E. coli, it appeared that GUG signifies the amino acid methionine when it initiates a chain and valine in the middle of the chain, a situation complicating the determination of nonsense polarity (the effect of a mutation on a neighboring gene). Degeneracy had been an accepted feature of the code since the 1950s, but "the skeleton in our cupboard is the possibility of ambiguity. It may be that even in the middle of a message a certain triplet may stand for more than one amino acid," Crick admitted. "Even when we know the genetic code, we will still not know what it is that signals the beginning of an operon and the end of an operon," he warned. Nevertheless, he predicted these difficulties would be resolved within the next few years. The same could not be claimed for the harder questions about the structure of the code, its universality, and its (evolutionary) origins. Crick stated, "Whereas in the Fifties we had to endure a whole lot of rather poor papers on the nature of the genetic code, in the last years there has been a rash of papers on its structure and origin. I am considering offering an annual prize for the worst paper published on this subject—I don't think there will ever be any lack of candidates for it."[138] These topics composed the subjects of the symposium's final session, led by Woese (just then completing his book, *The Genetic Code*, which represented protein synthesis as information processing directed through scriptural technologies). With ninety-four presentations parceled into thirteen sessions (Codons in Vitro; Direction of Reading; In Vivo Code and Polarity; Polarity; Punctuation; Control of Gene Expression; Transfer RNA: Chemistry; Transfer RNA: Function; Transfer RNA: Interactions; Transfer RNA and Ribosomes; Infidelity of Information Transfer; Origins of the Code; and Crick's Introduction), and with three hundred scientists attending, it was the largest Cold Spring Harbor meeting to date. The participants included a bright constellation of leaders in molecular biology and biochemistry: Watson, Crick, Jacob, Monod, Arthur Pardee, Stent, Gros, Lipmann, Ochoa, Khorana, Heppel, Tomkins, and Nirenberg. But the less visible multitudes—literally hundreds of participants (including Matthaei, who presented a paper)—composed the battalions of foot soldiers who had attacked the multifaceted intricacies of the coding problem in the 1960s.

Nirenberg's overview paper, "The RNA Code and Protein Synthesis," launched the Cold Spring Harbor presentations (though he also presented the results of more specific research in a later session). He used the occasion to announce another major finding from his laboratory: the demonstration of the (near) universality of the genetic code. "The results of many studies indicate that the RNA code is largely universal," he conceded. Still, his findings remained inconclusive. Drawing on his diverse biological knowledge, he pointed out that translation of the RNA code—at various stages and in different ways—could be altered in vivo and that different cells sometimes displayed variations in their "recognition apparatus" (namely, in their specificities for codons). In fact, Nirenberg speculated that these subtle differences in recognition might play a role in embryonic differentiation (a topic that had engaged him only five years earlier).[139]

Together with a junior collaborator, C. Thomas Caskey (later a major presence in the Human Genome Project), Nirenberg undertook the task of investigating the fine structure of the recognition apparatus in different organisms. Fifty RNA codons recognized by tRNA from amphibian (South African clawed toad) and mammalian (guinea pig) liver preparations were determined and compared with those recognized by E. coli tRNA. Almost identical translations were obtained with tRNA from the three species, though the response of E. coli tRNA diverged markedly from the other two; tRNA from mammalian liver, amphibian liver, and amphibian muscle responded similarly to RNA codons. Thus the code was "essentially universal," they concluded. "The remarkable similarity in codon base sequence recognized by bacterial, amphibian, and mammalian AA-sRNA suggest that most, if not all, forms of life on this planet use almost the same genetic language, and that the language has been used, possibly with few major changes, for at least 500 million years."[140] The nearly universal dictionary for an almost universal language was finally in place. "Genetic Language Called Universal. Researchers Find Various Organisms Use Same Code but Different 'Dialects.' Experts Hail Results," announced the *New York Times*. "He really nailed it down," exclaimed biochemist Ralph T. Hinegardner of Columbia University, who had been working on the same problem. "It was a real tour de force."[141]

The response of scientists and media to these findings was extraordinary, conveying both awe and alarm about the meanings and implications of the new biological knowledge. Geneticist and Nobel laureate George Beadle had just completed his book, *The Language of Life*. Its penultimate chapter, "Breaking the Code," retold the story of the Rosetta stone and explained:

> What has happened in genetics during the past decade has been the discovery of a Rosetta stone. The unknown language was the molecular one of DNA. Science can now translate at least a few messages written in DNAese into the chemical language of blood and bone and nerves and muscle. One might also say that the deciphering of the DNA code has revealed our possession of a language much older than hieroglyphics, a language as old as life itself, a language that is the most living language of all—even if its letters are invisible and its words are buried deep in the cells of our bodies.[142]

Robert Sinsheimer, Beadle's Caltech colleague (and who a decade and a half later was a progenitor of the Human Genome Project), had also prepared a little book, *The Book of Life*, where he analogized the genetic code to a Mayan Codex. "This," he wrote, pointing to the array of the Codex, "is a book. No man living now can fully decipher it." "This," he wrote, displaying the bundles of human chromosomes, "is another book—or perhaps more accurately an encyclopedia." He continued:

> It is a set of human chromosomes, the book of life. In this book are instructions, in a curious and wonderful code, for making a human being. In one sense—on a sub-conscious level—every human being is born knowing how to read this book in every cell of his body. But on the level of conscious knowledge it is a major triumph of biology in the past two decades that we have begun to understand the content of these books and the language in which they are written.[143]

No one speculated as to who wrote the book of life, but the possibilities of its rewriting elicited both excitement and fear. While in his final chapter Beadle carefully pointed to the Nazi experience and the dilemmas of racial betterment or breeding a Superman, he had little trouble envisioning a "brave new world" of medicine where errors could be erased from a gene pool besieged by accumulated genetic defects. Reviewing Beadle's book in *The New York Times Book Review*, his friend, the distinguished biologist Theodosius Dobzhansky, agreed that the "tremendous achievement" of breaking the genetic code should not be used to justify negative or positive eugenics; though he did underscore the danger of accumulating genetic defects (mainly due to radioactive fallout). "They are, however, a part of a more general and graver problem: accumulation of hereditary diseases and constitutional weakness," he warned.[144]

Linus Pauling, chemist and twice a Nobel laureate, suggested a policy of eugenic prophylaxis: "There should be tattooed on the forehead of every young person a symbol showing possession of the sickle-cell gene or whatever other similar gene. . . . It is my opinion that legislation along this line, compulsory testing for defective genes before marriage, and some form of semi-public display of this possession, should be adopted."[145] Meanwhile, Pauling's Caltech colleague Sinsheimer rejoiced in the powerful technologies for perfecting the remarkable product of two billion years of evolution: "The old eugenics was limited to a numerical enhancement of the best of our existing gene pool. The new eugenics would permit in principle the conversion of all the unfit to the highest genetic level."[146] And distinguished geneticist H. Bentley Glass, vice president of the State University of New York, Stoney Brook, forecasted to an audience of forty-five hundred school administrators that in the year 2000:

> Man will be free of hunger and infectious disease. A vigorous physical and mental life will be enjoyed by most people until the age of 90 or 100. Defective parts of the body will be replaced, even prenatally. The frozen reproductive cell, sometimes of people long dead, will be used to create life. . . . Here is our "brave new world" in full, with bottle babies in different kinds of solutions that condition their mental growth to suit a certain caste.

Though he did acknowledge that "this staggering power" over human evolution "provoked another great crisis in human affairs, the crisis of values and goals."[147]

Marshall Nirenberg, in a *Science* editorial, "Will Society Be Prepared?" was emphatic about the impending social dilemmas. He did not know how long it would take before cells programmed with synthetic messengers were available. And though the obstacles were formidable, he had little doubt they would eventually be overcome; he speculated probably within twenty-five years, perhaps less, if efforts were intensified.

> The point which deserves special emphasis is that man may be able to program his own cells with synthetic information long before he will be able to assess adequately the long-term consequences of such alterations, long before he will be able to formulate goals, and long before he can resolve the ethical and moral problems which will be raised. When man becomes capable of instructing his own cells, he must refrain from doing so until he has sufficient wisdom to use this knowledge for the benefit of mankind. I state this problem well in advance of the need to resolve it, because decisions concerning the application of this knowledge must ultimately be made by society, and only an informed society can make such decisions wisely.

These, and other expressions of concern, initiated a series of congressional hearings in 1968 to consider a national commission on health, science, and society, with the purpose of anticipating, examining, and reporting on the legal, ethical, and social implications of biomedical research, including genetic engineering.[148]

By then Nirenberg was one of the leading figures in molecular biology and recipient of numerous prestigious awards (Paul Lewis Award in Enzyme Chemistry, American Chemical Society, 1964; the National Medal of Science, 1965; the Research Corporation Award, 1966; the Hildenbrand Award, 1966; the Gairdner Foundation Award of Merit, 1967; the Prix Charles Leopold Meyer, French Academy of Sciences, 1967; the Joseph Priestly Award, 1968; and the Franklin Medal). In 1968 he shared the Nobel Prize for Physiology or Medicine with Robert Holley and H. Gobind Khorana; "a triplet of great sense," as Maxine Singer put it.[149] By then, too, Nirenberg, like Delbrück, Stent, Benzer, Pauling, Crick, and others, was on his way to the new scientific frontier of neurobiology. Adhering to his philosophy—"after a system is going well, turn it over to someone else"—he tackled the challenges of the nervous system, choosing to remain at NIH despite numerous other prestigious offers.

Thus, with the elucidation of the genetic code and even before the advent of the recombinant DNA technologies of the 1970s, a new era in biology had dawned. Molecular biology became a kind of information science aimed at rewriting the Book of Life, a field constituted through the efforts of large research teams representing different approaches. Literally hundreds of scientists from diverse backgrounds—senior researchers, junior collaborators, postdoctoral fellows, and laboratory technicians—contributed to the solution of the genetic code; however, heroic narratives have led to the inescapable conclusion that "after the Eighth Day Francis Crick rested." As Robert Martin observed, rather than a "world filled with giants and pygmies . . . most molecular biologists . . . have stood between four-foot-ten and six-foot-one."[150] But it was not only the story of individuals. The solution of the genetic code in the 1960s was a social and cultural process, representing the convergence of discursive, material, and social technologies. As a technoscientific imaginary, it was constituted through a continuous dialectic of altering and being altered by laboratory practices; by informational tropes and scriptural representations of protein synthesis; through shifting disciplinary reconfigurations; and the institutional and patronage structures supporting life-science research through the changing political landscape of the 1960s.

CHAPTER SEVEN

In the Beginning Was the Wor(l)d?

The history of the genetic code does not end with the completion of a "dictionary" filled with "words." Instead, the code's history has become intertwined and evolved with the history of linguistics. As I have traced in the previous chapters, the genetic code was constituted as a scientific object and scriptural technology in tandem with the emergence and dissemination of the information discourse. In conjunction with the communication technosciences of the postwar era, namely cybernetics, information theory, and computer science, the age-old Book of Life acquired its historically specific meaning as genetic information transfer. No longer was biological specificity captured solely within the viscous materiality of biological pattern; now the transmitted messages were constituted through alphabetic writing, as a form of "verbal heredity." The new discourse emerged as pure representation.

By the mid-1960s, with the spread of the vision of genomic writing, a sense of an astonishing discovery, of a transcendent scientific reality, arose. "La surprise, c'est que la spécifité génétique soit écrit, non avec des idéogrammes comme en chinois, mais avec un alphabet comme en français, ou plutôt en Morse," exclaimed François Jacob at his 1965 inaugural address to the Collège de France (that same year he received a Nobel Prize, shared with Andre Lwoff and Jacques Monod). Likewise, George Beadle celebrated the genetic Rosetta stone, which revealed "our possession of a language much older than hieroglyphics, a language as old as life itself." And Robert Sinsheimer provided vivid imagery of the genetic code as a latter-day Book of Life, or as an instruction manual. These are quintessential instances of the representation of DNA as the prime mover (as Max Delbrück had phrased it), the origin of information, writing, and life.[1] This reification of DNA as a verbal code received its highest legitimation from the internationally acclaimed linguist Roman Jakobson.

What was the special appeal of these scriptural representations of DNA? What unique grip did writing and language have on the imagination of postwar science? The significance of these historical issues, which are addressed in this chapter, rests with the remarkable endurance of the vision of the Book of Life. This vision persists up to present-day genomics, manifested most notably in human genome projects. Once the elementary of unit of life became informational, the imagery of the "word" served to reconfigure the larger biological terrain, including theories of the origin of life and evolution.

Theistic implications, though only implicit, have loomed large over the scriptural representations of the genetic code as the Book of Life. Originating in Ecclesiastes in the Old Testament, the Book of Life received its most potent reification (and also problematization) with the first verse of St. John's gospel: "In the beginning was the Word." This proclamation has led to generations of explications of both "the Word" and "the Book"; St. Augustine and St. Thomas Aquinas in particular meditated on definitions and interpretations. For Goethe's Faust this statement became the challenge in his quest for epistemic mastery and source of his subversive power, as he translated "Word" ("It seems absurd") to "Mind" (but "is mind the all-creating source?") to "Force" ("yet something warns me") to, finally, "In the beginning was the Act." As Friedrich Kittler argued, with Goethe's final move the word as signifier is submitted to the primacy of the (transcendental) signified; without this move, the word remains the aporic signified, as pure representation.[2] Thus, if the genome stands for the origin of life, then the Word—the first DNA sequence—has brought molecular biologists as close to the act of creation as could be experienced, invoking supernatural or Faustian powers (recall my discussion about Trifonov and Brendel's *Gnomics* in Chapter 1).

Indeed, language occupied a privileged position in both the theological history of life and in its counterpart, the theory of evolution. For both histories the distinction between bestiality and humanity resides in the acquisition of language and the development of writing. In fact, the status of animal communication, called biosemiotics, has been a key point of debate between linguists, semioticians, and biologists and has served to stretch linguists' narrow definition of language. Nobel laureate Manfred Eigen augmented this debate by inquiring into the rules of "the game of life": What order of information—letters, words, or even DNA sequences—has, since prebiotic time, enabled the acquisition of evolutionary meaning and thus ascension up the tree of life to the pinnacle of evolution, to the possession of language?[3] Language and writing have also formed a strategic site in the long-standing debates between anthropology and biology: Was language largely a product of nature or of culture? Nature's law or human's code?

If the possession of a genomic language and writing represented a form of secular transcendence, then its appeal was surely enhanced through an aura of secret knowledge (just recently made known as the genetic code). Indeed, secret writings, codes, and ciphers have long fascinated humans. The cryptographic imagination reveals its beginnings alongside that of writing in the third millennium B.C. But the appeal of genomic secrecy was even further enhanced through one of biology's oldest quests: the search for the key to the "secret of life" (a leitmotif in the field since the nineteenth century, when the soul as an explanatory category was purged from science). As with reconfigurations of the Book of Life, the code as a form of secret writing also acquired synchronic significance through its association with the intrigues of cold-war defense, communication technologies, and espionage. Indeed as James Bono has observed, the power of metaphors resides in their linkages to other discourses and texts. Such tacit and explicit linkages can both amplify and destabilize a metaphor's contemporary meanings.[4] Thus, in their travel through time, age-old metaphors, such as "code," "information," and the "Book of Life," are reconstituted; and in the postwar period their reconfiguration created the possibility of information without meaning, codes with no language, messages with no sender, and writing devoid of authorship.

Similar slippages, destabilizations, or unintended significations also arose from the metaphor of language, from the very notion of what language is, or is perceived to be. Undoubtedly, the seductive appeal of DNA as an ancient language derived mostly from biologists' mechanistic conceptions of language, which had been formulated in the Enlightenment (e.g., by Étienne de Condillac), ideas long ago discarded by modern linguists. Reacting against the system of linguistic signification of the Renaissance—when relations between words and things were mediated through occult alchemical knowledge, doctrine of signs, analogies, and similitudes—natural philosophers of the seventeenth and eighteenth centuries sought to rationalize phenomena through linguistic reforms. In their searches for a universal language and certitudes they attempted to purge all vestiges of discursive idols from the construction of knowledge. Condillac's *Logic* epitomized that kind of domestication, proclaiming that "well-made language" was a form of algebra, where words, just as mathematical symbols, referred to precisely defined, even quantifiable things. "For in the art of reasoning as in art of calculating, everything is reduced to compositions and decompositions; and it must not be thought that these [algebra and language] are two different arts."[5] (Condillac, in turn, became the guiding framework for Lavoisier's reform of chemistry, then severing its epistemological ties with alchemy through a nomenclature based on such algebraic reason rather than occult analogies.)

It seems this mechanistic notion of language was taken up by various molecular biologists in the 1960s: faith in the absolute nature of the sign; insistence on the exact correspondences between the signifier and the signified and on the possibility of unambiguous positive reading of texts, the printed word, or the Book of Life. The breaking of the genetic code thus acquired the dimension of revealed knowledge (as Beadle alluded): life could be unambiguously read from the genomic text, written in a DNA language. By then, however, linguistics, like modernist transformations in other realms of knowledge (not only the arts but also physics and mathematics), experienced an epistemic rupture from the absolutes of Newtonianism to the relational regimes of structuralism, with profound implications for the genomic Book of Life.

James Watson's mandate for molecular biology and the Human Genome Project makes evident that these problematic scriptural representations of the genome as the Book of Life have come to signify a kind of divine intervention, a new form of biopower and technologies of the self. He wrote, "For the genetic dice will continue to inflict cruel fates on all too many individuals and their families who do not deserve this damnation. Decency demands that someone must rescue them from genetic hells. If we don't play God, who will?"[6] Emerging in the late 1940s, this form of biopower was first articulated in Norbert Wiener's declaration, "We are forced nowadays to recognize individuality as something . . . that shares the nature of communication," thus literally recasting the individual as the word. Layered above the viscous materiality of organic substance, information—the memory of its form and its logos—has since promised other levels of controlling bodies and populations. The molecular vision of life was supplemented by the information gaze and empowered by the technologies of the DNA word. As molecular biologist and biotechnology champion David Jackson recently put it, "To be fluent in a language, one needs to be able to *read*, to *write*, to *copy*, and to *edit* in that language. The functional equivalents of each of those aspects of fluency have now been embodied in technologies to deal with the language of DNA."[7] This notion of biowriting was authorized in the 1960s and 1970s by invoking the rules of language and the game of life, with all their aporias, circularities, and loss of referentiality, until, as Derrida observed, their "own historically metaphysical character is also exposed."[8]

VERBAL CODE: ROMAN JAKOBSON AND MOLECULAR BIOLOGY

Roman Jakobson, the Russian-born Harvard linguist (1896–1982), was swept up by the cybernetic wave and shared in the widespread conviction that communication theory ushered in a new epoch in human knowledge, including the science of linguistics. He rushed a note to Wiener in 1949, soon after Wiener's *Cybernetics* appeared: "At every step I was again and again surprised at the extreme parallelism between the problems of modern linguistic analysis and the fascinating problems you discuss. The linguistic pattern fits excellently into the structures you analyze and it is becoming still clearer how great are the outlooks for a consistent cooperation between modern

FIGURE 40. Roman Jakobson at MIT, ca. 1960. Courtesy of the MIT Museum.

linguistics and the exact sciences."[9] The parallels Jakobson observed were grounded in the general similarities between the definition of information and Ferdinand de Saussure's theory of the sign at the turn of the century.

Saussure, as Robert Brain has demonstrated, placed the phoneme—the elementary unit of speech, materialized as scientific object through acoustical inscription—at the center of his investigations. He thus not only elevated the study of *parole* (spoken language) over *langue* (system of language) but also shifted the focus from the diachronic (philological) to the synchronic (structural) study of language. It was this focus on the material properties of speech, from his early studies in Leipzig to the later collaborations with French linguists, that led Saussure to his conceptual break with the past and, importantly, to his radical theory of the linguistic value of the sign. Analogizing linguistics to equilibrium (homeostatic) theories in economics (where the price of cotton in 1900 depended not on its price in 1890 but on the interactions of contemporary market conditions), he envisioned language as a system of relative and contrasting acoustic images—with all their attendant semantic and grammatical relations—distinguished from one another not through any absolute reference, but rather by their differences. The sign derived its meaning only from the context of the whole linguistic system.[10] Within the Saussurian notion of language, speaking, reading, writing, and the Book of Life lose their perceived positivities.

The Saussurian program, pursued around the world, was elaborated in the interwar years by the Prague school, led by Roman Jakobson and Nikolay Troubetskoy. Rather than taking the phoneme as the minimal unit of analysis, they viewed phonemes as sets of distinctive features: each of these distinctive features entails a choice between binaries, two terms of an opposition (e.g., /P/ and /t/ are phonemes representing grave and acute sounds, respectively, with contrasting physical and physiological properties). As with other prominent Jewish scholars, Jakobson's life was disrupted in 1939. After two years on an escape route from Czechoslovakia through Denmark, Norway, and Sweden, he found refuge in the United States in 1941. Following a couple of interim appointments, he pursued his work at Columbia University (1943–49) until he went to Harvard. Moving beyond phonology, in his characteristically panoramic view, Jakobson demonstrated that such binary oppositions could be also applied to other linguistic levels: syntax, semantics, and pragmatics.[11]

Jakobson had enormous impact on linguistics; his command of numerous languages, his remarkably broad scope of interests, his discipline-building skills, his charm and penchant for bold theorizing all extended his sphere of influence into other realms of knowledge, particularly the social sciences (most of all anthropology) and biology. In the early 1940s, as Claude Lévi-Strauss was searching for a framework within which to interpret his massive kinship and social organization data, Jakobson's structural methods of linguistic analysis opened up a new analytical approach. Linguistic structuralism became a major resource along with systems thinking in anthropology, where an element—a linguistic sign, codified behavior, ritual practice, or gift exchange—no longer possessed intrinsic meaning but derived its value from its interactions with other elements of the cultural system, from its context. The similarities between the Saussurian view of *La Langue* as a system and cybernetics seemed compelling. Just as linguistic structuralism is bound to contextuality and thus to the arbitrariness of reference, so Wiener's cybernetics sees communication as a stochastic process within a statistical universe, where the message is signified only in relation to other messages. For Saussure, Jakobson, and Wiener signification was about relations, not about objects, or the world as a thing in itself.[12]

Jakobson also conveyed his excitement to the officers of the Rockefeller Foundation's Humanities Division. In response, Warren Weaver sent to Jakobson a copy of his and Claude Shannon's book, *Mathematical Theory of Communication* (1949). Jakobson became equally enthusiastic about Weaver and Shannon's theory. He responded to Weaver, "Having returned from my lecture and study trip through seven European countries may I tell you that discussing the controversial problem of *Sound and Meaning* in various universities I frankly pointed out your and Shannon's *Mathematical Theory of Communication* as the most important among the recent American publications in the science of language."[13] Undoubtedly these new directions and the possibility of their funding offered prospects for the revitalization of linguistics, which at the time was in a state of stagnation and even crisis.[14] Outlining his new disciplinary visions, Jakobson—most likely encouraged by Weaver—attempted to stretch the limits of information theory well beyond its engineering domain to cover numerous aspects of linguistics. "I fully agree with W. Weaver," Jakobson proclaimed, "That one is now, perhaps for the first time, ready for a real theory of meaning, and of communication in general. The elaboration of this theory asks of an efficacious cooperation of linguists with representatives of several other fields such as mathematics, logic, communication engineering, acoustics, physiology, psychology and the social sciences." Despite Shannon's insistence that informational linguistic analysis applied only to syntax, Jakobson had grander visions. He proceeded to outline the areas in linguistics to be developed within this new information paradigm—morphemes (smallest units of meaning, e.g., prefixes and suffixes), phonemes, grammar (morphology and syntax), vocabulary and phraseology, the coding process with its redundancy and ambiguity, and, finally, semantics and pragmatics. A later version of the proposal also included semiotics, especially in relation to anthropology. Several of these investigations, particularly psychoacoustics, were to be developed collaboratively with his MIT colleagues.[15]

These were auspicious plans. In 1957 Jakobson received a joint appointment at MIT on the occasion of the creation in 1958 of the Center for Communication Sciences (he was, in fact, on its steering committee; among the center's members were Claude Shannon, Jerome Wiesner, Walter Rosenblith, Jerome Lettvin, Noam Chomsky, Marvin Minsky, and John McCarthy). Affiliated with the Research Laboratory of Electronics, the center offered a "unique environment in which to study communication processes in both natural and man-made systems" (e.g., the SAGE system). MIT President Julius Stratton announced:

> Development of a mathematical theory of communication and advances in computer technology have made possible the study of the principles of organization of such systems. . . . [Attracting] to the Institute scientists interested in speech communication, in learning and perception, in small group behavior, in the logical analysis of language and of other symbol systems, in translation, in sensory processes, and in a variety of problems dealing with the functioning of the nervous system. These scientists joined forces with mathematicians and engineers who were interested in information theory, game theory, and in the behavior of automata.

Although it was a center for basic research, its military relevance was implicit.[16] Like most academic disciplines, linguistics became embedded in cold-war knowledge production, especially after the Soviet launching of *Sputnik I* in 1957. Not only was mechanized linguistic analysis central to cryptanalysis (as in Shannon's work) but also it was essential for the national project of developing automatic mechanical translation—mainly from Russian to English (and particularly scientific publications)—sponsored by the National Science Foundation, National Bureau of Standards, the U.S. Army, Air Force, and the Central Intelligence Agency. By 1960 these agencies had collectively spent about $3 million on linguistic research related to automatic translation; by the end of fiscal year 1963 the figure would grow to about $8 million.[17] In the late 1950s, Jakobson, in collaboration with MIT protégé Morris Halle and the noted information theorist, Collin Cherry, ventured an information-theoretical approach to phonemic theory. That approach entailed a logical redescription of language, converting syntax into binary symbol chain. (For example, any letter in the alphabet may be identified by asking: "Is it in the first half, A to M; yes or no? If yes, "is it in the first half; yes or no?" and so on, until identified. Similarly, because language is based on binary oppositions, high-low, inside-outside, fact-fiction, it is easily convertible into binary code. There are many such historical examples.) A similar approach was used in the analysis of phonemes. Originally taken to represent the smallest linguistic units, Jakobson had already pushed the analysis of phonemes much further to argue for their constitutive attributes, or *distinctive features*, thus relating the phoneme to its articulatory production. In the 1950s, together with Morris Halle, they identified twelve such features, or binary oppositions (e.g., grave/acute, nasal/oral, tense/lax, voiced/unvoiced, etc.) across numerous languages. They also called for the information-theoretical representations of these distinctive features, where, presumably, their information content could be quantified in bits and the rules of language operationalized.[18] This could potentially contribute to automated informational systems.

By 1961 Jakobson had consolidated these findings. Drawing on the Wiener-Shannon theory, particularly on its corrective in the work of British information theorist Donald M. MacKay, Jakobson spelled out a vision for linguistics as a (mathematical) communication science. The Wiener-Shannon theory was too blunt a tool for the nuances of linguistic analysis. MacKay, by triangulating between information, semantics, and pragmatics, accounted for context and subjectivity of human communication. But precisely because of the technical complexities it introduced, MacKay's version of information theory never took root in the United States. As Jakobson saw it, linguistics, qua communication science, was dialectical: it was based on Bohr's complementarity principle. That is, linguistic communication was grounded in "the inseparability of objective content and observing subject" and on a feedback between the two irreducible opposites of speaking and hearing, encoder and decoder. Jakobson forecasted "the diverse possibilities for measurement of the amount of phonemic information, which are foreseen by the communication engineers." He highlighted the parallels between the concept of redundancy in communication theory and its proposed counterpart in phonemic analysis, where the goal was the elimination of redundancy.

He also extolled coding theory as a key tool for linguistic analysis.

> The code matches the *signan* with its *signatum* and the *signatum* with its *signans*. Today, with respect to the treatment of coding problems in communication theory, the Saussurian dichotomy *langue-parole* can be restated much more precisely and acquire a new operational value. Conversely, in modern linguistics communication theory may find illuminating information about stratified structure of the intricate language code in its various aspects. . . . Communication theory, after having mastered the level of phonemic information, may approach the task of measuring the amount of grammatical information, since the system of grammatical, particularly morphological categories, like the system of distinctive features, is ostensibly based on a scale of binary oppositions. . . . To sum up, there exists a wide range of questions calling for the cooperation of the two different and independent disciplines we are discussing.[19]

Information, coding, and encoding were the elements of the information discourse that reconfigured linguistics as an information science in the 1950s, as it did in molecular biology. Ironically, Saussurian linguistics, constructed along the differential material representations of speech (acoustic/physiological), was finally devocalized and dematerialized through its conversion to logical syntax. As in biology, content became form. The multidimensional organic attributes of speech and life were conceptually flattened, ostensibly captured on the imaginary tape, a logos studded with zeros and ones.

The Center for Communication Science never did fulfill its grand epistemic and interdisciplinary vision; instead, it disappeared into the annals of history. While generating considerable enthusiasm among some participants (e.g., Jakobson), it also attested—as critics, even at MIT, pointed out—to the discrepancies between an expansive scientific imagination and the technical limits of information theory. One such critic wrote:

> Scientists in the biological and social sciences and in the related applied fields connected with these disciplines had a sudden insight that the "simple" problems of the communication engineer were rather similar with their own more complicated ones. Under the influence of the cybernetic ground-swell, some of them were sufficiently dazzled to believe that they had only to borrow the recently developed concepts and research techniques from communication theorists to solve their problems.[20]

Specifically targeting linguistics, critics pointed out that in Shannon's information theory the salient features—code, transmitter, receiver, channel, and signal—are precisely defined entities and their attributes quantified. In speech communication, on the other hand, the knowledge of the (respectively) analogous features—language, speaker, listener, ambient air, and acoustic wave—is imprecise and fragmentary and thus cannot be quantified.

"Communication theory is now shaping the language. . . . There are dangers here," cautioned a reviewer of the edited volume *On Translation* (which included a contribution by Jakobson). The reviewer specifically highlighted the confusions in borrowing the terms *information* and *code*, repeating what by now had become Shannon and Weaver's refrain: "Semantic aspects of communication are irrelevant to the engineering problem" and "information must not be confused with meaning." The reviewer further argued, "*Code*, *encode*, *decode* present similar dangers."

> In communication theory, the message (say, a typewritten sentence) and the signal (say, a sequence of Morse) must be separately identifiable and the relation between them accord with specifiable *rules of transformation*. Where this is not so, talk of encoding and decoding of messages is a dangerously misleading metaphor, a thing linguists should, above all, be equipped to detect. Unfortunately such talk has become fashionable and is spreading.[21]

Can language (or mind) study itself and still deliver certitudes? Can it be, simultaneously, both the object and subject of positive knowledge? As Derrida and Michael Reddy would observe, the project of developing a language about language cannot escape the hall of mirrors of their referentialities and metaphoricities.

The young Noam Chomsky too was critical of the new cybernetic trend. Although he had a joint appointment in the Research Laboratory of Electronics and the Center for Communication Sciences he later recalled finding no relevance in it to his work on generative transformational grammar, then rapidly gaining primacy over linguistic structuralism. He recalled, "Virtually every engineer or psychologist with whom I had any contact, and many professional linguists as well, took for granted that the formal models of language proposed in the mathematical theory of communication provided the appropriate framework for general linguistic theory. But this assumption is not correct."[22] Perhaps this is so in the narrow technical sense, but the centrality of syntactic and transformational linguistic analyses to the larger project of automation of language and mechanical translation does point to technoepistemic resonances between the theory of generative grammar and the cybernetic culture.[23]

It was just at the ascent of Chomskian linguistics and the low ebb of structuralism that the aging Jakobson turned to biology and got swept up by the euphoria around breaking the genetic code. In a sense his visions of an information-based "verbal heredity" could be seen as an attempt to preserve the power of his declining version of linguistics by extending its sphere of influence to biology (much as the architects of the Rockefeller Foundation's molecular biology program, physicists Max Mason and Warren Weaver, sought to preserve classical physics in the 1930s within the domain of a new biology).[24] And as in molecular biology, even among biochemists, Jakobson found an audience willing to engage with his ideas.

The first public occasion for such explorations occurred soon after the genetic code was broken. The wide-ranging implications of the code—especially prospects of identifying a chemical memory code—captured the scientific imagination well beyond the terrain of molecular biology. These ideas became a focus for brainstorming at the first annual disciplinary meeting of MIT's Neuroscience Research Project (NRP) under the direction of Francis O. Schmitt (who championed the nascent field he christened "neuroscience"), held at the American Academy of Arts and Sciences in Boston, August 1962. With all the charm of a European statesman, Jakobson delivered his talk, "Phonemes as Linguistic Code," as a preamble to an extended discussion on "possible relation to molecular information code." Among the participants were biochemists Severo Ochoa, Richard Roberts, William H. Sweet, Gerhard Schramm, Albert L. Lehninger, biophysicist (and information theorist) Leroy Augenstine, and physical chemist Manfred Eigen, who was just then forging his links to molecular biology. (Marshall Nirenberg attended a subsequent meeting.) Though absent, Beadle had encouraged Jakobson to pursue the linguistic study of the language of life.[25]

"In language we deal with information," Jakobson began his morning presentation. "The messages are messages carrying information. . . . When we speak . . . there is an enormous amount of redundance, and there is enormous help in prompting context," he stated, providing a framework. He then proceeded to explain his work on distinctive features and his analysis centering primarily on the attributes of speech and articulatory mechanisms. After some questions of clarification about the notions of redundancy and noise in language (which led to no resolution), the formal discussion turned to their relations to the genetic code. Curiously no one addressed the obvious problem: What had speech to do with the so-called language of life, and thus, what was the relevance of phonemic analysis to the genetic code? (Eventually Jakobson would find ways to finesse this problem.) "Is there some fraction . . . of redundance, and if so, what is the redundance in molecular coding?" asked Eigen. Jakobson could not answer that question; even for language, redundancy was difficult to quantify (Shannon had estimated it to be between 50 and 75 percent). But "in the case of amino acid code, what would you refer to as redundancies?" queried Ochoa, venturing a guess that "If you compare it to language, you can say that there are a number of words which have the same meaning" (presumably analogizing codons as words).[26] At this time, one begins to encounter traces of the information discourse in Ochoa's writings.

Schramm wrestled with the concept of redundancy by recasting the tobacco mosaic virus (TMV) RNA as a three-page text of six thousand letters (each letter being a nucleotide). Drawing on his TMV inactivation studies, where nitrous acid techniques had led to the sensational findings (1956) that a single-base mutation inactivated the virus, Schramm now described these results as changing one letter in a textbook. But unlike language, it was not clear which letter, or whether letter choice even mattered. If you alter one, then the whole text cannot be figured out; it becomes senseless. Or the whole sense of this text is changed, and you get a different result ("missense," according to Delbrück). So the redundancies must be very small, if there are redundancies at all. While Ochoa surmised that such single-letter deletion would produce nonsense (in the technical sense of noncoding triplet?), as Schramm saw it, it was not nonsense but just unreadable. In any case this situation was clearly not comparable to linguistic redundancy, they concluded (since the ability to read was altered). Jakobson, on the other hand, thought "this is a redundancy in the sense of channels" and letter redundancy could be determined only through context (the context determining how far one could get with letter omissions). But one thing is clear on the phonemic level, he stressed: given equal context, one could distinguish distinctive elements (or features) from one another.[27] But, what would it mean to apply such phonemic-level analysis to the genome? What kind of articulatory practices would be captured within a syntactic chain of RNA composed of four symbols? And, coming back to Eigen's question, how was redundance represented in a genetic code?

At the session's closing the participants reached the conclusion that by "redundancy" they probably meant "degeneracy": the sixty-four codons specified twenty amino acids with about a 30 percent redundancy (a figure that clearly contradicted Schramm's findings of single-base inactivation). Sweet proposed that the function of the code's redundancy could be similar to that of language: a kind of safety feature that serves to increase the likelihood that the communication attempt will succeed. Could it be the same for the genetic code? There were no clear answers. As Ochoa reminded the participants, "In the genetic code, the last word has not been said."[28] Indeed, in 1962 there were several competing versions of the triplet code, as well as other kinds that seemed to fit the data, including Roberts's two-letter code, which was extremely nonredundant yet highly respectable.

While neither concrete nor conclusive findings emerged out of the session, the discussion seems to have inspired Jakobson to pursue in print, lectures, and on television some of these issues in subsequent years. Between 1962 and 1965 he visited twice the newly founded Salk Institute for Biological Studies in La Jolla, California, where several leaders of molecular biology worked. The institute was then also committed to Jonas Salk's humanistic vision of science, as epitomized by its deputy director, Jacob Bronowski. Like other European scholars, Jakobson was charmed by the physical beauty and intellectual intensity of the Salk Institute and planned to move there soon. "I am looking forward with great and joyous expectation to my research activities with your institute," he wrote to Bronowski in the spring of 1965. "My present research is increasingly oriented toward biological problems. Recently I delivered a lecture at the American Academy about the Quest for the Essence of Language which is to appear in Diogènes and is the first draft of a much more extensive study."[29] Several publications followed.

Though the move to the Salk Institute never materialized, Jakobson's interest in molecular biology intensified. In 1967, as Jakobson reported to

Bronowski, his ideas about the relation between linguistics and molecular biology circulated nationally and worldwide and generated lively discussion at, among other places, the Academy of Science in Moscow (where linkages between biology and cybernetics were particularly strong), the International Congress of Linguistics in Bucharest, and in Paris. Jakobson wrote:

> Finally, in Paris together with Lévi-Strauss, I had a thoroughly improvised TV discussion with the biologists F. Jacob and M. L. Heritier [*sic*]. All of us were fascinated by the striking isomorphisms between the code of language and molecular code which were revealed in this discussion. I am working now on the development of my study on the relation of linguistics with the sciences of man and the natural sciences.[30]

The Debate

The televised debate, "Vivre et Parler," aired in September 1967 and subsequently published under that title in *Les Lettres Françaises*, was apparently a unique scholarly encounter, even for the community of French public intellectuals (a "revolutionary discussion," the article put it). The discussion also created a remarkable historical document: for all the contemporary coverage about the epistemic and social significance of the genetic code, it was the first time (and probably the last) that leaders of linguistics, anthropology, genetics, and molecular biology convened in a public forum to explore the convergences in these representations and to grapple with their ramifications with considerable intensity. And though no consensus was reached about the linguistic status of DNA, the debate and its publication served to spread the gospel of the world as communication system, according to which both nature and culture imploded at their shared site—language—and each used language simultaneously as tool and as object, *episteme* and *techne*.[31]

Initiating the discussion with an examination of "Genetic Information and the Function of Language," Jacob explained how "what we call genetic information . . . is genuinely inscribed in the chromosomes [via permutations of elements] . . . exactly like in a phrase in a text." This view, he argued, had enormous consequences for the central problem of biology: organization. And this concern with the informational integration of levels, from molecules to cells to specialized structures to whole organisms and society, was with "systems of communication." Following with an exploration of "Linguistics and Biology," Jakobson recalled that he had been interested in the intersections between linguistics and biology since his youth and had followed earlier studies linking linguistic differences to theories of human evolution. But until recently, he had "tended to avoid the biological analogies since they were grounded in premature theories, biological theories of language that had been shown to be completely false," which had subsequently fueled racial studies and Nazi ideology. He also previously opposed the "very mechanical application of Mendelism to theory of evolution of language." But his numerous encounters with molecular biologists and his work at the Salk Institute in the last couple of years had convinced him that "one could find not only isomorphisms but also profound and important convergences between linguistics and biology."[32]

Directing his query to both Jacob and Jakobson, the moderator (Michel Treguer) suggested that, perhaps, Jacob's representation of organisms could be analogized to the Saussurian paradigm shift in linguistics at the turn of the century, when language came to be regarded as a system. Jakobson wholeheartedly agreed. His views were buttressed by Lévi-Strauss, who pointed out that structuralism affected not only thinking in biology but also reoriented the ways in which anthropologists studied human societies.[33] Philippe L'Héritier concurred that heredity was a system of information transfer but proposed to push the analysis much further, into a kind of protosociobiology. Could it be, therefore, that the possession of language in the human species "introduced into the biological world a genuinely novel form of heredity, which one can call social heredity, or perhaps, the heredity of language?" Jakobson called this possibility "verbal heredity." Verbal heredity, L'Héritier reasoned, would be subject to the same rules of evolution and natural selection as other forms of heredity but would operate not on the level of individuals but of groups and civilization. That, of course, opened up the fascinating problems of children's language acquisition (Was it physiologically hard-wired or culturally shaped learning process?) and raised the contested issue of animal versus human communication (Could bird calls, bee dances, or dolphin clicks be considered languages? Are they genetically determined?). In any case, these intriguing and still unresolved issues seemed tangential to the central question: Is DNA a language?[34] Later Jakobson would insist that it was a natural language and even went so far as to suggest that human DNA as a source of verbal heredity was somehow tied to the emergence of language.

The moderator queried the group: What was the scope of the convergences between biology and linguistics? If phonemes were the elements of communication, on what level of biological organization would phonemic analysis (or its analogue) operate? Was there a true link in the manner in which mechanisms of language and biological organization functioned? Jacob observed, "It was particularly striking to find that genetic information is formed through the juxtaposition and the sequence of four units, and that language is equally formed through the organization, combination, permutation and the sequences of very small units." Undoubtedly, the question of level was only part of the problem, Jacob noted, for structure (or organization) meant something quite different in physics, biology, and social science. Did Jakobson agree with Jacob's observations? "Absolutely," exclaimed Jakobson with the conviction of Glaukone in Plato's dialogues.

> Since the first time I encountered the linguistic terms in the literature of the biologists, I said to myself: one must check whether this is simply fashionable talk, a metaphoric usage, or whether there is here something exceedingly profound. I must say that biologists' talk is totally legitimate from a linguistic view point and that one could go much further. What do the system of molecular genetics and the linguistic system have in common? Primarily it is this: that it is perhaps a most extraordinary thing and most important—it is the same architecture, these are the same principles of construction, a fully hierarchical principle.[35]

Jakobson spelled out the hierarchy: phonemic, word, then syntactic levels. Circumventing the obvious difficulties with a notion of phonemic analysis of the genetic code, and echoing Jacob, he explained that the words of the genetic code were like those of a linguistic code (conflating—as many others have done—the notion of the genetic code as relation, with that of a DNA code as thing). And he reiterated how the laws of composition in the genetic code are the same as those for the Indo-European and Semitic roots. On the syntactic level he pointed to punctuation signs, such as commas (forgetting that the genetic code was comma free) and equated punctuation with initiation and termination codons; thus Jakobson insisted on the parallels. These punctuations applied also on the levels of phrases, which, like life forms, displayed infinite variability. Jacob agreed wholeheartedly. In their discussion "Are There Linguistic Models in Biology?" Jacob added that for a long time, before being viewed as beads on a string, genes had been regarded as patterns, or idiograms, and not as phrases written with very simple combinations. He found the comparison between the linearity of the gene and linearity of alphabetical language compelling.[36] His conviction would wane in the coming years.

Lévi-Strauss, however, was not convinced by the linguistic view of life. He saw major obstacles for such analogies, mostly in the form of difficulties centering on the issue of *signification*, which had not been properly defined. The claim that signification emerged out of combination of elements, themselves devoid of signification—phonemes, letters of the alphabet, or DNA bases—was problematic in biology, he argued, since signification was not employed here in the same sense. In language, signification applied to us, the speakers, but in molecular biology, it applied to inert molecules (and the decoder, the biologists added); there was a slippage between object and subject. L'Héritier pursued Lévi-Strauss's train of thought: being a symbolic language, human language presupposes an interlocutor and a comprehending brain, but in genetic language there is nothing but information transfer between molecules (and even this is metaphorically speaking). "In what sense, finally, is this signification?" L'Héritier demanded. But Jakobson thought that the similarities between the two systems far outweighed their differences, especially if one broadened the notion of communication to include semiotics: nonverbal messages and other sign systems (in this sense anthropology becomes a science of communication, he argued).

Lévi-Strauss was not swayed. For him the whole discussion finally led to a fundamental philosophical conundrum: Can there be a prediscursive knowledge of language existing prior to its construction by humans? Could there be something, as biologists claim, that resembles the structure of language but which involves neither consciousness nor subject? Could one really retrieve the phenomenon of communication, as occurring prior to the consciousness of the members of the social group, which does not intervene in the claims of the speaking subjects? This point he saw as the greatest challenge of the human sciences.[37] Indeed, it remains a fundamental critique of the idea that a genomic Book of Life existed prior to the material and scriptural technologies that brought it into being.

BETWEEN ONTOLOGIES AND ANALOGIES: CHIMERA OF THE BOOK OF LIFE

Jakobson was inspired by the debate, because it highlighted the salient features binding biology and linguistics and pointed to areas for further interdisciplinary study. As he had announced to Bronowski, these ideas now occupied much of his energies. As a visiting professor at the Salk Institute (1969) Jakobson participated in a September meeting on biology as a bridge between the "two cultures." Speaking on the "Language of Life and the Life of Language," he proposed that language itself was the yoke between science and the humanities. In his talk, he explored the links between language and thought, the biological mechanisms of human language, and the validity of the genetic code as a language; the latter point accounting for much of the subsequent discussion. By that time he had already written extensively about the language of life.[38] One such fine-grained study appeared in 1970 in the volume *Main Trends of Research in the Social and Human Sciences*. "The spectacular discoveries of the last few years in molecular genetics are presented by the explorers themselves in terms borrowed from linguistics and communication theory," he pointed out approvingly.

> The title of the book by George and Muriel Beadle, *The Language of Life*, is not mere figurative expression, and the extraordinary degree of analogy between the systems of genetic and verbal information fully justifies the guiding statement of the volume "The deciphering of the DNA code has revealed our possession of a language much older than hieroglyphics, a language as old as life itself, a language that is the most living language of all."

The validity of that statement, according to Jakobson, was based on several isomorphic features of the two informational systems, the verbal and genetic code: their elemental and phonetic structures; their hierarchies, context-dependence, colinearity, and feedback regulation; and the balance between stability and variability.[39]

Drawing on the writings of Francis Crick and Charles Yanofsky, Jakobson explained how genetic messages, like verbal ones, consist of words, or codons, which in turn consist of the four letters of the code alphabet, combined three at a time, and that the dictionary of the genetic code encompassed sixty-four distinct words, three of which are apparently termination signals. This had enormous significance, Jakobson claimed, quoting Jacob's 1965 inaugural address at the Collège de France about the alphabetic and Morse-like attributes of genetic writing (versus ideographic forms). He also seemed to have overcome the obstacles to phonemic analysis. Undeterred by the muteness of the genome, Jakobson argued syllogistically that phonemic analysis could be performed through a process of double abstraction.

> Since our letters are mere substitutes for the phonemic pattern of language and Morse alphabet is but a secondary substitute for letters, the subunits of the genetic code are to be compared directly with phonemes. We may state that among all the information-carrying systems, the genetic code and the verbal code are the only ones based upon the use of discrete components [phonemes] which, by themselves, are devoid of inherent meaning but serve to constitute the minimal senseful units.

He never specified what "sense" (or semantics) could mean in molecular biology. This devocalization of phonemes enabled Jakobson to push the analogy much further, exposing what he saw as a striking feature: just as all the interrelations of phonemes were decomposable into the binary oppositions of distinctive features, so too the two molecular binary oppositions—the paired DNA bases, A and T, and G and C—underlay the four letters of the code.[40] His excitement over this coincidental surface similarity seemed to have masked the obvious problem that distinctive features and nucleotide bases were operationally (in their functions) totally unrelated.

Jakobson further argued that beyond elemental structures, there were the precise analogies between hierarchical design of the verbal and genetic messages. And although he never directly confronted the thorny question of semantics, he pursued such higher-order analysis. He wrote, "The transition from lexical to syntactical units of different grades is paralleled by the ascent from codons to 'cistrons' [Benzer's coinage for a genetic unit of function] and 'operons' [Jacob and Monod's term for coordinated genes], and the latter two ranks of genetic sequence are equated by biologists [e.g., Russian biologist V. A. Ratner] with ascending syntactic constructions." Jakobson readily conceded that such ascending syntactic constructions depended on punctuation marks for the formation of meaning, and the code had none. However, he also agreed that initiation and termination codons ("words") could be described *metaphorically* as punctuation marks, thereby conflating the notion of mark with word. Drawing on recent findings that an initiation codon had a different meaning at the middle, rather than the beginning, of a DNA sequence, Jakobson inferred a generality: as in natural language and in contrast to formal language, the meaning of a genetic message was context-sensitive, so that a word could display a variety of dissimilar contextual meanings. Moreover, he argued, a strict colinearity of the time sequence in the encoding and decoding operations characterized both the verbal and genetic language. But how did this colinearity accord with the observation that, unlike communication between speakers, the genetic code has directionality, that its machinery can translate in one direction only? Jakobson's solution was to invoke the notion of feedback regulation: regulatory mechanisms of molecular genetics and language corresponded to the dialogic nature of speech. These common attributes of the two informational systems ensured stability, speciation, and boundless individuation, he proclaimed.[41]

Speciation led to the ascendance of humans. Indeed, for Jakobson these homologies carried profound significance for humanity (since, he believed, they were not shared by animal communication). He wrote, "The genetic code, the primary manifestation of life, and, on the other hand, language, the universal endowment of humanity and its momentous leap from genetics to civilization, are the two fundamental stores of information transmitted from the ancestry to the progeny, the molecular heredity and the verbal legacy as necessary prerequisite of cultural tradition." He ventured what he considered a legitimate and important question: Was the isomorphism exhibited by the two informational systems merely a convergence of two independent developments guided by similar natural constraints, or could it be a manifestation of a common phenomenon, the genetic bases of language? He favored the second hypothesis: "the possibility of genetic endowment" in human speech.[42] As Jacob would later remark, Jakobson had come a long way since his youthful suspicion of the biologization of language and its premature theories. Through these suggestive analogies the code acquired an ontological status as language; some would say a cosmological principle.

Jacob, on the other hand, became increasingly more circumspect about the scope of these analogies with the passage of time, qualifying his views with rejoinders that biology deals merely with models and representations of life rather than with preexisting realities. In contrast to his 1965 exclamation at the Collège de France about the surprise that genetic specificity "is written" (not in ideograms, as in Chinese, but with a linear alphabet, as in French), his 1970 book, *The Logic of Life*, speaks of representations and models: "The representations of the genes . . . Like a string of beads"; and "The model that describes our knowledge of heredity is indeed that of a chemical message. Not a message written in ideograms like Chinese, but with an alphabet like that of the Morse code." His parting words leave the readers to ponder the dilemmas of scientific representation: "But science is enclosed in its explanatory system, and cannot escape from it. Today the world is message, codes and information. Tomorrow what analysis will break down our objects to reconstitute them in a new space? What new Russian doll will emerge?"[43] By 1974, in his article, "Le Modele Linguistique en Biologie," Jacob made his reservations explicit. Examining the roles of models in biology, he conveyed his critique with a sense of irony and even challenge to Jakobson.[44] In fact, he began by quoting Jakobson (who had originally quoted Jacob to make the very point). He observed, "We may state that among all the information-carrying systems, the genetic code and verbal code are the only ones based upon the use of discrete components which, by themselves are devoid of inherent meaning but serve to constitute the minimal senseful unit, i.e. entities endowed with their own intrinsic meaning in a given code." The analogies were compelling but the differences between biological and social "language" were equally striking. It merits particular attention, Jacob pointed out, that this assessment comes from someone who had always sought to demonstrate that linguistics belonged to the social sciences, not the natural sciences.

The notion of language in heredity was, of course, not entirely new. Back in 1943, Schrödinger had pointed out the possibility of some Morse-like code-script to account for genetic diversity. But, as Jacob argued, it was through the theory of information that biologists transformed their views of heredity. Within a few years heredity became information, message, and codes. Outlining the salient features of the code, he recapitulated Jakobson's analysis of the suggestive homology between the two informational systems: the verbal and linguistic code. In the final analysis, what do these remarkable analogies amount to? Jacob wondered. Does the description of heredity in terms of program, instructions, and code express simply the thoughts of an epoch dominated by the theory of information or does it recover a most profound "reality"? Examining Jakobson's proposal that the isomorphisms of the two systems were not merely convergences of contingencies but a kind of filiation, he pointed to the fundamental limits of these parallels:

> Language studies the messages transmitted from an emitter to a recipient. Now there is nothing of the kind in biology: no emitter, no recipient. The famous message of heredity transmitted from one generation to the other, *no one* has ever written it [personne ne l'a jamais écrit]; it is constituted by itself, slowly, painfully traversing the vicissitudes of reproductions subtended by evolution. Moreover, no one has received a true message [emphasis added].[45]

The two salient features of the isomorphism—combinatorial elements and the strict linearity—did not prove the identity of the two systems. There were many examples in science where diverse and complex structures arose from combinatorial arrangements (e.g., periodic table and atomic structure), Jacob argued. However, unlike language, genetic structures were three-dimensional products of organismic evolution. He could have added, of course, two other severe objections to the linguistic-genetic analogy based on the finding in the 1950s, first by Quastler and his collaborators and soon after by George Gamow and Alexander Rich: these researchers found that, unlike any known language, the so-called genetic language displayed no intersymbol restriction and hence the code could neither be overlapping nor be solved cryptographically. Similarly, in no natural language would words comprise consecutive triplet letters. Even without invoking these key structural discrepancies Jacob came to the conclusion that the remarkable analogies between heredity and language were an expression of exigencies born out of proximate functions (e.g., development of vocal and auditory apparatus that imposed linearity on language). And, he judged, while linguistics could aid genetic analysis, genetics had little to contribute to linguistics.[46] Powerful as it was, linguistics served merely as a model in biology.

Echoing Georges Canguilhem, Jacob distinguished biology from the physical sciences: "Lacking perhaps the means of supplying itself with mathematically-based theories, biology functions most frequently through the use of models. There exist, in effect, in biology numerous generalizations but few genuine theories." In the necessary dialogue between theory and experiment in the progression of the natural sciences, models have often played the role of theory in guiding experiment in biology. But frequently there was a tendency to take the model of an explanation and the analogies as identities, he observed. If the value of a model was measured by the operational efficacy, then the linguistic model has been remarkably effective. “Rarely could a model imposed on the conceptions of an epoch find more faithful applications,” he conceded. But how seriously can we take an analogy, even if it appears so exact? To drive the point, Jacob recounted the homologies between the genetic code and the ancient Chinese Book of Life (the I Ching, or Book of Changes), which were unearthed in the late 1960s; the correspondences were far more astonishing than those of linguistics and the genetic code. “It is perhaps the I Ching which one should study in order to grasp the relations between heredity and language,” he concluded (one has to wonder if he was being sarcastic) based on the writings of Gunther Stent.[47]

The Genetic Code and the I Ching: A Serious Joke?

Around 1969 several individuals in Europe and the United States observed, from very different professional vantage points, that the ancient Chinese I Ching and the newly completed genetic code shared remarkable similarities.[48] The three-thousand-year-old Book of Changes—a symbolic system for comprehending human experience—and the genetic Book of Life exhibited striking correspondences. Both symbolic systems claimed to account for the patterns governing diversity through permutations of four basic elements taken three at a time, producing sixty-four building blocks (codons in the genetic code; hexagrams in the I Ching). This diversity arose from the interaction of the two binary oppositions, or antithetical principles, Yang (the active, or male, principle represented by an unbroken line) and Yin (the passive, or female, principle represented by a broken line). In Chinese philosophy the cosmos is organized into principles of unity, duality, and change. Yang and Yin are viewed as complementary polarities; they are not dualities. Yin and Yang combine to form four digrams: Old Yang (⚌), Old Yin (⚏), New Yang (⚎), and New Yin (⚍). These four digrams are combined, three at a time, to form $4^3=64$ hexagrams. Read from bottom to top, each hexagram symbolizes one of the sixty-four fundamental states of life and its polar opposite: for example, abundance and limitation; deliverance and obstruction; before completion and after completion. All of these are generated through the interaction of the three digrams composing the hexagram and derive from the transmutation of complementarities: Yin to Yang, or vice versa. In this manner, the arrangements of the hexagrams (varying over the millennia) supposedly capture the changing cosmic patterns, which is the ceaseless flow of life.[49] (See Fig. 41.)

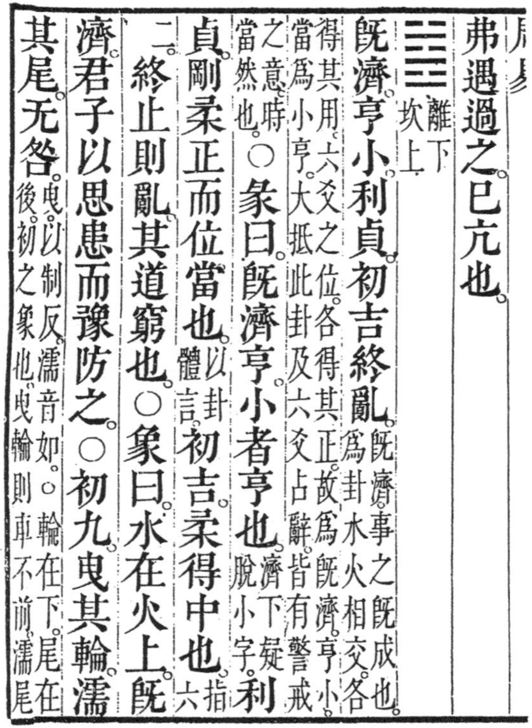

63.

弗遇過之巳亢也。

離下 坎上

既濟亨小利貞初吉終亂。既濟事之既成也。爲卦水火相交各得其用六爻之位各得其正故爲既濟亨小當爲小亨大抵此卦及六爻占辭皆有警戒之意時當然也。○象曰既濟亨小者亨也。濟下疑脫小字利貞剛柔正而位當也。以卦體言初吉柔得中也。指六二。終止則亂其道窮也。○象曰水在火上既濟君子以思患而豫防之。○初九曳其輪濡其尾无咎。曳以制反濡音如。○輪在下尾在後初之象也曳輪則車不前濡尾

63. Chi Chi–After completion

64. Wei Chi–Before completion

The final two complementary states (63, 64) in the I Ching code, "Before Completion" and "After Completion": their meanings and combinatorial potential.

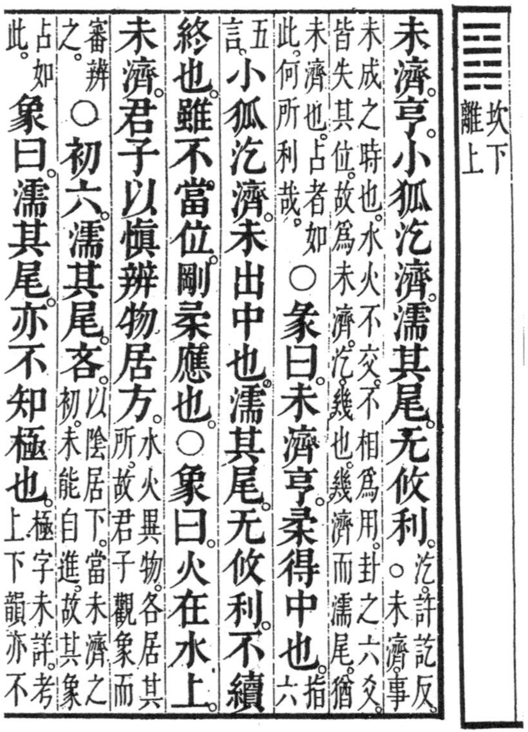

64.

坎下 離上

未濟亨小狐汔濟濡其尾无攸利。汔許訖反。○未濟事未成之時也水火不交不相爲用卦之六爻皆失其位故爲未濟汔幾也幾濟而濡尾猶未濟也占者如此何所利哉。○象曰未濟亨柔得中也。指六五言。小狐汔濟未出中也濡其尾无攸利不續終也雖不當位剛柔應也。○象曰火在水上未濟君子以愼辨物居方。水火異物各居其所故君子觀象而審辨之。○初六濡其尾吝。以陰居下當未濟之初未能自進故其象占如此。象曰濡其尾亦不知極也。極字未詳考上下韻亦不

FIGURE 41. Complementary hexagrams in the I Ching. From *Zhou yi benyi* (*The Fundamental Meaning of the I Ching*) (Taipei, Taiwan: Huailan chubanshe, 1975).

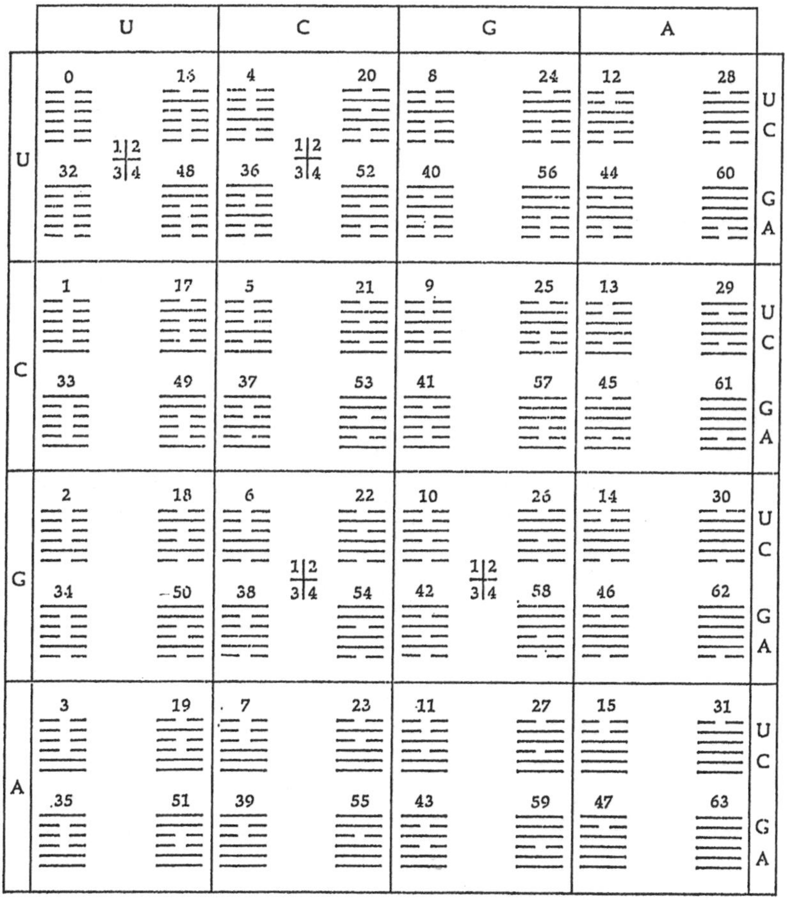

	U		C		G		A		
U	0	16	4	20	8	24	12	28	U C
	32	48	36	52	40	56	44	60	G A
C	1	17	5	21	9	25	13	29	U C
	33	49	37	53	41	57	45	61	G A
G	2	18	6	22	10	26	14	30	U C
	34	50	38	54	42	58	46	62	G A
A	3	19	7	23	11	27	15	31	U C
	35	51	39	55	43	59	47	63	G A

1|2
3|4

FIGURE 42. Martin Schönberger, *The I Ching and the Genetic Code*, p. 72. Reprinted by permission of the Aurora Press.

If one of four DNA bases is assigned to one of the digrams (note that the assignment is arbitrary), then each of the sixty-four hexagrams comes to represent one of the codons. In this manner the "natural" order of the I Ching states can generate the full array of the genetic code (see Fig. 42). Stent felt that "the congruence between it [I Ching] and the genetic code is nothing short of amazing. . . . Perhaps students of the presently still mysterious origins of the genetic code might consult the extensive commentaries on the I Ching to obtain clues to the solution of their problem."[50] Martin Schönberger, the first to work out in detail these homologies, extrapolated their meanings in a manner similar to Jakobson: rather than observing contingent convergences, he extracted ontological significance from the two systems:

the Book of Changes and the Book of Life. He saw both as a manifestation of universal flow of information representing a cosmological principle.

> And yet we shall not be able to evade the question: Are both "books" a manifestation of a common principle? Is what is involved here perhaps one universal code which was discovered 5,000 years ago by the Chinese—and 10 years ago by Watson and Crick? In other words: Is there only one spirit whose manifestation (=information?) must of necessity find its expression in the 64 words of the genetic code on the one hand or in the 64 possible states and development of the I Ching . . . on the other?[51]

As with Jakobson, the answer was affirmative and pointed to a universe fundamentally different from that portrayed in Jacques Monod's *Chance and Necessity*. Rather than viewing DNA-based life as a product of chance, it would be chance subject to the structures and patterns of the I Ching. And rather than being a gypsy living on the edge of an alien world, as Monod decried, a human being would enjoy a deep sense of security that emerged from being planted physically and spiritually in an internal natural order.[52] True, scientists might discard such spiritual claims, but they cannot easily ignore a clear double standard of warrant: if one were unwilling to consider the validity of the striking analogy between the I Ching and the genetic code, then how could one embrace the far weaker analogy between language and DNA, and even treat it as an ontology?

This tension between ontology and analogy may be neutralized through the notion of "chimera," proposed by the semiotician Françoise Bastide, in her essay "Linguistique et Génétique." She analyzed critically the exchanges between Jacob and Jakobson, demonstrating (as Canguilhelm had argued) that models do not act unilaterally: the act of transporting models across disciplines simultaneously subverts their respective objects of inquiry. Thus, in assigning "intrinsic meaning" to the code Jakobson negated Saussure's theory of arbitrariness of the sign; Jacob, on the other hand, undermined the very idea of linguistic communication by admitting to a code without senders or receivers. But rather than expect fidelities and in order to cope with "our spaces of ignorance," Bastide suggests that modern biology might view its objects as *chimera*: a hybrid mythological creature like a centaur, an animal body with human head. The animal body is nature, subject to the laws of elemental interactions, while the head represents the indeterminate level of signification, "a mixed denomination, fruitful but ambiguous," as Anne-Marie Moulin put it. "Nature does not spontaneously fabricate chimeras, it is man's way to integrate nature and culture . . . it is the greatest source of productivity," Bastide insists.[53] It is in this sense that the genomic Book of Life may be viewed as a chimera, a production of nature and culture. As Jacob wrote, "The genetic message, the programme of the present-day organism, therefore, resembles a text without an author, that a proof-reader has been correcting for more than two billion years, continually improving, refining and completing it, gradually eliminating all imperfections."[54] Part nature, part culture, the chimera of the book of life has accommodated the overload of meanings through which that book became an authorless creation.

Thus the simultaneous transportation of cybernetic and informational representations into both linguistics and molecular biology in the 1950s propelled the striking analogies between the two fields. As in other disciplines, through the circulation of the information discourse their objects of study were (separately) reconfigured anew, and then emerged, not entirely surprisingly, with some parallel features. And it is this simultaneous dematerialization of both language and life that soon formed the conditions of possibility for envisioning the word (information of the DNA sequence) as the origin of self-organization, the ontological unit of life and evolution. This vision, elaborated by Manfred Eigen in the 1970s, set the stage for simulating and engineering life with computer-generated mathematical modeling, creating the theoretical possibility of an evolutionary biotechnology, a postgenomic future. On the more pragmatic, short-term level, DNA linguistics has promised to fashion a powerful tool for uplifting the coding sequences from the morass of so-called junk DNA (the 97 percent of noncoding sequences), for extracting semantics out of syntax. In the 1970s, as Jakobson's influence waned, the study of the linguistic properties of DNA would be revisited with the Chomskian paradigm, but still subject to the same critique about the unwarranted extrapolation from linguistic analogy to ontology.

EVOLUTION OF THE WOR(L)D

Manfred Eigen's (b. 1927) excursions into life science began in the early 1960s, soon after the breaking of the genetic code, and were stimulated by his regular participation in Schmitt's Neuroscience Research Program (NRP) gatherings. Even before his Nobel Prize (shared in 1967 with Ronald Norrish and George Porter) and directorship of the Max-Planck Institute for Biophysical Chemistry in Göttingen he began applying his expertise in ultrashort chemical reactions to biomolecular systems. He treated nucleic acid and enzyme reactions as information transfer systems, as molecular learning, storage, and retrieval. By the end of the decade these excursions had coalesced into a focused research program on the origin of life as information: self-organization of matter; molecular evolution; and the beginning of what

FIGURE 43. Manfred Eigen, 1985. Courtesy of Manfred Eigen.

would grow into DNA linguistics. Eigen's project bestowed new meanings on Gamow's originary random sequence, Shannon's statistical communications, Wiener's vision of the individual as the word, and von Neumann's dream of self-reproducing automata: they now merged into biological algorithms grounded in the rules of neo-Darwinian evolution, reconfigured as information-based game theory.[55]

Eigen argued there is no need to get snagged up in chicken-and-egg conundrums about the origin of life—which came first, nucleic acids or proteins? (DNA cannot replicate without enzymes; enzymes cannot be made without DNA.) If "information" is substituted for nucleic acids and "function" for proteins, then the relation between the two is a closed loop: function is predicated on information; information acquires meaning only through function. Fully aware of the absence of meaning in information theory, Eigen viewed semantics as the dynamic and functional properties resulting from a prebiotic interplay between nucleic acids and proteins. Consequently, there was no need for viewing the emergence of life as one enormously unlikely accident (some estimate it on the order of 10^{-255}) but as random effects that were able to feed back to their origin and became themselves the cause of amplified action. As Eigen put it, they formed a communication system of sufficient legislative and executive powers. Under certain external conditions such a multiple interplay between cause and effect built up—through hypercycles—to a macroscopic functional organization, including self-reproduction, selection, and evolution to a level of sophistication where the system could escape the prerequisite of its origin and change the environment to its advantage. Thus genetic information arose selectively by self-organization of a material system. In Eigen's cosmogony, "Life is neither creation nor revelation. It is neither the one nor the other, because it is both at once." His theory was not entirely novel; Henry Quastler had proposed an information-theoretical Darwinian model for biological organization in the early 1960s (Eigen and his colleagues did not seem to be aware of it).[56] But Eigen went much further, elaborating and quantifying the features of this evolutionary model and its putative linguistic properties toward futuristic visions of postgenomic biopower.

"Von Neumann's idea of a self-reproducing automaton has stirred mathematicians' interest [among them S. Ulam and J. H. Conway] in a particular category of games simulating proliferation and growth," Eigen recounted.[57] But he modified these statistical "life games"—colored glass beads and dice shaped as platonic solids, played on specially constructed boards—by introducing an element of randomness. Any molecular reproduction process was subject to randomly occurring "errors" and these "mutations," if properly selected, would be the source of new information. Beginning with a sequence of letters (e.g., AGUUCCGCAGGCU), the game's objective is to arrive at a specified sequence (say, GCUGGCUACUAGC) by random variation of single digits under the guidance of certain selection rules that favor survival, namely, the conservation of information or "ur-semantics." Sequences supplying further information for enhancement of speed or fidelity of reproduction or protection against decomposition possess selective advantage. In this stochastic universe of molecular Darwinism the analogue of species is a "quasi-species," defined as a given distribution of macromolecular species with closely interrelated sequences. External constraints enforce the selection of the best-adapted distribution, referred to as the *wild-type*.[58]

The lessons Eigen drew from those glass bead games were that Darwinian selectional behavior was attained by fulfillment of certain preconditions, which became unalterable once the complexity of the system had become so great that all the alternatives could not be represented at the same time. As long as the game started with simple preconditions to be fulfilled for large classes of substance in unlimited concentration ranges, only one definite combination of strategies—metabolism, self-reproduction, and mutageneity—applied for Darwinian selectional values. In this manner natural selection yielded a quality grounded in the properties of matter and readily checked by independent experiments; the model thus circumvented the proverbial tautology inherent in the Spencerian dictum "survival of the fittest," where fitness was assessed through survival rates, thus leading to the "survival of the survivor." Beyond theoretical prowess, the model suggested the possibility of building an "evolution machine"—von Neumann's dream of genetic simulacra extended even beyond Baudrillard's apocalypse—automatically controlling and maintaining the specified condition and leading to self-evolving molecular systems. A decade later Eigen would outline the kinetics of evolutionary molecular engineering of RNA replication in terms of models tested by computer simulations; he also sketched the basic features of an evolution reactor. Not purely of a theoretical import, this work has attracted the support of biotechnology companies (e.g., Bayer, Hoechst, and Hoffmann La Roche). An implosion of technologic and biologic, evolutionary biotechnology has since become a symbol of future biological machines and industrial ecology in what Kevin Kelly forecasted as the rise of neobiological civilization.[59]

To ascend from survival at the macromolecular to the protocellular level Eigen postulated an ordering principle in the form of the emergence of the hypercycle (closed loop of nucleic acids and proteins). As a prerequisite for self-organization, nucleic acids provide legislative powers: complementary instructions for code formation using binary or quaternary digit system. But their recognition power is not sufficiently high for the accumulation of a large and still reproducible information content in single chains. Proteins, on the other hand, possess enormous executive powers, namely, functional and recognitive diversity and specificity. Via catalytic couplings they can link many information carriers and build up large information capacity; however, they lack the prerequisite for evolution—instructions. A combination of legislative and executive powers, or complementary instruction with catalytic coupling, will lead to nonlinear selection behavior as the simplest mechanism of functional coupling, or a self-reproductive hypercycle. Those fluctuations in the system that lead to unique translation and its reinforcement via the formation of a hypercycle offer an enormous selective advantage. Thus the emergence of life occurs through selective reiterations. Rather than being a fluke, the origin of life's information turns out to be an inevitable event.

By the mid-1970s this information-theoretical molecular Darwinism acquired a linguistic dimension. Jakobson had articulated a point of departure, but with the shift from structuralism to Chomsky's program—syntactic structures, generative grammar, and transformational rules—a handful of molecular biologists began exploring the so-called genetic language within the new paradigm. Rather than being a system grounded in differential phonemic attributes, "language" now meant a set of sentences, meaningful constructions of finite length but infinite possibilities, built from a finite set of elements by means of concatenation. It was defined by applying to its alphabet (set of elements) a finite set of rules (grammar) for generating all sentences and by imposing a set of punctuation marks, or strings of alphabetical signs, functioning as punctuations.[60]

Drawing on such preliminary explorations Eigen and his collaborator, Ruthilde Winkler, outlined a rationale for studies of evolutionary biomolecular linguistics. They observed:

> The existence of a "language" is equally important to the material self-organization of living beings, to the communication between men, and the evolution of ideas. The essential precondition for the development of a language is the assignment of unambiguous meanings to symbols. In molecular languages, this assigning takes the form of definite physical and chemical interactions; in communication between human beings, it is based on the assignment of meaning to phonemes and on their graphic representation. The assignment of meaning to combinations of symbols, as well as the interrelationships between such combinations, arises from an evolutionary process based on functional evaluation. According to Noam Chomsky, the deep structures of all languages—just like the genetic language that has emerged from molecular mechanics—have common elements that reflect the functional logic inherent in the operations of the central nervous system. The parallels between molecular genetics and the generative grammar of linguistic communication make the rules affecting evolutionary processes eminently clear.

With this dialectic of language and matter, word and act, Eigen circumvented the Faustian dilemma raised by St. John's gospel: life is neither act nor word, neither creation nor revelation; it is both at once. In accordance with the view of emergent life as a concatenation of hypercycles, as a communication between legislative and executive powers, semantics was assigned to proteins. As in human languages twenty alphabetical symbols (amino acids), having specific functions, formed cooperative units of words and sentences, the authors argued; the legislative language of nucleic acids was analogized to formal machine language, pure (syntactic) information processing.[61]

But Eigen's linguistic distinction between DNA syntax and protein semantics did not take root. After all, what would be the incentive of investing efforts in linguistic analysis of known entities instead of operationalizing the predictive process for deriving biological meaning (functionality) from opaque genomic syntax? The quest for genetic meaning intensified as it became clear in the 1980s that only a small fraction of human DNA (about 3 percent) specified the manufacture of proteins; this search gained urgency as the visions of sequencing the entire human genome took shape. Edward Trifonov and Volker Brendel in the mid-1980s first began applying operationally the rules of Chomskian grammar to genomic language, christening it "Gnomic" as their predictive tool. Words treated as internally correlated strings of limited size were the basis for future linguistic analysis of nucleotide sequences, they argued. "Bearing in mind the importance of information contained in these texts on living matter, its functions and malfunctioning, one could envisage that Gnomic will soon become a most intensively studied language, and most intriguing reading as it is already."[62]

DNA linguistics did not become a scientific movement, but it did gather momentum and emerged as a visible subspecialty within theoretical, or computational, molecular biology. For example, inspired by Jakobson and Jacob, Julio Collado-Vides has been a champion of DNA linguistics. According to Collado-Vides, one of the biggest problems in biology is the accumulation of enormous amounts of data in the absence of appropriate theoretical frameworks. Generative grammar, he explained in 1989, could provide a broad and flexible framework for constructing a global paradigm for understanding genomic organization and gene expression regulation. Linguists' criticisms of the validity of this project notwithstanding, theoretical biologists continue to sharpen their linguistic tools on prokaryotic and eucariotic systems in quest of biological meaning.[63]

"Biologists Seek the Words in DNA's Unbroken Text," announced a 1991 *New York Times* article. The reporter explained:

> Now, in an effort to decipher the great helical string of biochemical letters that make up the *book of life*, a handful of particularly imaginative biologists are applying the techniques of linguistics to the study of DNA. . . . The idea of thinking of genes as language is not really new. After all, said Dr. Konopka

> [mathematical biologist at the National Cancer Institute], the science of molecular biology first burst to life in the 1940s, which happened to be the time when social scientists were exploring the nature of communication and language. "This obviously influenced our thinking. . . . Most biologists were mentally ready to think of the genome as a communication system." . . . Biolinguists are trying to find a method for picking up the core three percent from the biochemical background noise and they are trying to spot the words without having to worry about what those words say.[64]

No longer taken as a metaphor, the chimera of the Book of Life, with all its incongruities and aporias, has become the dominant icon in the quest of biopower, genomic mastery predicated on "DNA literacy," and control of the word; it professes both creation and revelation.

Conclusion

The imagery of information written in the genomic Book of Life, which awaits reading and editing, has proved to be scientifically productive and culturally compelling. Each day on the average a new gene is identified; genetic sequence data is cascading exponentially as market stocks soar. But what does all this information mean? As several scientists have argued, these genomic visions are simplistic, promising a great deal more than can be reasonably delivered. For even if one grants these slippery scriptural analogies, the Book of Life cannot be read or edited unambiguously. As in literary creations, transliteration differs from translation and cannot capture the nuances of meaning. Transliterated DNA sequences would be polysemic and context-dependent; even "context" is by no means simple to define. Many biologists acknowledge that these large-scale sequencing initiatives, although useful, are based on faith in the predictive powers of genomic sequences, a view presupposing straightforward correspondence between genes, structures, and functions; a "genetic program." But with transposons, exons, and introns, and with splicing and posttranslation modification, the relation is plastic, context-dependent, and contingent. In several laboratories around the world, genomics is now moving beyond monogenetic and polygenetic determinism, even beyond functional genomics, toward a phase concerned with nonlinear, adaptive properties of complex dynamic systems, where visions of linear causality would be replaced by analyses of networks interacting with the environment and operating across levels of regulations: genetic, epigenetic, morphogenetic, and organismal.

Indeed, most known human disorders (about 98 percent) are polygenetic (involving the participation of several genes) and multifactorial (influenced by somatic and environmental interactions). Only about 2 percent of known human disorders are monogenetic, as in the simplest paradigmatic case of cystic fibrosis where to date nearly five hundred mutations of the CF gene have been tracked down, although some mutations may never express, or manifest in only a mild form. And gene therapy has been forbiddingly costly and difficult to effect. Experts acknowledge that as a standard medical intervention, gene therapy, even if eventually successful for a limited number of disorders, lies far in the future. At a rate of one per year, recombinant-DNA drugs (variously referred to as "gene therapy") have been slow to reach the market. In the meantime, driven by global capital, the human genome projects are generating voluminous "raw" genetic information, only some of it useful, and copious diagnostics of genetic predispositions, which are beginning to alter employment practices, family planning, educational policies, insurance practices, investment portfolios, and cultural attitudes. A fount for journalistic hype, human genome projects offer only little in the way of therapeutics. Their current medical prowess and their economic and cultural potency inheres mainly in their geneticization of society, in the ways genetic information is reconfiguring our notions of self, health, and disease. Well ahead of medical technologies, social technologies have already been set in motion.

Indeed, the Human Genome Project is the vision of biopower of the information age. The possession of a genetic map and the DNA sequence of a human being will transform our lives, so we are told. For Leroy Hood, working at the vanguard of the Human Genome Project, the creation of the encyclopedia of life is essentially a technological process, which begets more powerful technologies, especially computer techniques for inputting, storing, and accessing the three billion base pairs, and fast microchips for pattern recognition in the hunt for anomalies. An article, "Hacking the Mother Code," in the September 1995 issue of *Wired* magazine reports on Hood's expansive genomic visions, his forecasts backed by the fortunes of software magnate Bill Gates. Gates's own excitement derives from the belief in the enormous possibilities inhering in the most sophisticated program of all: the genome. Molecular biologist and Nobel laureate Walter Gilbert too sees the essence of ourselves in terms of genetic information, predicting that soon one will be able to identify one's self by the information contained on a single compact disk (CD). Thus beyond control of bodies and populations, in all their material messiness and physical contingencies, genomic biopower promises new levels of control over life through the pristine metalevel of information: through control of the word, or the DNA sequence.

While these human genome projects (in the United States, Europe, and Japan) were launched only in the past decade, the technoscientific imaginary and the discursive practices that have animated them, specifically the textual and linguistic representations of the genome, are quite old. In their (post?)modern form they first emerged in the late 1940s and were then fully elaborated within the work on the genetic code in the 1950s and 1960s. Through that work DNA was conceptualized as programmed information, as a Book of Life of the nascent information age. Although these informational representations of genetic phenomena were imprecise, sometimes self-negating, and often metaphorical, they proved remarkably seductive and productive both operationally and culturally. They aided the scientific imagination in the process of meaning making, in and beyond the laboratory. And they linked molecular biology with other realms of postwar technoculture shaped by the new communication sciences.

Thus this study could be viewed as a kind of genealogy of the future, tracing the material, discursive, and social practices that contributed to the emergence and instantiations of the informational/scriptural vision of life, representations of heredity which animate the genomic future. But it is also a study of an epistemic rupture from purely material and energetic to an informational view of nature and society. We have seen that genes did not always transfer information, that these informational modes of reasoning were historically contingent. Up until around 1950 molecular biologists (supported mainly by the Rockefeller Foundation) described genetic mechanisms without ever using the term *information* (some, in fact, resisted its usage well into the 1960s and beyond); what had been transferred across biological space and time earlier was biological and chemical specificity. An overarching theme in the life sciences, specificity originated in an earlier historical epoch, a different biological world picture, and within the discourse of organization. Though often interchangeable, the two concepts—*specificity* and *information*—did not directly map onto each other; being historically situated, discourses seldom do. The discrepancies resided in the categorical difference between the two: specificity denoting material and structural properties; information denoting nonmaterial attributes, such as soul, potentialities, and form (telos), previously captured by the notion of organization and plan (logos). The genetic code had been widely viewed as the key to life's secret logos.

The early attempts to explain genetic specificity through the permutations of nucleic acids (which some have viewed as protocodes), were formulated without notions of information. In fact, and at the risk of posing a historical counterfactual, had the coding problem been studied in the 1930s, its representations would most likely look very different. After all, the twenty amino acids were known since the turn of the century, and the four DNA (and RNA) bases were identified already in the 1920s. That the theoretical correlations of the nucleic acid bases with amino acid did not even constitute an interesting biological problem derived from the prevailing belief in the protein view of life: from the conviction that the genetic material was a protein. Had DNA and RNA been of genetic interest in the 1930s, the problem of their correlations with amino acids would have probably assumed central importance. But then the terminology and modes of reasoning would have not been informational and scriptural, since the information discourse had not yet come into being.

This picture changed radically at the end of the 1940s; World War II and the subsequent militarization of science and culture in the cold war played a significant role in this shift. Several leaders of the information revolution had a major impact on the biological sciences and social sciences, including molecular biology (still within the protein paradigm of heredity). Though the mathematical aspects of these works did not influence the technical content and experimental agenda of molecular genetics (as many information theorists had predicted) the discursive framework—the information discourse—that it stimulated did endure. Information, partially displacing the concept of specificity, became a guiding metaphor, or rather a metaphor of a metaphor, in molecular biology and in the work on the genetic code.

We have seen how astrophysicist George Gamow, following up on Watson and Crick's DNA structure, articulated what became known as the coding problem: how the DNA bases, taken three at a time, specified the assembly of twenty amino acids. Gamow thus initiated the first phase of the genetic code: the formalistic phase, 1953–61. Envisioning the code as a military cryptogram and as a system of command and control, he enlisted some of the leading physicists, mathematicians, and communications experts (several working at the hub of weapons design) for its solution. The work consisted of viewing protein synthesis as a black box, and decoding the DNA input based on the protein output. Through these contributions and those of Gamow's RNA Tie Club, notably the studies of Francis Crick and Sydney Brenner, the genetic code was constituted as an information system and linguistic communication. This approach did not lead to "breaking the code," since from linguistic and cryptanalytic standpoints the genetic code is not a code.

It was during the formalistic phase (as studies moved from overlapping to nonoverlapping codes) that the scriptural representation of the genetic code as text, reading, alphabet, and words was introduced; they served as the conceptual framework and as analytical tools for establishing the correlations between nucleotide triplets and amino acids. And despite the definitional slippages (down to the very definition of *code*), tautologies, empirical contradictions, and against the objections of information theorists these communication tropes were instantiated as the discursive framework of molecular biology. And as we have seen, by 1959 it even reoriented the thinking of biochemists. By 1960 the genetic code was viewed as the arbiter of genetic information, the central problem in molecular biology; many researchers in American and European laboratories were racing to crack the code of life.

The information discourse also assumed a pivotal role in the researches of Jacques Monod and François Jacob at the Pasteur Institute in the late 1950s. These intricate studies of genetic regulation of enzyme synthesis in E. coli recast these processes as a cybernetic communication system (information-driven negative feedback). It led to the messenger hypothesis, visualized within a system of information transfer, and to the subsequent (1961) identification of the messenger as an unstable RNA intermediate. The postulation and preparations of synthetic messenger RNA in combination with refinements of the cell-free E. coli system completely reoriented the approach to the coding problem.

Indeed, Marshall Nirenberg at NIH (partly inspired by the Pasteur Institute work) had been thinking about cracking the genetic code using synthetic messenger RNA in a cell-free protein synthesizing system well before the two groups identified messenger RNA. Early in the summer of 1961, together with postdoctoral fellow Heinrich Matthaei, they "cracked the code" by using synthetic RNA messenger poly-U in their fine-tuned E. coli cell-free system, showing that it specified the synthesis of polyphenylalanine. This was one of the most stunning events in the history of modern science; and it was a major upset for the theorists and those molecular biologists (e.g., Crick and Brenner) who had tried—and failed—to solve the code by black-boxing the problem of protein synthesis. Research on the genetic code entered the second phase: the biochemical phase (supplemented by genetic analyses), 1961–67.

The conceptual and discursive framework laid down during the first phase guided the efforts in the second. That work was marked by fierce competition to complete all the "words" of the code (to establish the "dictionary"), with Nirenberg's laboratory and the laboratory of Nobel laureate Severo Ochoa leading the race and running neck-and-neck. The ingenious genetic recombination studies in phage, spearheaded by Crick and Brenner, contributed significantly to the elucidation of the structure of the code. But it took several additional biochemical feats to determine all the "code words." By 1967 the code was essentially completed; its momentous significance captured by the many scientific writings and media coverage which announced an impending revolution in biology, both its epic promises and perils. The Nobel Prize for the work was awarded in 1968 to Marshall Nirenberg, Har Gobind Khorana, and Robert Holley. Some of the same prophecies delivered by the champions of the Human Genome Project in the 1980s could be heard already in the mid-1960s, all based on the biopower derived from decoding the book of life.

By then, the view of DNA as a universal language was ubiquitous both in biology and the culture at large, becoming a source of inspiration for Roman Jakobson's promotions of DNA linguistics. Having come under the spell of information theory in the 1950s, linguistic structuralism, like molecular biology, had emerged in the 1960s with striking similarities. The equally strong divergences were ignored. Taken to its extreme by Manfred Eigen, information came to be seen as the ontological unit of life, evolution, and natural selection; the "word" (the first DNA sequence) as both revelation and (re)creation. Genomic textuality had become a fact of life and commercial futures, a metaphor literalized, with all the humbling limits that this conflation of analogy and ontology entails for textual and material mastery of the "book of life."

REFERENCE MATTER

NOTES

Some of the documents used throughout this book are in French or German. The translations are mine, although I have sometimes retained foreign titles and specific expressions.

Archival sources are cited using the following abbreviations:

AIP	American Institute of Physics (George Gamow Oral History; Alexander Rich–George Gamow Correspondence; Leon Brillouin Papers; Henry Quastler–Henry Dancoff Correspondence)
APS	American Philosophical Society Library (Erwin Chargaff Papers; Warren S. McCulloch Papers; Robert Olby Collection)
AT&T	AT&T/Bell Laboratory Archive (Claude Shannon Papers)
CIT	California Institute of Technology (Max Delbrück Papers; George Beadle Papers; Biology Division Papers)
CIW	Carnegie Institution of Washington (George Gamow's 1946 conference)
JLF	Joshua Lederberg Files
LC	Library of Congress (George Gamow Papers; John von Neumann Papers; Vannevar Bush Papers; Gertrud Quastler Papers)
MIT	Massachusetts Institute of Technology (Norbert Wiener Papers; Roman Jakobson Papers; Julius Stratton Papers)
MND	Marshall Nirenberg Diaries
OSU	Oregon State University (Linus and Ava Helen Pauling Papers)
RAC	Rockefeller Archive Center (Detlev Bronk Papers; MIT series; Rockefeller Brothers Fund; Warren Weaver Diaries; General Correspondence; Henry Quastler; Heinz Fraenkel-Conrat; and Jacques Monod fellowship records)
SAIP	Service des Archives de L'Institut Pasteur (Fonds Monod [Jacques Monod Papers])
SBF	Sydney Brenner Files
UCB	University of California at Berkeley, Archives of The Bancroft Library (Wendell M. Stanley Papers)
UCSD	University of California at San Diego, Archives (Leo Szilard Papers)
UIA	University of Illinois Archive (Henry Quastler administrative records)

CHAPTER 1. THE GENETIC CODE

1. PBS's *Nova*, "Decoding the Book of Life."
2. W. Gilbert, p. 96.
3. D. Jackson, p. 358.
4. Pollack.
5. Notably, Aquinas; Derrida, *Of Grammatology*. The most complete study of the Book of Nature is Blumenberg's *Die Lesbarkeit*, but he accepts, rather than interrogates, the concept of nature's writing.
6. Bono, "Science, Discourse, and Literature"; idem, "Locating Narratives."
7. It is well known that Erwin Schrödinger, in his acclaimed book, *What Is Life?* spoke of the notion of a code, and there is a long-standing historiographic debate about Schrödinger's role in the history of molecular biology (see Chap. 2, n. 7, this volume). There is no doubt that Schrödinger was the first to use the term *code-script* in relation to heredity and that this played a role in the history of the genetic code, which Richard M. Doyle has eloquently analyzed in "On Beyond Living," Chap. 2. But I agree with Edward Yoxen's assessment in "Where Does Schrödinger's *What Is Life?* Belong in the History of Molecular Biology?" that the myths of Schrödinger's code were constructed mainly in the 1960s; neither Crick nor Gamow referred to Schrödinger when they began work on the genetic code in the early 1950s.
8. Public Papers of the Presidents of the United States: Dwight D. Eisenhower, Washington, 1960–61, p. 1045; quoted in Sherry, *In the Shadow of War*, p. 235.
9. Forman, "Behind Quantum Electronics." "The Cold War and Expert Knowledge: New Essays on the History of the National Security State," *Radical History Review* 63 (fall 1995); the entire issue is dedicated to the impact of the cold war in the academy. See also P. Novick. I use the term *cultural hegemony* in the Gramscian sense as developed by T. J. Jackson Lears and by Ernesto Laclau and Chantal Mouffe: a noncoercive power circulating through discursive practices. See Gramsci, p. 12; Lears, pp. 567–93; Laclau and Mouffe, p. 109. For a superb discussion of cultural hegemony, see Comaroff and Comaroff, pp. 19–27. See also Kay, "Rethinking Institutions."
10. McCormick, pp. 7–15; on Britain, France, and Germany, see Chap. 3–6. On Britain's key role in America's global hegemony, see Howard Whidden, "Europe at the Crossroads: The Next 10 to 15 Years," December 1956, Box 3.35, Special Studies Project Records, Rockefeller Brothers Fund, RAC. See also Pestre.
11. McCormick, p. 98. See also Leffler, *A Preponderance of Power*; and Wittner, Chap. 4–7.
12. Forman, "Behind Quantum Electronics"; Leslie, *The Cold War and American Science*; idem, "Science and Politics"; Edwards. See also Noble, *Forces of Production*, esp. Chap. 3; and idem, "Command Performance." See also Herken, Chap. 3–7; J. Wang; Reingold, "Science and Government"; Aaserud; and Z. Wang.
13. Haraway, "Signs of Dominance." Attenuated because by 1960 there were multiple sources of support, federal as well as private, for life science. Several scholars are now working on the impact of the cold war in the academy (see special issue of *Radical History Review*, fall 1995). The session on "The Cold War and the Shape of

Science" at the 1994 meeting of the History of Science Society included works on economics and social science. See also Simpson. Many historians have been studying the culture of the cold war. For example, Lipsitz; D'Emilio; May; Whitfield; Carmichael; and several unpublished papers were presented at the Landmarks Conference on the Cold War and American Culture at the American University, Washington, D.C., 17–19 March 1994, among them Appy, which begins to deal with historical silences (I am grateful to Peter Kuznick for availing these papers to me).

14. Tabulated from information provided annually for the years 1950–69 by the National Science Foundation, *Federal Funds for Science*; on the national laboratories and life science, see Chap. 3, this volume.

15. Beatty, "Opportunities for Genetics"; idem, "Origins of the U.S. Human Genome Project." Jean-Paul Gaudillière, in "Biologie Moléculaire," has examined in detail the political dynamics of French molecular biology (though not directly in relation to the cold war); and Soraya de Chadarevian is currently completing the book *The Making of a New Science*, where she examines explicitly the relation of molecular biology to postwar policies. On the relation of British microbiology to the military context, see Bud, "Bugs and Institutes"; Fulbright. Admiral Rickover referred to the "Military-Scientific Complex." See Nelson, "Research Probe."

16. Biagioli, *Galileo, Courtier*, Intro. and Epi. Hacking, "Weapons Research." See also Leslie, *The Cold War and American Science*, Intro. The preoccupations with national security and the influence of the cold war extended also to the writing of American history. On the broader cognitive and cultural impact of the cold war, see D. Campbell, Intro. and Chap. 6; and P. Novick, Chap. 10, esp. pp. 281–82.

17. Baudrillard, pp. 103–4.

18. There is currently a growing interest in standardization practices in science, in general, and biology, in particular; see, for example, Clarke and Fujimura; Kohler, *Lords of the Fly*; and Rader.

19. Rheinberger, "Experiment, Difference, and Writing I and II"; idem, *Toward a History of Epistemic Things*.

20. Burroughs; Fredrickson; J. A. Shannon; McElheny, "Research in Biology."

21. C. Shannon and W. Weaver.

22. Medawar, pp. 56–57.

23. On the influence of cybernetics and information theory on the social sciences, see Heims, *The Cybernetics Group*; Edwards; see also Chap. 3, this volume. For a contemporary overview on information in biology, see Elsasser. On the cybernetic influence in evolutionary biology, see Haraway, "The High Cost of Information"; and idem, "Signs of Dominance." On cybernetics' influence in ecology, see Mitman, p. 5. An extensive literature search conducted (by my research assistant Evan Ingersoll) in journals of endocrinology and immunology for the period 1950–70 revealed numerous articles that deployed the concepts of cybernetics and information; probably the best known is Burnet, in which he devotes considerable space to discussion of the information concept in immunology. Embryology was particularly responsive to the information discourse; see, for example, Raven; Waddington; and Keller, "The Body of a New Machine." On Soviet cybernetics, see the work of Gerovitch. Adams, in

"Molecular Answers in Soviet Genetics," showed that Soviet molecular biology derived its institutional legitimation from cybernetics. My search through the journal *Problems of Cybernetics*, edited by A. A. Lyapunov, for the years 1960–65 reveals that, as in the United States, the widespread cybernetic influence was mainly rhetorical. I am grateful to Slava Gerovitch for pointing me in this direction.

24. R. G. Martin, "A Revisionist View," p. 283.

25. Woese, *The Genetic Code*, p. 5.

26. Jacob, *The Logic of Life*, Intro. and Chap. 5, quotes on pp. 1 and 254, respectively.

27. Monod, *Chance and Necessity*, esp. Chap. 3–4, quotes on pp. 45 and 68, respectively.

28. Beadle and Beadle, p. 207.

29. Sinsheimer, *The Book of Life*, pp. 5–6.

30. Jacob, *The Logic of Life*, p. 287.

31. Foucault, *The Archeology of Knowledge*, p. 38.

32. Ibid., p. 48.

33. Foucault, *Power/Knowledge*, p. 93; and Smart. For an excellent discussion, see Lenoir. In response to criticisms that Foucault's power is everywhere and thus nowhere, he has explained the central role of institutions in the formation of a knowledge/power nexus. See Dreyfus and Rabinow, pp. 222–24; and Foucault, "Politics and the Study of Discourse." See also Rouse.

34. Foucault, *The History of Sexuality*, pp. 135–59. See also Hacking, "Biopower and the Avalanche of Numbers"; idem, *The Taming of Chance*. See also Kay, "Problematizing Basic Research"; and idem, "Rethinking Institutions."

35. "Information," *Compact Oxford English Dictionary*, p. 847. For discussion on the three levels of linguistic communications (proposed by Colin Cherry), see Chap. 3, this volume.

36. Aspray, "The Scientific Conceptualization of Information."

37. C. Shannon and W. Weaver, p. 8.

38. Ibid., quoted by Weaver.

39. Cherry, p. 40; von Foerster.

40. Bar-Hillel, quoted in Cherry, p. 219.

41. Lakoff and Johnson, pp. 3, 124. Lakoff has attempted to extend his claim that all knowledge is experiential and metaphorical to mathematics: the backbone of information theory. Mathematics—the Platonic ideal—usually undergirds the claim that the universe possesses rationality, which transcends human experience. Drawing on writings of the noted mathematician Saunders MacLane, Lakoff asserts that Platonic characterizations of pure mathematics do not account for *which* ideals are realized in *which* particulars; they do not provide a pairing of applicable branches of mathematics and given phenomena. According to Lakoff's brand of realism—experientialism—mathematics is based on structures within the human conceptual system, structures that people use to comprehend ordinary experiences. See Lakoff, Chap. 20.

42. The shift in the philosophy of science away from theory-construction to ex-

perimental practice was signaled by Hacking in *Representing and Intervening*. The most detailed analyses of how language has shaped laboratory practice in molecular biology is given by Rheinberger, "Experiment, Difference, and Writing I and II"; and idem, *Toward a History of Epistemic Things*.

43. Black; Hesse, "The Explanatory Function of Metaphor"; idem, *Models and Analogies*; and idem, *Revolutions and Reconstructions*, pp. 111–12.

44. Arbib and Hesse, p. 156.

45. Reddy, p. 165. On the linguistic/informational crisis of representation, see Derrida, *Of Grammatology*, esp. p. 9; Baudrillard; and Lyotard.

46. Reddy, p. 182.

47. Von Foerster, p. 20; Galison, "The Ontology of the Enemy"; Noble, *Forces of Production*; Winner; and MacKenzie.

48. Boyd, Chap. 21, p. 486. On the other hand, Evelyn Fox Keller, in "The Body of a New Machine," has claimed that the "cybersciences" (cybernetics, information theory, operations research, and computing) and molecular biology may have been products of the same historical moment, but with respect to their models and causal structures, they were running on two separate tracks, and that molecular biologists modeled their organisms on the machines of yesteryear. As we will see, there is overwhelming evidence—published and archival—that molecular biologists were influenced by the Wiener-Shannon theory, through formal and informal transactions and in various and complicated ways. This includes Max Delbrück, whose 1973 letter is cited by S. J. Heims in *The Cybernetics Group* (and Keller in "Body of a New Machine") nearly twenty years after the fact, after Delbrück had taught a graduate course on information theory. We will also see that cold-war machines (such as guidance systems and computers) supplied direct and indirect models in analyses of genetic codes in the 1950s—including its information-based textualization—as did the institutional and discursive practices used in weapons research. In fact, these exchanges provide powerful examples for Keller's observation that physicists (mathematicians and communication engineers) reconfigured biology's agenda within the new knowledge/power nexus of the postwar era.

49. Simon to Yčas, 5 December 1956, p. 2, Yčas fld. (1956), Gamow Papers, LC. Burnet, esp. Chap. 5. See also Boyd.

50. *Science Citation Index* (1955–63) contained nearly four hundred references to Quastler's work. Several researchers in molecular biology, e.g., Gamow, Sinsheimer, and Delbrück, drew on his work. For detailed examination, see Chap. 3, this volume.

51. Canguilhem, "The Role of Analogies and Models." See also Delaporte, esp. Chap. 4, which deals with the role of information in biology.

52. Medawar, pp. 56–57.

53. Woese, *The Genetic Code*, p. 17.

54. Canguilhem, "The Role of Analogies and Models."

55. Chargaff, "A Few Remarks on Nucleic Acids," Chap. 8, quote on p. 113.

56. Florkin, p. 13.

57. Fruton, pp. 200–201.

58. Edge; the railway metaphor was studied in depth by Leo Marx in *The Machine in the Garden.*

59. Rosenberg, pp. 4–6; Stepan. These are just a few isolated examples out of the myriad of metaphors in science. There is also vast literature on gendered metaphors in science; see, for example, Keller, "Molecules, Messages, and Memory"; idem, *Secrets of Life*; and idem, "Gender and Science," which provide a good overview.

60. Bono, "Science, Discourse, and Literature," p. 61. On the historicities of discourses, see Kusch, esp. Chap. 4.

61. Crick, "On Protein Synthesis," quote on pp. 152–53.

62. Rheinberger, *Toward a History of Epistemic Things*, esp. Chap. 10.

63. Fussell, p. 187. See also Pynchon.

64. Aquinas, pp. 287–88.

65. Ibid. On the logos of the world soul, see Plato. On the separation of theology and philosophy in the medieval university, see Ben-David, Chap. 4.

66. Derrida, *Of Grammatology*, p. 18.

67. Blumenberg.

68. Poster, pp. 6–11. See also H-J. Martin. For a nuanced view of epochal changes of books, see J. Martin.

69. Lucretius, verses 170 and 195, pp. 29–30. I am grateful to Matthew Meselson for calling my attention to this source.

70. Stock, esp. pp. 315–25.

71. Eisenstein; Derrida, *Of Grammatology*, p. 15. See also Kay, "Who Wrote the Book of Life?" in *Science in Context.*

72. On the scriptural representations of nature and the Book of Nature metaphor in the seventeenth century, see Arbib and Hesse, Chap. 8, esp. p. 149. Bacon; Bonnet's quote (undocumented) in Derrida, *Of Grammatology*, pp. 15–16. See also Bono, *The Word of God*; and Mario Biagioli, "Stress in the Book of Nature."

73. Immanuel Kant, *Briefwechsel* (Hamman, 1759), Akademie-Ausgabe X 28, quoted in Blumenberg, p. 190.

74. Von Goethe to Charlotte von Stein, 15 June 1786 (Werke XVIII 931), quoted in Blumenberg, p. 216. On Schrödinger's code-script, see Chap. 2, this volume. Von Goethe, pp. 1224–37. On romanticism and the Book of Nature, see Blumenberg, Chap. 16; K. Hartley; and Steigerwald.

75. Sinsheimer, *The Book of Life*, pp. 5–6; see also his autobiography, *The Strands of Life*, Chap. 16.

76. On the critiques of genomic "language," see transcript of "Un Débat Entre François Jacob, Roman Jakobson, Claude Lévi-Strauss et Philippe L'Héritier: Vivre et Parler," 20 September 1967, pp. 17–18, 31, Box 18.48, MC 72, Jakobson Papers, MIT. See Chap. 7, this volume. Baudrillard.

77. On the rise of structuralist linguistics, see Chap. 7, this volume; see also Pollack, esp. Intro.

78. On the shift from structuralism to poststructuralism, see Derrida, *Writing and Difference*, esp. Chap. 10. On autopoiesis and the redefinition of system, see Maturana and Varela; Varela, Thompson, and Rosch; Luhmann; Rasch and Wolfe.

79. Heidegger; Hacking, *Representing and Intervening*; Rheinberger, "Genetic Engineering."

80. Derrida, *Of Grammatology*, Part I.

81. See, for example, Delbrück's Festschrift by Cairns, Stent, and Watson; Lwoff and Ullman; and Stent and Calendar. See also Nelkin and Lindee.

82. Trifonov and Brendel, Preface. I am very grateful to Manfred Eigen for making this text available to me.

83. Watson, "Values from Chicago Upbringing," p. 197.

CHAPTER 2. MOLECULAR BIOLOGY BEFORE THE AGE OF INFORMATION

1. Delbrück, "Aristotle-totle-totle."

2. Ibid., pp. 54–55. A comparison between several key passages from Aristotle's *Generation of Animals* that Delbrück cited (I, 21, 730a, 24–30; I, 21, 729b, 5–8; I, 22, 730b, 10–19) and their counterparts in a Standard English translation reveals some of the liberties Delbrück took with the text. His modernized usages of terms facilitated the (post)modernist interpretations. See also Canguilhem, "Epistemology of Biology," Chap. 4 and Intro.

3. Delbrück and Stent, quote on p. 730.

4. Canguilhem, "Epistemology of Biology"; Foucault, *The Order of Things*, pp. 228, 266. Foucault did not insist on the primacy of the term *organization* and used it interchangeably with *hierarchical order* and *plan*. It was Karl M. Figlio's important essay, "The Metaphor of Organization: An Historiographical Perspective on the Bio-Medical Sciences of the Early Nineteenth Century," that elaborated Foucault's ideas. Figlio finds the discourse of organization to be more useful in historical studies of modern life science than Foucault's broad notion of *episteme*.

5. Jacob, *The Logic of Life*, p. 74. Michel Foucault, "La Logique du Vivant," p. 9. Foucault's praise appears also on the back cover. The "Integron" is the book's conclusion.

6. That reconfiguration (knowledge/power nexus) has been discussed by Haraway, for example, "A Semiotics of the Naturalistic Field," Chap. 5; idem, "A Cyborg Manifesto," Chap. 8; and Edwards. For elaboration on the military influence on the form and organization of biological knowledge, see Chap. 3–4, this volume. The many linkages of systems theories, both in the United States and Europe, were discussed at the Dibner Institute Systems Workshop, held in May 1996 (published proceedings in Hughes and Hughes).

7. On the place of the protein paradigm in the history of molecular biology, see Olby, "The Protein Version of the Central Dogma"; idem, *The Path to the Double Helix*; and Judson. See also Kay, *The Molecular Vision of Life*, Interlude I.

8. On these contributions, see Olby, *The Path to the Double Helix*, Sec. 3–4; and Judson, Part I.

9. Schrödinger. Among those who have accorded Schrödinger this formative role are Stent and Calendar; Jacob, *The Logic of Life*, Chap. 5; Olby, *The Path to the*

Double Helix, Sec. IV; idem, "Schrödinger's Problem: What Is Life?"; Moore, pp. 394–404; J. A. Witkowski; Doyle, "Mr. Schrödinger Inside Himself"; and idem, "On Beyond Living," Chap. 2. Some arguments against such interpretations have been advanced by Yoxen, "Where Does Schrödinger's *What Is Life?* Belong in the History of Molecular Biology?"; and to some extent by Symonds. See also Sigurdsson.

10. François Jacob, "Le Modele Linguistique en Biologie."

11. Judson, in his remarkable book, *The Eighth Day of Creation*, pp. 608–12, pointed to the centrality of the theme of biological specificity in the history of molecular biology. But he did not probe the broader epistemic context, material space, and meanings of specificity, nor its apparent equivalence to "information." Because of this narrow hindsight, Judson concluded that in the 1930s "specificity was really a term almost empty of meaning. . . . Forty years later, biological specificity is richly stuffed with meaning" (p. 12). According to him, it was Francis Crick's "sequence hypothesis" and the "Central Dogma" that infused specificity with meaning. As we shall see, with the introduction of the notion of "information," Crick followed the discursive trend of displacing "specificity" by a more metaphorical term. Olby too has written about biological specificity in "The Recasting of the Sciences." See also Sarkar, "Biological Information: A Skeptical Look at Some Central Dogmas of Molecular Biology"; and Thieffry and Sarkar. But the discursive significance of the concept of specificity in shaping the objects of biological research and their subsequent reshaping by the information discourse have not been previously explored.

12. Silverstein, "History of Immunology"; idem, *A History of Immunology*, Chap. 5–6; Tauber and Chernyak; and Tauber, esp. Part I. See also Gilbert and Greenberg; and Mazumdar, "The Antigen-Antibody Reaction and the Physics and Chemistry of Life." Landsteiner's work is elaborated in his noted *The Specificity of Serological Reactions*.

13. Frank R. Lillie, quoted in Gilbert and Greenberg, p. 27.

14. On the heated controversy between Lillie and Loeb, see Gilbert and Greenberg, p. 31; Manning; and Pauly, *Controlling Life*. On the relations of Loeb and Arrhenius, see Kay, *Molecules, Cells, and Life*, p. 64.

15. For example, Nuttall; Reichert and Brown. See also Mazumdar, "Karl Landsteiner and the Problem of Species, 1838–1968"; and idem, *Species and Specificity*.

16. Reichert, quoted in Reichert and Brown, p. iv.

17. For contemporary discussion on these studies, see Loeb, pp. 63–68. See also discussion in Olby, "The Recasting of the Sciences."

18. Loeb, p. 61.

19. Ibid., p. 70. On these debates, see Sapp, *Beyond the Gene*, Chap. 1.

20. Thomas H. Morgan, *The Physical Basis of Heredity*, pp. 225–26. See also Allen, *Thomas Hunt Morgan*.

21. Morgan, *The Theory of the Gene*, p. 306. On Morgan's thoughts about the materiality of the gene, see Kay, *The Molecular Vision of Life*, Interlude I.

22. On the different approach in European genetics, see Burian, Gayon, and Zallen, "The Singular Fate of Genetics in the History of Biology"; Sapp, *Beyond the Gene*; Harwood, "National Styles in Science"; and idem, *Styles of Scientific Thought*.

23. Report of the Committee on Appraisal and Plan, 11 December 1934, p. 25,

Box 24.184, RG3, 900, RAC. On the Rockefeller Foundation's support of science and molecular biology, see Kohler, *Partners in Science*; idem, "The Management of Science"; Yoxen, "Giving Life a New Meaning"; Abir-Am, "The Discourse of Physical Power"; and Kay, *The Molecular Vision of Life*.

24. Weaver's Diary, 1934, pp. 98–110, Box 68, RG12.1, RAC; see also Kay, *The Molecular Vision of Life*, Intro. and Chap. 1, for discussion on the relationship between the new biology and the old eugenics. On the theme of biopower, see Foucault, *The History of Sexuality*, pp. 135–59; Hacking, "Biopower and the Avalanche of Numbers"; and idem, *The Taming of Chance*.

25. Morgan to Mason, 15 May 1933, pp. 2–3, Box 5.71, RG1.1, RAC. For discussion on the role of scientific elites (e.g., Morgan, Pauling, or Beadle) in building cultural hegemonies, see Kay, *The Molecular Vision of Life*, Intro.; and Kay, "Rethinking Institutions."

26. See, for example, Irwin and Cumley; and Boyden. See also Standskot. For a review of Haldane's researches, see "John Burdon Sanderson Haldane," *Biographical Memoirs of the Royal Societies* 12 (1966): 219–49.

27. On the role of metaphors in science, see Chap. 1, this volume. For an illuminating analysis of metaphors, with particular reference to immunology, see Moulin, "Text and Context in Biology."

28. Foucault, *The Order of Things*, Chap. 8; Jacob, *The Logic of Life*, Chap. 2; Figlio, pp. 17–53.

29. Weiss, *Principles of Development*, p. 102. See also idem, "Principles of Development."

30. Foucault, *The Order of Things*, Chap. 8. On the relation of biology to modernist culture, see Pauly, "Modernist Practice in American Biology."

31. See, for example, Spencer; Durkheim, Book III, Chap. 1; Weber, "Science as a Vocation"; and idem, "Politics as a Vocation." See also Crook, Pakulski, and Waters, Chap. 7. For an extensive bibliography on naturalism and the human sciences, see Cross and Albury. On "Fordism," see Harvey, Part II. On the rhetorical and ideological dimensions of "Co-operation" in American culture in the interwar era, see Kay, *The Molecular Vision of Life*, esp. Chap. 1. On the role of discursive economy in natural science, see Lenoir.

32. Cannon, pp. 287–92. See also Cross and Albury.

33. Huxley; Geison. See also Kay, *The Molecular Vision of Life*, Interlude I.

34. Chamberlin and Gilman; Ludmerer; Haller; Pickens; Kevles, *In the Name of Eugenics*; Allen, "The Eugenics Record Office"; Paul; Adams, *The Wellborn Science*; Weingart, Kroll, and Bayertz. On biopower, see Foucault, *The History of Sexuality*; idem, *Power/Knowledge*.

35. Kohler, "The Enzyme Theory."

36. Stanley, "Isolation of Crystalline Protein Properties." See also Kay, "W. M. Stanley's Crystallization of the Tobacco Mosaic Virus"; and Olby, *The Path to the Double Helix*, Chap. 9–10.

37. General Correspondence, Warren Weaver, 28 August 1939, Box 170.1235, RG2, 100, RAC. See also Kay, *The Molecular Vision of Life*, pp. 111–12.

38. On the managerial view in molecular biology, see Yoxen, "Life as a Produc-

tive Force"; on the managerial view in the human sciences, see Haraway, "A Pilot Plant for Human Engineering"; and Kay, *The Molecular Vision of Life*, Intro. and Chap. 1.

39. Weaver; Kay, "Problematizing Basic Research"; idem, "Rethinking Institutions." On the relationship between representation and intervention (first formulated by Heidegger), see Hacking, *Representing and Intervening*, pp. 130–46. On their ideological implications, see Merchant, Chap. 7–9; Keller, "Critical Silences in Scientific Discourse," pp. 73–92; idem, "Physics and the Emergence of Molecular Biology"; and Kay, "Life as Technology."

40. On Pauling's visions of social control, see Kay, "Life as Technology"; and idem, *The Molecular Vision of Life*, Chap. 8.

41. Mirsky and Pauling. See also Kay, *The Molecular Vision of Life*, Chap. 5.

42. Pauling and Delbrück, pp. 78–79.

43. See, for example, Delbrück. See also Kay, *The Molecular Vision of Life*, Chap. 8.

44. Haldane, quoted in Pollock, "From Pangens to Polynucleotides," p. 467.

45. Olby, *The Path to the Double Helix*, pp. 115–18; Kay, *The Molecular Vision of Life*, Interlude I, Sec. 1.

46. Pauling, "A Theory of the Structure." See also Kay, "Molecular Biology and Pauling's Immunochemistry"; and idem, *The Molecular Vision of Life*, Chap. 6.

47. Pauling, "Antibodies and Specific Biological Forces," p. 53.

48. Grant in Serological Genetics, 14 June 1940, Box 7.91, RG1.1, 205D, RAC. See Kay, *The Molecular Vision of Life*, Chap. 6.

49. Pauling, in fact, applied for a patent for a production of artificial antibodies. See Kay, *The Molecular Vision of Life*, pp. 174–75.

50. Burnet, Chap. 5. See also Moulin, Part II, Chap. 1.

51. Schultz.

52. On Beadle's *Neurospora* research, see Kay, "Selling Pure Science in Wartime"; and idem, *The Molecular Vision of Life*, Chap. 4 and 7. See also Kohler, "Systems of Production."

53. Beadle, "The Genetic Control of Biochemical Reactions," quote on p. 192; and idem, "Biochemical Genetics."

54. Report on Serological Genetics, March 1943; and Sturtevant (the paper was written in 1940), Box 7.94, RG1.1, 205D, RAC. See Kay, *The Molecular Vision of Life*, p. 191. Sterling Emerson's publications were well received by life scientists.

55. "The Structure and Function of the Gene," 14 October 1955, p. 7, Box 31.2, Beadle Papers, CIT. See also Beadle and Beadle.

56. Lederberg, "Comments on the Gene-Enzyme Relationship," quote on p. 167. I am grateful to Sahotra Sarkar for making this material available to me.

57. Gamow, Rich, and Yčas, pp. 23–67, definition on p. 66.

58. Fantini, "Monod, Jacques Lucien." The term *enzyme adaptation* was coined by Henning Karstron, who distinguished between constitutive and adaptive enzymes.

59. The official convergence of genetics, bacteriology, and virology took place around 1946. See, for example, Delbrück and Bailey; Hershey; Lederberg and Tatum. See also Bud, *The Uses of Life*, Chap. 8.

60. Lwoff, "Jacques Lucien Monod"; Judson, pp. 354–58. On the place of genetics in France, see Burian, Gayon, and Zallen, "The Singular Fate of Genetics"; and Sapp, *Beyond the Gene*, Chap. 5.

61. Monod, "The Phenomenon of Enzymatic Adaptation," *Growth* 2 (1947): 224.

62. Weiss, quoted in Monod, "The Phenomenon of Enzymatic Adaptation" in *Growth Symposium* XI (1947): 2.

63. Monod, "The Phenomenon of Enzymatic Adaptation," pp. 260–61, 280. See also Burian, "Technique, Task Definition, and the Transition from Genetics to Molecular Genetics."

64. For example, Brachet; Caspersson. See also Kay, *Molecules, Cells, and Life*, p. 4; and idem, *The Molecular Vision of Life*, Interlude I. See Olby, *The Path to the Double Helix*, Sec. II.

65. Avery, MacLeod, and McCarty. See also McCarty; Olby, *The Path to the Double Helix*, Sec. III. For historiographical discussion, see Stent, *Molecular Genetics*; Wyatt; Lederberg, "Genetic Recombination in Bacteria"; idem, "The Transformation of Genetics by DNA."

66. Avery, MacLeod, and McCarty, p. 152.

67. Ibid., pp. 154–55. See also Olby, *The Path to the Double Helix*, pp. 189–90; Amsterdamska, "Stabilizing Instability"; and idem, "Between Medicine and Science."

68. James Watson, *The Double Helix*, Chap. 18; Chargaff, *Heraclitean Fire*, Part II; Olby, *The Path to the Double Helix*, Chap. 14; Judson, pp. 142–44; Abir-Am, "From Biochemistry to Molecular Biology."

69. Chargaff, "On the Nucleoproteins and Nucleic Acids of Microorganisms," quote on p. 32.

70. Chargaff, "On the Nucleoproteins and Nucleic Acids of Microorganisms," pp. 32–33.

71. Chargaff, "Chemical Specificity of Nucleic Acids." See also Stent and Calendar, Chap. 8.

72. Stent and Calendar, p. 209. See also Chargaff, "Some Recent Studies of the Composition and Structure of Nucleic Acids."

73. Chargaff, "The Chemistry and Function of Nucleoproteins and Nucleic Acids," Chap. 4, quote on pp. 72–73.

74. Ibid.; idem, "First Steps toward a Chemistry of Heredity," Chap. 8, quote on p. 113.

75. Ibid.; idem, "A Few Remarks on Nucleic Acids," Chap. 10, quote on p. 163; and idem, "Amphisbaena," Chap. 11; imaginary dialogue quote on pp. 188–89.

76. Watson, *The Double Helix*, Chap. 18; Judson, Part I.

77. Watson and Crick, "Molecular Structure of Nucleic Acids." For Watson's earlier works, see, for example, Watson, "The Biological Properties of X-ray Inactivated Bacteriophage."

78. Ephrussi et al., p. 701.

79. Watson and Crick, "Genetical Implications of the Structure of Deoxyribonucleic Acid," quote on p. 966.

80. Stent and Calendar, p. 26.

81. Jacob, *The Logic of Life*, p. 260.

82. See note 9 above.

83. Symonds, p. 226. See also Yoxen, "Where Does Schrödinger's *What Is Life?* Belong?" Sec. I; Kay, "Conceptual Models and Analytical Tools"; and Keller, "Physics and the Emergence of Molecular Biology."

84. Schrödinger, p. 5.

85. Ibid., pp. 21–22.

86. Ibid., p. 23.

87. Ibid., p. 66.

88. Teich.

89. Semon; see W. Moore, pp. 46–49.

90. Yoxen, "Where Does Schrödinger's *What Is Life?* Belong?" pp. 31–36; and idem, "The Social Impact of Molecular Biology," pp. 139–41; Sigurdsson, esp. pp. 61–64.

91. Schrödinger to E. I. Conway, 25 October 1942, quoted in Yoxen, "The Social Impact of Molecular Biology," p. 152.

92. These materials were listed in Schrödinger's lectures' file entitled "Biologica I" and included in Sinnott and Dunn. Haldane, *New Paths in Genetics*; Darlington; Sherrington; and Timofeff-Ressovsky, Zimmer, and Delbrück. See Yoxen, "Where Does Schrödinger's *What Is Life?* Belong?" pp. 31–36; and Sigurdsson, p. 64.

93. Sherrington, p. 163; also quoted in Yoxen, "Where Does Schrödinger's *What Is Life?* Belong?" p. 35.

94. Schrödinger, p. 22. See also Blumenberg, Chap. 22.

95. This conclusion was also reached by Yoxen in "Where Does Schrödinger's *What Is Life?* Belong?" p. 37.

96. Schrödinger, p. 82.

97. Szilard; English translation by Rapoport and Knoller. See also Hayles, *Chaos Bound*, Chap. 2; Leff and Rex; and Keller, "Molecules, Messages, and Memory," Chap. 2. Von Neumann was familiar with Szilard's paper since the early 1930s and brought it to the attention of Claude Shannon around 1947. But it was Brillouin who gave it a broader audience. Warren Weaver (who collaborated with Shannon, see Chap. 3, this volume) brought Szilard's papers to the attention of his friend Brillouin in 1951 and also introduced the two. See Weaver Diary, 11 September 1950, Box 68, RG12.1, RAC; and Weaver to Szilard, 15 September 1950, Box 20.21, MSS.32, Szilard Papers, UCSD. This sequence of events explains why Brillouin, who had been writing about information and cybernetics since 1949, first discussed Szilard's paper in 1951. See Brillouin, "Maxwell's Demon Cannot Operate." See also idem, *Science and Information Theory*. For biographical materials on Brillouin, see Box 7.14, Leon Brillouin Papers, AIP; and Warren Weaver Diary, Boxes 68–68, RG12.1, RAC, for many entries on Brillouin's professional trajectory throughout the 1940s and 1950s.

98. Schrödinger to Brillouin, 9 October 1953, pp. 3–4, Box 7.6, Brillouin Papers, AIP.

99. For a perceptive discussion, see Kusch, esp. Chap. 4.

100. The various accounts include Crick, "The Genetic Code"; Woese, *The Genetic Code*, Chap. 2; Yčas, *The Biological Code*, Chap. 2; Judson, pp. 245–47; and Sarkar, "Reductionism and Molecular Biology," Chap. 2. Olby, *The Path to the Double Helix*, pp. 217–21; both Judson and Sarkar also included the ideas of Stern in their individual historical reconstructions of coding.

101. See, for example, Stern's contributions to electrophoresis in Kay, "Laboratory Technology and Biological Knowledge."

102. Stern, quote on p. 943. On the significance of Stern's chemical structure, see Olby, *The Path to the Double Helix*, pp. 217–21.

103. Stern, p. 945.

104. See, for example, Hinshelwood, pp. 3, 206.

105. Caldwell and Hinshelwood, quote on p. 3157.

106. On biochemists' response to Schrödinger's book, see Fruton, *A Skeptical Biochemistry*, p. 198.

107. Dounce, "Nucleoproteins."

108. Dounce, "Duplicating Mechanism for Peptide Chain."

109. Ibid., pp. 253–54.

110. Dounce, "Nucleic Acid Template Hypothesis," p. 541.

111. Ibid.

CHAPTER 3. CYBERNETICS, INFORMATION, LIFE

1. Wiener, "A Scientist Rebels."

2. D. C. Cronemeyer to Wiener, 2 February 1947, Box 2.75, MC 22, Wiener Papers, MIT.

3. Ibid.

4. On the relation between regimes of signification, discourse, and cultural hegemony, see Laclau and Mouffe, p. 139. See also Comaroff and Comaroff, Intro.

5. For early official histories of science in World War II, see Baxter; and Stewart. See also Gray. For later histories, see Cochrane, pp. 382–432; Kevles, *The Physicists*, pp. 102–38; and Noble, *Forces of Production*, Chap. 1. For specific case studies, see Hevly; Dennis.

6. Noble, *Forces of Production*, Chap. 1; Cochrane, pp. 382–432; Stewart, Chap. 6–8.

7. Kevles, "The National Science Foundation and the Debate"; Noble, *Forces of Production*, pp. 11–12; Hevly, Chap. 2; Dennis, Part II. Similar patterns can be discerned in the life sciences. See Kay, *The Molecular Vision of Life*, Chap. 6.

8. Bush. For different interpretations of this document, see Kevles, "The National Science Foundation"; Reingold, "Vannevar Bush's New Deal for Research"; and Kay, *The Molecular Vision of Life*, pp. 223–25 and Interlude II.

9. For a cultural study of the bipolar world, see Edwards. See also Wittner, Chap. 1–4.

10. On the Marshall Plan, see Leffler, "The American Concept of National Security"; Wexler. See also Kay, *The Molecular Vision of Life*, Interlude II.

11. Wittner, Chap. 1–4; Boyer; McDougall.

12. "USAF Establishes Broad Research Policy," 5 March 1949, Box 28.18, RG 303U, Bronk Papers, RAC. General Hoyt S. Vandenberg, U.S. Air Force Chief of Staff, approved a regulation that established a broad policy for the support of an R&D program. See also Edwards, pp. 68–69; Noble, *Forces of Production*, pp. 15–16. On

the navy's support of science, see Allison; Rees; Sapolsky, *The Polaris System Development*; and idem, *Science and the Navy*; and Forman's provocative review of the latter in *IEEE Annals of the History of Computing.*

13. Noble, *Forces of Production*, p. 16; Hewlett and Anderson; Hewlett and Duncan; Weart, Part II. On the AEC support of genetics, see Beatty, "Genetics in the Atomic Age"; and idem, "Opportunities for Genetics in the Atomic Age."

14. Leslie, *The Cold War and American Science*, pp. 6–8; see also Sherry, *Planning for the Next War*; Melman, pp. 231–34.

15. Forman, "Behind Quantum Electronics"; Leslie, *The Cold War and American Science*; Hevly. See also Smith, Chap. 5–8; and Mendelsohn, Smith, and Weingart, which also include European and Russian perspectives.

16. Leslie, *The Cold War and American Science*, p. 8; Forman, "Behind Quantum Electronics," Intro.

17. Hacking, "Weapons Research." This article has provided a point of departure for Forman's "Behind Quantum Electronics," for Leslie's *The Cold War and American Science*, and for Keller's chapter "Critical Silences in Scientific Discourse," in her *Secrets of Life, Secrets of Death*. See also Simpson.

18. Aspray, "The Scientific Conceptualization of Information."

19. While the term *military-industrial complex* was coined by Dwight Eisenhower, the later term, *military-industrial-academic complex*, was coined by Senator J. William Fulbright.

20. Wiener, *Cybernetics: Or Control and Communication*, pp. 3–5. See also Aspray, "The Scientific Conceptualization of Information," pp. 124–25. This cybernetic turn did not mean that its substance was all new. Recently David A. Mindell, in "'Datum for Its Own Annihilation,'" has demonstrated in great detail the many theoretical and technological antecedents to Wiener's theory; see esp. Chap. 9.

21. Owens. See also Baxter, pp. 409–11.

22. Owens, p. 296 and p. 291, respectively; Bigelow to Weaver, 22 April 1944, AMP. On Wiener's war activities, see also Heims, *John von Neumann and Norbert Wiener*, Chap. 9.

23. Wiener, "Cybernetics"; Heims, *John von Neumann*, pp. 182–86. On the cultural role of cyborgs, see Haraway, "A Cyborg Manifesto," Chap. 8; and Edwards, esp. Chap. 3. For a critique of cyborg phenomenology, see Kay, "Who Wrote the Book of Life?" in *Science in Context*. See also Galison, "The Ontology of the Enemy"; and Pickering. Again, despite the impact of these representations their novelty has been persuasively challenged by Mindell's "'Datum for Its Own Annihilation.'"

24. Wiener to Haldane, 22 June 1942, Box 2.62, MC22, Wiener Papers, MIT.

25. Holton, "Ernst Mach"; idem, "The Joys and Sorrows of the Vienna Circle in Exile."

26. West, quote on p. 4. Wiener's secret report was later published as the book *Extrapolation, Interpolation and Smoothing of Stationary Time Series*. For a detailed analysis of the tracking problem, see Galison, "The Ontology of the Enemy."

27. Rosenblueth, Wiener, and Bigelow.

28. Ibid., p. 18.

29. Ibid., p. 21. On the origins of servomechanisms, see notes 44 and 45 below.

30. Rosenblueth, Wiener, and Bigelow, p. 22.

31. Wiener to von Neumann, 17 October 1944, and von Neumann to Wiener, 16 December 1944, Box 2.66, MC22, Wiener Papers, MIT.

32. Wiener to Rosenblueth, 24 January 1945, Box 2.67, MC22, Wiener Papers, MIT.

33. Von Neumann to Wiener, 21 April 1945, Box 2.68, MC22, Wiener Papers, MIT.

34. On the Macy Foundation, see Heims, *The Cybernetics Group*. "Mathematical Biology, 1945–46"; five-year grant (1947–52) for $27,500, RG1.1, 224 MIT, fld., MIT. Beginning in 1944, a grant of $4,000 and an appropriation of $18,000 for equipment was made to Rosenblueth at the Institute of Cardiology in Mexico City. In 1946 Wiener received a grant for $850 to collaborate with Rosenblueth in Mexico City. See also Hayles, *How We Became Posthuman*.

35. Scheffler; Nagel, Chap. 12; Becker.

36. "Mathematical Biology"; Morison to Rosenblueth, 17 January 1947, RG1.1, 224 MIT, fld., RAC.

37. Wimsatt.

38. "Mathematical Biology"; Weaver to Wiener, 28 January 1949, RG1.1, 224 MIT, fld., RAC.

39. For their responses, see Rosenblueth to Morison, 25 January 1947; and Wiener to Weaver, 4 April 1949, RG1.1, 224 MIT, fld., RAC.

40. Weaver Diary, 19 May 1947, Box 68, RG12.1, RAC.

41. Wiener, *Cybernetics*, p. 11. The first edition was published in France by Hermann & Cie and appeared a few months later in the United States. On the negotiations of the American publication, see Wiener Papers, Box 2, passim, MC22, MIT.

42. Nowinski to Wiener, 7 January 1952; Wiener to Nowinski, 20 February 1952, Box 4.145, MC22, Wiener Papers, MIT. In fact, Ampère had proposed *cybernétique* as a term for the science of government (Part II, pp. 140–41).

43. Wiener, *Cybernetics*, p. 39; and on the relation between technology and epistemology, p. 12. Foucault, *The Order of Things*.

44. Wiener, *Cybernetics*, pp. 8–11, quote on p. 11. He also credited Ronald A. Fisher, who had developed the notion of statistical information in his book, *The Design of Experiments*.

45. O. Mayr. Maxwell's often-cited paper "On Governors" (in *Proceedings of the Royal Society* (London) 16 [1868]: 270–83) is of special historical significance.

46. Bennet, Chap. 1.

47. For example, Hazen; Ivanoff; and A. V. Mikhailov, "The Method of Harmonic Analysis in Regulation Theory," in *Automatika i Telemekhania* (cited in West, "Forty Years of Control," p. 4). See also Mindell, "'Datum for Its Own Annihilation.'"

48. West, p. 1.

49. Wiener, *Cybernetics*, p. 11.

50. Ibid., pp. 41–42.

51. Ibid., pp. 93–94.

52. The difficulty was compounded by the poor editing of the first edition of the manuscript, which was due to Wiener's poor eyesight and related surgery. Wiener put

the blame on his junior collaborator, Walter Pitts, who was in charge of proofreading.

53. Haldane to Wiener, 12 November 1948, Box 2.86, MC22, Wiener Papers, MIT.

54. Haldane to Wiener, 13 July 1950, Box 3.121, MC22, Wiener Papers, MIT; and Haldane to Wiener, 6 May 1952, Box 4.150, MC22, Wiener Papers, MIT.

55. Kalmus.

56. Mandelbrot to Wiener, [n.d.] July 1948, Box 2.84, MC22, Wiener Papers, MIT. Wiener went there in 1951.

57. Science Service release, 22 October 1948, Box 2.85, MC22, Wiener Papers, MIT.

58. Sturtevant to Wiener, 8 November 1948; Feller to Wiener, 18 November 1948, Box 2.86; and McCulloch to Wiener, 9 December 1948, Box 2.87, MC22, Wiener Papers, MIT.

59. Freymann to Wiener, 29 December 1948, Box 2.87; Wallman to Wiener, 4 January 1949, Box 2.89; and Krozybski to Wiener, 19 January 1949, Box 2.90, MC22, Wiener Papers, MIT.

60. Jakobson to Wiener, 24 February 1949, Box 2.92, MC22, Wiener Papers, MIT.

61. "Mathematical Biology"; Weaver to Lovitt, 28 January 1949, Box 4, fld., RG1.1, 224, MIT.

62. R. S. Morison, interview, 21 February 1949, Box 4, fld., RG1.1, 224 MIT.

63. Wiener, *The Human Use of Human Beings*.

64. Ibid., pp. 103–9, quote on p. 103.

65. Ibid., pp. 109–11.

66. Noble, *Forces of Production*, p. 54. See also idem, "Command Performance."

67. It is not possible to record here all the responses that may be found in Wiener Papers, Boxes 3–5, MC22, MIT. The selected vignettes capture the diversity and spirit of the reactions.

68. Rathe to Wiener, 7 August 1950, Box 3.122, MC22, Wiener Papers, MIT.

69. Wheeler to Wiener, October 1953, Box 4.174; and Wiener to Deutsch, 7 April 1953, Box 4.168, MC22, Wiener Papers, MIT. Deutsch had been converted to cybernetics through the "Unity of Science" group. His classic text, *Nerves of Government*, summarized his cybernetic vision in political science. On Talcott Parsons's studies of social control in the interwar era (*The Structure of Social Action*), see Buxton, Chap. 5; on Parsons's reconceptualization of social control in cybernetic terms, see Yoxen, "The Social Impact of Molecular Biology," Chap. 4.

70. Boulding to Wiener, 12 January 1954, Box 4.186; and Kepes to Wiener, 1 August 1951, Box 3.140, MC22, Wiener Papers, MIT.

71. Ashby to Wiener, 2 April 1953, Box 4.168, MC22, Wiener Papers, MIT.

72. Still to Wiener, 29 December 1952, Box 4.161; and Wiener to Still, 5 January 1953, Box 4.162, MC22, Wiener Papers, MIT.

73. Vonnegut. For the context and significance of the novel, see Noble, *Forces of Production*, Appendix II and passim.

74. Wiener to English, 17 July 1952; and Vonnegut to Wiener, 26 July 1952, Box 4.150, MC22, Wiener Papers, MIT.

75. See, for example, Hague to Wiener, 3 January 1953; and Wiener to Hague, 12 January 1953, Box 4.162, MC22, Wiener Papers, MIT; and Greely to Wiener, 3 April 1953, Box 4.165, MC22, Wiener Papers, MIT.

76. Wiener to Rabinowitch, 22 June 1951, Box 3.138, MC22, Wiener Papers, MIT; and Rabinowitch to Wiener, 18 July 1951, Box 3.139, MC22, Wiener Papers, MIT.

77. J. Jackson. See also Noble, *Forces of Production*, Chap. 1, for figures on Bell Labs's military contracts. To date, there are no scholarly studies of Bell Labs.

78. "Claude E. Shannon," curriculum vitae, file copies, AT&T Archives; and Liversidge. See also Kahn, pp. 743–44; Fagen, p. 165; and Edwards, pp. 251–52. Shannon received his Ph.D. at the age of twenty-four. While a doctoral student he worked during the summer of 1937 at Bell Labs, where he demonstrated the application of Boolean algebra to relay circuit analysis. That work was immediately credited as a turning point in the history of the field. See C. Shannon, "A Symbolic Analysis of Relay and Switching Circuits." On Shannon's study of genetics, see Claude Shannon fld., "Mathematical Theory of Genetics," 1938, pp. 1–42, Box 12, Vannevar Bush Papers, LC; and Roch, who argues that Shannon's 1939 genetics work shaped his conceptualization of information theory.

79. Shannon, telephone interview by David Kahn, 27 November 1961; David Slepian, interview by David Kahn, 28 October 1962; both quoted in Kahn, p. 744.

80. Edwards, pp. 251–55. See also Fagen, p. 317; and Millman, pp. 104, 405–6.

81. C. Shannon, "Communication Theory"; idem, "The Mathematical Theory." In human languages the rules of syntax introduce redundancy into messages, thereby raising the likelihood of their correct reception; Shannon estimated the redundancy of ordinary English to be 50 percent. In communication theory the syntax is described as a set of conditional probabilities. Redundancy increases the channel capacity in the presence of noise and is defined as one minus the relative entropy. Kahn, p. 744; Cherry, pp. 180–87; and Aspray, *John von Neumann*, pp. 198–99.

82. Shannon to Wiener, 13 October 1948, Box 2.85, MC22, Wiener Papers, MIT.

83. Weaver to Wiener, 21 December 1948, Box 2.87, MC22, Wiener Papers, MIT.

84. Wiener to Bello (technology editor of *Fortune Magazine*), 13 October 1953, Box 4.179, MC22, Wiener Papers, MIT.

85. Nyquist.

86. Ibid., pp. 332–33. See also Aspray, "The Scientific Conceptualization of Information," p. 121.

87. R. V. Hartley; Aspray, "The Scientific Conceptualization of Information," pp. 121–22; and Cherry, pp. 170–76.

88. Aspray, "The Scientific Conceptualization of Information," pp. 122–24.

89. C. Shannon, "The Mathematical Theory," p. 379.

90. Cherry, Chap. 5–6. Cherry is especially instructive in offering a non-American perspective on the development of communication theory.

91. Ibid.; and Aspray, "The Scientific Conceptualization of Information," pp. 123–24.

92. C. Shannon and W. Weaver, pp. 33–34; Edwards, p. 256; and Aspray, "The Scientific Conceptualization of Information," p. 123.

93. See note 92 above. The visibility of Shannon's work was also due to his participation in the Macy conferences on cybernetics, see note 103 below.

94. C. Shannon and W. Weaver, pp. 3–9.

95. Ibid., pp. 8 and 28, respectively.

96. Cherry, Chap. 6. See also Edwards, pp. 253–54; and Heims, *The Cybernetics Group*, pp. 74–75. See also C. Shannon, "Prediction and Entropy of Printed English."

97. Wiener, "Relation of Cybernetics to Semantics," 3 January 1951, Box 15.830, MC22, Wiener Papers, MIT. For Derrida's critique of the relation between words and deeds, see Chap. 1, this volume.

98. Cherry, p. 225. See also note 81 above.

99. Warren Weaver Diary, 27 September 1951, Box 68, RG12.1, RAC. See, for example, Carnap and Bar-Hillel; Bar-Hillel, "Linguistic Problems"; and idem, "Logical Syntax." See also Sarkar, "The Boundless Ocean." Bar-Hillel had been in communication with MIT faculty and with Warren McCulloch (see below) since the late 1940s. He also spent significant time at MIT. For example, his paper "A Logician's Reaction to Recent Theorizing on Information Search Systems" was written in 1956 while he was at the Research Laboratory of Electronics, and the work was supported by the army (signal corps), the air force (Office of Scientific Research, Air Research and Development Command), and the navy (Office of Naval Research); and in part by Eastman Kodak Company. See General Correspondence (1957), Box 25.201, RG2, 200, RAC.

100. Quoted in Cherry, p. 219.

101. Ibid., p. 40.

102. Ibid., p. 61.

103. McCulloch-Shannon Correspondence, 1949–50, B M139 No. 1, McCulloch Papers, APS. Kahn, p. 744. See also Millman, pp. 58–61. For a detailed analysis of these relations, see Edwards, Parts II and III, passim; Heims, *The Cybernetics Group*, pp. 74–77; and Mirowski, "When Games Grow Deadly Serious"; and idem, "What Were von Neumann and Mogenstern Trying to Accomplish?"

104. Biographical note, von Neumann Papers, LC; Aspray, *John von Neumann*, Chap. 9; Heims, *John von Neumann*, Chap. 11.

105. Ceruzzi, esp. Chap. 6; I. B. Cohen. Aspray, *John von Neumann*, esp. Chap. 2; Edwards, pp. 95–97; and Cortada.

106. Edwards, pp. 102–7.

107. Ibid., quote on p. 169.

108. Heims, *John von Neumann and Norbert Wiener*, Chap. 11, quote on p. 247.

109. Aspray, *John von Neumann*, pp. 241–45.

110. Von Neumann to Burington (Department of the Navy), 19 January 1951, quote on p. 1, Box 2, fld. "B" Misc., von Neumann Papers, LC.

111. Rose, p. 36, quoted in Edwards, p. 106.

112. McCulloch and Pitts. For an extended description and assessment of McCulloch's project see Warren Weaver Diary, 10 January 1951, Box 68, RG12.1, RAC. An interesting retrospective by McCulloch about these developments can be found in "Biological Computers," [n.d.] ca. 1957, B M139, No. 2, McCulloch Papers, APS.

See also Aspray, *John von Neumann*, pp. 180–81; and idem, "The Scientific Conceptualization of Information," pp. 127–30.

113. Northrop to Wiener, 5 May 1947, B M139, No. 1, McCulloch Papers, APS.

114. B M139, No. 1, McCulloch Papers, APS, contains a significant number of correspondence items concerning McCulloch's ties to the navy, army, and air force. Sec. No. 2 of the collection contains records that document the relation of cybernetics to military agencies. The quote was McCulloch's first sentence in his "Why the Mind Is in the Head."

115. Lt. Col. Callahan, Jr., to Lt. Col. Sieber, Jr., 12 December 1962, B M139, No. 1, McCulloch Papers, APS.

116. Aspray, *John von Neumann*, pp. 181–89; Heims, *The Cybernetics Group*, pp. 93–94.

117. "The Ninth Washington Conference on Theoretical Physics," 18 November 1946, Theoretical Physics Conferences Series, Department of Terrestrial Magnetism Archive, CIW. There are no proceedings of the conference.

118. On Delbrück's phage work and his relation to physicists, see Kay, "Conceptual Models and Analytical Tools"; idem, "The Secret of Life"; and idem, *The Molecular Vision of Life*, Chap. 4 and 8. On Spiegelman, see Sapp, *Beyond the Gene*, esp. p. 103. See also Gaudillière, "J. Monod, S. Spiegelman et l'adaptation enzymatique."

119. Spiegelman to von Neumann, 3 December 1946, Box 7.1, von Neumann Papers, LC.

120. Von Neumann to Spiegelman, 10 December 1946, von Neumann Papers, LC.

121. Ibid., passim (six letters).

122. Von Neumann to Wiener, 29 November 1946, Box 2.72, MC 22, Wiener Papers, MIT.

123. Ibid.

124. On the protein paradigm in life science, see Kay, *The Molecular Vision of Life*, Interlude I.

125. Langmuir to von Neumann, 11 November 1946, Box 5.5; and von Neumann to Langmuir, 12 November 1946; and von Neumann to Harker (General Electric), 16 December 1946, Box 4.8, von Neumann Papers, LC. See also Aspray, *John von Neumann*, p. 186. On Langmuir's support of Wrinch's theory, see Fruton, "Early Theories of Protein Structure." On the professional and personal saga of Wrinch, see Abir-Am, "Synergy or Clash." And on the controversy around her cyclol structure, see Serafini, Chap. 6. For further discussion on Wrinch, see Kay, *The Molecular Vision of Life*, Interlude I and Chap. 5.

126. Correspondence between von Neumann and Edsall, 1951–54, Box 3.9; and Szent-Gyorgyi to von Neumann, 22 June 1949, Box 6.14, von Neumann Papers, LC.

127. Jeffress to von Neumann, 6 November 1947, Box 19.19, von Neumann Papers, LC. On the significance of the symposium, see H. Gardner, pp. 10–16. For a different interpretation of the move from behaviorism to cognitivism, see Edwards, Chap. 8–10.

128. Von Neumann to Jeffress, 11 November 1947, Box 19.19, von Neumann

Papers, LC. The Hixon Committee members are listed in the *Cerebral Mechanisms in Behavior*, p. ix.

129. Von Neumann to Jeffress, "Abstract of Paper by Von Neumann," [n.d.] ca. 1948, Box 19.19, von Neumann Papers, LC.

130. Ibid.

131. Ibid. See the rest of the correspondence for reactions, expectations, and von Neumann's discomfort in committing his thoughts to print.

132. Von Neumann.

133. Ibid., pp. 25–31.

134. Ibid., quote on pp. 30–31.

135. Kemeny. On these conferences, see Aspray, *John von Neumann*, pp. 198–206. The fragments, notes, and unfinished manuscripts for these lectures were compiled and completed by Burks. See also C. Shannon and J. McCarthy.

136. Kemeny.

137. Millman, Chap. 9.3; and E. F. Moore.

138. Penrose, "Self-Reproducing Machines"; and idem, "Mechanics of Self-Reproduction." The interest in biological automata in relation to developmental biology was later sustained by Michael Arbib. See Arbib, "Automata Theory and Development: Part I"; and idem, "Self-Reproducing Automata."

139. Lederberg to von Neumann, 10 March 1955. The von Neumann–Lederberg correspondence is in Box 58, von Neumann Papers, LC. However, that correspondence is incomplete; I am very grateful to Joshua Lederberg for providing me with the missing letters.

140. Lederberg to von Neumann, 3 April 1955, JLF. See Lederberg, "Infection and Heredity."

141. Lederberg to von Neumann, 10 August 1955, JLF.

142. Von Neumann to Lederberg, 15 August 1955, JLF.

143. Lederberg to von Neumann, 3 September 1955, JLF.

144. See, for example, Spiegelman, "On the Nature of the Enzyme-Forming System"; and Lederberg, "A View of Genetics." I am grateful to Sahotra Sarkar for bringing this important volume to my attention.

145. McLuhan. For Jean Baudrillard's later critique, relating the culture of simulation to the genetic code, see Baudrillard, p. 54.

146. Coggeshall to Bates, 21 June 1939, Box 1.6, RG11, Series 704I, RAC. Letters between Bates and Sawyer 15 July and 1 August 1939, RG Personnel File. Staff Appointments Papers, R.S. 2/5/15, UIA. "Records of Training and Professional Experience" forms for Henry Quastler, 1947–55. Curtis, pp. vii–viii. The unsuccessful searches to find the Quastler Papers through the relevant institutions and family suggest that his papers are lost. I have been able to reconstruct his trajectory through several interviews: personal interview with Henry Linschitz, 16 July 1993; and with Heinz von Foerster, 26 June 1994; telephone communications with Quastler's niece, Joan Zimmerman, 2 June 1994, and with his sister, Mrs. Johanna Zimmerman, 6 and 14 June 1994; and J. Hastings, E. P. Krankeit, and Maurice Goldhaber, 15 and 17 November 1994.

147. Curtis. R.S. 2/5/15, UIA. Hewlett and Duncan, pp. 228–30. Von Foerster, interview, 26 June 1994. On hiring in molecular biology at the University of Illinois, see Kay, *The Molecular Vision of Life*, p. 249. Warren Weaver Diary, 11 February 1949, Box 68, RG12.1, RAC.

148. Dancoff-Quastler Correspondence, AIP; Memorials for Sydney Dancoff, August 1951, fld. 3; Dancoff's notes and incomplete manuscripts. Much of that correspondence revolves around their joint paper and the conceptual issues it raised. For a description of that collection, see also Robert C. Olby, AIP.

149. Dancoff and Quastler, pp. 263–73. Dancoff-Quastler Correspondence, fld. 1–2, passim, AIP. Quote in Quastler to Dancoff, fld. 1, 4 July 1950, AIP.

150. Dancoff-Quastler Correspondence, fld. 1; Dancoff to Quastler, 31 July 1950, AIP. In the final version of the paper Quastler offered the corrective that there were at least two genetic alleles.

151. Dancoff to Quastler, 17 August 1950, p. 6, AIP.

152. Dancoff and Quastler. Fruton, "Early Theories."

153. Quastler, *Information Theory in Biology*, pp. 1–4.

154. Quastler, "The Measure of Specificity." In actuality enzyme-substrate reactions display analog behavior, which through proper axiomatization may be theoretically "digitalized."

155. Branson, quote on pp. 84–85.

156. Augenstine, Branson, and Carver, "A Search for Intersymbol Influences in Protein Structure," pp. 105–18.

157. Lederberg to Quastler, 3 May 1951; and Lederberg to Kay, 7 July 1993. I am grateful to Joshua Lederberg for making this material available to me.

158. Haurowitz; Irwin; Quastler, "The Specificity of Elementary Biological Functions," quote on p. 188.

159. Bragdon, Nalbandov, and Osborne; and Tweedell, quote on p. 215.

160. Linschitz; and Linschitz, interview, 16 July 1993.

161. Quastler, "Feedback Mechanisms"; and von Foerster, interview, 26 June 1994.

162. Birthday file (1954), Box 3, Gertrud Quastler Papers, LC; and "In Memorium," (anonymous), 1963, Box 2, Gertrud Quastler Papers, LC.

163. Curtis. On the national laboratories, see Hewlett and Duncan, pp. 223–27; and Hewlett and Holl, pp. 252–70.

164. "Remarks by Commissioner Henry D. Smyth," 25 October 1949, pp. 4–5, Box 28.2, Series 303u, Detlev Bronk Papers, RAC.

165. Hewlett and Duncan, pp. 4, 224–25, 242–51; and Hewlett and Holl, pp. 253–54; Ramsey; and Rowe.

166. All the symposia proceedings since 1948 were published by Brookhaven National Laboratory.

167. Yockey, *Symposium on Information Theory in Biology*; McCulloch to Yockey, 25 April 1956, B M139, McCulloch Papers, APS.

168. Quastler, "A Primer on Information Theory." According to von Foerster that primer was one of the best-written treatments of information theory; von Foer-

ster, interview, 26 June 1994. On his contributions to information theory in psychology, see Quastler, *Information Theory in Psychology.*

169. Yockey, "Some Introductory Ideas."

170. Gamow and Yčas, "The Cryptographic Approach."

171. Yčas, "The Protein Text."

172. Quastler, "The Status of Information Theory in Biology," quote on p. 399.

173. Ibid., p. 402.

174. Yčas, "Biological Coding and Information Theory," quote on p. 256. I am grateful to Martynas Yčas for making these materials available to me.

175. Cherry, p. 169; see Chap. 1, this volume.

CHAPTER 4. GENETIC CODES IN THE 1950S

1. These narratives include Stent and Calendar; Judson; and Crick, *What Mad Pursuit.* For more detailed (purely) technical analysis of genetic codes in the 1950s, see Sarkar, "Reductionism in Molecular Biology," Chap. 3. Quoted by R. G. Martin, "A Revisionist View," p. 282.

2. Woese, *The Genetic Code*, p. 17. This point is also demonstrated in the way that the work on in vitro protein synthesis and tRNA was assimilated into the coding problem within the information discourse. See Rheinberger, *Toward a History of Epistemic Things*, esp. Chaps. 10–12.

3. Fleming; Olby, *The Path to the Double Helix*, Sec. IV; Kohler, "The Management of Science"; Yoxen, "Giving Life a New Meaning"; Haraway, "The Biological Enterprise"; Abir-Am, "The Discourse of Physical Power"; idem, "From Multidisciplinary Collaboration to Transnational Objectivity"; Kay, "Conceptual Models and Analytical Tools"; idem, "The Secret of Life"; Keller, "Critical Silences in Scientific Discourse," pp. 56–72. Keller's observation is in her "Physics and the Emergence of Molecular Biology."

4. Richard Doyle, in "On Beyond Living," has provided an excellent analysis of Gamow's contributions to scriptural representations of the genetic code (Chap. 3). However, he has based his study of the code essentially on Gamow's first note in *Nature* (1954), rather than on his collaborative work on various codes over a period of several years. Furthermore, his analysis does not address the relation of the information discourse to linguistics and to the new semiotics of codes, nor to the influences of cold-war military technoculture, notably communication technologies, on representations of heredity. See also Kay, "Who Wrote the Book of Life? Information and the Transformation of Molecular Biology," in Hagner, Rheinberger, and Wahrig-Schmidt; and the English version "Who Wrote the Book of Life?" in *Science in Context.*

5. Gamow to Delbrück, 13–22 April 1941, Box 8.21, Delbrück Papers, CIT. The proposed book was either never written or left uncompleted. Gamow, *Mr. Tompkins Learns the Facts of Life.*

6. Interview by Charles Weiner, 4 April 1968, p. 76, Oral History, George Gamow,

AIP. Gamow to Watson and Crick, 8 July 1953, Archive Box 2, SBF. See also Judson, Chap. 5.

7. Gamow, p. 80, AIP.

8. For sources on U.S. foreign policy in the 1950s, see Leffler, *A Preponderance of Power*, Chap. 10–11; McCormick, Chap. 5–6; Wittner, Chap. 4–7. On McCarthyism and national security, see Griffith and Theoharis; Caute; D. Campbell, *Writing Security*, Chap. 6; Carmichael, Part I. For a general overview, see Halberstam.

9. Beadle, "Science and Security," [n.d. 1954 or 1955], p. 1, Box 31.7, Beadle Papers, CIT. Scientists' response to the loyalty oath in 1954 is documented in Resolution of the American Physiological Society, 29 April 1954, Box 6.12, RG 303-U, Bronk Papers, RAC. For earlier reactions, see J. Wang.

10. On the development of science and technology in the 1950s, see Forman, "Behind Quantum Electronics," pp. 224–25; Leslie, "Science and Politics in Cold War America," pp. 200–233; Noble, *Forces of Production*, esp. Chap. 3; idem, "Command Performance," Chap. 8; and Edwards.

11. Kahn; Pratt.

12. Kahn, Chap. 22.

13. Ibid., Chap. 12, 15, and 17. On cryptography in World War II, see Welchman; and Hinsley and Stripp; on postwar developments, see Bamford; Richelson. See also O'Toole. On Friedman's contributions, see Rosenheim.

14. C. Shannon, "Communication Theory"; idem, "Prediction and Entropy of Printed English." See also Cherry, Chap. 3; and Kahn, pp. 743–52.

15. On use of electronic computers in cryptanalysis, see Kahn, p. 725.

16. Crick, *What Mad Pursuit*, p. 94.

17. Resume of Gamow's career, January 1968, p. 2, Box 8.21, Delbrück Papers, CIT.

18. As documented in the von Neumann Papers, as well as the Gamow Papers, passim, LC.

19. January 1968, p. 2, Box 8.21, Delbrück Papers, CIT.

20. Box 7, Yčas fld., passim, Gamow Papers, LC; Gamow's television appearances are mentioned in several places in the correspondence. His selected popular books include: *Mr. Tompkins in Wonderland*; *The Birth and Death of the Sun*; *Mr. Tompkins Explores the Atom*; *One, Two, Three . . . Infinity*; *The Creation of the Universe*; *Mr. Tompkins Learns the Facts of Life*; *The Moon*; *Puzzle-Math*, with M. Stern; *Mr. Tompkins in Paperback*; *Thirty Years that Shook Physics*; *Mr. Tompkins Inside Himself*, with M. Yčas.

21. Gamow, "Possible Relation" (communicated 22 October 1943); Doyle, "On Beyond Living," p. 78.

22. Gamow, "Possible Relation," p. 318. On the choice of twenty amino acids (since there are actually twenty-two), see Judson, pp. 253–54; and Crick, *What Mad Pursuit*, pp. 91–92.

23. Gamow to Pauling, 22 October 1953; Pauling to Gamow, 9 December 1953, Box 263.39, Ava Helen and Linus Pauling Papers, OSU.

24. Crick, *What Mad Pursuit*, p. 94. Sanger's results were made available to

Crick upon their completion in 1952. Sanger and Thompson, pp. 353–74. On the relation of Sanger (and biochemists) to Crick and Brenner, see de Chadarevian, "Sequences, Conformation, Biochemists."

25. Interview by Charles Weiner, 1968, pp. 76–77, Oral History, George Gamow, AIP.

26. Gamow, "Possible Mathematical Relation"; (Emil Fischer's quote from *Sitzungsber. der Kgl. Preuss. Akad. der Wissenschaft*, p. 990, my translation).

27. Gamow, "Possible Mathematical Relation," pp. 10–13.

28. Chargaff to Gamow, 2 March 1954, Gamow fld., B:C37, Chargaff Papers, APS.

29. Chargaff to Gamow, 3 March 1954, Gamow fld., B:C37, Chargaff Papers, APS. On the gradual acceptance of RNA's role in protein synthesis, see Judson, Part II; Burian, "Technique, Task Definition, and the Transition"; idem, "Underappreciated Pathways Toward Molecular Genetics"; Rheinberger, "Experiment and Orientation: Early Systems of in Vitro Protein Synthesis"; idem, "Experiment, Difference, and Writing: I. Tracing Protein Synthesis," and "II. The Laboratory Production of Transfer RNA"; and Denis Thieffry, "Contributions of the 'Rouge-Clôitre Group'" (special issue on the Rouge-Clôitre group).

30. Yčas, interview, 6 October 1993. I am grateful to Yčas for supplying a detailed CV, reprints, and other relevant materials.

31. Delbrück's quote in his "Experiments with Bacterial Viruses (Bacteriophages)."

32. Gamow to W. G. Parks (director of Gordon Research Conferences, AAAS), 5 April 1956, Yčas fld. (1956), Gamow Papers, LC.

33. Yčas to Gamow, 16 March 1954; and Gamow to Yčas, 31 March 1954, Box 7, Yčas fld. (1954), Gamow Papers, LC.

34. Gamow to Yčas, 28 April 1954, Box 7, Yčas fld. (1954), Gamow Papers, LC.

35. On Gamow's visiting professorship at Berkeley, see "Scientists in the News," *Science* 119 (1954): 540.

36. Gamow to Yčas, 28 April 1954, Box 7, Yčas fld. (1954), Gamow Papers, LC; and Stent to Delbrück, 27 March 1954, Box 20.20, Delbrück Papers, CIT.

37. Interview by Charles Weiner, 1968, pp. 81–82, Oral History, George Gamow, AIP. Anecdote on scientists and drinking in Press Conference, U.S. Atomic Energy Commission, 9 May 1950, p. 19, Box 28.3, RG 303-U, Bronk Papers, RAC. On Gamow's drinking, see Crick, *What Mad Pursuit*, p. 95.

38. Gamow to Chargaff, 6 May 1994, p. 2, Box 7, Yčas fld. (1954), Gamow Papers, LC. A somewhat more formal version went to Vincent du Vigneaud, 6 May 1954, p. 1.

39. RNA Tie Club, Box 7.30, Gamow Papers, LC; Yčas flds. passim, Gamow Papers, LC. See also Crick, *What Mad Pursuit*, p. 95. There were a few variations in the names during the period of the club's activities.

40. Yčas to Gamow, 15 August 1954, Box 7, Yčas fld. (1954); and Yčas to Gamow, 13 October 1954, Gamow Papers, LC. Also Gamow to Bailey (Chief, Research & Development Division, U.S. Army, Philadelphia), 23 September 1954, B:C37,

Gamow fld., Chargaff Papers, APS; and Bailey to Gamow, 8 October 1954, SBF. Perhaps some RNA Tie Club members objected to military sponsorship, or perhaps some could not receive clearance.

41. Oral History, p. 82, Gamow, AIP. This is also Yčas's assessment; Yčas, interview, 6 October 1993. Abir-Am, "From Multidisciplinary Collaboration to Transnational Objectivity," focuses on informal international networks in the cold war but does not examine the RNA Tie Club.

42. Gamow to Chargaff, 6 May 1954, Box 7, Yčas fld. (1954), Gamow Papers, LC. The same day a similar letter went to du Vigneaud.

43. Gamow, Rich, and Yčas, "The Problem of Information Transfer," pp. 41–51. For technical analyses of these codes, see Sarkar, "Reductionism in Molecular Biology," Chap. 3.

44. Gamow, Rich, and Yčas, "The Problem of Information Transfer," pp. 51–53. On Dounce's model, see Chap. 1, this volume.

45. Ibid., pp. 53–54; and Gamow to Yčas, 27 May and 11 June 1954, quote on p. 2, Box 7, Yčas fld. (1954), Gamow Papers, LC. On Teller's role in the Oppenheimer affair, see Halberstam, Chap. 24.

46. Gamow and Metropolis. See also Galison, *Image and Logic*, Chap. 8. "Computer Used to Probe Protein Structure," *Science Service*, 11 October 1954, Archive Box 2, SBF.

47. Gamow and Metropolis. Gamow, Rich, and Yčas, "The Problem of Information Transfer," pp. 59–61. On how that language trope leads to its own deconstruction, see Kay, "Who Wrote the Book of Life?" in *Science in Context*.

48. Letters between Gamow and Yčas, 23 July to 18 August 1954, Box 7, Yčas fld. (1954), Gamow Papers, LC. See also Judson, p. 277; and Crick, *What Mad Pursuit*, p. 95. Stent to Brenner, 6 December 1954; Crick to Brenner, 12 January 1955; and Gamow to Brenner, 22 January 1955, Archive Box 3; "Short Biography," SBF.

49. Gamow to Yčas, 27 November 1954, Box 7, Yčas fld. (1954), Gamow Papers, LC; Ledley, quotes on p. 498.

50. Ledley, p. 511.

51. Gamow, Rich, and Yčas, "Information Transfer in Biology," pp. 55–59. Also Yčas to Crick, 26 April 1955, Box 7, Yčas fld. (1954), Gamow Papers, LC.

52. Gamow to Watson, 17 Dec. 1954, Archive Box 2, SBF. Gamow to Yčas, [n.d. but probably 2] Dec. 1954, p. 2, Box 7, Yčas fld. (1954), Gamow Papers, LC.

53. Yčas to Rich, 23 December 1954; Yčas fld. (1955); and Yčas to Crick, 15 February 1955, Gamow Papers, LC.

54. Gamow, Rich, and Yčas, "The Problem of Information Transfer," p. 66.

55. Ibid., p. 24.

56. Kahn, pp. xiv–xv; Pratt, pp. 12–13.

57. Pratt, pp. 12–13.

58. Crick, *What Mad Pursuit*, p. 90. See also Judson, p. 278.

59. Gamow, Rich, and Yčas, "The Problem of Information Transfer," p. 40.

60. Ibid., pp. 64–65.

61. Ibid., pp. 65–66.

62. Ibid., p. 66.

63. Gamow, "Information Transfer in the Living Cell," p. 70.

64. Ibid., p. 78.

65. Gamow, Rich, Yčas, "The Problem of Information Transfer," pp. 23–67. Gamow to Yčas, 29 January 1955, Box 7, Yčas fld. (1955), Gamow Papers, LC. For discussion see Chap. 3, this volume.

66. Francis Crick, "On Degenerate Templates," an unpublished note to the RNA Tie Club, [n.d.] mid-January 1955. I thank Francis Crick for sending me a copy of this paper. For discussion of this paper, see Judson, pp. 287–93.

67. Crick, "On Degenerate Templates," p. 4.

68. Ibid., pp. 4–5. As Judson points out in *The Eighth Day of Creation*, the epigram is from a minor Persian writer (p. 287). On nature as text, see Doyle, "On Beyond Living," Chap. 3.

69. Crick, "On Degenerate Templates," p. 17. Yčas to Crick, 15 February 1955, Box 7, Yčas fld. (1955), Gamow Papers, LC. Crick to Brenner, 6 July 1955, Archive Box 3, SBF. On Britain in the postwar decade, see Howard Whidden, "Europe at the Crossroads: The Next 10 to 15 years," December 1956, Box 3.35, Special Studies Project Records, Rockefeller Brothers Fund, RCA; and McCormick, Chaps. 3–5. Yčas to Crick, 26 April 1955, Box 7, Yčas fil. (1955), Gamow Papers, LC.

70. Gamow to Rich, 15 November 1955, p. 1, MPC, George Gamow, AIP.

71. Gamow to von Neumann, 8 July 1955, Box 4, Gamow fld., von Neumann Papers, LC. George Gamow, "Information Transfer"; Gamow and Yčas, "Statistical Correlation of Protein."

72. Gamow to von Neumann, 8 July 1955; von Neumann to Gamow, 25 July 1955, Box 4, Gamow fld., von Neumann Papers, LC.

73. Letters between Gamow and Yčas, July 1955, Gamow Papers, LC. Gamow and Yčas, "Statistical Correlation," p. 1013.

74. Gamow and Yčas, "Statistical Correlation," p. 1013; Ulam's proof on pp. 1017–19. Elson and Chargaff.

75. Gamow to Yčas, 1 August 1955; Yčas to Gamow, 5 August 1955, Yčas fld. (1955), Gamow Papers, LC.

76. Yčas to Crick, 26 April 1955, Gamow Papers, LC. Gamow and Yčas, "Statistical Correlation," pp. 1013–14. See also Sarkar, "Reductionism in Molecular Biology," Chap. 3.

77. Delbrück to Rich, 9 November 1955, Box 8.21, Delbrück Papers, CIT. Crick to Yčas, 17 November 1955, Yčas fld. (1955), Gamow Papers, LC.

78. Schwartz, "Speculations on Gene Action"; idem, "Coding Problem in Proteins."

79. Dounce, "Duplicating Mechanism"; idem, "Role of Nucleic Acid and Enzymes." Yčas to Dounce, 19 July 1955, Yčas fld. (1955), Gamow Papers, LC.

80. Simon to Yčas, 5 December 1956, p. 2, Yčas fld. (1956), Gamow Papers, LC. On Simon's role in the information discourse, see Poster, p. 148.

81. Gamow to Yčas, 3 May, 29 May 1956, Yčas fld. (1956), Gamow Papers, LC.

82. "Short Biography of S. Brenner," SBF. Brenner, "On the Impossibility of All

Overlapping Triplet Codes," note to the RNA Tie Club, September 1956, Box 30.5, Delbrück Papers, CIT. For discussions of Brenner's proof, see Woese, *The Genetic Code*, p. 19; Yčas, *The Biological Code*, pp. 43–46; Judson, pp. 329–30; and Sarkar, "Reductionism in Molecular Biology," Chap. 3.

83. Brenner, "On the Impossibility of All Overlapping Triplet Codes," p. 3 (see note 82 above).

84. Brenner, "On the Impossibility of All Overlapping Triplet Codes."

85. Notes on Correspondence, 17 October 1972, Yčas fld. (1972), Gamow Papers, LC. Gamow and Yčas, "The Cryptographic Approach"; and Yčas, "The Protein Text," pp. 70–101.

86. Yčas, "The Protein Text," p. 71.

87. Ibid., pp. 88–90. See also Kay, "Who Wrote the Book of Life?" in *Science in Context*.

88. Yčas, *The Biological Code*, p. 31.

89. Ibid.

90. Crick, Griffith, and Orgel, quote on pp. 417–18.

91. Ibid., p. 419.

92. Crick, "On Degenerate Templates," see note 66 above.

93. Crick, Griffith, and Orgel, p. 420. Crick, *What Mad Pursuit*, pp. 99–100.

94. Yčas, "The Protein Text," pp. 91–92 (see note 85); Crick, *What Mad Pursuit*, p. 100.

95. On the attendance at Cold Spring Harbor, see Demerec; see also Kay, "Conceptual Models and Analytical Tools." Szilard's career is documented in detail in Szilard Papers, Box 1.2 and Box 2.9, MSS.32, UCSD. Lanquette. For personal impressions, see A. Novick, "Phenotypic Mixing."

96. Jacob, *The Statue Within*, p. 293; Curriculum Vitae (including list of publications), Box 1.2, MSS.32, Szilard Papers, UCSD; and Lanquette, Chap. 25.

97. Application to the National Science Foundation, 26 July 1956, pp. 2–3, Box 67.22, Biology Division Papers, CIT.

98. Davis to Conzolazio (NSF), 6 July 1956; and "Confidential Memorandum: On a Roving Professorship for Leo Szilard," by George W. Beadle, Bernard D. David, and Theodore Puck, 25 May 1956, quote on pp. 1–2, Box 67.22, Biology Division Papers, CIT.

99. Szilard to Crick, 14 June 1957, Box 6.38, MSS.32, Szilard Papers, UCSD.

100. Leo Szilard, "How Amino Acids Read the Nucleotide Code," 7 June 1957, pp. 1–14, plus appendix, quote on pp. 2–4, Box 26.3, MSS.32, Szilard Papers, UCSD.

101. Ibid., pp. 8–13.

102. Crick to Szilard, 20 June 1957; Szilard to Crick, 24 June 1957; Szilard to Delbrück, 24 June 1957, Box 6.38, MSS.32, Szilard Papers, UCSD; and Szilard to Novick, 25 June 1957, Box 14.21, MSS.32, Szilard Papers, UCSD.

103. Leslie, *The Cold War and American Science*, passim; MacKenzie; and McDougall.

104. Beadle to Weaver, 9 December 1947, Box 4.24, RG1.1, 205D, RAC. See also Kay, *The Molecular Vision of Life*, Chap. 8.

105. Golomb, Gordon, and Welch.

106. Ibid., p. 209; and McMillan. On this two-way traffic, see Golomb, where he refers to Jayne and to "Recent Results in Comma-Free Codes," Research Summary, Jet Propulsion Laboratory, CIT, 15 February 1961, pp. 36–37. On the meaning and significance of boundary objects, see Fujimura; and Star and Griesemer.

107. Golomb, Welch, and Delbrück.

108. Ibid., n.p.

109. Ibid., p. 11. See also Yčas, *The Biological Code*, p. 33. On Delbrück's earlier articulation of this feature, see Crick, "On Degenerate Templates," pp. 4–5 (see note 66).

110. Golomb, Welch, and Delbrück, p. 1.

111. Delbrück and Stent, quote on p. 730.

112. Delbrück to Sinsheimer, 21 November 1955, Box 20.3, Delbrück Papers, CIT. The DNA data Delbrück referred to was Sinsheimer, "The Action of Pancreatic Desoxyribosnuclease," Parts I and II.

113. Osmundsen, "New Way to Read Life's Code Found." The report refers to Golomb's paper, "Efficient Coding for the Desoxyribonucleic Channel" (see note 106).

114. De Chadarevian, "Sequences, Conformation, Information: Biochemists and Molecular Biologists in the 1950s."

115. Ibid.; and Abir-Am, "The Politics of Macromolecules."

116. Crick, "On Protein Synthesis," pp. 138–63; quote on p. 144. Wiener, *Cybernetics: or Control and Communication in the Animal and the Machine*, pp. 41–42. Judson, pp. 333–40. Sarkar, "Biological Information."

117. Quastler, "The Measure of Specificity," p. 41; and Augenstine, Branson, and Carver, "A Search for Intersymbol Influences in Protein Structure."

118. Crick, "On Protein Synthesis," pp. 152–53. See also Olby, "The Protein Version of the Central Dogma." For an extensive critique of the Central Dogma, see Sarkar, "Biological Information"; and Thieffry and Sarkar.

119. Crick, "On Protein Synthesis," p. 160.

120. Belozersky and Spirin; Lee, Wahl, and Barbu.

121. Sinsheimer, "Is the Nucleic Acid Message in a Two-Symbol Code?" quote on p. 219.

122. Sinsheimer to Kay, 19 July 1994, personal communication. I am grateful to Sinsheimer for providing me with this information. See also Sinsheimer, *The Strands of Life*; and review by Kay, *Bulletin of the History of Medicine* 69 (1995): 318–19.

123. Sinsheimer, "Is the Nucleic Acid Message in a Two-Symbol Code?" p. 219; Elson and Chargaff; Zubay.

124. Brenner and Crick, unpublished results, 1958, on the quadruplet code, reported in Brenner, "The Mechanism of Gene Action," Box 30.5, Delbrück Papers, CIT; Crick and Brenner, "Some Footnotes on Protein Synthesis: A Note for the RNA Tie Club," December 1959, Yčas fld. (1957–59), Gamow Papers, LC; Gamow to Yčas, February to April 1957, Gamow Papers, LC. Gamow to Brenner, 4 June 1957, Archive 1 (personal collection), SBF.

125. Crick, "The Present Position of the Coding Problem."

126. Brenner, "The Mechanism of Gene Action"; Benzer; Brenner and Barnett.

127. See Brenner, "The Mechanism of Gene Action," pp. 317–18, for Kalmus's "Discussion."

128. Kahn, pp. 917–37, quote on p. 937.

129. Brenner, "Mechanism of Gene Action," pp. 318–19.

130. Ibid., p. 38. Ingram, "A Specific Chemical Difference"; idem, "How Do Genes Act?" Ingram was following the hemoglobin studies first reported by Pauling et al. Neel. See also Judson, pp. 300–308; Kay, *The Molecular Vision of Life*, Chap. 8.

131. Olby, *The Path to the Double Helix*, pp. 235–37; Stent and Calendar, p. 578; Kay, "W. M. Stanley's Crystallization of the Tobacco Mosaic Virus"; Van Helvoort, "What Is a Virus?"; and idem, "History of Virus Research in the 20th Century."

132. Kay, "W. M. Stanley's Crystallization of the Tobacco Mosaic Virus," p. 470. Correspondence with geneticists during that period refers largely to the logistics of fitting genetics meetings into Stanley's crowded schedule; 78/18c, Ct. 1 (1937–40), Stanley Papers, UCB.

133. Knight, quote on p. 307.

134. For biographical accounts, see Edsall; and Kay, "Wendell Meredith Stanley." On Stanley's prestige and connections, see Kay, "W. M. Stanley's Crystallization of the Tobacco Mosaic Virus," p. 470; idem, "The Tiselius Electrophoresis Apparatus"; idem, "The Politics of Fame"; and idem, "The Intellectual Politics of Laboratory Technology."

135. Creager. Gaudillière, "Oncogenes as Metaphors for Human Cancer"; idem, "Circulating Mice and Viruses"; idem, "Norms and Practices of Molecular Medicine."

136. Creager.

137. Application to the National Science Foundation, 24 February 1959, Box 4.59; and Stanley to the Nobel Committee, 22 December 1960, Box 5.12, 78/18c, Stanley Papers, UCB, both include detailed biographical accounts of Fraenkel-Conrat.

138. Fraenkel-Conrat to Stanley, 4 July 1947, Box 8.58, 78/18c, Stanley Papers, UCB.

139. Fraenkel-Conrat to Stanley, 8 November 1947, Box 8.58, 78/18c, Stanley Papers, UCB.

140. Application to the National Science Foundation, 24 February 1959, Box 4.59; and Stanley to the Nobel Committee, 22 December 1960, Box 5.12, 78/18c, Stanley Papers, UCB.

141. Harris and Knight; Fraenkel-Conrat, "Protein Chemists Encounter Viruses."

142. Fraenkel-Conrat and Williams; Fraenkel-Conrat, "The Role of the Nucleic Acid," quote on p. 883; idem, "Rebuilding a Virus"; Fraenkel-Conrat and Singer; and Kay, "Matter of Information." The competition between Stanley's group and Gerhard Schramm's group in Tübingen goes back to the 1930s. On the prewar virus research of Schramm, see Olby, *The Path to the Double Helix*, Chap. 10. On their competition in the 1950s, see Stanley to the Nobel Committee, 12 December 1960, Box 5.12, 78/18c, Stanley Papers, UCB.

143. Gamow to Yčas, [n.d.] mid-June 1955, Box 7, Yčas fld. (1955), Gamow Papers, LC; Kaempffert. Similar experiments were being conducted by Barry Commoner's group at Washington University in St. Louis.

144. Stanley to the Nobel Committee, 12 December 1960, p. 2, Box 5.12, 78/18c, Stanley Papers, UCB.

145. Niu and Fraenkel-Conrat; Narita; and Fraenkel-Conrat, "Protein Chemists Encounter Viruses," p. 311.

146. Rich to Crick, 17 December 1957, p. 1, Olby Collection, APS.

147. Stent to Delbrück, 30 October 1957, Box 20.20, Delbrück Papers, CIT.

148. Stent to Delbrück, 7 November 1957, Box 20.20, Delbrück Papers, CIT.

149. Gamow to Stanley, 13 November 1946, Box 8.77, 78/18c, Stanley Papers, UCB.

150. Gamow to Stanley, 24 February 1947, Box 8.77, 78/18c, Stanley Papers, UCB.

151. Heinz Fraenkel-Conrat, interview, 29 June 1994. His recollections are supported by his publications.

152. Schuster and Schramm; Gierer and Mundry.

153. Schmeck.

154. Tsugita and Fraenkel-Conrat.

155. Ibid., quote on p. 641. The complication was pointed out by the German biochemist Wittman in "Comparison of the Tryptic Peptides." On the new amino acid sequence technique, see Moore and Stein.

156. Osmundsen, "Scientists Find Clue to Heredity's Code."

157. Tsugita et al., quote on p. 1468.

158. Gamow to Stanley, 30 November 1960; and Stanley to Gamow, 5 December 1960, Box 8.77, 78/18c, Stanley Papers, UCB.

159. Stanley to the Nobel Committee, 22 December 1960, p. 5, Box 5.12, 78/18c, Stanley Papers, UCB.

160. "Genetic Rosetta Stone," *Time*, 23 May 1960, p. 50. See also Stent and Calendar, pp. 540–43.

161. Kahn, pp. 905–12.

162. Ibid., p. 910.

163. Cryptic references to Yčas's African trip can be found in his correspondence with Gamow, Box 7, Yčas fld.(1956, passim), Gamow Papers, LC. It is recounted by Gamow, "What Is Life?"

164. Yčas, "Correlation of Viral Ribonucleic Acid"; idem, "Replacement of Amino Acids in Proteins"; idem, *The Biological Code*, pp. 63–66; and Stent and Calendar, pp. 541–43.

165. Eck, "Non-Randomness," where he used arguments from information theory to make his point; Woese, "Composition of Various Ribonucleic Acid Fractions"; idem, "Coding Ratios"; idem, "A Nucleotide Triplet Code"; idem, *The Genetic Code*.

166. For example, Stanley, "The Regulation and Transfer of Biological Information." Several slightly different versions of the same presentation for different occasions are found in the Stanley Papers, UCB. Fraenkel-Conrat, "Synthetic Mutants," quote on p. 200.

167. Fraenkel-Conrat, "Synthetic Mutants," p. 200.

168. Ibid., p. 205.

CHAPTER 5. 'CYBERNÉTIQUE ENZYMATIQUE,' 'GÈNE INFORMATEUR,' AND MESSENGER RNA

1. This broad, but useful, binary characterization was given by Brenner, "RNA, Ribosomes, and Protein Synthesis," literally just a few weeks before the announcement of the breaking of the code by Marshall Nirenberg and Heinrich Matthaei, whose studies clearly belonged in the second approach. Brenner's view of competition between groups to break the code is supported by the contemporary coverage of "the race" in the *New York Times* and *Time* magazine. On the peripherality of genetics at Berkeley's Virus Lab, see Creager. See also Chap. 4, this volume.

2. Gaudillière, "Biologie Moléculaire et Biologistes"; idem, "Molecular Biology in the French Tradition?"; and Burian, "Technique, Task Definition, and the Transition." On the notion of "trading zones" in science, see Galison, "Context and Constraints"; idem, *Image and Logic*, Intro. and Chap. 9.

3. On the Pasteur group and molecular biology, see Yoxen, "The Social Impact of Molecular Biology," Chap. 4 and 8. Lwoff and Ullman; Judson; Sapp, *Beyond the Gene*; Gaudillière, "Biologie Moléculaire et Biologistes"; idem, "Molecular Biology in the French Tradition?"; Burian, "Technique, Task Definition, and the Transition" (see note 2); Abir-Am, "From Multidisciplinary Collaboration to Transnational Objectivity"; Doyle, "On Beyond Living," Chap. 4; Morange, *Histoire de la Biologie Moléculaire*, Chap. 6; and idem, *A History of Molecular Biology*, Chap. 14.

4. Yoxen, "The Social Impact of Molecular Biology," (see note 3); Fantini, "Utilisation par la Génétique Moléculaire"; Canguilhem, "The Role of Analogies and Models"; Delaporte, esp. Chap. 4.

5. Monod, *Chance and Necessity*; Jacob, *The Logic of Life*; both first appeared in French in 1970. Yoxen, "Social Impact of Molecular Biology," has examined the widespread reception of these works and several examples out of the hundreds of French and English reviews of these books.

6. Cohn et al. See also Cohn; Gaudillière, "J. Monod, S. Spiegelman et l'adaption Enzymatique"; idem, "Biologie Moléculaire et Biologistes," Chap. 1. On microbial genetics, enzyme adaptation, and molecular biology, see Sapp, *Beyond the Gene*, esp. Chap. 5; and idem, *Where the Truth Lies*. See also Thieffry, "Escherichia coli as a Model System."

7. Luria and Delbrück, a paper Monod read soon after it appeared; Stent and Calendar, Chap. 6. For an important critique, see Keller, "Between Language and Science."

8. Fantini, *Jacques Monod*, Preface; idem, "Monod, Jacques Lucien"; Morange, "L'oeuvre scientific de J. Monod"; Gaudillière, "Biologie Moléculaire et Biologistes," Chap. 1.

9. Monod, "The Phenomenon of Enzymatic Adaptation," in *Growth*, quote on p. 280; Sapp, *Beyond the Gene*, Chap. 5. See also Chap. 2, this volume.

10. On the protein paradigm of life, see Kay, *The Molecular Vision of Life*, Interlude I; on its relation to antibody research and molecular genetics, see ibid., Chap. 6–7, describing the promise of serological genetics for the elucidation of gene

action; idem, "Molecular Biology and Pauling's Immunochemistry"; and Moulin, *Le Dernier Langage de la Médecine*, Part II, Chap. 1.

11. Gaudillière, "Biologie Moléculaire et Biologistes," Chap. 1; Cohn (see note 6 above).

12. Monod, "La Technique de Culture Continue"; Novick and Szilard, "Experiments with the Chemostat"; A. Novick, "Phenotypic Mixing"; idem, "Introductory Essay." Szilard to Schrödinger, 9 November 1950, and subsequent exchanges, Box 17.12, MSS.32, Szilard Papers, UCSD. Although this invention seemed to be independently conceived, the constant exchanges of ideas, materials, and methods between these researchers could easily lead to like-minded projects.

13. Novick and Szilard, "II. Experiments with the Chemostat." On the patent negotiations see "Minutes of Meetings Held Between Representatives of American Sterilizer and Marc Wood International, Inc., to Discuss Licensing of the Monod Patents," 12 December 1958, and related correspondence, Box 14.21, MSS.32, Szilard Papers, UCSD; Novick and Monod, 13 and 18 September 1958, Correspondence, Box Mon.Cor 12, Fond Monod, SAIP. As Hans-Jörg Rheinberger has detailed for the case of protein synthesis, the graphamatic representations of the workings of experimental systems (here the E. coli system)—generated through instruments (such as the bactogen and chemostat), manipulations, measurements, calculations, pictures, graphs, and discursive practices—reconfigured the experimental space of enzyme regulation research. On the role of experimental systems in life science, see Rheinberger, "Experiment, Difference, and Writing: I. Tracing Protein Synthesis"; and idem, "Experiment, Difference, and Writing: II. The Laboratory Production of Transfer RNA"; idem, *Experiment, Differenz, Schrift*; idem, "Experiment and Orientation; idem, *Toward a History of Epistemic Things*.

14. Monod, "Foreword" (see note 12 above). Monod to Mrs. Szilard, 22 March 1968, pp. 1–5 [archival earlier version], Correspondence, Fonds Monod, SAIP. Monod's Notebooks (1953–63), Box MON.Cor 16, contains references to Maxwell demons, MON.MSS.01, No. 1, p. 4, January 1953. See also Abir-Am, "From Multidisciplinary Collaboration to Transnational Objectivity" (see note 3 above); and Chap. 4, this volume. The Pasteur group's formal exposure to coding was in early May 1955, when Crick first gave a seminar there.

15. Cohn and Monod; Cohn, p. 78 (though it appears that the expression "theater of the absurd" was initiated by Monod); Fantini, "Monod, Jacques Lucien," p. 642; Lederberg to Monod, 8 May 1952, Box MON.Cor.09, SAIP.

16. Cuénot, as quoted in Sapp, *Beyond the Gene*, p. 126. See also Burian, Gayon, and Zallen, "The Singular Fate of Genetics"; Burian and Gayon; Burian, Gayon, and Zallen, "Boris Ephrussi"; Zallen.

17. Monod's Notebooks (1953–63), January 1953, p. 3, Box MON.MSS.01, SAIP.

18. Cohn et al., "Terminology of Enzyme Formation"; Cohn, p. 79.

19. Jacob, *The Logic of Life*, p. 3.

20. Fantini, "Monod, Jacques Lucien"; Gaudillière, "Biologie Moléculaire et Biologistes," Chap. 2, where he details the Caen Colloquium (1956) and the formation of a politico-scientific network for the modernization of French science, in which

Monod played a central role. See also Gaudillière, "Molecular Biology in the French Tradition."

21. Ibid.; and Monod to Pomerat, 7 July 1954, Box 6.52, RG1.2, 500D, Pasteur Institute, RAC.

22. GRP Diary, 9 May 1955, Box 6.52, RG1.2, 500D, Pasteur Institute, RAC.

23. Fantini, "Monod, Jacques Lucien"; G. N. Cohen.

24. On the cold war in genetics, see Sapp, *Beyond the Gene*, Chap. 6. On the Marshall Plan and science, see Kay, *The Molecular Vision of Life*, Interlude II. Letter to Monsieur Larkin (consul des Etats-Unis), Monod to Szilard, [n.d.] 1952, Box 13.20, MSS.32, Szilard Papers, UCSD. See also the published letter, in "Passports and Visas," *Science* 116 (1952): 178–79. On the historiographic significance of Monod's political activities, see Abir-Am, "How Scientists View Their Heroes." On Szilard's relation to Teller and Oppenheimer and on his political activism in the 1950s, see Lanquette, Chap. 23–24. See also Chap. 4, this volume.

25. "Letter to the editor" (*Bulletin of Atomic Scientists*), Monod to Szilard, 17 July 1953, quote on p. 4, Box 13.20, MSS.32, Szilard Papers, UCSD.

26. Novick to Szilard, 8 February 1954, p. 1, Box 14.21, MSS.32, Szilard Papers, UCSD.

27. GRP interview with DR, 16 September 1954; and GRP Diary, 2 February 1954, RG1.2, 500D, Pasteur Institute, RAC.

28. Cohen and Monod; Monod, "Remarks on the Mechanism of Enzyme Induction"; Lederberg, "Gene Control"; idem, "Genetic Studies"; Spiegelman and Landman, for a different perspective; and G. N. Cohen; see also an overview by Stent, "Induction and Repression."

29. Cohen and Monod, p. 190.

30. Monod's Notebooks (1953–63), January 1953, p. 4, Box MON.MSS.01, No. 1, SAIP; on Szilard's contribution to the problem of the Maxwell demon, see Chap. 2, this volume.

31. On their relations to the phage group, see Lwoff, "The Prophage and I"; and Wollman.

32. Jacob, "Biography (1965)"; and idem, *The Statue Within*. On Elie Wollman's parent, see Wollman; and Judson, pp. 373–74.

33. Wollman and Jacob, poem on p. 109; see also Jacob and Wollman.

34. Jacob, "Genetics of the Bacterial Cell"; Jacob, "The Switch." Whether a bacterium was a donor or recipient was determined by a fertility factor F.

35. Pardee, pp. 109–10.

36. Litman and Pardee; Umbarger.

37. Yates and Pardee, "Pyrimidine Biosynthesis"; idem, "Control of Pyrimidine Biosynthesis." *Repression* came to denote inhibition of enzyme synthesis, while *inhibition* blocked (extant) enzyme activity. In 1961 Monod and Jacob would call the discovery of end-product inhibition "the Novick-Szilard-Umbarger Effect." See also Maas, for another historical perspective. A couple of years later, Pardee was involved in a priority dispute with Monod's laboratory (specifically with Jean-Pierre Changeux) over the discovery of allostary. See Creager and Gaudillière.

38. Pardee, Jacob, and Monod, "Sur l'expression"; idem, "The Genetic Control";

Jacob, "Genetics of the Bacterial Cell"; Monod, "From Enzymatic Adaptation"; Jacob, "The Switch"; Pardee, "The PaJaMa Experiment"; and Jacob, *The Statue Within*, pp. 290–99. There is also sizable secondary literature on the PaJaMa experiment: Schaffner; Judson, pp. 402–18; Grmek and Fantini; Fantini, "Jacques Monod et la Biologie Moléculaire"; Burian, "On the Cusp between Biochemistry and Molecular Biology"; Gaudillière, "How Biochemical Regulation Held"; and idem, "Molecular Biologists, Biochemists and Messenger RNA."

39. Pardee, p. 112; Jacob, "The Switch," p. 97.

40. Monod, "From Enzymatic Adaptation to Allosteric Transitions," p. 270. See also Maas.

41. On coding collaborations, see Szilard to Crick, 14 June 1957, Box 6.38, MSS.32, Szilard Papers, UCSD; on the restructuring of German biology, see Leo Szilard, "Welche Methoden Eignen Sich Zur Forderung Der Biologischen Forschung In Deutschland," pp. 1–6, Box 34.10, MSS.32, Szilard Papers, UCSD; Box 29.7 contains his unpublished manuscripts, "Control of Enzyme Production, Suppressor Genes and Enzyme Induction in Microorganisms," as well as "Drug Tolerance and Antibody Formation in Mammals," 13 May 1957, pp. 1–7, MSS.32, Szilard Papers, UCSD; and "On the Formation of Adaptive Enzymes," 30 August 1957, pp. 1–14, Box 29.7, MSS.32, Szilard Papers, UCSD. Lanquette, pp. 371–72; Melvin Cohn, interview with Lanquette, 11 August 1987, quoted on p. 391. Monod warmly credited Szilard's contributions both in the publication of the PaJaMa experiment and in his Nobel Lecture, though Pardee was less enthusiastic about these attributions.

42. Pardee, Jacob, and Monod, "Sur l'Éxpression"; idem, "The Genetic Control."

43. Monod, "An Outline of Enzyme Induction," quote on p. 571, paper presented early that summer at the Dutch Chemical Society meeting.

44. Jacob, *The Statue Within*, pp. 297–99.

45. Jacob, "Genetic Control of Viral Functions," pp. 1–39, quote on p. 24.

46. Jacob, "Genetics of the Bacterial Cell," p. 225; idem, "The Switch," pp. 99–100; idem, *The Statue Within*, p. 301; Judson, pp. 416–17. Jacob remembered hitting on the idea of the switch while watching his son play with his electric train.

47. Jacob, *The Statue Within*, though in his Nobel Lecture he used the more neutral analogy of automatic controlled doors; Monod, *Chance and Necessity*, p. 68. Keller, "Between Language and Science," p. 177.

48. Manuscripts, drafts of "Dunham Lectures," delivered 16–23 November 1958, Box MON.MSS.03, No. 10, SAIP; Monod to Stanier, 5 December 1957, related correspondence, Box MON.Cor.16, SAIP; and Monod to Cohn, 22 September 1958, Box MON.Cor.03, SAIP.

49. Manuscripts, "Dunham II," pp. 1–2, Box MON.MSS.03, No. 10, SAIP.

50. Manuscripts, "Dunham II," pp. 12–27, Box MON.MSS.03, No. 10, SAIP; and "Dunham III," pp. 1, 15, Box MON.MSS.03, No. 10, SAIP.

51. Monod to Krayer, 3 December 1958, "Dunham Lectures" Correspondence, Box MON.Cor.09, SAIP; Manuscripts, Cohn to Monod, 15 December 1958, Box MON.Cor.03, SAIP. *Cybernétique Enzymatique*, Intro., pp. 6–7 (text dictated 15 June–7 July 1959), Box MON.MSS.08, No. 2, SAIP.

52. Manuscripts, Berkeley lecture notes, "Induction, Repression, Genes," 17–25 November 1958, Box MON.MSS.08, No. 2, SAIP.

53. Monod, "Information, Induction, Répression," quotes on pp. 127, 137.

54. Ibid., pp. 129, 137.

55. Manuscripts, Monod's Notebooks, 1953–63, May 1959, pp. 42–43, Box MON.MSS.01, No. 1, SAIP.

56. Medawar, passim, but esp. pp. 56–57. See also Layzer, pp. 28–32.

57. Canguilhem, "The Role of Analogies," pp. 515, 519 (see note 4 above).

58. Manuscripts, Monod's Notebooks, 1953–63, pp. 41–42, Box MON.MSS.-01, No. 1, SAIP. The term *teleonomy* was coined by Pittenridgh (see "Adaptation") and formed the basis for the analysis on teleonomy by E. Mayr, "Teleological and Teleonomic." According to Mayr, it was he who introduced Monod and Jacob to the new concept of teleonomy (personal communication with author, 1995), but while adopting it, they failed to appreciate the role of natural selection in teleonomic mechanisms. See E. Mayr, *The Growth of Biological Thought*, Chap. 1 and p. 516, for a critique of Monod's usage. See also below.

59. On Kendrew's role in the institutionalization of molecular biology, see Abir-Am, "The Politics of Macromolecules," pp. 164–94; de Chadarevian, "Sequences, Conformation, Information"; and Gaudillière, "Molecular Biologists, Biochemists, and Messenger RNA."

60. Pardee, Jacob, and Monod, "Genetic Control," pp. 175–77.

61. Jacob and Monod, "Gènes de structure."

62. Jacob et al.; Jacob and Monod, "Genetic Regulatory Mechanisms," quote on p. 354. For literary analysis of the operon, see Doyle, "On Beyond Living," Chap. 4. On earlier related contributions from NIH, see Ames and Garry; and Chap. 6 in this volume.

63. Crick to Monod, 4 April 1955, Correspondence, Box MON.Cor.04, SAIP.

64. Crick to Monod, 14 April 1955, Box MON.Cor.04, SAIP; Crick, "Sailing with Jacques"; Jacob, *The Statue Within*, p. 287; Crick, *What Mad Pursuit*, p. 98.

65. Brenner to Jacob, 19 February 1959, SBF; I am deeply indebted to Sydney Brenner for availing to me his correspondence.

66. Jacob to Brenner, 1 April 1959, SBF; and passim, 1959.

67. Manuscripts; *Cybernétique Enzymatique*, Chap. VI, p. 3, Box MON.MSS.08, SAIP.

68. Misc. notes: Operon 1959/60, Box MON.MSS.01, No. 17, SAIP.

69. Jacob and Monod, p. 346; Jacob to Brenner, 7 February 1961, SBF. Thieffry, "Contributions of the 'Rouge-Clôitre Group'"; and Thieffry and Burian. See also Gros, "Code et Messenger," Chap. V.

70. Riley et al.; Jacob, "Genetics of the Bacterial Cell," p. 223; Gros, "Code et Messenger," Chap. V.

71. Riley et al.; Naono and Gros; Watson, "The Involvement of RNA"; Gros, "The Messenger"; Gaudillière, "Molecular Biologists, Biochemists, and Messenger RNA."

72. Riley et al., p. 225.

73. Jacob, "Genetics of the Bacterial Cell," p. 224.

74. Hershey, Dixon, and Chase; Volkin and Astrachan, "Intracellular Distribution"; idem, "RNA Metabolism in T2-Infected Escherichia Coli"; Volkin, Astrachan, and Countryman; Watson, "The Involvement of RNA in the Synthesis of Proteins," pp. 190–92; Jacob, *The Statue Within*, p. 312; Judson, pp. 426–27.

75. Nomura, Hall, and Spiegelman; Hall and Spiegelman; Watson, "The Involvement of RNA in the Synthesis of Proteins," pp. 190–91.

76. Watson, "The Involvement of RNA in the Synthesis of Proteins," p. 191; Davern and Meselson; Jacob, "Genetics of the Bacterial Cell," p. 224; Judson, pp. 436–41; Jacob, *The Statue Within*, pp. 313–18. Brenner to Meselson, 7 May 1960, Olby Collection, APS.

77. Brenner, Jacob, and Meselson; Watson, "The Involvement of RNA in the Synthesis of Proteins," pp. 190–92; Jacob, "The Genetics of the Bacterial Cell," pp. 222–24; Jacob, *The Statue Within*, quote on p. 315.

78. Gros et al.; Gros, "The Messenger," pp. 122–23; Gros, "Code et Messenger," Chap. V; Watson, "The Involvement of RNA in the Synthesis of Proteins," p. 191.

79. Jacob to Brenner, 12 September 1960, SBF.

80. Jacob to Brenner, 16 February 1961, SBF; Meselson to Brenner and Jacob, 15 February 1961, Olby Collection, APS. As Hans-Jörg Rheinberger pointed out (personal communication with author), Spiegelman was an interesting case. He was one of the first to adopt the information discourse and was one of the first (1954–56) to show that induced enzyme synthesis is mandatorily coupled to the de novo synthesis of RNA—in a sense he had the "short-lived RNA" at hand in 1955. It seemed that he possessed all the material and discursive ingredients to "invent" the messenger RNA, but he did not. Spiegelman's papers are deposited at the University of Wyoming, American Heritage Center, and have not been processed for scholarly use.

81. Brenner, Jacob, and Meselson, p. 580.

82. Gros et al., p. 585; Yčas and Vincent. Vincent was one of those who worked on RNA's role in protein synthesis already in 1956. See Rheinberger, *Toward a History of Epistemic Things*, pp. 146, 206.

83. Spiegelman, "The Relation of Informational RNA to DNA," quote on pp. 87–88. On his early interests in cybernetics, see Chap. 3, this volume. Spiegelman's friendship with Henry Quastler and the strong interest in control and communication sciences at the University of Illinois sustained his interests in information theory, echoes of which may found in Spiegelman's articles in the 1950s.

84. Jacob to Brenner, 2 March 1961, SBF.

85. Davis, quote on p. 1.

86. Jacob and Monod, "Elements of Regulatory Circuits in Bacteria," quotes on p. 1; symposium held at Varenna, September 1962.

87. Monod and Jacob, quote on p. 389. Reverberations of this polemical statement may be found in Monod to Ephrussi, 23 February 1967, Correspondence, Box MON.Cor.06, SAIP.

88. Monod and Jacob, p. 393.

CHAPTER 6. WRITING GENETIC CODES IN THE 1960S

1. The rise of NIH stemmed from several sources, including original legislation establishing the NIH, the New Deal, the Manhattan Project, and congressional support, especially Senator Lister Hill and Representative John Forgarty. See Harden, *Inventing the NIH*; Strickland, pp. 80–91; Steelman. For the postwar period, see Mullan, Chap. 5–7; Pursell; Burroughs.

2. Fredrickson; J. A. Shannon; Harden, "National Institutes of Health." The *NIH Almanac, 1993–94*, NIH Publication No. 94–95, cites slightly less dramatic figures.

3. Stetten; Kornberg, pp. 129–34, quote on p. 130.

4. R. G. Martin, "A Revisionist View," quotes on pp. 286–87 (see note 3 above).

5. Ibid., pp. 288–89.

6. Ibid., p. 288.

7. Ibid., pp. 284–86, quote on p. 284. Even in the commemorative volume, *Origins of Molecular Biology: A Tribute to Jacques Monod*, edited by Lwoff and Ullman, several of the contributors commented on Monod's autocratic and, at times, abrasive style. See also Ames and Garry.

8. Nirenberg, interview, 18 July 1994; R. G. Martin, "A Revisionist View," p. 287.

9. Nirenberg, "Biography."

10. Nirenberg and Jacoby, "Enzymatic Utilization"; idem, "On the Sites of Attachment"; and idem, "Constraints in the Determination." For discursive contrast, see Nirenberg, "The Inducation of Two Enzymes." On the role of the information idiom as a Derridian *supplement*, see Rheinberger, *Toward a History of Epistemic Things*, Chap. 9 and 11.

11. *New York Times*, "Biographies of 3 Nobel Laureates." "That's really funny," Nirenberg responded, "but the truth is I have never had trouble driving a car and I work more like ten hours a day. Except usually I view it more as playing than working." Nirenberg, interview, 4 October 1996.

12. I am deeply indebted to Marshall Nirenberg not only for providing me with complete photocopies of these diaries but also for clarifying many of the entries and, generally, for giving so generously of his time for several lengthy interview sessions conducted in 1994, 1995, and 1996.

13. On DNA as the transforming principle in tumor cells, see Book IV A, 31 December 1957–11 January 1958, pp. 139–48, MND; on his scientific philosophy, see Book V A, 22 August 1958, pp. 121–22, MND.

14. Nirenberg, interviews, 18 July 1994, and 19 July 1996.

15. Book V A, 10 October 1958, p. 153, MND; McElroy and Glass; R. G. Martin, "A Revisionist View," pp. 289–90.

16. Book V A, passim, quote on 9 November 1958, p. 171, MND.

17. Book V A, [n.d.] end of November 1958, p. 189, MND.

18. Nirenberg, interviews, 18 July 1994, 18 November 1995, and 19 July 1996; Book VI A, 9 April 1959, pp. 84, 89, MND.

19. Nirenberg, "The Genetic Code," pp. 336–37; Pollock, "An Exciting but Exasperating Personality" (see note 7 above).

20. Nirenberg, "The Genetic Code," pp. 336–37; Book VI A, 29 April 1959, p. 100, MND; for the references to the PaJaMa experiments, 2 August 1959, p. 181, MND. See also Eagle (see note 3 above).

21. Book VI A, [n.d.] end of April 1959, p. 106, MND; and 6 June 1959, p. 125, MND; and [n.d.] early February 1959, p. 38, MND.

22. Nirenberg, "The Genetic Code," p. 337; Book VI A, 25 August 1959, MND; R. G. Martin, "A Revisionist View," p. 290, Nirenberg, interview, 4 October 1996.

23. Rheinberger, "Experiment, Difference, and Writing: I. Tracing Protein Synthesis"; idem, "Experiment, Difference, and Writing: II. The Laboratory Production of Transfer RNA"; idem, "Experiment and Orientation," pp. 443–71; see also Burian, "Technique, Task Definition, and the Transition from Genetics to Molecular Genetics"; Rheinberger, "From Microsomes to Ribosomes."

24. On Crick's "adaptors" and the genetic code, see Chap. 3, this volume. For example, Stent and Calendar; Judson; and Burian, "Technique, Task Definition," have each equated Crick's hypothetical "adaptor" with tRNA.

25. Rheinberger, *Toward a History of Epistemic Things*, Chap. 10.

26. Lamborg was then a postdoctoral fellow in Zamecnik's laboratory. Lamborg and Zamecnik. See also Judson, pp. 472–73.

27. Tissières, Schlesinger, and Gros. Other reports included Kameyama and Novelli; and Nisman and Fukuhara.

28. Book VII A, [n.d.] end of April 1960, p. 13 (unnumbered pages), MND; and Book VII A, 26 June 1960, MND. Martin and Ames.

29. Book VII A, 5 August 1960, p. 113, MND.

30. Book IX A, 30 September 1960, p. 3 (of unnumbered section), MND.

31. Matthaei, interview, 3 March 1992; I deeply indebted to Matthaei for giving so generously of his time and for providing me with a complete collection of his publications and with copies of his laboratory notebooks for 1960–62. Interview with Heinrich Matthaei, 2 April 1968, pp. 1–12, Robert C. Olby Collection, APA. That Matthaei was informed of Brachet's and Caspersson's work is evident in his publication "Vergleichende Untersuchungen des Eiweiss-Haushalts beim Streckungswachstum von Bluttenblattern und Anderen Organen."

32. "N.A.T.O—Application, Dr. Johann Heinrich Matthaei," for academic year 1960 [n.d.], and attached letter to the German Academic Exchange Service, 14 September 1960, pp. 1–3, Olby Collection, APS. See also the account in Judson, pp. 470–71.

33. Book IX A, 11 November 1960, p. 21, MND; [n.d. the following week], pp. 30–31; R. G. Martin, "A Revisionist View," p. 92.

34. Book IX A, 13 January 1961, p. 89, MND.

35. Nirenberg, interview, 18 July 1994, and 19 July 1996; Matthaei and Nirenberg, "Some Characteristics of a Cell-Free DNAase Sensitive System."

36. Matthaei and Nirenberg, "The Dependence of Cell-Free Protein Synthesis," quotes on pp. 404, 407, respectively. References to messenger RNA appear frequently in the diaries since the fall of 1960. On the messenger work see Chap. 5, this volume.

37. "Interview with Matthaei," pp. 7–8, Olby Collection, APS.

38. Book IX, 8 May 1961, p. 186, MND.

39. See Chap. 5. Quote, *Time* magazine, 23 May 1960, p. 50 (quoting Wendell M. Stanley); Stent and Calendar, pp. 540–41. See Chap. 4, this volume.

40. Nirenberg, interview, 19 July 1996. Book IX, 12–14 May 1961, pp. 189–94, MND.

41. Matthaei, interview, 3 March 1992; Nirenberg, interview, 19 July 1996 (the resentment eventually culminated in Matthaei's explosion and the termination of their friendship). Matthaei's Notebooks, M1, experiment 29G, p. 104; and experiment 27N, 25 May 1961. Rheinberger, *Toward a History of Epistemic Things*, Chap. 13. See also Judson, pp. 476–79. Interview with Matthaei, pp. 10–11, Olby Collection, APS.

42. Matthaei Notebooks, M1, experiment 27Q, 27 May 1961 (the reported figure of 38,000 cpm/ml is an order of magnitude higher to adjust from microgram amounts in the reaction mixture). Quoted in Judson, p. 478. See also Rheinberger, *Toward a History of Epistemic Things*, Chap. 13.

43. Interview with Matthaei, pp. 11–12, Olby Collection, APS.

44. The paper by Tsugita, Fraenkel-Conrat, and Nirenberg turned out to be flawed and was later retracted. Book X, 11 June 1961, p. 21, MND. Matthaei, interview, 3 March 1992; Nirenberg, interview, 19 July 1996, and 4 October 1996.

45. Matthaei and Nirenberg, "Characterization and Stabilization"; Nirenberg and Matthaei, "The Dependence of Cell-Free Protein Synthesis," both in *Proceedings of the National Academy of Sciences, U.S.A*, 47 (1961), pp. 1580–88 and 1588–1602, respectively; quotes on pp. 1589, 1601.

46. Crick to Delbrück, 19 June 1961, Box 5.41, Delbrück Papers, CIT. For Nirenberg's Moscow presentation, see Nirenberg and Matthaei. On the Moscow congress and reactions to Nirenberg, see Borek, pp. 199–200; and Judson, pp. 463–69, 480–82.

47. Interview with Charles Wiener, 4 April 1968, p. 78, Oral History, George Gamow, APS. On Jacob's attempts, see Chap. 5, this volume; Jacob and Tissières, as quoted in Judson, p. 482; and letter from Alexander Rich to Francis Crick, 21 October 1961, Olby Collection, APS.

48. Stent and Calendar, p. 545.

49. Zamecnik.

50. R. G. Martin, "A Revisionist View," p. 291. Nirenberg, interview, 19 July 1996. Interview with Matthaei, p. 8, Olby Collection, APS.

51. Woese, *The Genetic Code*, p. 17.

52. R. G. Martin, "A Revisionist View," p. 293. See also the autobiographical essay by Ochoa, "The Pursuit of a Hobby," esp. pp. 20–21.

53. R. G. Martin, "A Revisionist View," p. 294. Nirenberg, interview, 19 July 1996.

54. Interview with Matthaei, p. 12, Olby Collection, APS.

55. Ochoa, "The Pursuit of a Hobby"; idem, "Enzymatic Synthesis"; and idem, "Biography"; Kornberg, pp. 49–69, quote on p. 50. Ochoa's post–Nobel Prize work included Beljanski and Ochoa, "Protein Biosynthesis by Cell-Free Bacterial System, I," and idem, "Protein Biosynthesis by Cell-Free Bacterial System, II."

56. Grunberg-Manago; and Grunberg-Manago, Ortiz, and Ochoa. See also

Ochoa, "Enzymatic Synthesis"; idem, "The Pursuit of a Hobby," pp. 19–20; and Borek, pp. 202–5.

57. Letters from Alexander Rich to Francis Crick and to George Gamow, from 6 July 1956 to 24 April 1963, Olby Collection, APS; Ochoa, "The Pursuit of a Hobby," p. 20; and idem, "Enzymatic Mechanism" (I thank Robert Olby for pointing out this important volume). On the collaboration of Ochoa and Heppel, see also Singer, Heppel, and Hilmoe (Hilmoe worked in Ochoa's laboratory), "Oligonucleotides as Primers"; Heppel, Ortiz, and Ochoa; Singer, Heppel, and Hilmoe, "Oligonucleotides as Primers."

58. Ochoa, "The Pursuit of a Hobby," p. 20. French visitor in Ochoa's lab, Lengyel, on Ochoa's suggestion had tried a poly-A experiment before Nirenberg's but did not recover an effect since polylysine is soluble in the TCA precipitating reagent.

59. Lengyel, Speyer, and Ochoa. On the relation between technical and epistemic things, see Rheinberger, *Toward a History of Epistemic Things*, Chap. 2.

60. Ochoa, "The Pursuit of a Hobby," p. 20.

61. Lengyel, Speyer, and Ochoa, p. 1941.

62. Speyer et al., *Synthetic Polynucleotides II.*

63. Nirenberg, Matthaei, and Jones.

64. R. G. Martin, "A Revisionist View," p. 293.

65. Nirenberg, interview, 19 July 1996.

66. R. G. Martin et al., quotes on pp. 411–12.

67. Matthaei et al. On Gamow's paper, see Chap. 4, this volume.

68. Yčas, "The Protein Text," quotes on pp. 70 and 91, respectively. For further analysis, see Kay, "Who Wrote the Book of Life? Information and the Transformation of Molecular Biology, 1945–1955" in *Science in Context*; and Chap. 3–4, this volume.

69. Book IX, 16 September 1961, pp. 60–61, MND; 6 October 1961, p. 73, MND; and early October [n.d.] 1961, p. 86, MND.

70. Book IX, 3 December to end of 1961, pp. 131–54, MND.

71. Lengyel et al.

72. *New York Times*, "Gain Is Reported in Heredity Study."

73. Crick et al.

74. Jacob to Brenner, 13 December 1961, SBF. I am deeply indebted to Brenner for availing to me his correspondence.

75. Crick et al., p. 1227.

76. Ibid.

77. Ibid., p. 1228.

78. Ibid., p. 1231.

79. Ibid., p. 1232.

80. Quoted in Judson, p. 488.

81. Crick, "The Genetic Code—Yesterday, Today, and Tomorrow," quote on p. 6. Eck, "A Simplified Strategy"; and idem, "Genetic Code," quote p. 480; *New York Times*, "New Model Given for Genetic Code." See also Chap. 4, this volume.

82. Wall.

83. Hersh, p. 328.

84. Roberts, quotes, p. 897. See also Sueoka; Rychlik and Šorm; and Crick, "The Genetic Code—Yesterday Today and Tomorrow," p. 6.

85. Woese, "Nature of the Biological Code."

86. Medvedez. On Medvedez, see Judson, pp. 464–67.

87. Ageno.

88. Speyer et al., "Synthetic Polynucleotides"; Basilio et al.; Wahba, Basilio et al.; Gardner et al.; and Wahba, Gardner et al.

89. Jones and Nirenberg.

90. Crick to Delbrück, 6 November 1962, Box 5.41, Delbrück Papers, CIT.

91. Crick, "The Genetic Code," in *Nobel Lectures*; idem, "The Genetic Code," in *Scientific American*; and idem, "Towards the Genetic Code," quote on p. 16. See also Nirenberg, "The Genetic Code: II."

92. Crick, "The Recent Excitement in the Coding Problem." For an even more detailed review, see Lani.

93. Crick, "The Recent Excitement in the Coding Problem," pp. 176, 180.

94. Ibid., p. 185.

95. Ibid., pp. 213–14.

96. Ochoa, "Enzymatic Mechanisms," p. 159 (see note 57 above).

97. Platt, quotes on pp. 167–68.

98. Ibid., p. 179. See also Chap. 4, this volume.

99. Vogel, Bryson, and Lampen, p. xv.

100. Fruton, *A Skeptical Biochemist*, pp. 200–201. See also Chap. 1, this volume.

101. Chantrenne.

102. Laurence, "Structure of Life"; and idem, "Biochemists Wary on Life's Secrets."

103. *New York Times*, "New Gains Cited on Genetic Code"; also idem, "The Code of Life."

104. Ibid., "Biologists Hopeful of Solving Secrets of Heredity This Year." Baudrillard, p. 65.

105. *New York Times*, "Code of Genetics Proves Stubborn."

106. Ibid., "The Genetic Code Held Universal."

107. Lederberg, "Biological Future of Man"; the volume of proceedings from the CIBA symposium.

108. Using Thomas Kuhn's terminology of "normal science." Stent, "That Was the Molecular Biology That Was."

109. Mullan, Chap. 7; Fredrickson, p. 643.

110. McElheny, "Research in Biology." Quote on p. 908. Gottweis.

111. McElheny, "Research in Biology"; quote on pp. 909–10. The Pasteur group's criticisms and calls for reform of French biology (and science) are detailed in various publications. See, for example, "Nobel Winner Monod Criticizes French 'Scientific Backwardness,'" *Washington Post*; McElheny, "France Considers Significance of Nobel Awards"; and idem, "Pasteur Institute Scientists Demand Sweeping Reform." See also Gaudillière, "Biologie Moléculaire et Biologistes"; idem, "Molecular Biologists, Biochemists, and Messenger RNA"; Gottweis; and Chapter 5 in this volume.

112. Wolstenholme; *New York Times*, "Probing Heredity's Secrets"; idem, "Geneticists Meet to Review Gains"; and idem, "Gains in Genetics."

113. O'Connor. See also Weiner.

114. Sonneborn, *The Control of Human Heredity*, p. 47; quoted in Weiner, p. 34.

115. *New York Times*, "Hereditary Control by Man Is Foreseen."

116. The politicization of science during the Vietnam era and the concerns with these issues is treated in the pages of *Science* from 1965 onward. Organized efforts of scientists for social responsibility began in the 1930s; see Kuznick, *Beyond the Laboratory*. See also idem, "The Ethical and Political Crisis of Science."

117. Nirenberg's concern appeared in the editorial "Will Society Be Prepared?"

118. Khorana, "Biography."

119. Kornberg, pp. 138–39.

120. Nirenberg, interview, 19 July 1996. He pointed out that by then Khorana's book, *Some Recent Developments*, had become the definitive text in the field.

121. Khorana, "Nucleic Acid Synthesis," p. 306; Stent and Calendar, pp. 546–47.

122. Khorana, "Nucleic Acid Synthesis," pp. 306–7.

123. Ibid., p. 307; Kornberg, pp. 166–67; Falaschi, Adler, and Khorana.

124. Leder, "Biographical Statement" (courtesy of Department of Genetics, Harvard Medical School); R. G. Martin, "A Revisionist View," p. 294; *Cold Spring Harbor Symposia on Quantitative Biology* (including Delbrück's "A Theory of Autocatalytic Synthesis"). Several historical studies of Cold Spring Harbor are now in progress.

125. Nirenberg et al., "On the Genetic Code," pp. 549–57; and a longer version, idem, "Cell-Free Peptide Synthesis."

126. Nirenberg and Leder; Leder and Nirenberg, "RNA Codewords and Protein Synthesis, II." They credited three laboratories with the findings that aminoacyl-tRNA bound to ribosomes: Schweet's at the University of Kentucky; Lipmann's at Rockefeller University; and Kaji's at the University of Pennsylvania.

127. Ibid.; R. G. Martin, "A Revisionist View," p. 295. See also Nirenberg, "The Genetic Code"; and Singer.

128. Osmundsen, "Breaking the Code."

129. Leder and Nirenberg, "RNA Codewords and Protein Synthesis, III"; Bernfield and Nirenberg; Pestka, Marshall, and Nirenberg; Trupin, Rottman, and Brimacombe; Nirenberg et al., "RNA Codewords and Protein Synthesis, VII"; Brimacombe et al. See also Nirenberg, "The Genetic Code."

130. Nirenberg et al., "RNA Codewords, VII"; Sarabhai, Stretton, and Brenner; Yanofsky et al.; Weigert and Garen; and Sambrook, Fan, and Brenner.

131. Nirenberg et al., "RNA Codewords, VII," pp. 1166–67; quote on p. 1167.

132. Holley, "Biography."

133. Holley et al.; Holley, "The Nucleotide Sequence"; idem, "Alanine Transfer RNA"; and Singer (see note 127 above).

134. Holley et al., p. 1464; Crick, "Codon-Anticodon Pairing"; on the reception of Holley's work, see Singer; and Sonneborn, "Nucleotide Sequence of a Gene" (a letter calling attention to Holley's findings).

135. For a review of his work from 1961 to 1965, see Khorana, "Polynucleotide Synthesis"; for subsequent studies, see Khorana, "Nucleic Acid Synthesis, pp. 312–18.

136. Kellogg et al.; Pestka and Nirenberg; and Rottman and Nirenberg. See also Stent and Calendar, pp. 546–47.

137. Cairns; and Crick, "The Genetic Code—Yesterday, Today, and Tomorrow."

138. Crick, "The Genetic Code—Yesterday, Today, and Tomorrow," p. 8.

139. Nirenberg et al., "The RNA Code and Protein Synthesis"; and Marshall, Caskey, and Nirenberg.

140. Nirenberg et al., "The RNA Code," p. 19.

141. *New York Times*, "Genetic Language Called Universal."

142. Beadle and Beadle, Chap. 23. On scientists' popularization and media response, especially in Europe, see Yoxen, "The Social Impact of Molecular Biology."

143. Sinsheimer, *The Book of Life*, pp. 5–6.

144. Dobzhansky, p. 3.

145. Pauling, "Reflections on the New Biology," quoted in Duster, p. 46. See also Kay, *The Molecular Vision of Life*, esp. "Epilogue"; on the question of discontinuities versus cleavages in vision of biological engineering, see Kay, "Problematizing Basic Research in Molecular Biology."

146. Sinsheimer, "The Prospect of Designed Genetic Change," quoted in Keller, "Nature, Nurture, and the Human Genome Project."

147. Farber, p. 35.

148. Nirenberg, "Will Society Be Prepared?" See also a response, *New York Times*, "Geneticist Predicts Man Will Manipulate Heredity"; and Weiner, "Anticipating the Consequences of Genetic Engineering," p. 37.

149. Singer, p. 433.

150. R. G. Martin, "A Revisionist View," pp. 282–83.

CHAPTER 7. IN THE BEGINNING WAS THE WOR(L)D?

1. Jacob, "Inaugural Lecture"; quoted in Jakobson, *Main Trends of Research*, p. 438. Beadle and Beadle, p. 207; Sinsheimer, *The Book of Life*, pp. 5–6; Delbrück, "Aristotle-totle-totle."

2. Von Goethe, pp. 1224–37; Kittler, pp. 3–14.

3. Eigen and Winkler (the German edition is titled *Das Spiel*; the English edition is *Laws of the Game*).

4. Kahn; Eamon; and Rosenheim. See also Daston for a historicization of curiosity. Bono, "Science, Discourse, and Literature."

5. Foucault, *The Order of Things*, Chap. 2. Condillac, *Logic*, p. 305; idem, *Essai sur l'origine des connoissances humaines*. See also Aarsleff.

6. Watson, "Values from Chicago Upbringing."

7. For Wiener's scriptural vision of life, see Chap. 3, this volume; quote in Wiener, *The Human Use of Human Beings*, p. 103; D. Jackson.

8. Derrida, *Of Grammatology*, p. 9. For discussion see Chap. 1, this volume.

9. Jakobson to Wiener, 24 February 1949, Box 2.92, MC22, Wiener Papers, MIT. See also Chap. 3, this volume.

10. Brain; Piaget, Chap. 5; Hayles, *How We Became Posthuman*, Chap. 4. I am

grateful to Kate Hayles for sharing with me this material prior to publication. See also Aarsleff, *From Locke to Saussure*, pp. 372–400; Culler. Interestingly, seven centuries earlier Aquinas had pointed out that received writing, which possesses material potency, could only signify distinction and difference (see Chap. 1, this volume); at the end of the eighteenth century Wilhelm von Humboldt argued for the phonocentric economy of language and to the meaninglessness of the sign (p. 63). See Helmut Müller-Sievers, *Self-Generation: Biology, Philosophy, and Literature around 1800*, p. 109.

11. On Jakobson's career, see MC 72, Jakobson Papers, MIT. "Roman Jakobson: A Brief Chronology" by Stephen Rudy. Gardner, pp. 196–205.

12. See, for example, Levi-Strauss to Jakobson, 29 March and 5 May 1951, Box 12.45, MC 72, Jakobson Papers, MIT; Jakobson, Voegelin, and Sebeok, Chaps. 1–2. Gardner, pp. 202, 236–40; Lévi-Strauss. For other influences on the social sciences, see, for example, "Psycholinguistics: A Survey of Theory and Research Problems," in Supplement to *Journal of Abnormal and Social Psychology*; I am indebted to Henning Schmitgen for availing to me these materials. See also Hayles, *How We Became Posthuman*, Chap. 4.

13. Weaver to Jakobson, 15 December 1949; Jakobson to Weaver, 30 July 1950, Box 6.37, MC 72, Jakobson Papers, MIT.

14. Gardner, pp. 205–7.

15. Jakobson to Fahs, 22 February 1950, Box 6.37, MC 72, Jakobson Papers, MIT.

16. Stratton to Jakobson, 2 December 1957, Box 3.63, MC 72, Jakobson Papers, MIT; and Stratton to "Members of the Faculty and Staff," 18 April 1958, Box 3.63, MC 72, Jakobson Papers, MIT. See also Leslie, *The Cold War*; Edwards; and Hughes and Hughes.

17. See pp. 621–31.

18. Cherry, Halle, and Jakobson; Jakobson and Halle; Cherry, Chap. 3; Fehr. See also Gardner, pp. 200–202.

19. Jakobson, "Linguistics and Communication Theory," quote on p. 247. See, for example, MacKay, "The Epistemological Problem of Automata"; idem, *Information, Mechanisms and Meaning*. On MacKay, see Heims, *The Cybernetics Group*, pp. 111–12; Hayles, *How We Became Posthuman*, Chap. 3; Bohr; and Bowker.

20. "A Center for Information Sciences," [n.d.] ca. 1959, p. 4, Box 3.65, MC 72, Jakobson Papers, MIT.

21. M. Halle and K. N. Stevens from "A Survey of the Communication Sciences," edited by Wiesner and Rosenblith, 10 December 1959, Box 3.64, MC 72, Jakobson Papers, MIT.

22. Chomsky, pp. 39–40.

23. For an alternative view, see See; and Saumjan (contributors to Saumjan's edited volume include Jakobson, Chomsky, and Emile Benveniste); I am grateful for Henning Schmidgen for availing to me these materials.

24. Kay, *The Molecular Vision of Life*, p. 43.

25. "Phonemes as Linguistic Code," 22 August 1962, pp. 1–57, Box 34.65, MC

72, Jakobson Papers, MIT. See also Schmitt; Olby, "The Impact of Molecular Biology"; and Landecker.

26. "Phonemes as Linguistic Code," 22 August 1962, quotes on pp. 1, 50, and 52, respectively, Box 34.65, MC 72, Jakobson Papers, MIT.

27. Ibid., pp. 52–57.

28. Ibid., pp. 27–28.

29. Jakobson to Bronowski, 15 April 1969 (for the Diogenes publication, see note 23 above), Box 6.39, MC 72, Jakobson Papers, MIT.

30. Jakobson to Bronowski, 5 October 1967, Box 6.39, MC 72, Jakobson Papers, MIT.

31. Transcript of "Un Débat Entre François Jacob, Roman Jakobson, Claude Lévi-Strauss et Philippe L'Héritier: Vivre et Parler," 20 September 1967; "Vivre et Parler, I and II," *Les Lettres Francaises, No. 1221 and 1222*, 14–20 and 21–28 February 1968, pp. 1–7 and 1–6, respectively, Box 18.48, MC 72, Jakobson Papers, MIT. For an overview on this debate, see Doyle, *On Beyond Living*, Chap. 5.

32. Transcript, pp. 5–7, Box 18.48, MC 72, Jakobson Papers, MIT.

33. Ibid., pp. 7–8.

34. Ibid., pp. 8–11.

35. Ibid., pp. 11–14, Jakobson's quote on pp. 12–13.

36. Ibid., pp. 14–17.

37. Ibid., quotes on pp. 17–18 and 31, respectively.

38. On the Salk Institute conference, see Slater to Ross, 10 September 1969, Box 6.39, MC 72, Jakobson Papers, MIT.

39. Jakobson, *Main Trends*, Chap. VI, quote on p. 437.

40. Ibid., p. 438. See also Roman Jakobson, "Linguistics and Adjacent Sciences," Tenth International Congress of Linguistics, Bucharest, Romania, 1967, Box 18.63, MC 72, Jakobson Papers, MIT; and "Linguistics in Its Relation to Other Sciences," *Actes Du Xe Congres International des Linguistes*, Bucharest, 1967, Box 18.63, MC 72, Jakobson Papers, MIT.

41. Jakobson, *Main Trends*, pp. 439–40.

42. Ibid.

43. Jacob, *The Logic of Life*, p. 324.

44. Jacob, "Le Model Linguistique en Biologie."

45. Ibid., p. 200.

46. Ibid., pp. 202–3.

47. Ibid., pp. 203–5.

48. Stent, *The Coming of the Golden Age*, p. 64, where he credits two individuals for the discovery; Schönberger, pp. 29–31, originally published as *Verborgener Schlussel zum Leben*. Schönberger cites two authors—von Franz and Grafe. See also Walter, esp. Sec. II.

49. Stent, *The Coming of the Golden Age*, pp. 64–65.

50. Ibid., p. 65.

51. Schönberger, p. 34.

52. Ibid., p. 106; Monod, *Chance and Necessity*, pp. 172–73.

53. Bastide, quote p. 28; Moulin, "Text and Context in Biology," pp. 146–61, quote on p. 157.

54. Jacob, *The Logic of Life*, p. 287.

55. See, for example, Eigen, "Chemical Means"; and Eigen and De Maeyer. Eigen's research project on the origin of life and information was initiated in his classic paper, "Selforganization of Matter."

56. Eigen, "Selforganization of Matter"; idem, "The Origin of Biological Information"; Eigen and Winkler, *Das Spiel* (English translation by Robert and Rita Kimber); idem, *Laws of the Game*; Eigen, "How Does Information Originate?" See also Gatlin; Atlan; for a thoughtful critique, see Oyama. For a comprehensive overview of Eigen's project, see Kuppers, *Information and the Origin of Life*. For the earliest study of information-based self-organization, see Quastler, *The Emergence of Biological Organization*. I am grateful to Leslie Orgel for availing this book to me.

57. Eigen, "The Origin of Biological Information," pp. 601–3.

58. While Eigen defined the general properties of a "quasi-species" in his first paper, "Selforganization of Matter," as well as in subsequent ones (see note 56 above), the concept was elaborated and refined by his collaborator, Peter Schuster. See, for example, Eigen and Schuster, "The Hypercycle: A Principle of Natural Self-Organization," Parts A, B, and C.

59. Eigen mentioned the possibility of an "evolution machine" already in his 1971 paper, "Selforganization of Matter," but elaborated on this "evolution reactor" a decade later. See Eigen and Gardiner. See also Coghlan; on industrial interests in evolutionary biotechnology, see Eigen, *Selection*; and Kelly, passim.

60. See, for example, Masters and Broda; and Ratner. See also Sereno; and J. Campbell, esp. Chap. 14. For an overview on Chomsky's work, see H. Gardner, Chap. 7; and Chomsky.

61. Quoted in Eigen and Winkler, *Laws of the Game*, p. 259; their discussion on creation and revelation, see pp. 165–72. See also Eigen, "Sprache und Lernen auf molekularer Ebene"; and Kuppers, "The Context-Dependence of Biological Information"; idem, "Der semantische Aspekt von Information."

62. Trifonov and Brendel, p. 8.

63. Collado-Vides. Collado-Vides has since published several papers on the subject. See also Dong and Searls; I am grateful to Denis Thieffry for bringing these publications to my attention. For a critique, see Dresher. I am grateful to Wayne O'Neil for availing this article to me and for sharing with me his own critiques of DNA linguistics.

64. Angier.

WORKS CITED

Aarsleff, Hans. *From Locke to Saussure: Essays on the Study of Language and Intellectual History*. Minneapolis: Minnesota University Press, 1982.

Aaserud, Finn. "*Sputnik* and the 'Princeton Three': The National Security Laboratory that Was Not to Be." *Historical Studies in the Physical and Biological Sciences* 25 (1995): 185–240.

Abir-Am, Pnina. "The Discourse of Physical Power and Biological Knowledge in the 1930s: A Reappraisal of the Rockefeller Foundation's 'Policy' in Molecular Biology." *Social Studies of Science* 12 (1982): 341–82.

———. "From Biochemistry to Molecular Biology: DNA and the Acculturated Journey of the Critic of Science Erwin Chargaff." *History and Philosophy of the Life Sciences* 2 (1980): 3–60.

———. "From Multidisciplinary Collaboration to Transnational Objectivity: International Space as Constitutive of Molecular Biology, 1930–1970." In Elizabeth Crawford, Terry Shinn, and Sverker Sorlin, eds., *Denationalizing Science: The Contexts of International Scientific Practice*, pp. 153–86. Dordrecht: Kluwer Academic Publishers, 1992.

———. "How Scientists View Their Heroes: Some Remarks on the Mechanisms of Myth Construction." *Journal of the History of Biology* 15 (1982): 281–315.

———. "The Politics of Macromolecules: Molecular Biologists, Biochemists, and Rhetoric." *Osiris* 7 (1992): 210–37.

———. "Synergy or Clash: Disciplinary and Marital Strategies in the Career of Mathematical Biologist D. M. Wrinch (1894–1976)." In Pnina Abir-Am and Dorinda Outram, eds., *Uneasy Careers and Intimate Lives: Women in Science, 1789–1979*, pp. 338–94. New Brunswick, N.J.: Rutgers University Press, 1987.

Adams, Mark B. "Molecular Answers in Soviet Genetics." Paper presented at the Second Mellon Workshop, "Building Molecular Biology: Comparative Studies of Ideas, Institutions, and Practices," MIT, Cambridge, Mass., April 1992.

———, ed. *The Wellborn Science: Eugenics in Germany, France, Brazil, and Russia*. New York: Oxford University Press, 1990.

Ageno, Mario. "Deoxyribonucleic Acid Code." *Nature* 195 (1962): 998–99.

Allen, Garland E. "The Eugenics Record Office of Cold Spring Harbor, 1910–1940." *Osiris* 2 (1986): 225–65.

———. *Thomas Hunt Morgan: The Man and His Science*. Princeton: Princeton University Press, 1979.

Allison, David K. "U.S. Navy Research and Development since World War II." In Merritt Roe Smith, ed., *Military Enterprise and Technological Change: Perspectives on the American Experience*, pp. 289–328. Cambridge: MIT Press, 1985.
Ames, Bruce, and Barbara Garry. "Coordinated Repression of the Synthesis of Four Histidine Biosynthetic Enzymes by Histidine." *Proceedings of the National Academy of Sciences (U.S.A.)* 45 (1959): 1453–61.
Ampère, André-Marie. In *Essai sur la Philosophie des Sciences*, Pt. II, pp. 140–41. Paris, 1834.
Amsterdamska, Olga. "Between Medicine and Science: The Research Career of Oswald T. Avery." In Ilana Löwy, ed., *Medicine and Change: Historical and Sociological Aspects*, pp. 181–212. London: John Libbey, 1993.
———. "Stabilizing Instability: The Controversy over Cyclogenic Theories of Bacterial Variation during the Interwar Period." *Journal of the History of Biology* 24 (1991): 191–222.
Angier, Natalie. "Biologists Seek Words in DNA's Unbroken Text." *New York Times*, 9 July 1991, pp. C1, C11.
Appy, Christian. "'We'll Follow the Old Man': Sentimental Militarism and Cold War Films of the 1950s." Paper presented at the Landmarks Conference on the Cold War and American Culture at the American University, Washington, D.C., 17–19 March 1994.
Aquinas, St. Thomas. *Truth*. Vol. I. Trans. Robert W. Mulligan, S.J. Chicago: Henry Regency, 1952.
Arbib, Michael. "Automata Theory and Development: Pt. I." *Journal of Theoretical Biology* 14 (1967): 131–56.
———. "Self-Reproducing Automata—Some Implications for Theoretical Biology." In C. H. Eddington, ed., *Toward a Theoretical Biology*, Vol. 2, pp. 204–26. Edinburgh: Edinburgh University Press, 1968.
Arbib, Michael A., and Mary Hesse. *The Construction of Reality*. Cambridge: Cambridge University Press, 1986.
Aristotle. *Generation of Animals*. Ed. A. L. Peck. Loeb ed. Cambridge: Harvard University Press, 1979.
Aspray, William. *John von Neumann and the Origins of Modern Computing*. Cambridge: MIT Press, 1990.
———. "The Scientific Conceptualization of Information: A Survey." *Annals of the History of Computing* 7, no. 2 (1985): 117–40.
Atlan, Henri. *L'Organisation Biologique et la Theorie de L'Information*. Paris: Hermann, 1972.
Augenstine, Leroy, Herman R. Branson, and Eleanore B. Carver. "A Search for Intersymbol Influences in Protein Structure." In Henry Quastler, ed., *Information Theory in Biology*, p. 105. Urbana: University of Illinois Press, 1953.
———. "A Search for Intersymbol Influences in Protein Structure." In H. Yockey, ed., *Information Theory in Biology*, pp. 105–18. New York: Pergamon, 1956.
Avery, Oswald T., Colin M. MacLeod, and Maclyn McCarty. "Studies on the Chemical Transformation of Pneumococcal Types." *Journal of Experimental Medicine* 79 (1944): 137–58.

Bacon, Francis. *Natural and Experimental History*. In Richard Foster Jones, ed., *Essays, Advancement of Learning, New Atlantis, and Other Pieces*. New York: Odyssey Press, 1937.

Bamford, James. *The Puzzle Palace: A Report on America's Most Secret Agency*. Boston: Houghton Mifflin Co., 1982.

Bar-Hillel, Yehoshua. "Linguistic Problems Connected with Machine Translation." *British Journal of Philosophy of Science* 20, no. 3 (1953): 217–25.

———. "Logical Syntax and Semantics." *Language* 30, no. 2, 1954: 230–37.

Basilio, Carlos, Albert J. Wahba, Peter Lengyel, Joseph Speyer, and Severo Ochoa. "Synthetic Polynucleotdies and the Amino Acid Code, V." *Proceedings of the National Academy of Sciences* 48 (1962): 613–16.

Bastide, Françoise. "Linguistique et Génétique." *Bulletin du Groupe de Recherches Semio-Linguistiques de L'Ecole de Hautes Etudes en Sciences Sociales* 33 (1985): 21–28.

Baudrillard, Jean. *Simulations*. New York: Semiotext(e), 1983.

Baxter, James P. *Scientists against Time*. Boston: Little, Brown, 1946.

Beadle, George. "Biochemical Genetics." *Chemical Reviews* 37 (1945): 15–96.

———. "The Genetic Control of Biochemical Reactions." *The Harvey Lectures* Series XL (1945): 179–94.

Beadle, George, and Muriel Beadle. *The Language of Life*. New York: Doubleday, 1966.

Beatty, John. "Genetics in the Atomic Age: The Atomic Bomb Casualty Commission, 1947–1956." In K. R. Benson, J. Maienschein, and R. Rainger, eds., *The American Expansion of Biology*, Chap. 10. New Brunswick, N.J.: Rutgers University Press, 1991.

———. "Opportunities for Genetics in the Atomic Age." Paper presented at the Fourth Mellon Workshop, "Institutional and Disciplinary Contexts of the Life Sciences," MIT, Cambridge, Mass., April 1994.

———. "Origins of the U.S. Human Genome Project: The Changing Relationship of Genetics to National Security." In Phillip Sloan, ed., *Controlling Our Destinies: Historical, Philosophical, Social, and Ethical Perspective on the Human Genome Project*. Notre Dame: University of Notre Dame Press, 1999.

Becker, M. "Function and Teleology." *Journal of the History of Biology* 2 (1969): 151–64.

Beljanski, Mirko, and Severo Ochoa. "Protein Biosynthesis by Cell-Free Bacterial System, I." *Proceedings of the National Academy of Sciences* 44 (1958): 498–501.

———. "Protein Biosynthesis by Cell-Free Bacterial System, II." *Proceedings of the National Academy of Sciences* 44 (1958): 1157–61.

Belozersky, A. N., and A. S. Spirin. "Correlation between the Composition of Deoxyribonucleic and Ribonucleic Acids." *Nature* 182 (1958): 111–12.

Ben-David, Joseph. *The Scientist's Role in Society: A Comparative Study*. Chicago: University of Chicago Press, 1971.

Bennet, S. *A History of Control Engineering, 1800–1930*. London: Institution of Electrical Engineers, 1979.

Benzer, Seymour. "The Elementary Units of Heredity." In William D. McElroy and Bentley Glass, eds., *The Chemical Basis of Heredity*, pp. 70–93. Baltimore: Johns Hopkins University Press, 1957.

Bernfield, Merton R., and Marshall Nirenberg. "RNA Codewords and Protein Synthesis. The Nucleotide Sequences of Multiple Codewords for Phenylalanine, Serine, Leucine, and Proline." *Science* 147 (1965): 479–84.

Biagioli, Mario. *Galileo, Courtier: The Practice of Science in the Culture of Absolutism*. Chicago: University of Chicago Press, 1993.

———. "Stress in the Book of Nature: Galileo's Realism and Its Supplements." Unpublished manuscript, 1996.

Black, Max. *Models and Metaphors*. Ithaca: Cornell University Press, 1962.

Blumenberg, Hans. *Die Lesbarkeit der Welt*. Frankfurt am Mein: Suhrkamp Verlag, 1981.

Bohr, Niels. "Quantum Physics and Biology." *Symposia of the Society for Experimental Biology* (1960): 5.

Bono, James J. "Locating Narratives: Science, Metaphor, Communities, and Epistemic Styles." In Peter Weingart, ed., *Grensuberschreitungen in der Wissenschaft: Crossing Boundaries in Science*, pp. 119–51. Baden-Baden: Nomos Verlagsgesellschaft, 1995.

———. "Science, Discourse, and Literature: The Role/Rule of Metaphor in Science." In Stuart Peterfreund, ed., *Literature and Science: Theory and Practice*, pp. 59–90. Boston: Northeastern University Press, 1990.

———. *The Word of God and the Languages of Man: Interpreting Nature in Early Modern Science and Medicine*. Vol. I. Madison: University of Wisconsin Press, 1995.

Borek, Ernest. *The Code of Life*. New York: Columbia University Press, 1965/69.

Bowker, Geoffrey. "The Age of Cybernetics or How Cybernetics Aged." Paper presented at the First Mellon Workshop on "Comparative Perspectives on the History and Social Studies of Modern Life Science," MIT, Cambridge, Mass., April 1991.

Boyd, Richard. "Metaphor and Theory Change: What Is a 'Metaphor' a Metaphor For?" In Andrew Ortony, ed., *Metaphor and Thought*, Chap. 21. Cambridge: Cambridge University Press, 1993.

Boyden, A. "Serology and Animal Systematics." *American Naturalist* 57 (1934): 234–49.

Boyer, Paul. *By the Bomb's Early Light: American Thought and Culture at the Dawn of the Atomic Age*. New York: Pantheon Books, 1985.

Brachet, Jean. "Recherches sur le synthèse de l'acide thymonuclèique pendant le developement de l'oeuf d'oursin." *Archive Biologie* 44 (1933): 519–76.

Bragdon, Douglas E., Olga Nalbandov, and James W. Osborne. "The Control of the Blood Sugar Level." In Hubert P. Yockey, ed., *Symposium on Information Theory in Biology*, pp. 191–214. New York: Pergamon Press, 1956.

Brain, Robert M. "Standards and Semiotics: The Laboratory in Modern French Linguistics." In Timothy Lenoir, ed., *Inscribing Science*. Stanford: Stanford University Press, 1998.

Branson, Herman R. "Information Theory and the Structure of Proteins." In *Symposium on Information Theory in Biology*, pp. 84–104. New York: Pergamon Press, 1956.

Brenner, Sydney. Interview by author. Cambridge, Eng., 9 July 1992.

———. "The Mechanism of Gene Action." In G. E. Wolstenholme and Cecilia M. O'Connor, eds., *CIBA Foundation Symposium on Biochemistry of Human Genetics*, pp. 304–28. Boston: Little, Brown, 1959.

———. "On the Impossibility of All Overlapping Triplet Codes in Information Transfer from Nucleic Acid to Proteins." *Proceedings of the National Academy of Sciences* 43 (1957): 687–94.

———. "RNA, Ribosomes, and Protein Synthesis." *Cold Spring Harbor Symposia on Quantitative Biology* 26 (1961): 101–10.

Brenner, Sydney, and Leslie Barnett. "Genetic and Chemical Studies on the Head Protein of Bacteriophages T_2 and T_4." *Brookhaven Symposia* 12 (1959): 86–94.

Brenner, Sydney, François Jacob, and Matthew Meselson. "Unstable Intermediate Carrying Information from Genes to Ribosomes for Protein Synthesis." *Nature* 190 (1961): 576–81.

Brillouin, Leon. "Maxwell's Demon Cannot Operate: Information and Entropy. I." *Journal of Applied Physics* 22, no. 3 (1951): 334–37.

———. *Science and Information Theory*. New York: Academic Press, 1956.

Brimacombe, R., J. Trupin, M. Nirenberg, P. Leder, M. Bernfield, and T. Jaouni. "RNA Codewords and Protein Synthesis, VIII. Nucleotide Sequences of Synonym Codons for Arginine, Valine, Cysteine, and Alanine." *Proceedings of the National Academy of Sciences* 54 (1965): 954–60.

Bud, Robert. "Bugs and Institutes." Paper presented at the University of Manchester, Eng., March 1992.

———. *The Uses of Life: A History of Biotechnology*. Cambridge: Cambridge University Press, 1993.

Burian, Richard M. "On the Cusp between Biochemistry and Molecular Biology: The Pyjama [or PaJaMo] Experiment." Unpublished paper, 1994.

———. "Technique, Task Definition, and the Transition from Genetics to Molecular Genetics: Aspects of the Work on Protein Synthesis in the Laboratories of J. Monod and P. Zamecnik." *Journal of the History of Biology* 26, no. 3 (1993): 387–407.

———. "Underappreciated Pathways toward Molecular Genetics as Illustrated by Jean Brachet's Chemical Embryology." In Sahotra Sarkar, ed., *The Philosophy and History of Molecular Biology: New Perspectives*, pp. 67–85. Dordrecht: Kluwer Publishers, 1996.

Burian, Richard M., and Jean Gayon. "Genetics after World War II: The Laboratories at Gif." *Cahiers pour l'Histoire du CNRS* 7 (1990): 25–43.

Burian, Richard, Jean Gayon, and Doris Zallen. "Boris Ephrussi and the Synthesis of Genetics and Embryology." In Scott Gilbert, ed., *Conceptual History of Embryology*, pp. 207–27. New York: Plenum Press, 1991.

———. "The Singular Fate of Genetics in the History of Biology, 1900–1940." *Journal of the History of Biology* 21 (1988): 357–402.

Burks, Arthur W., ed. *Theory of Self-Reproducing Automata.* Urbana: University of Illinois Press, 1966.

Burnet, F. Macfarlane. *Enzyme, Antigen, and Virus: A Study of Macromolecular Pattern in Action.* Cambridge: Cambridge University Press, 1956.

Burroughs, Mider G. "The Federal Impact on Biomedical Research." In John Z. Bowers and Elizabeth Purcell, eds., *Advances in American Medicine: Essays at the Bicentennial,* 2, pp. 806–71. New York: Josiah Macy, Jr., Foundation, 1976.

Bush, Vannevar. *Science: The Endless Frontier.* 1980. Reprint, Washington, D.C.: National Science Foundation, 1945.

Buxton, William. *Talcott Parsons and the Capitalist Nation-State: Political Sociology as Strategic Vocation.* Toronto: University of Toronto Press, 1985.

Cairns, John. "Foreword." *Cold Spring Harbor Symposia on Quantitative Biology* 31 (1966): v.

Cairns, John, Gunther S. Stent, and James D. Watson, eds. *Phage and the Origins of Molecular Biology.* Cold Spring Harbor: Cold Spring Harbor Laboratory of Quantitative Biology, 1966.

Caldwell, P. C., and Cyril N. Hinshelwood. "Some Considerations on Autosynthesis in Bacteria." *Journal of the Chemical Society* 4 (1950): 3156–59.

Campbell, David. *Writing Security: United States Foreign Policy and the Politics of Identity.* Minneapolis: University of Minnesota Press, 1992.

Campbell, Jeremy. *Grammatical Man: Information, Entropy, Language, and Life.* New York: Simon and Schuster, 1982.

Canguilhem, Georges. "Epistemology of Biology." In François Delaporte, *A Vital Rationalist: Selected Writings from Georges Canguilhem.* Trans. Arthur Goldhamer. New York: Zone Books, 1994.

———. "The Role of Analogies and Models in Biological Discovery." In A. C. Crombie, ed., *Historical Studies in the Intellectual, Social, and Technical Conditions for Scientific Discovery and Technical Invention from Antiquity to the Present*, pp. 507–20. London: Heinemann, 1962.

Cannon, Walter. *The Wisdom of the Body.* New York: W. W. Norton, 1932.

Carmichael, Virginia. *Framing History: The Rosenberg Story and the Cold War.* Minneapolis: University of Minnesota Press, 1993.

Carnap, Rudolph, and Yehoshua Bar-Hillel. *Technical Reports of the Research Laboratory of Electronics*, no. 247. Cambridge: MIT, 1952.

Caspersson, Torbjorn O. "Über den chemischen Aufbau der Strukturen des Zellkernes." *Acta Med. Skand.* 73, Suppl. 8 (1936): 1–151.

Caute, David. *The Great Fear: The Anti-Communist Purge under Truman and Eisenhower.* New York: Simon and Schuster, 1978.

Ceruzzi, Paul E. *Reckoners: The Prehistory of the Digital Computer, from Relays to the Stored Program Concept, 1935–1945.* Westport, Conn.: Greenwood Press, 1983.

Chamberlin, Edward J., and Sander L. Gilman, eds. *Degeneration: The Dark Side of Progress.* New York: Columbia University Press, 1985.

Chantrenne, H. "Information in Biology." *Nature* 197 (1963): 27–30.

Chargaff, Erwin. "Amphisbaena." *Essays on Nucleic Acids*. New York: Elsevier, 1963.

———. "Chemical Specificity of Nucleic Acids and Mechanism of Their Enzymatic Degradation." *Experientia* VI (1950): 201–40.

———. "The Chemistry and Function of Nucleoproteins and Nucleic Acids." 1955. Reprint in his *Essays on Nucleic Acids*, New York: Elsevier, 1963.

———. "A Few Remarks on Nucleic Acids, Decoding, and the Rest of the World." 1962. Reprint in his *Essays on Nucleic Acids*, New York: Elsevier, 1963.

———. "First Steps toward a Chemistry of Heredity." 1958. Reprint in his *Essays on Nucleic Acids*, New York: Elsevier, 1963.

———. *Heraclitean Fire: Sketches from a Life before Nature*. New York: Rockefeller University Press, 1978.

———. "On the Nucleoproteins and Nucleic Acids of Microorganisms." *Cold Spring Harbor Symposia on Quantitative Biology* XII (1947): 28–34.

———. "Some Recent Studies of the Composition and Structure of Nucleic Acids." *Journal of Cellular and Comparative Physiology* 38 (1951): 41–59.

Cherry, Colin. *On Human Communications*. Cambridge: MIT Press, 1957; and New York: John Wiley and Sons, 1957.

Cherry, Colin, Morris Halle, and Roman Jakobson. "Toward the Logical Description of Languages and Their Phonemic Aspect." *Language* 29 (1953): 34–46.

Chomsky, Noam. *The Logical Structure of Linguistic Theory*. New York: Plenum Press, 1975.

Clarke, Adele E., and Joan H. Fujimura. *The Right Tools for the Job: At Work in Twentieth-Century Life Science*. Princeton: Princeton University Press, 1992.

Cochrane, R. G. *The National Academy of Sciences*. Washington, D.C.: National Academy of Sciences, 1978.

Coghlan, Andy. "Survival of the Fittest Molecules." *New Scientist* 136, no. 1841 (1992): 37–40.

Cohen, Georges N. "Permeability as an Excuse to Write What I Feel." In Andre Lwoff and Agnes Ullmann, eds., *Origins of Molecular Biology: A Tribute to Jacques Monod*, pp. 89–94. New York: Academic Press, 1979.

Cohen, George N., and Jacques Monod. "Bacterial Permeases." *Bacteriological Reviews* 21 (1957): 169–94.

Cohen, I. Bernard. "The Computer: A Case Study of the Support by Government, Especially the Military, of a New Science and Technology." In Everett Mendelsohn, Merritt Roe Smith, and Peter Weingart, eds., *Science, Technology, and the Military*, pp. 119–54. Dordrecht: Kluwer Academic Publishers, 1988.

Cohn, Melvin. "In Memorium." In Andre Lwoff and Agnes Ullmann, eds., *Origins of Molecular Biology: A Tribute to Jacques Monod*, pp. 75–88. New York: Academic Press, 1979.

Cohn, Melvin, and Jacques Monod. "Adaptation in Microorganisms." *London Symposium, April 1953*, pp. 132–49. Cambridge: Cambridge University Press, 1953.

Cohn, M., J. Monod, M. R. Pollock, S. Spiegelman, and R. Y. Stanier. "Terminology of Enzyme Formation." *Nature* 172 (1953): 1096.

Cold Spring Harbor Symposia on Quantitative Biology 28 (1963).

Collado-Vides, Julio. "A Transformational-Grammar Approach to the Study of the Regulation of Gene Expression." *Journal of Theoretical Biology* 136 (1989): 403–25.

Comaroff, Jean, and John Comaroff. *Of Revelation and Revolution: Christianity, Colonialism, and Consciousness in South Africa*. Vol. I. Chicago: University of Chicago Press, 1991.

Compact Oxford English Dictionary. 2d ed. Oxford: Clarendon Press, 1994.

Condillac, Ettiene de. *Essai sur l'origine des connaissances humaines*. 1746. Reprint, in George le Roy, ed., *Oeuvres Philosophiques de Condillac*, Vol. I. Paris: Presses Universitaires de France, 1947.

———. *Logic*. Trans. W. R. Albury. New York: Abaris Books, 1979.

Cortada, James W. *The Computer in the United States: From Laboratory to Market, 1930–1960*. New York: M. E. Sharp, 1993.

Creager, Angela N. H. "Wendell Stanley and the Dream of a Free-Standing Biochemistry Department at the University of California, Berkeley." *Journal of the History of Biology* 29 (1996): 331–60.

Creager, Angela N. H., and Jean-Paul Gaudillière. "Meanings in Search for Experiments and Vice-Versa: The Invention of Allosteric Regulation in Paris and Berkeley, 1959–1968." *Historical Studies in the Physical and Biological Sciences* 27 (1996): 1–89.

Crick, Francis. "Codon-Anticodon Pairing. The Wobble Hypothesis." *Journal of Molecular Biology* 19 (1966): 548–55.

———. "The Genetic Code." In David Baltimore, ed., *Nobel Lectures in Molecular Biology, 1933–1975*, pp. 205–13. New York: Elsevier North-Holland, 1977.

———. "The Genetic Code." *Scientific American* 207 (1962): 66–75.

———. "The Genetic Code—Yesterday, Today, and Tomorrow." *Cold Spring Harbor Symposia on Quantitative Biology* 31 (1966): 3–9.

———. "On Degenerate Templates and the Adaptor Hypothesis." Unpublished note to the RNA Tie Club, n.d. but mid-Jan. 1955.

———. "On Protein Synthesis." In *Symposium of the Society for Experimental Biology, 12*, pp. 138–63. New York: Academic Press, 1958.

———. "The Present Position of the Coding Problem." *Brookhaven National Laboratory Symposia* June (1959): 35–39.

———. "The Recent Excitement in the Coding Problem." *Progress in Nucleic Acids Research* 1 (1963): 163–217.

———. "Sailing with Jacques." In Andre Lwoff and Agnes Ullmann, eds., *Origins of Molecular Biology: A Tribute to Jacques Monod*, pp. 225–30. New York: Academic Press, 1979.

———. "Towards the Genetic Code." *Discovery* 23, no. 3 (1962): 8–16.

———. *What Mad Pursuit: A Personal View of Scientific Discovery*. New York: Basic Books, 1988.

Crick, Francis H., Leslie Barnett, S. Brenner, and R. J. Watts-Tobin. "General Nature of the Genetic Code for Proteins." *Nature* 192 (1961): 1227–32.

Crick, Francis, John S. Griffith, and Leslie E. Orgel. "Codes Without Commas." *Proceedings of the National Academy of Sciences* 43 (1957): 416–21.
Crook, S., J. Pakulski, and M. Waters, eds. *Postmodernization: Change in Advanced Society*. London: Sage Publications, 1992.
Cross, Stephen J., and William R. Albury. "Walter B. Cannon, L. J. Henderson, and the Organic Analogy." *Osiris* 2d series, 3 (1987): 165–92.
Cuénot, Lucien. *Invention et Finalité en Biologie*. Paris: Flammarion, 1941.
Culler, Jonathan. *Ferdinand de Saussure*. Ithaca: Cornell University Press, 1976.
Curtis, H. J., ed. "Henry Quastler, 1908–1965." In *The Emergence of Biological Organization*. New Haven: Yale University Press, 1964.
Dancoff, Sydney M., and Henry Quastler. "The Information Content and Error Rate of Living Things." In Henry Quastler, ed., *Essays on the of Information Theory in Biology*, pp. 263–73. Urbana: University of Illinois Press, 1953.
Darlington, C. D. *The Evolution of Genetic Systems*. Cambridge: Cambridge University Press, 1939.
Daston, Lorraine. "Curiosity in Early Modern Science." *Word & Image* II, no. 4 (1995): 391–404.
Davern, C. I., and Matthew Meselson. "The Molecular Conservation of Ribonucleic Acid during Bacterial Growth." *Journal of Molecular Biology* 2 (1960): 153–60.
Davis, Bernard. "The Teleonomic Significance of Biosynthetic Control Mechanisms." *Cold Spring Harbor Symposia on Quantitative Biology* 26 (1961): 1–10.
de Chadarevian, Soraya. *The Making of a New Science: Molecular Biology in Britain, 1945–1975*. Cambridge: Cambridge University Press, forthcoming.
———. "Sequences, Conformation, Information: Biochemists, and Molecular Biologists in the 1950s." *Journal of the History of Biology* 29 (1996): 361–86.
Delaporte, François, ed. *A Vital Rationalist: Selected Writings from Georges Canguilhem*. Trans. Arthur Goldhamer. New York: Zone Books, 1994.
Delbrück, Max. "Aristotle-totle-totle." In Jacques Monod and E. Borek, eds., *Microbes and Life*, pp. 50–55. New York: Columbia University Press, 1971.
———. "Experiments with Bacterial Viruses (Bacteriophages)." *Harvey Lectures* 41 (1945–46): 161.
———. "A Theory of Autocatalytic Synthesis of Polypeptides and Its Application to the Problem of Chromosome Reproduction." *Cold Spring Harbor Symposia on Quantitative Biology* IX (1941): 122–26.
Delbrück, Max, and W. T. Bailey, Jr. "Induced Mutations in Bacterial Viruses." *Cold Spring Harbor Symposia on Quantitative Biology* XI (1946): 33–37.
Delbrück, Max, and Gunther Stent. "On the Mechanism of DNA Replication." In William D. McElroy and Bentley Glass, eds., *The Chemical Basis of Heredity*, pp. 699–736. Baltimore: Johns Hopkins University Press, 1957.
Demerec, Milislav. "Annual Report." *Carnegie Institution of Washington Yearbook, 1946–1947*, p. 127. Baltimore: Lord Baltimore Press, 1947.
D'Emilio, John. *Sexual Politics, Sexual Communities: The Making of a Homosexual Minority in the United States, 1940–1970*. Chicago: University of Chicago Press, 1983.

Dennis, Michael A. "A Change of State: The Political Cultures of Technical Practice at the MIT Instrumentation Laboratory and the Johns Hopkins University Applied Physics Laboratory, 1930–1945." Ph.D. diss., Department of the History of Science, Johns Hopkins University, 1991.
Derrida, Jacques. *Of Grammatology*. Trans. Gayatri Chakravorty Spivak. Baltimore: Johns Hopkins University Press, 1976.
———. *Writing and Difference*. Chicago: University of Chicago Press, 1978.
Deutsch, Karl W. *Nerves of Government*. New York: Free Press, 1967.
Dobzhansky, Theodosius. "The Code Was Broken." *New York Times*, 17 April 1966, Sec. VII, p. 3.
Dong, Shan, and David B. Searls. "Gene Structure Prediction by Linguistic Methods." *Genomics* 23 (1994): 540–51.
Dounce, Alexander L. "Duplicating Mechanism for Peptide Chain and Nucleic Acid Synthesis." *Enzymologia* 15 (1952): 251–58.
———. "Nucleic Acid Template Hypothesis." *Science* 172 (1952): 541.
———. "Nucleoproteins." *Journal of Cellular Comparative Physiology* 47 (1956): 103–6.
———. "Role of Nucleic Acid and Enzymes in Peptide Chain Synthesis." *Nature* 176 (1955): 597–98.
Doyle, Richard M. "Mr. Schrödinger Inside Himself." *Qui Parle* 5 (1992): 1–20.
———. "On Beyond Living: Rhetorics of Vitality and Post-Vitality in Molecular Biology." Ph.D. diss., Department of Rhetoric, University of California at Berkeley, 1993.
———. *On Beyond Living: Rhetorics of Vitality and Post-Vitality in Molecular Biology*. Stanford: Stanford University Press, 1998.
Dresher, Elan. "Recent Issues in Linguistics: Functional Categories in DNA." *Glot International* 1, no. 7 (1995): 8.
Dreyfus, Hubert, and Paul Rabinow. *Michel Foucault: Beyond Structuralism and Hermaneutics*. Chicago: University of Chicago Press, 1982.
Durkheim, Emil. *The Division of Labor in Society*. New York: The Free Press, 1964.
Duster, Troy. *Backdoor to Eugenics*. New York: Routledge, 1990.
Eagle, Harry. "Studies in Cell Biology, NIAID." In DeWitt Stetten, Jr., ed., *NIH: An Account of Research in Its Laboratories and Clinics*, pp. 99–107. Orlando, Fla.: Academic Press, 1984.
Eamon, William. *Science and the Secrets of Nature: Books of Secrets in Medieval and Early Modern Culture*. Princeton: Princeton University Press, 1994.
Eck, Richard V. "Genetic Code: Emergence of a Symmetrical Pattern." *Science* 140 (1963): 477–80.
———. "Non-Randomness in Amino-Acid 'Alleles.'" *Nature* 191 (1961): 1284–85.
———. "A Simplified Strategy for Sequence Analysis of Large Proteins." *Nature* 193 (1962): 241–43.
Edge, David. "Technological Metaphors and Social Control." *New Literary History* 6 (1974): 135–48.

Edsall, John D. "Wendell Meredith Stanley." *The American Philosophical Society Year Book* (1971): 184–90.

Edwards, Paul. *The Closed World: Computers and the Politics of Discourse in Cold War America.* Cambridge: MIT Press, 1996.

Eigen, Manfred. "Chemical Means of Information Storage and Readout in Biological Systems." *Neurosciences Research Program Bulletin* II (1964): 11–22.

———. "How Does Information Originate? Principles of Biological Self-Organization." *Advances in Chemical Physics* (ed. Ilya Prigogine) 38 (1978): 211–62.

———. Eigen, Manfred. Interviews by author. Göttingen, Ger., March–May 1992.

———. "The Origin of Biological Information." In Jagdish Mehra, ed., *The Physicist's Conception of Nature.* Dordrecht: D. Reidel Publishing Co., 1973.

———. "Selforganization of Matter and the Evolution of Biological Macromolecules." *Die Naturwissenschaften* 10 (1971): 465–523.

———. "Sprache und Lernen auf molekularer Ebene." In Anton Peisl and Armin Mohler, eds., *Der Mensch und seine Sprache*, pp. 181–218. Berlin: C. F. von Siemens Stiftung, 1979.

———, ed. *Selection—Natural and Unnatural in Biotechnology.* Report on the International Workshop, 18–20 April 1991, Max-Planck Institut für Biophysikalische Chemie, Göttingen, Ger.

Eigen, Manfred, and Leo C. M. De Maeyer. "Summary of Two NRP Work Sessions on Information Storage and Processing in Biomolecular Systems." *Neurosciences Research Program Bulletin* III (1965): 244–66.

Eigen, Manfred, and William Gardiner. "Evolutionary Molecular Engineering Based on RNA Replication." *Pure and Applied Chemistry* 56 (1984): 967–78.

Eigen, Manfred, and Peter Schuster. "The Hypercycle: A Principle of Natural Self-Organization. Part A." *Naturwissenschaften* 64 (1977): 541–65.

———. "The Hypercycle: A Principle of Natural Self-Organization. Part B." *Naturwissenchaften* 65 (1978): 7–41.

———. "The Hypercycle: A Principle of Natural Self-Organization. Part C." *Naturwissenschaften* 65 (1978): 341–69.

Eigen, Manfred, and Ruthilde Winkler. *Das Spiel.* Munich: R. Riper Verlag, 1975.

———. *Laws of the Game: How the Principles of Nature Govern Chance.* Trans. Robert and Rita Kimber. New York: Alfred A. Knopf, 1981.

Eisenstein, Elizabeth L. "The Advent of Printing and the Problem of the Renaissance." *Past and Present* 45 (1969): 19–89.

Elsasser, Walter M. *The Physical Foundation of Biology: An Analytical Study.* New York: Pergamon Press, 1958.

Elson, David, and Erwin Chargaff. "Regularities in the Composition of Pentose Nucleic Acid." *Nature* 173 (1954): 1037.

Ephrussi, B., U. Leopold, James D. Watson, and J. J. Weigle. "Terminology in Bacterial Genetics." *Nature* 171 (1953): 701.

Fagen, M. D., ed. *A History of Engineering and Science in the Bell System.* Murray Hill, N.J.: Bell Telephone Laboratories, 1978.

Falaschi, Arthuro, Julius Adler, and H. G. Khorana. "Chemically Synthesized De-

oxypolynucleotides as Templates for Ribonucleic Acid Polymerase." *Journal of Biological Chemistry* 238 (1963): 3080–85.
Fantini, Bernardino. "Jacques Monod et la Biologie Moléculaire." *La Recherche* 218 (February 1990): 180–87.
———. "Monod, Jacques Lucien." In Frederick L. Holmes, ed., *Dictionary of Scientific Biography*, Vol. 18, Suppl. II, pp. 636–49. New York: Charles Scribner, 1990.
———. "Utilisation par la Génétique Moléculaire du Vocabulaire de la Théorie de L'Information." In Martin Groult, ed., *Transfert De Vocabulaire Dans Les Sciences*, pp. 159–70. Paris: Editions du Centre National de la Recherche Scientifique, 1988.
———, ed. *Jacques Monod. Pour Une Éthique De La Connaissance*. Paris: Édition La Decouverte, 1988.
Farber, M. A. "Geneticist Looks at the Year 2000." *New York Times*, 13 February 1967, p. 35.
Fehr, Johannes. "Code-Transfer." *Annual Report of the Max-Planck Institute for the History of Science, Berlin* (1996): 92–102.
———. "Der 'Code" (in) der Sprachwissenschaft: Zur Geschichte eines Konzeots." Unpublished paper, ETH University of Zürich, 1998.
———. *Untersuchen über Kohlenhydraten und Fermente 1884–1908*. Berlin: Springer Verlag, 1909.
Figlio, Karl M. "The Metaphor of Organization: An Historiographical Perspective on the Bio-Medical Sciences of the Early Nineteenth Century." *History of Science* XIV (1976): 17–53.
Fischer, Emil. "Einflus der Konfiguration auf die Wirkung der Enzyme, I." *Berichte der Deutsch. chem. Gesellschaft* 27 (1894): 2985–93.
———. "Bedentung der Stereochemie für die Physiologie." *Zeitschrift für Physiologische Chemie* 26 (1898): 60–87.
———. *Sitzungsber. der Kgl. Preuss. Akad. der Wissenschaft.*
Fisher, Ronald A. *The Design of Experiments*. London: Oliver and Boyd, 1942.
Fleming, Donald. "Émigré Physicists and the Biological Revolution." *Perspectives in American History* 2 (1968): 152–89.
Florkin, Marcel. *Concepts of Molecular Biosemiotics and Molecular Evolution*. New York: Elsevier Scientific, 1974.
Forman, Paul. "Behind Quantum Electronics: National Security as the Basis for Physical Research in the United States, 1940–1969." *Historical Studies in the Physical and Biological Sciences* 18 (1987): 149–229.
———. Review of Sapolsky's *Science and the Navy: The History of the Office of Naval Research* (Princeton: Princeton University Press, 1990). In *IEEE Annals of the History of Computing* 14 (1992): 60–62.
Foucault, Michel. *The Archeology of Knowledge and the Discourse on Language*. New York: Pantheon Books, 1972.
———. *The History of Sexuality, Vol. I: An Introduction*. Trans. Robert Hurley. New York: Pantheon, 1978.

———. "'La Logique du Vivant' by François Jacob." *Le Monde*, 15 November 1970.
———. *The Order of Things: An Archeology of the Human Sciences.* 1966. Reprint, New York: Vintage Books, 1973.
———. "Politics and the Study of Discourse." In Graham Burchell, Colin Gordon, and Peter Miller, eds., *The Foucault Effect: Studies in Governmentality*, Chap. 2. Chicago: University of Chicago Press, 1991.
———. *Power/Knowledge: Selected Interviews and Other Writings, 1972–1977.* Ed. Colin Gordon. New York: Pantheon Books, 1980.
Fraenkel-Conrat, Heinz. Interview by author. Berkeley, Cal., 29 June 1994.
———. "Protein Chemists Encounter Viruses." *Annals of the New York Academy of Sciences* 325 (1979): 309–18.
———. "Rebuilding a Virus." *Scientific American* 194 (1956): 42–47.
———. "The Role of the Nucleic Acid in the Reconstitution of Active Tobacco Mosaic Virus." *Journal of the American Chemical Society* 78 (1956): 882–83.
———. "Synthetic Mutants." In Wendell M. Stanley and Evans G. Valens, eds., *Viruses and the Nature of Life*, pp. 191–204. New York: E. P. Dutton, 1961.
Fraenkel-Conrat, Heinz, and Bea Singer. "Virus Reconstitution, II. Combination of Protein and Nucleic Acid from Different Strains." *Biochimica et Biophysica Acta* 24 (1957): 540–48.
Fraenkel-Conrat, Heinz, and Robley C. Williams. "Reconstitution of Active Tobacco Mosaic Virus from Its Inactive Protein and Nucleic Acid Components." *Proceedings of the National Academy of Sciences* 41 (1955): 690–98.
Fredrickson, Donald S. "The National Institutes of Health Yesterday, Today, and Tomorrow." *Public Health Service* 93 (1978): 642–47.
Fruton, Joseph. "Early Theories of Protein Structure." *Annals of the New York Academy of Sciences* 325 (1979): 1–15.
———. *A Skeptical Biochemist.* Cambridge: Harvard University Press, 1992.
Fujimura, Joan. "The Molecular Bandwagon in Cancer Research: Where Social Worlds Meet." *Social Problems* 35 (1988): 261–83.
Fulbright, J. William. "The War and Its Effects: The Military-Industrial-Academic Complex." In Herbert I. Schiller, ed., *Super-State: Readings in the Military-Industrial Complex*, pp. 171–78. Urbana: University of Illinois Press, 1970.
Fussell, Paul. *The Great War and Modern Memory.* New York: Oxford University Press, 1975.
Galison, Peter. "Context and Constraints." In Jed Buchwald, ed., *Scientific Practice: Theories and Stories of Physics*, pp. 13–41. Chicago: University of Chicago Press, 1995.
———. *Image and Logic: The Material World of Microphysics.* Chicago: University of Chicago Press, 1997.
———. "The Ontology of the Enemy: Norbert Wiener and the Cybernetic Vision." *Critical Inquiry* 21 (1994): 228–65.
Gamow, George. "Information Transfer in the Living Cell." *Scientific American* 193 (1955): 70–78.

———. *Mr. Tompkins Learns the Facts of Life.* Cambridge: Cambridge University Press, 1953.
———. "Possible Mathematical Relation between Deoxyribonucleic Acid and Proteins." *Det Kongelige Danske Videnskabernes Selkab, Biologiske Meddelelsker* 22, no. 3 (1954): 1–13.
———. "Possible Relation between Deoxyribonucleic Acid and Protein Structure." *Nature* 173 (1954): 318.
———. "What Is Life?" *Transactions of the Bose Research Institute* 24 (1961): 185–92.
Gamow, George, and Nicholas Metropolis. "Numerology of Polypeptide Chains." *Science* 120 (1954): 779–80.
Gamow, George, Alexander Rich, and Martynas Yčas. "The Problem of Information Transfer from Nucleic Acids to Proteins." *Advances in Biological and Medical Physics,* 4, pp. 23–68. New York: Academic Press, 1956.
Gamow, George, and Martynas Yčas. "The Cryptographic Approach to the Problem of Protein Synthesis." In Hubert P. Yockey, ed., *Symposium on Information Theory in Biology*, pp. 63–69. New York: Pergamon Press, 1956.
———. "Statistical Correlation of Protein and Ribonucleic Acid Composition." *Proceedings of the National Academy of Sciences* 41 (1955): 1011–19.
Gardner, Howard. *The Mind's New Science: A History of the Cognitive Revolution.* New York: Basic Books, 1987.
Gardner, Robert S., Albert J. Wahba, Carlos Basilio, Robert S. Miller, Peter Lengyel, and Joseph Speyer. "Synthetic Polynucleotdies and the Amino Acid Code, VII." *Proceedings of the National Academy of Sciences* 48 (1962): 2087–94.
Gatlin, Lila L. *Information Theory and the Living System.* New York: Columbia University Press, 1972.
Gaudillière, Jean-Paul. "Biologie Moléculaire et Biologistes dans les Années Soixante: La Naissance d'une Discipline. Le cas Francais." Ph.D. diss., University of Paris, 1991.
———. "Circulating Mice and Viruses: The Jackson Memorial Laboratory, the National Cancer Institute, and the Genetics of Breast Cancer." In Michael Fortun and Everett Mendelsohn, eds., *The Practices of Human Genetics.* Sociology of the Sciences, Yearbook XXI, 1997/99. Dordrecht: Kluwer Academic Publishers, 1997 [1999].
———. "How Biochemical Regulation Held? The Practice and Rhetoric of the PaJaMa Experiment." Paper delivered at the 1995 meeting of the International Society for History, Philosophy, and Social Studies of Biology, Leuven, Belgium, mid-July 1995.
———. "J. Monod, S. Spiegelman et l'adaptation enzymatique. Programmes de recherche, cultures locales et traditions disciplinaires." *History and Philosophy of Life Science* 14 (1992): 23–71.
———. "Molecular Biologists, Biochemists and Messenger RNA: The Birth of a Scientific Network." *Journal of the History of Biology* 29 (1996): 417–45.
———. "Molecular Biology in the French Tradition? Redefining Local Traditions

and Disciplinary Patterns." *Journal of the History of Biology* 26 (1993): 474–98.

———. "Norms and Practices of Molecular Medicine: The Singular Fate of Cancer Viruses in Post War United States." Paper presented at the Fifth Mellon Workshop, "From Molecular Power to Biological Wisdom: Challenges and Needs in Historical and Social Studies of 20th-Century Life Sciences," MIT, Cambridge, Mass., May 1995.

———. "Oncogenes as Metaphors for Human Cancer: Articulating Laboratory Practices and Medical Demands." In Ilana Lowy, ed., *Medicine and Change: Historical and Sociological Aspects*, pp. 213–48. Paris: J. Libbey-Inserm Editions, 1992.

Geison, Gerald L. "The Protoplasmic Theory of Life and the Vitalist-Mechanist Debate." *Isis* 60 (1969): 273–92.

"Genetic Rosetta Stone." *Time*, 23 May 1960, p. 50.

Gerovitch, Slava. "Beyond the Rhetoric: The Construction of Soviet Cybernetics." Second-year paper, Program in Science, Technology, and Society, MIT, 1994.

———. "Speaking Cybernetically: The Soviet Remaking of an American Science." Ph.D. diss., MIT, 1999.

Gierer, Alfred, and Karl-Wolfgang Mundry. "Production of Mutants of Tobacco Mosaic Virus by Chemical Alteration of Its Ribonucleic Acid in Vitro." *Nature* 182 (1958): 1457–58.

Gilbert, Scott F., and Jason P. Greenberg. "Intellectual Traditions in the Life Sciences. II. Stereocomplementarity." *Perspectives in Biology and Medicine* 28, no. 1 (1984): 18–34.

Gilbert, Walter. "A Vision of the Grail." In Daniel J. Kevles and Leroy Hood, eds., *The Code of Codes: Scientific and Social Issues in the Human Genome Project*, p. 96. Cambridge: Harvard University Press, 1990.

Goldhaber, Maurice. Telephone conversation with author. Brookhaven National Laboratory, 15 and 17 November 1994.

Golomb, Solomon W. "Efficient Coding for the Desoxyribonucleic Channel." *Proceedings of Symposia in Applied Mathematics* 14 (1962): 87–100.

Golomb, Solomon W., Basil Gordon, and Lloyd R. Welch. "Comma-Free Codes." *Canadian Journal of Mathematics* 10 (1958): 202–9.

Golomb, Solomon W., Lloyd R. Welch, and Max Delbrück. "Construction and Properties of Comma-Free Codes." *Biologiske Meddelelsker Det Kongelige Danske Videnskabernes Selskab* 23 (1958): 1–34.

Gottweis, Herbert. *Governing Molecules*. Cambridge: MIT Press, 1998.

Grafe, E. H. "I Ching." *Zeitschrift fur Allgemeinmedizin-Der Landarzt* 5 and 16 (1969).

Gramsci, Antonio. *Selections from the Prison Notebooks*. Ed. Q. Hoare and G. Nowell Smith. New York: International, 1971.

Gray, George W. *Science at War*. New York: Harper, 1943.

Griffith, Robert, and Athan Theoharis, eds. *The Specter: Original Essays on the Cold War and the Origins of McCarthyism*. New York: New View Points, 1974.

Grmek, Mirko D., and Bernadino Fantini. "Le Rôle du Hasard dans la Naissance du Modèle de l'Opéron." *Revue d'histoire des Sciences* 35 (1982): 193–215.

Gros, François. "Code et Messenger." *Les Secrets Du Gène*. Paris: Editions Odile Jacob, 1986.

———. "The Messenger." In Andre Lwoff and Agnes Ullmann, eds., *Origins of Molecular Biology: A Tribute to Jacques Monod*, pp. 117–24. New York: Academic Press, 1979.

Gros, François, H. Hiatt, Walter Gilbert, C. G. Kurland, R. W. Risebrough, and James B. Watson. "Unstable Ribonucleic Acid Revealed by Pulse Labelling of Escherichia Coli." *Nature* 190 (1961): 581–85.

Grunberg-Manago, Marianne. "Enzymatic Synthesis and Breakdown of Polynucleotide Phosphorylase." *Journal of the American Chemical Society* 77 (1955): 3165–66.

Grunberg-Manago, Marianne, Priscilla J. Ortiz, and Severo Ochoa. "Enzymatic Synthesis of Polynucleotides: I. Polynucleotide Phosphorylase of *Azotobacter Vinelandii*." *Biochimica et Biophysica Acta* 20 (1956): 269–85.

Hacking, Ian. "Biopower and the Avalanche of Numbers." *Humanities in Society* V (1982): 279–95.

———. *Representing and Intervening: Introductory Topics in the Philosophy of Natural Science*. Cambridge: Cambridge University Press, 1983.

———. *The Taming of Chance*. Cambridge: Cambridge University Press, 1989.

———. "Weapons Research and the Form of Scientific Knowledge." *Canadian Journal of Philosophy*, Suppl. 12 (1986): 237–62.

Halberstam, David. *The Fifties*. New York: Ballantine Books, 1993.

Haldane, John B. S. *New Paths in Genetics*. London: George Allen and Unwin, 1941.

Hall, Benjamin D., and Sol Spiegelman. "Sequence Complementarity of T2 DNA and T2 Specific RNA." *Proceedings of the National Academy of Sciences, U.S.A.* 47 (1961): 137–46.

Haller, Mark H. *Eugenics: Hereditarian Attitudes in American Thought*. New Brunswick, N.J.: Rutgers University Press, 1963.

Haraway, Donna J. "The Biological Enterprise: Sex, Mind, and Profit from Human Engineering to Sociobiology." *Radical History Review* 29 (1979): 206–37.

———. "A Cyborg Manifesto: Science, Technology, and Socialist-Feminism in the Late Twentieth Century." In *Simians, Cyborgs, and Women: The Reinvention of Nature*. New York: Routledge, 1991.

———. "The High Cost of Information in Post–World War II Evolutionary Biology: Ergonomics, Semiotics, and the Sociobiology of Communication Systems." *The Philosophical Forum* XIII (winter-spring 1981–82): 244–78.

———. "A Pilot Plant for Human Engineering: Robert Yerkes and the Yale Laboratories of Primate Biology, 1924–1942." *Primate Visions: Gender, Race, and Nature in the World of Modern Science*, pp. 59–83. New York: Routledge, 1989.

———. "A Semiotics of the Naturalistic Field: From C. R. Carpenter to S. A. Altmann, 1930–55." In *Primate Visions: Gender, Race, and Nature in the World of Modern Science*. New York: Routledge, 1989.

———. "Signs of Dominance: From a Physiology to a Cybernetics of Primate Society, C. R. Carpenter, 1930–1970." *Studies in the History of Biology* 6 (1983): 129–219.

Harden, Victoria. *Inventing the NIH*. Baltimore: Johns Hopkins University Press, 1986.

———. "National Institutes of Health: Celebrating 100 Years of Medical Progress." In *Encyclopedia Britannica*, pp. 158–75. Chicago: Encyclopedia Britannica Inc., 1988.

Harris, J. I., and C. A. Knight. "Action of Carboxypeptidase on TMV." *Nature* 170 (1952): 613.

Hartley, Keith, ed. "The Romantic Spirit in German Art 1790–1990." Exhibit text. Scottish National Gallery of Modern Art, Edinburgh, summer 1994.

Hartley, R. V. "Transmission of Information." *Bell System Technical Journal* 7 (1928): 535–53.

Harvey, David. *The Condition of Postmodernity*. Oxford: Basil Blackwell, 1989.

Harwood, Jonathan. "National Styles in Science: Genetics in Germany and the United States Between the World Wars." *Isis* 78 (1987): 390–414.

———. *Styles of Scientific Thought: The German Genetics Community, 1900–1933*. Chicago: University of Chicago Press, 1993.

Hastings, Julius. Telephone conversation with author. Brookhaven National Laboratory, 15 and 17 November 1994.

Haurowitz, Felix. "Protein Synthesis and Immunochemistry." In Hubert P. Yockey, ed., *Symposium on Information Theory in Biology*, pp. 125–46. New York: Pergamon Press, 1956.

Hayles, N. Katherine. *Chaos Bound: Orderly Disorder in Contemporary Literature and Science*. Ithaca: Cornell University Press, 1990.

———. *How We Became Posthuman: Virtual Bodies in Cybernetics, Information, and Literature*. Chicago: University of Chicago Press, 1998.

Hazen, H. L. "Theory of Servo-mechanisms." *Journal of the Franklin Institute* 218 (1934): 279–303.

Heidegger, Martin. *The Question Concerning Technology and Other Essays*. New York: Harper and Row, 1977.

Heims, Steve J. *The Cybernetics Group*. Cambridge: MIT Press, 1991.

———. *John von Neumann and Norbert Wiener: From Mathematics to the Technologies of Life and Death*. Cambridge: MIT Press, 1980.

Heppel, L. A., P. J. Ortiz, and S. Ochoa. "Small Polyribonucleotides with 5′-Phosphomonoester End-Groups." *Science* 123 (1956): 415.

Herken, Gregg. *Cardinal Choices: Presidential Science Advising from the Atomic Bomb to SDI*. New York: Oxford University Press, 1992.

Hersh, R. T. "Mutants of TMV and the Commaless Code." *Journal of Theoretical Biology* 2 (1962): 326–28.

Hershey, Alfred D. "Spontaneous Mutations in Bacterial Viruses." *Cold Spring Harbor Symposia on Quantitative Biology* XI (1946): 66–77.

Hershey, Alfred D., June Dixon, and Martha Chase. "Nucleic Acid Economy in

Bacteria Infected with Bacteriophage T_2." *Journal of General Physiology* 36 (1953): 777–89.

Hesse, Mary. "The Explanatory Function of Metaphor." In Yehoshua Bar-Hillel, ed., *Logic, Methodology, and the Philosophy of Science*. Amsterdam: Elsevier North-Holland, 1965.

———. *Models and Analogies in Science*. Notre Dame: Notre Dame University Press, 1966.

———. *Revolutions, and Reconstructions in the Philosophy of Science*. Bloomington: Indiana University Press, 1980.

Hevly, Bruce W. "Basic Research within a Military Context: The Naval Research Laboratory and the Foundations of Extreme Ultraviolet and X-ray Astronomy, 1923–1960." Ph.D. diss., Department of the History of Science, Johns Hopkins University, 1987.

Hewlett, Richard G., and Oscar E. Anderson, Jr. *The New World, 1939/46*. Philadelphia: University of Pennsylvania Press, 1962.

Hewlett, Richard G., and Francis Duncan. *Atomic Shield, 1947/52. A History of the United States Atomic Energy Commission*. Vol. II. Philadelphia: University of Pennsylvania Press, 1969.

Hewlett, Richard G., and Jack M. Holl. *Atoms for Peace and War, 1953–1961: A History of the United States Atomic Energy Commission*. Vol. III. Berkeley and Los Angeles: University of California Press, 1989.

Hinshelwood, Cyril N. *Chemical Kinetics of the Bacterial Cell*. Oxford: Clarendon Press, 1946.

Hinsley, F. H., and Alan Stripp, eds. *Code Breakers: The Inside Story of Bletchley Park*. New York: Oxford University Press, 1993.

Holley, Robert W. "Alanine Transfer RNA." In David Baltimore, ed., *Nobel Lectures in Molecular Biology, 1933–1975*, pp. 285–300. New York: Elsevier North-Holland, 1977.

———. "Biography." In David Baltimore, ed., *Nobel Lectures in Molecular Biology, 1933–1975*, pp. 300–301. New York: Elsevier North-Holland, 1977.

———. "The Nucleotide Sequence of a Nucleic Acid." *Scientific American* 214 (1966): 31–39.

Holley, Robert W., Jean Apgar, George A. Everett, James T. Madison, Mark Marquisee, Susan H. Merrill, John Robert Penswick, and Ada Zamir. "Structure of Ribonucleic Acid." *Science* 147 (1965): 1462–65.

Holton, Gerald. "Ernst Mach and the Fortunes of Positivism in America." *Isis* 83 (1992): 27–60.

———. "The Joys and Sorrows of the Vienna Circle in Exile." Paper presented at the Boston University Colloquium for Philosophy of Science, Boston, Mass., December 1993.

Hughes, Thomas, and Agatha Hughes, eds. *Systems, Experts, and Computers*. Cambridge: MIT Press, 1999.

Huxley, Thomas. "On the Physical Basis of Life." 1864. Reprint, in *Collected Essays*, Vol. 1, New York: D. Appleton, 1894.

Ingram, Vernon M. "How Do Genes Act?" *Scientific American* 198 (1958): 68–74.
———. "A Specific Chemical Difference between the Globins of Normal Human and Sickle-Cell Anaemia Haemoglobin." *Nature* 178 (1956): 792–94.
Irwin, M. R. "Genes and Antigens." In Hubert P. Yockey, ed., *Symposium on Information Theory in Biology*, pp. 147–69. New York: Pergamon Press, 1956.
Irwin, M. R., and R. W. Cumley. "Immunogenetics Studies of Species Relationships." *American Naturalist* 57 (1934): 211–33.
Ivanoff, A. "Theoretical Foundation of Automatic Regulation of Temperature." *Journal of the Institute of Fuel* 7 (1934): 117–30.
Jackson, David. "Template for an Economic Revolution." In Donald Chambers, ed., *DNA, The Double Helix: Perspective and Prospective at Forty Years*, p. 358. New York: New York Academy of Science, 1995.
Jackson, Janet. "AT&T Bell Laboratories." In Fritz E. Froehlich and Allen Kent, eds., *The Froehlich/Kent Encyclopedia of Telecommunications*, Vol. 1, pp. 397–406. New York: Marcel Dekker, 1989.
Jacob, François. "Biography." In David Baltimore, ed., *Nobel Lectures in Molecular Biology, 1933–1975*, pp. 243–44. New York: Elsevier North-Holland, 1977.
———. "Genetic Control of Viral Functions." *Harvey Lectures, 1958–1959*. New York: Academic Press, 1960.
———. "Genetics of the Bacterial Cell." In David Baltimore, ed., *Nobel Lectures in Molecular Biology, 1933–1975*, pp. 219–21. New York: Elsevier North-Holland, 1977.
———. "Inaugural Lecture." Collège de France, Paris, 7 May 1965.
———. *The Logic of Life: A History of Heredity*. 1970. Reprint, trans. Betty E. Spillmann, New York: Pantheon Books, 1974/82.
———. "Le Modele Linguistique en Biologie." *Critique* 322 (1974): 197–205.
———. *The Statue Within: An Autobiography*. New York: Basic Books, 1988.
———. "The Switch." In Andre Lwoff and Agnes Ullmann, eds., *Origins of Molecular Biology: A Tribute to Jacques Monod*, pp. 96–97. New York: Academic Press, 1979.
———. "Transfer and Expresssion of Genetic Information in Escherichia Coli K12." *Experimental Cell Research*, Suppl. 6 (1958): 51–68.
Jacob, François, and Jacques Monod. "Elements of Regulatory Circuits in Bacteria." In R. J. C. Harris, ed., *Biological Organization at the Cellular and Supercellular Level*, pp. 1–23. New York: Academic Press, 1963.
———. "Gènes de structure et gènes de regulation dans la biosynthèse des protéines." *Comptes Rendus de l'Académie des Sciences* 249 (1959): 1282–84.
———. "Genetic Regulatory Mechanisms in the Synthesis of Proteins." *Journal of Molecular Biology* 3 (1961): 318–59.
Jacob, François, David Perrin, Carmen Sanchez, and Jacques Monod. "L'opéron: groupe de gènes à expression coordonnée par un opérateur." *Comptes Rendus de l'Académie des Sciences* 250 (1960): 1727–29.
Jacob, François, and Elie Wollman. "Genetic Aspects of Lysogeny." In William D.

McElroy and Bentley Glass, eds., *The Chemical Basis of Heredity*, pp. 468–98. Baltimore: Johns Hopkins University Press, 1957.

———. *Main Trends of Research in the Social and Human Sciences*. The Hague: Mouton/Unesco, 1979.

Jakobson, Roman, and Morris Halle. *Fundamentals of Language*. The Hague: Mouton, 1956.

Jakobson, Roman, C. F. Voegelin, and Thomas A. Sebeok. "Results of the Conference of Anthropologists and Linguists." *International Journal of American Linguistics* Memoir 8 (1953): Chaps. 1–2.

Jayne, E. T. "Note on Unique Decipherability." *IRE Transactions on Information Theory* IT-5 (1959): 98–102.

Jeffress, Lloyd, ed. *Cerebral Mechanisms in Behavior: The Hixon Symposium*. New York: Hafner Publishing Company, 1967.

Jones, Oliver W., Jr., and Marshall W. Nirenberg. "Qualitative Survey of RNA Codewords." *Proceedings of the National Academy of Sciences* 48 (1962): 2115–23.

Judson, Horace F. *The Eighth Day of Creation: The Makers of the Revolution in Biology*. New York: Simon and Schuster, 1979.

Kaempffert, Waldemar. "Reconstitution of Virus in Laboratory Reopens the Question: What Is Life?" *New York Times*, 30 October 1955, Sec. IV, p. E9.

Kahn, David. *The Code Breakers: The Story of Secret Writing*. New York: Macmillan, 1967.

Kalmus, H. "A Cybernetical Aspect of Genetics." *Journal of Heredity* 41 (1950): 19–22.

Kameyama, T., and G. D. Novelli. *Biochemical and Biophysical Research Communication* 2 (1959): 2240.

Kay, Lily E. "Conceptual Models and Analytical Tools: The Biology of Physicist Max Delbrück." *Journal of the History of Biology* 18 (1985): 207–46.

———. "The Intellectual Politics of Laboratory Technology: The Protein Network and the Tiselius Apparatus." In Svante Linquist, ed., *Historical Aspects of 20th-Century Swedish Physics*, pp. 398–423. Canton, Mass.: Science History Publications, 1993.

———. "Laboratory Technology and Biological Knowledge: The Tiselius Electrophoresis Apparatus, 1930–1945." *History and Philosophy of the Life Sciences* 10 (1988): 51–72.

———. "Life as Technology: Representing, Intervening, and Molecularizing." *Rivista di Storia della Scienza* Series II, I (1993): 85–103.

———. "Matter of Information: Changing Meanings of the Tobacco Mosaic Virus." Paper presented at the XIXth International Congress of History of Science, Zaragoza, Spain, August 1993.

———. "Molecular Biology and Pauling's Immunochemistry: A Neglected Dimension." *History and Philosophy of the Life Sciences* 11, no. 2 (1989): 51–72.

———. *The Molecular Vision of Life: Caltech, the Rockefeller Foundation and the Rise of the New Biology*. New York: Oxford University Press, 1993.

———. *Molecules, Cells, and Life: An Annotated Bibliography of Manuscript Sources on Physiology, Biochemistry, and Biophysics, 1900–1960, in the Library of the American Chemical Society*. Philadelphia: American Philosophical Society, 1989.

———. "The Politics of Fame: The Protein Network and the Tiselius Apparatus." Paper presented at the XIXth International Congress of History of Science, Zaragoza, Spain, August 1993.

———. "Problematizing Basic Research in Molecular Biology." In Arnold Thackray, ed., *Private Science: Biotechnology and the Rise of the Molecular Sciences*. Philadelphia: Chemical Heritage Foundation Penn Series, 1997.

———. "Rethinking Institutions: Philanthropy as an Historiographic Problem of Knowledge and Power." *Minerva* 35 (1997): 283–93.

———. "The Secret of Life: Niels Bohr's Influence on the Biology Program of Max Delbrück." *Rivista di Storia della Scienza* 2 (1985): 487–510.

———. "Selling Pure Science in Wartime: The Biochemical Genetics of G. W. Beadle." *Journal of the History of Biology* 22 (1989): 85–98.

———. "The Tiselius Electrophoresis Apparatus and the Life Sciences, 1930–1945." *History and Philosophy of the Life Sciences* 10 (1988): 51–72.

———. "Wendell Meredith Stanley." Unpublished contribution to the *Encyclopedia of American Science*, 1994.

———. "Who Wrote the Book of Life? Information and the Transformation of Molecular Biology." In Michael Hagner, Hans-Jörg Rheinberger, and Bettina Wahrig-Schmidt, eds., *Objekte, Differenzen und Konjunkturen: Experimentalsysteme im Historischen Kontext*, pp. 151–79. Berlin: Akademie Verlag, 1994.

———. "Who Wrote the Book of Life? Information and the Transformation of Molecular Biology, 1945–1955." *Science in Context* 8 (1995): 609–34.

———. "W. M. Stanley's Crystallization of the Tobacco Mosaic Virus, 1930–1940." *Isis* 77 (1986): 450–72.

Keller, Evelyn Fox. "Between Language and Science: The Question of Directed Mutation in Molecular Biology." *Perspectives in Biology and Medicine* 35 (1992): 292–305.

———. "The Body of a New Machine: Situating the Organism between Telegraphs and Computers." In her *Refiguring Life: Changing Metaphors in Twentieth-Century Biology*. New York: Columbia University Press, 1995.

———. "Critical Silences in Scientific Discourse: Problems of Form and Re-form." In her *Secrets of Life, Secrets of Death: Essays on Language, Gender, and Science*. New York: Routledge, 1992.

———. "Gender and Science." *Osiris* 10 (1995): 27–38.

———. "Molecules, Messages, and Memory: Life and the Second Law." In her *Refiguring Life: Changing Metaphors in Twentieth-Century Biology*. New York: Columbia University Press, 1995.

———. "Nature, Nurture, and the Human Genome Project." In Daniel J. Kevles and Leroy Hood, eds., *The Code of Codes: Scientific and Social Issues in the Human Genome Project*, pp. 281–99. Cambridge: Harvard University Press, 1992.

———. "Physics and the Emergence of Molecular Biology: A History of Cognitive and Political Synergy." *Journal of the History of Biology* 23 (1990): 389–410.

———. *Secrets of Life, Secrets of Death: Essays on Language, Gender, and Science.* New York: Routledge, 1992.

Kellogg, D. A., B. P. Doctor, J. F. Loebel, and M. W. Nirenberg. "RNA Codons and Protein Synthesis, IX. Synonym Codon Recognition by Multiple Species of Valine-, Alanine-, and Methionine-sRNA." *Proceedings of the National Academy of Sciences* 55 (1966): 912–19.

Kelly, Kevin. *Out of Control: The Rise of Neo-Biological Civilization.* Reading, Mass.: Addison-Wesley, 1992.

Kemeny, John G. "Man Viewed as Machine." *Scientific American* 196 (April 1955): 58–67.

Kepes, Gyorgy. *The New Landscape in Art and Science.* Chicago: P. Theobald Publishers, 1956.

Kevles, Daniel J. *In the Name of Eugenics: Genetics and the Uses of Human Heredity.* New York: Alfred A. Knopf, 1985.

———. "The National Science Foundation and the Debate over Postwar Research Policy, 1942–45." *Isis* 68 (1977): 5–26.

———. *The Physicists: A History of a Scientific Community in America.* New York: Alfred A. Knopf, 1979.

Khorana, Har Gobind. "Biography." In David Baltimore, ed., *Nobel Lectures in Molecular Biology, 1933–1975*, pp. 332–33. New York: Elsevier North-Holland, 1977.

———. "Nucleic Acid Synthesis in the Study of the Genetic Code." In David Baltimore, ed., *Nobel Lectures in Molecular Biology, 1933–1975*, pp. 306–7. New York: Elsevier North-Holland, 1977.

———. "Polynucleotide Synthesis and the Genetic Code." *Federation Proceedings* 24 (1965): 1473–87.

———. *Some Recent Developments in the Chemistry of Phosphate Esters of Biological Interest.* New York: John Wiley and Sons, 1961.

Kittler, Friedrich A. *Discourse Networks 1800/1900.* Trans. Michael Metteer. Stanford: Stanford University Press, 1990.

Knight, C. Arthur. "The Nature of Some of the Chemical Differences among Strains of Tobacco Mosaic Virus." *Journal of Biological Chemistry* 171 (1947): 297–309.

Kohler, Robert E. "The Enzyme Theory and the Origins of Biochemistry." *Isis* 64 (1973): 181–96.

———. *Lords of the Fly: Drosophila Genetics and the Experimental Life.* Chicago: University of Chicago Press, 1994.

———. "The Management of Science: The Experience of Warren Weaver and the Rockefeller Foundation Programme in Molecular Biology." *Minerva* 14 (1976): 249–93.

———. *Partners in Science: Foundations and Natural Scientists, 1900–1945.* Chicago: University of Chicago Press, 1991.

———. "Systems of Production: Drosophila, Neurospora, and Biochemical Genetics." *Historical Studies in the Biological and Physical Sciences* 22 (1991): 87–130.

Kornberg, Arthur. *For the Love of Enzymes: The Odyssey of a Biochemist.* Cambridge: Harvard University Press, 1989.

Krankeit, Eugene P. Telephone conversation with author. Brookhaven National Laboratory, 15 and 17 November 1994.

Kuppers, Bernd-Olaf. "The Context-Dependence of Biological Information." In K. Kornwachs and K. Jacoby, eds., *Information: New Questions to a Multidisciplinary Concept.* Berlin: Akademie Verlag, 1995.

———. "Der semantische Aspekt von Information und seine evolutionsbiologische Bedeutung." *Nova Acta Leopoldina* 72, no. 294 (1996): 195–219.

———. *Information and the Origin of Life.* Cambridge: MIT Press, 1990.

Kusch, Martin. *Foucault's Strata and Fields: An Investigation into Archeological and Genealogical Science Studies.* Boston: Kluwer Academic, 1991.

Kuznick, Peter J. *Beyond the Laboratory: Scientists as Political Activists in the 1930s.* Chicago: University of Chicago Press, 1987.

———. "The Ethical and Political Crisis of Science: The AAAS Confronts the War in Vietnam." Paper presented at the Annual Meeting of the History of Science Society, New Orleans, La., October 1994.

Laclau, Ernesto, and Chantal Mouffe. *Hegemony and Socialist Strategy.* New York: Verso Press, 1985.

Lakoff, George. *Women, Fire, and Dangerous Things: What Categories Reveal about the Mind.* Chicago: University of Chicago Press, 1987.

Lakoff, George, and Mark Johnson. *Metaphors We Live By.* Chicago: University of Chicago Press, 1980.

Lamborg, Marvin R., and Paul C. Zamecnik. "Amino Acid Incorporation into Protein by Extracts of E. Coli." *Biochemica et Biophysica Acta* 42 (1960): 206–11.

Landecker, Hannah. "Molecular Memory as Allegory at the Inception of the Neurosciences." Second-year paper, Program in Science, Technology, and Society, MIT, 1996.

Landsteiner, Karl. *The Specificity of Serological Reactions.* 2d ed. Cambridge: Harvard University Press, 1945.

Lani, Frank. "The Biological Coding Problem." *Advances in Genetics* 12 (1964): 2–141.

Lanquette, William, with Bela Szilard. *Genius in the Shadows: A Biography of Leo Szilard, The Man behind the Bomb.* New York: Macmillan, 1992.

Laurence, William L. "Biochemists Wary on Life's Secrets." *New York Times,* 13 March 1962, p. 23.

———. "Structure of Life." *New York Times,* 14 January 1962, Sec. IV, p. 7.

Layzer, David. *Cosmogenesis: The Growth of Order in the Universe.* New York: Oxford University Press, 1990.

Lears, T. J. Jackson. "The Concept of Cultural Hegemony: Problems and Possibilities." *American Historical Review* 9 (1985): 567–93.

Leder, Philip, and Marshall Nirenberg. "RNA Codewords and Protein Synthesis, II. Nucleotide Sequences of a Valine RNA Codeword." *Proceedings of the National Academy of Sciences* 52 (1964): 420–27.

———. "Biographical Statement." Department of Genetics, Harvard Medical School, Cambridge, Mass.

———. "RNA Codewords and Protein Synthesis, III. On the Nucleotide Sequence of a Cysteine and a Leucine RNA Codeword." *Proceedings of the National Academy of Sciences* 52 (1964): 1521–29.

Lederberg, Joshua. "Biological Future of Man." In Gordon Wolstenholme, ed., *Man and His Future*, pp. 264–65. Boston: Little, Brown, 1963.

———. "Comments on the Gene-Enzyme Relationship." In Oliver H. Gaebler, ed., *Enzymes: Units of Biological Structure and Function*, pp. 161–69. New York: Academic Press, 1956.

———. "Gene Control of β-Galactosidase in E. coli." *Genetics* 33 (1948): 617–18.

———. "Genetic Recombination in Bacteria: A Discovery Account." *Annual Reviews of Genetics* 21 (1987): 23–46.

———. "Genetic Studies of Bacteria." In L. C. Dunn, ed., *Genetics in the Twentieth Century*, pp. 281–92. New York: Macmillan, 1951.

———. "Infection and Heredity." In *Cellular Mechanisms in Differentiation and Growth*, pp. 101–24. Princeton: Princeton University Press, 1956.

———. "The Transformation of Genetics by DNA: An Anniversary Celebration of Avery, MacLeod and McCarty (1944)." *Genetics* 136 (1994): 423–26.

———. "A View of Genetics." In David Baltimore, ed., *Nobel Lectures in Molecular Biology*, pp. 81–106. New York: Elsevier North-Holland, 1977.

Lederberg, Joshua, and E. L. Tatum. "Novel Genotypes in Mixed Cultures of Biochemical Mutants of Bacteria." *Cold Spring Harbor Symposia on Quantitative Biology* XI (1946): 139–55.

Ledley, Robert S. "Digital Computational Methods in Symbolic Logic, with Examples in Biochemistry." *Proceedings of the National Academy of Sciences* 41 (1955): 498–511.

Lee, Ki Yong, R. Wahl, and E. Barbu. "Contenu en Basses Puriques et Pyrimidiques des Acides Desoxyribonucleiques des Bacteries." *Annales de L'Institut Pasteur* 91 (1956): 212–24.

Leek, John M. "Biographies of 3 Nobel Laureates." *New York Times*, 17 October 1968, pp. 1, 42, col. 3.

———. "Biologists Hopeful of Solving Secrets of Heredity This Year." *New York Times*, 2 February 1962, pp. 1, 14.

———. "Code of Genetics Proves Stubborn." *New York Times*, 9 September 1962, Sec. IV, p. 11.

———. "The Code of Life." *New York Times*, 28 January 1962, Sec. IV, p. 8.

———. "Gain Is Reported in Heredity Study." *New York Times*, 21 December 1961, p. 18.

———. "Gains in Genetics." *New York Times*, 8 September 1963, Sec. IV, p. E9.

———. "The Genetic Code Held Universal." *New York Times*, 12 October 1962, p. 33.

———. "Genetic Language Called Universal." *New York Times*, 22 February 1967, p. 31.

———. "Geneticists Meet to Review Gains." *New York Times*, 2 September 1963, p. 17.

———. "Geneticist Predicts Man Will Manipulate Heredity." *New York Times*, 12 August 1967, p. 14.

———. "Hereditary Control by Man Is Foreseen." *New York Times*, 20 October 1963, p. 44.

———. "New Gains Cited on Genetic Code." *New York Times*, 24 January 1962, p. 35.

———. "New Model Given for Genetic Code." *New York Times*, 5 May 1963, p. 56.

———. "Probing Heredity's Secrets." *New York Times*, 12 September 1963, p. 36.

Leff, Harvey S., and Andrew F. Rex. "Maxwell Demon: Entropy Historian." Paper presented at the Dibner Institute Workshop, "The Meaning and Use of Entropy," MIT, Cambridge, Mass., 15–16 April 1994.

Leffler, Melvyn P. "The American Concept of National Security and the Beginning of the Cold War, 1945–1948." *American Historical Review* 89 (1984): 346–81.

———. *A Preponderance of Power: National Security, The Truman Administration, and the Cold War*. Stanford: Stanford University Press, 1992.

Lengyel, Peter, Joseph F. Speyer, Carlos Basilio, and Severo Ochoa. "Synthetic Polynucleotides and the Amino Acid Code, III." *Proceedings of the National Academy of Sciences* 48 (1962): 282–84.

Lengyel, Peter, Joseph F. Speyer, and Severo Ochoa. "Synthetic Polynucleotides and the Amino Acid Code." *Proceedings of the National Academy of Science* 47 (1961): 1936–42.

Lenoir, Timothy. "The Discipline of Nature and the Nature of Disciplines." In idem, *Instituting Science*. Stanford: Stanford University Press, 1997.

Leslie, Stuart W. *The Cold War and American Science: The Military-Industrial-Academic Complex at MIT and Stanford*. New York: Columbia University Press, 1993.

———. "Science and Politics in Cold War America." In Margaret Jacob, ed., *The Politics of Western Science, 1640–1990*, pp. 200–33. New York: Humanities Press, 1994.

Lévi-Strauss, Claude. *Structural Anthropology*. 1958. Reprint, New York: Basic Books, 1963.

Lillie, Frank R. "Studies of Fertilization. VI. The Mechanism of Fertilization in *Arbacia*." *Journal of Experimental Zoology* 9 (1914): 523–90.

Linschitz, Henry. "The Information Content of a Bacterial Cell." In Hubert P. Yockey, ed., *Symposium on Information Theory in Biology*, pp. 251–62. New York: Pergamon Press, 1956.

———. Interview with author. Waltham, Mass., 16 July 1993.

Lipsitz, George. *Class and Culture in Cold War America: "A Rainbow at Midnight."* New York: Praeger, 1981.

Litman, Rose M., and Arthur B. Pardee. "Production of Bacteriophage Mutants by a Disturbance of Deoxyribonucleic Acid Metabolism." *Nature* 179 (1956): pp. 529–31.

Liversidge, Anthony. "Profile of Claude Shannon." In N. Sloane and A. Wyner, eds., *Claude Elwood Shannon: Collected Papers*, pp. xix–xxxiii. Piscataway, N.J.: IEEE Press, 1993.

Loeb, Jacques. *The Organism as a Whole*. New York: G. B. Putnam, 1916.

Lucretius. *De Rerum Natura*. Baltimore: Johns Hopkins University Press, 1993.

Ludmerer, Kenneth L. *Eugenics and American Society: A Historical Survey*. Baltimore: Johns Hopkins University Press, 1972.

Luhmann, Niklas. "The Cognitive Program of Constructivism and a Reality that Remains Unknown." In Wolfgang Krohn, Gunther Kuppers, and Helga Nowotny, eds., *Self-Organization: Portrait of a Scientific Revolution*, pp. 30–52. Dordrecht: Kluwer Publishers, 1991.

Luria, Salvador, and Max Delbrück. "Mutations of Bacteria from Virus Sensitivity to Virus Resistance." *Genetics* 28 (1943): 491–511.

Lwoff, Andre. "Jacques Lucien Monod." In Andre Lwoff and Agnes Ullmann, eds., *Origins of Molecular Biology: A Tribute to Jacques Monod*, pp. 1–24. New York: Academic Press, 1979.

———. "The Prophage and I." In John Cairns, Gunther S. Stent, and James D. Watson, eds., *Phage and the Origins of Molecular Biology*, pp. 88–99. Cold Spring Harbor: Cold Spring Harbor Laboratory of Quantitative Biology, 1966.

Lwoff, Andre, and Agnes Ullmann, eds. *Origins of Molecular Biology: A Tribute to Jacques Monod*. New York: Academic Press, 1979.

Lyapunov, A. A., ed. *Problems of Cybernetics*. New York: Pergamon, 1960–65.

Lyotard, Jean-Françoise. *The Postmodern Condition: A Report on Knowledge*. Minneapolis: University of Minnesota Press, 1984.

Maas, Werner K., "The Regulation of Arginine Biosynthesis: Its Contribution to Understanding the Control of Gene Expression." *Genetics* 128 (1991): 489–94.

MacKay, Donald M. "The Epistemological Problem of Automata." In John McCarthy and Claude Shannon, eds., *Automata Studies*, pp. 235–53. Princeton: Princeton University Press, 1956.

———. *Information, Mechanisms and Meaning*. Cambridge: MIT Press, 1969.

MacKenzie, Donald. *Inventing Accuracy: A Historical Sociology of Nuclear Missiles Guidance*. Cambridge: MIT Press, 1990.

Manning, Kenneth. *Black Apollo in Science*. New York: Oxford Univ. Press, 1983.

Marshall, Richard E., C. Thomas Caskey, and Marshall Nirenberg. "Fine-Structure of RNA Codewords Recognized by Bacterial, Amphibian, and Mammalian Transfer RNA." *Science* 155 (1967): 820–26.

Martin, Henri-Jean. *The History and Power of Writing*. Trans. Lydia G. Cochrane. Chicago: University of Chicago Press, 1994.

Martin, Julian. "Why Manuscripts Matter: Reception and Mutable Mobiles in 17th-Century England." Paper presented at Harvard University, Cambridge, Mass., February 1998.

Martin, Robert G. "A Revisionist View of the Genetic Code." In DeWitt Stetten,

Jr., ed., *NIH: An Account of Research in Its Laboratories and Clinics*, p. 283. New York: Academic Press, 1984.

Martin, Robert G., and Bruce N. Ames. "A Method for Sedimentation Behavior of Enzymes: Application to Protein Mixtures." *Journal of Biological Chemistry* 236 (1961): 1372–79.

Martin, Robert G., J. Heinrich Matthaei, Oliver W. Jones, and Marshall Nirenberg. "Ribonucleotide Composition of the Genetic Code." *Biochemical and Biophysical Research Communications* 6 (1961/62): 410–14.

Marx, Leo. *The Machine in the Garden: Technology and the Pastoral Ideal in America*. New York: Oxford University Press, 1964.

Masters, M., and P. Broda. "New Biology." *Nature* 232 (1971): 137–40.

Matthaei, Heinrich. Interview with author. Göttingen, Ger., 3 March 1992.

———. "Vergleichende Untersuchungen Des Eiweiss-Haushalts Beim Streckungswachstum Von Bluttenblattern Und Anderen Organen." *Planta* 48 (1958): 468–522.

Matthaei, J. Heinrich, Oliver W. Jones, Robert G. Martin, and Marshall Nirenberg. "Characteristics and Composition of RNA Coding Unit." *Proceedings of the National Academy of Sciences* 48 (1962): 667–77.

Matthaei, Heinrich, and Marshall W. Nirenberg. "Characterization and Stabilization of DNAase-Sensitive Protein Synthesis in E. Coli Extracts." *Proceedings of the National Academy of Sciences, U.S.A.* 47 (1961): 1580–88.

———. "The Dependence of Cell-Free Protein Synthesis in *E. Coli* upon RNA Prepared from Ribosomes." *Biochemical and Biophysical Research Communications* 4 (1961): 404–8.

———. "Some Characteristics of a Cell-Free DNAase Sensitive System Incorporating Amino Acids into Protein." *Federation Proceedings* 29 (1961): 391.

Maturana, Humberto R., and Francisco J. Varela. *The Tree of Knowledge: The Biological Roots of Human Understanding*. Boston: Shambhala Press, 1992.

Maxwell, James C. "On Governors." *Proceedings of the Royal Society of London* no. 100 (1868): 105–20.

May, Larry, ed. *Recasting America: Culture and Politics in the Age of the Cold War*. Chicago: University of Chicago Press, 1989.

Mayr, Ernst. *The Growth of Biological Thought: Diversity, Evolution, and Inheritance*. Cambridge, Mass.: Belknap Press, 1982.

———. "Teleological and Teleonomic: New Analysis." *Boston Studies in Philosophy of Science* 14 (1974): 91–117.

Mayr, Otto. *The Origins of Feedback Control*. Cambridge: MIT Press, 1970.

Mazumdar, Pauline H. M. "The Antigen-Antibody Reaction and the Physics and Chemistry of Life." *Bulletin of the History of Medicine* 48 (1974): 1–21.

———. "Karl Landsteiner and the Problem of Species, 1838–1968." Ph.D. diss., Department of the History of Science, Johns Hopkins University, 1976.

———. *Species and Specificity: An Interpretation of the History of Immunology*. Cambridge: Cambridge University Press, 1995.

McCarty, Maclyn. *The Transforming Principle*. New York: W. W. Norton, 1985.

McCormick, Thomas J. *America's Half-Century: United States Foreign Policy in the Cold War*. Baltimore: Johns Hopkins University Press, 1989.
McCulloch, Warren S. "Why the Mind Is in the Head." In Lloyd Jeffress, ed., *Cerebral Mechanisms in Behavior*, pp. 42–57. New York: Hafner, 1951.
McCulloch, Warren S., and Walter Pitts. "A Logical Calculus of the Ideas Immanent in Nervous Activity." *Bulletin of Mathematical Biophysics* 5 (1943): 115–33.
McDougall, Walter. *The Heavens and the Earth: A Political History of the Space Age*. New York: Basic Books, 1985.
McElheny, Victor K. "France Considers Significance of Nobel Awards." *Science* 150 (1965): 1013–15.
———. "Pasteur Institute Scientists Demand Sweeping Reform." *Science* 151 (1966): 809.
———. "Research in Biology: New Pattern of Support Is Developing." *Science* 145 (1963): 908–12.
McElroy, William D., and Bentley Glass, eds. *The Chemical Basis of Heredity*. Baltimore: Johns Hopkins University Press, 1957.
McLuhan, Marshall. *Understanding Media*. New York: McGraw-Hill, 1965.
McMillan, Brockway. "Two Inequalities Implied by Unique Decipherability." *IRE Transactions on Information Theory* 2 (1956): 115–16.
Medawar, Peter B. *The Art of the Solvable*. London: Methuen, 1967.
Medvedev, Z. A. "A Hypothesis Concerning the Way of Coding Interaction Between Transfer RNA and Messenger RNA at the Later Stages of Protein Synthesis." *Nature* 195 (1962): 39.
Melman, Seymour. *Pentagon Capitalism: The Political Economy of War*. New York: McGraw-Hill, 1971.
Mendelsohn, Everett, Merritt Roe Smith, and Peter Weingart, eds. *Science, Technology, and the Military*. Vols. I and II. Dordrecht: Kluwer Academic Publishers, 1988.
Merchant, Carolyn. *The Death of Nature*. San Francisco: Harper and Row, 1980.
Millman, S., ed. *A History of Engineering and Science in the Bell System*. Murray Hill, N.J.: AT&T Laboratories, 1984.
Mindell, David A. "'Datum for Its Own Annihilation': Feedback, Control, and Computing, 1916–1945." Ph.D. diss., Program in Science, Technology, and Society, MIT, 1996.
Mirowski, Philip. "What Were von Neumann and Mogenstern Trying to Accomplish?" In Roy Weintraub, ed., *Towards a History of Game Theory* 24, Suppl. 11 (1992): 111–47.
———. "When Games Grow Deadly Serious: The Military Influence on the Evolution of Game Theory." In Crawfurd Goodwin, ed., *Annual Supplement to Vol. 23, History of Political Economy* (1991): 227–55.
Mirsky, Alfred E., and Linus Pauling. "On the Structure of Native, Denatured, and Coagulated Proteins." *Proceedings of the National Academy of Science* 22 (1936): 439–47.
Mitman, Gregg. *The State of Nature: Ecology, Community, and American Social Thought, 1900–1950*. Chicago: University of Chicago Press, 1992.

Monod, Jacques. *Chance and Necessity: An Essay on the Natural Philosophy of Modern Biology*. Trans. Austryn Wainhouse. New York: Vintage Books, 1972.
———. "Foreword." In Bernard Feld and Gertrud Weiss Szilard, eds., *The Collected Works of Leo Szilard: Scientific Papers*, pp. vi–viii. Cambridge: MIT Press, 1972.
———. "From Enzymatic Adaptation to Allosteric Transition." In David Baltimore, ed., *Nobel Lectures in Molecular Biology, 1933–1975*, pp. 259–84. New York: Elsevier North-Holland, 1977.
———. "Information, Induction, Répression dans la Biosynthèse d'un Enzyme." *Colloquium der Gesellschaft für Physiologische Chemie* (April 1959): 120–45.
———. "An Outline of Enzyme Induction." *Recueil des Travaus Chimiques des Pays-Bas* 77 (1958): 569–85.
———. "The Phenomenon of Enzymatic Adaptation and Its Bearings on Problems of Genetics and Cellular Differentiation." *Growth* 2 (1947): 223–89.
———. "The Phenomenon of Enzymatic Adaptation and Its Bearings on Problems of Genetics and Cellular Differentiation." *Growth Symposium* XI (1947): 68–289.
———. "Remarks on the Mechanism of Enzyme Induction." In Oliver H. Gaebler, ed., *Enzymes: Units of Biological Structure and Function*, pp. 7–28. New York: Academic Press, 1956.
———. "La Technique de Culture Continue. Théorie et Applications." *Annales de l'Institut Pasteur* 79 (1950): 390–412.
Monod, Jacques, and François Jacob. "Teleonomic Mechanism in Cellular Metabolism, Growth, and Differentiation." *Cold Spring Harbor Symposia on Quantitative Biology* 26 (1961): 389–401.
Moore, E. F. "Artificial Living Plants." *Scientific American* 195, no. 4 (1956): 118–20.
Moore, L., and W. H. Stein. "Procedures for the Chromatographic Determination of Amino Acids on Four Percent Cross-Linked Sulfonated Polystyrene Resins." *Journal of Biological Chemistry* 211 (1954): 893–906.
Moore, Walter. *Schrödinger: Life and Thought*. New York: Cambridge University Press, 1989.
Morange, Michel. *Histoire de la Biologie Moléculaire*. Paris: Editions La Decouverte, 1994.
———. *A History of Molecular Biology*. Trans. Matthew Cobb. Cambridge: Harvard University Press, 1998.
———. "L'oeuvre Scientific de J. Monod." *Fundamenta Scientae* 3 (1982): 396.
Morgan, Thomas H. *The Physical Basis of Heredity*. Philadelphia: J. B. Lippincott, 1919.
———. *The Theory of the Gene*. New Haven: Yale University Press, 1926.
Moulin, Anne Marie. *Le Dernier Langage de la Medecine: Histoire de l'immunologie de Pasteur au Sida*. Paris: Press de Universitaires de France, 1991.
———. "Text and Context in Biology: In Pursuit of the Chimera." *Poetics Today* 9 (1988): 145–61.
Mullan, Fitzhugh. *Plagues and Politics: The Story of the United States Public Health Service*. New York: Basic Books, 1989.

Müller-Sievers, Helmut. *Self-Generation: Biology, Philosophy, and Literature around 1800*. Stanford: Stanford University Press, 1997.

Nagel, E. *The Structure of Science*. New York: Harcourt, 1961.

Naono, S., and François Gros. "Synthèse par E. Coli d'une Phosphatase Modifiée en Presence d'un Analogue Pyrimidique." *Comptes Rendes de l'academie de Sciences* 250 (1960): 3889.

Narita, K. "Isolation of Acetylpeptide from Enzymic Digests of TMV-Protein." *Biochimia et Biophysica Acta* 28 (1958): 184–91.

National Science Foundation. *Federal Funds for Science*. Washington, D.C.: United States Government Printing Office, n.d.

Neel, James V. "Inheritance of Sickle-Cell Anemia." *Science* 110 (1949): 64–66.

Nelkin, Dorothy, and M. Susan Lindee. *The DNA Mystique: The Gene as a Cultural Icon*. New York: W. H. Freeman and Co., 1995.

Nelson, Brice. "Research Probe: Rickover Broadside." *Science* 161 (1968): 446–48.

Nirenberg, Marshall. "Biography." In David Baltimore, ed., *Nobel Lectures in Molecular Biology, 1933–1975*, pp. 359–60. New York: Elsevier North-Holland, 1977.

———. "The Genetic Code." In David Baltimore, ed., *Nobel Lectures in Molecular Biology, 1933–1975*, pp. 335–58. New York: Elsevier North-Holland, 1977.

———. "The Genetic Code: II." *Scientific American* 208 (1963): 80–94.

———. "The Induction of Two Enzymes by One Inducer: A Test Case for Shared Genetic Information." *Federation Proceedings* 19 (1960): 42.

———. Interview with author. Bethesda, Md., 18 July 1994; 18 November 1995; 19 July 1996; and 4 October 1996.

———. "Will Society Be Prepared?" *Science* 157 (1967): 633.

Nirenberg, M., C. T. Caskey, R. Marshall, R. Brimacombe, D. Kellog, B. Doctor, D. Hatfield, J. Levin, F. Rotman, S. Pestka, M. Wilcox, and F. Anderson. "The RNA Code and Protein Synthesis." *Cold Spring Harbor Symposia on Quantitative Biology* 31 (1966): 11–24.

Nirenberg, Marshall W., and William B. Jacoby. "Constraints in the Determination of Active-Center Topography." *Nature* 188 (1960): 747–48.

———. "Enzymatic Utilization of β-hydroxybutyric acid." *Journal of Biological Chemistry* 235 (1960): 954–60.

———. "On the Sites of Attachment and Reaction of Aldehyde Dehydogenases." *Proceedings of the National Academy of Science, USA* 46 (1960): 206–12.

Nirenberg, M. W., O. W. Jones, P. Leder, B. F. C. Clark, W. S. Sly, and S. Pestka. "Cell-Free Peptide Synthesis Dependent Upon Synthetic Oligodeoxynucleotides." *Proceedings of the National Academy of Sciences* 50 (1963): 1135–43.

———. "On the Genetic Code." *Cold Spring Harbor Symposia on Quantitative Biology* 28 (1963): 549–57.

Nirenberg, Marshall, and Philip Leder. "RNA Codewords and Protein Synthesis. The Effect of Trinucleotides upon the Binding of sRNA to Ribosomes." *Science* 145 (1964): 1399–1407.

Nirenberg, Marshall, Philip Leder, M. Bernfield, R. Brimacombe, J. Trupin, F. Rottman, and C. O'Neal. "RNA Codewords and Protein Synthesis, VII. On the

General Nature of the RNA Code." *Proceedings of the National Academy of Sciences* 53 (1965): 1161–68.

Nirenberg, Marshall W., and Heinrich Matthaei. "The Dependence of Cell-Free Protein Synthesis in E. Coli upon Naturally Occurring or Synthetic Template RNA." *Proceedings of the Fifth International Congress of Biochemistry* 1 (1961): 184–89.

———. "The Dependence of Cell-Free Protein Synthesis in E. Coli upon Naturally Occurring or Synthetic Polyribonucleotides." *Proceedings of the National Academy of Sciences* 47 (1961): 1588–1602.

Nirenberg, Marshall W., J. Heinrich Matthaei, and Oliver W. Jones. "An Intermediate in the Biosynthesis of Polyphenylalanine Directed by Synthetic Template RNA." *Proceedings of the National Academy of Sciences* 48 (1962): 104–9.

Nisman, B., and H. Fukuhara. "Incorporation des aminés et synthèse de la β-galactosidase par les fraction enzymatique de Escherichia coli." *Comptes Rendes de l'academie de Sciences* (1959): 2240–42.

Niu, C.-I., and Heinz Fraenkel-Conrat. "C-Terminal Amino Acid Sequence of Tobacco Mosaic Virus Protein." *Biochimia et Biophysica Acta* 16 (1955): 597–98.

"Nobel Winner Monod Criticizes French 'Scientific Backwardness.'" *Washington Post*, 23 November 1965, p. 14.

Noble, David. "Command Performance: A Perspective on Military Enterprise and Technological Change." In Merritt R. Smith, ed., *The Military Enterprise: Perspectives on the American Experience*, Chap. 8. Cambridge: MIT Press, 1985.

———. *Forces of Production: A Social History of Industrial Automation*. New York: Oxford University Press, 1984.

Nomura, Masayasu, Benjamin D. Hall, and Sol Spiegelman. "Characterization of RNA Synthesized in E. coli after Bacteriophage T2 Infection." *Journal of Molecular Biology* 2 (1960): 306–26.

Novick, Aaron. "Introductory Essay." In Bernard Feld and Gertrud Weiss Szilard, eds., *The Collected Works of Leo Szilard: Scientific Papers*, pp. 389–92. Cambridge: MIT Press, 1972.

———. "Phenotypic Mixing." In John Cairns, Gunther Stent, and James B. Watson, eds., *Phage and the Origins of Molecular Biology*, pp. 133–41. Cold Spring Harbor: Cold Spring Harbor Laboratory of Quantitative Biology, 1966.

Novick, Aaron, and Leo Szilard. "Experiments with the Chemostat on Spontaneous Mutations in Bacteria." *Proceedings of the National Academy of Sciences* (U.S.A.) 36 (1950): 706–19.

———. "II. Experiments with the Chemostat on the Rates of Amino Acid Synthesis in Bacteria." In Edgar J. Boell, ed., *Dynamics of Growth Processes*, pp. 21–32. Princeton: Princeton University Press, 1954.

Novick, Peter. *That Noble Dream: The "Objectivity Question" and the American Historical Profession*. Pt. III. Cambridge: Cambridge University Press, 1988.

Nuttall, George H. F. *Blood Immunity and Blood Relationship*. Cambridge: Cambridge University Press, 1904.

Nyquist, Harry. "Certain Factors Affecting Telegraphy Speed." *Bell System Technical Journal* 3 (1924): 324–26.

Ochoa, Severo. "Biography." In David Baltimore, ed., *Nobel Lectures in Molecular Biology, 1933–1975*, pp. 125–26. New York: Elsevier North-Holland, 1977.
———. "Enzymatic Mechanism in the Transmission of Genetic Information." In Michael Kasha and Bernard Pullman, eds., *Horizons in Biochemistry*, pp. 158–66. New York: Academic Press, 1962.
———. "Enzymatic Synthesis of Ribonucleic Acid." In David Baltimore, ed., *Nobel Lectures in Molecular Biology, 1933–1975*, pp. 107–25. New York: Elsevier North-Holland, 1977.
———. "The Pursuit of a Hobby." *Ann. Rev. Biochem.* 49 (1989): 1–30.
O'Connor, Basil. "Where Genetics May Lead." *New York Times*, 20 September 1963, p. 32.
Olby, Robert C. "The Impact of Molecular Biology upon Neurobiology: Memory Molecules." *Journal for the History of Biology*, in press.
———. *The Path to the Double Helix*. London: Macmillan, 1974.
———. "The Protein Version of the Central Dogma." *Genetics* 79 (1975): 3–27.
———. "The Recasting of the Sciences: The Case of Molecular Biology." In G. Battimelli, M. de Maria, and A. Rossi, eds., *La Ristrutturazione della Scienze tra le Due Guerre Mondiale*, pp. 275–308. Rome: La Giolardica Editrice Universitaria di Roma, 1986.
———. "Schrödinger's Problem: What Is Life?" *Journal of the History of Biology* 4 (1971): 119–48.
Orgel, Leslie. Interview with author. La Jolla, Cal., 8 July 1994.
Osmundsen, John A. "Breaking the Code." *New York Times*, 2 August 1964, p. E7.
———. "New Way to Read Life's Code Found." *New York Times*, 7 April 1961, p. 12.
———. "Scientists Find Clue to Heredity's Code." *New York Times*, 16 May 1960, p. 1.
O'Toole, G. J. A. *Honorable Treachery*. New York: Atlantic Monthly Press, 1991.
Owens, Larry. "Mathematicians at War: Warren Weaver and the Applied Mathematics Panel, 1942–1945." In Howe and Mclearty, eds., *The History of Modern Mathematics, Vol. II*, pp. 287–305. Boston: Academic Press, 1988.
Oyama, Susan. *The Ontogeny of Information: Developmental Systems and Evolution*. Cambridge: Cambridge University Press, 1985.
Pardee, Arthur B. "The PaJaMa Experiment." In Andre Lwoff and Agnes Ullman, eds., *Origins of Molecular Biology: A Tribute to Jacques Monod*, pp. 108–10. New York: Academic Press, 1979.
Pardee, Arthur B., F. Jacob, and Jacques Monod. "The Genetic Control and Cytoplasmic Expression of 'Inducibility' in the Synthesis of β-galactosidase by Escherichia coli." *Journal of Molecular Biology* 1 (1959): 165–78.
———. "Sur l'expression et le rôle des allèles 'inductible' et 'constituitive' dans la synthèse de la β-galactosidase ches des zygotes d'Escherichia coli." *Comptes Rendus de l'Académie des Sciences* 246 (1958): 3125–28.
"Passports and Visas." *Science* 116 (1952): 178–79.
Paul, Diane. "The Rockefeller Foundation and Origins of Behavior Genetics." In Keith R. Benson, Jane Maienschein, and Ronald Rainger, eds., *The Expansion*

of American Biology, pp. 262–83. New Brunswick, N.J.: Rutgers University Press, 1991.

Pauling, Linus. "Antibodies and Specific Biological Forces." *Endeavour* VII, no. 26 (1948): 52–53.

———. "Reflections on the New Biology." *UCLA Law Review* 15 (1968): 269.

———. "A Theory of the Structure and Process of Formation of Antibodies." *Journal of the American Chemical Society* 62 (1940): 2643–57.

Pauling, Linus, and Max Delbrück. "The Nature of the Intermolecular Forces Operative in Biological Process." *Science* 92 (1940): 77–79.

Pauling, Linus, H. A. Itano, S. J. Singer, and I. C. Wells. "Sickle-Cell Anemia, a Molecular Disease." *Science* 110 (1949): 543–48.

Pauly, Philip J. *Controlling Life: Jacques Loeb and the Engineering Ideal in Biology*. New York: Oxford University Press, 1987.

———. "Modernist Practice in American Biology." In Dorothy Ross, ed., *Modernist Impulses in the Human Sciences*, Chap. 12. Baltimore: Johns Hopkins University Press, 1994.

PBS's *Nova*, "Decoding the Book of Life," 1 November 1989.

Penrose, Lionel S. "Mechanics of Self-Reproduction." *Annals of Human Genetics* 23 (1958–59): 59–72.

———. "Self-Reproducing Machines." *Scientific American* 200 (1959): 105–17.

Pestka, Sidney, Richard Marshall, and Marshall Nirenberg. "RNA Codewords and Protein Synthesis, V. Effects of Streptomycin on the Formation of Ribosome-sRNA Complexes." *Proceedings of the National Academy of Sciences* 53 (1965): 639–46.

Pestka, Sidney, and Marshall Nirenberg. "Regulatory Mechanisms and Protein Synthesis, X. Codon Recognition on 30s Ribosomes." *Journal of Molecular Biology* 21 (1966): 145–71.

Pestre, Dominique. "Science and the Military in France after WWII: A Chronological Overview and a First Interpretation." Paper presented at Harvard University, Cambridge, Mass., 21 January 1998.

Piaget, Jean. *Structuralism*. New York: Basic Books, 1970.

Pickens, Donald K. *Eugenics and the Progressives*. Nashville, Tenn.: Vanderbilt University Press, 1968.

Pickering, Andrew. "Cyborg History and the WWII Regime." *Perspectives on Science* 3 (1995): 1–48.

Pittendridgh, Colin S. "Adaptation, Natural Selection, and Behavior." In Gaylord G. Simpson and A. Roe, eds., *Behavior and Evolution*. New Haven: Yale University Press, 1958.

Plato. *Timaeus*. Trans. Francis M. Cornford. New York: Macmillan, 1959.

Platt, John R. "A 'Book Model' of Genetic Information—Transfer in Cells and Tissues." In Michael Kasha and Bernard Pullman, eds., *Horizons in Biochemistry*, pp. 167–87. New York: Academic Press, 1962.

Pollack, Robert. *Signs of Life: The Language and Meaning of DNA*. Boston: Houghton Mifflin, 1994.

Pollock, Martin R. "An Exciting but Exasperating Personality." In Andre Lwoff

and Agnes Ullmann, eds., *Origins of Molecular Biology: A Tribute to Jacques Monod*, pp. 61–74. New York: Academic Press, 1979.
———. "From Pangens to Polynucleotides: The Evolution of Ideas on the Mechanism of Biological Replication." *Perspectives in Biology and Medicine* 19, no. 4 (1976): 455–73.
Poster, Mark. *The Mode of Information: Poststructuralism and Social Context*. Chicago: University of Chicago Press, 1990.
Pratt, Fletcher. *Secret and Urgent: The Story of Codes and Ciphers*. New York: Blue Ribbon Books, 1942.
"Psycholinguistics: A Survey of Theory and Research Problems." Supplement to Pt. 2 of *Journal of Abnormal and Social Psychology* 49, no. 4 (1954).
Pursell, Carrol W., Jr. "Research in the United States: A Historical Perspective." In National Science Foundation, *Science at the Bicentennial: A Report from the Research Community*. Washington, D.C.: Government Printing Office, 1976.
Pynchon, Thomas. *Gravity's Rainbow*. New York: Collins Viking Press, 1973.
Quastler, Henry. *The Emergence of Biological Organization*. New Haven: Yale University Press, 1964.
———. "Feedback Mechanisms in Cellular Biology." In Heinz von Foerster, ed., *Cybernetics, 9th Conference*, pp. 167–81. New York: Josiah Macy Foundation, 1953.
———. "The Measure of Specificity." In Hubert P. Yockey, ed., *Symposium on Information Theory in Biology*, pp. 41–74. New York: Pergamon Press, 1956.
———. "A Primer on Information Theory." In Hubert P. Yockey, ed., *Symposium on Information Theory in Biology*, pp. 3–49. New York: Pergamon Press, 1956.
———. "The Specificity of Elementary Biological Functions." In Hubert P. Yockey, ed., *Symposium on Information Theory in Biology*, pp. 170–90. New York: Pergamon Press, 1956.
———. "The Status of Information Theory in Biology." In Hubert P. Yockey, ed., *Symposium on Information Theory in Biology*, pp. 399–402. New York: Pergamon Press, 1956.
———, ed. *Information Theory in Psychology*. Glencoe, Ill.: Free Press, 1955.
Rader, Karen. "Making Mice: The Standardization of *Mus Musculus* for American Biological Research, 1910–1965." Ph.D. diss., Department of History and Philosophy of Science, Indiana University, 1995. (Also forthcoming, Princeton University Press.)
Radical History Review 63, Special Issue (fall 1995).
Ramsey, Norman F. "Early History of Associated Universities and Brookhaven National Laboratory." *Brookhaven Lecture Series* 55 (1966): 1–16.
Rapoport, Anatol, and Mechthilde Knoller, trans. "On the Decrease of Entropy in a Thermodynamic System by the Intervention of Intelligent Beings." *Behavioral Sciences* 9, no. 4 (1964): 301–10. (English translation of Szilard, 1929.)
Rasch, William, and Cary Wolfe, eds. "Special Issue: The Politics of Systems and Environments, Part I." *Cultural Critique* 30 (1995).

Ratner, Vadim A. "The Genetic Language." In Robert Rosen, ed., *Progress in Theoretical Biology*, pp. 143–228. New York: Academic Press, 1974.

Raven, P. *Oogenesis: The Storage of Developmental Information.* New York: Pergamon Press, 1961.

Reddy, Michael J. "The Conduit Metaphor: A Case of Frame Conflict in Our Language about Language." In Andrew Ortony, ed., *Metaphor and Thought*, Chap. 10. Cambridge: Cambridge University Press, 1993.

Rees, Mina. "The Computing Program of the Office of Naval Research, 1946–1953." *Annals of the History of Computing* 4 (1982): 102–20.

Reichert, E. T., and A. P. Brown. "The Differentiation and Species Specificity of Corresponding Proteins and Other Vital Substances in Relation to Biological Classification and Organic Evolution." *Carnegie Institution Publication* no. 116 (Washington, D.C., 1909).

Reingold, Nathan. "Science and Government in the United States Since 1945." *History of Science* 32 (1994): 361–86.

———. "Vannevar Bush's New Deal for Research: Or the Triumph of the Old Order." *Historical Studies in the Physical and Biological Sciences* 17 (1987): 299–344.

Rheinberger, Hans-Jörg. "Experiment, Difference, and Writing: I. Tracing Protein Synthesis." *Studies in the History and Philosophy of Science* 23 (1991): 305–31.

———. "Experiment, Difference, and Writing: II. The Laboratory Production of Transfer RNA." *Studies in the History and Philosophy of Science* 23 (1991): 389–422.

———. *Experiment, Differenz, Schrift: Zur Geschichte Epistemischer Dinge.* Marburg: Basiliskenpresse, 1992.

———. "Experiment and Orientation: Early Systems of In Vitro Protein Synthesis." *Journal of the History of Biology* 26 (1993): 441–71.

———. "From Microsomes to Ribosomes: 'Strategies' of 'Representation.'" *Journal of the History of Biology* 28 (1995): 49–89.

———. "Genetic Engineering and the Practice of Molecular Biology." Paper presented at the Fourth Mellon Workshop, "Genetic Engineering: Transformation in Science, Politics, and Culture," MIT, Cambridge, Mass., May 1993.

———. *Toward a History of Epistemic Things: Synthesizing Proteins in the Test Tube.* Stanford: Stanford University Press, 1997.

Richelson, Jeffrey. *The U.S. Intelligence Community.* 2d ed. Cambridge, Mass.: Ballinger, 1989.

Riley, Monica, Arthur B. Pardee, François Jacob, and Jacques Monod. "On the Expression of a Structural Gène." *Journal of Molecular Biology* 2 (1960): 216–25.

Roberts, Robert B. "Alternative Codes and Templates." *Proceedings of the National Academy of Sciences* 48 (1962): 897–900.

Roch, Axel. "Mendels Message: Genetik und Informations Theorie." In Erika Keil and Verner Oeder, eds., *Versuchskaninchen: Bilder und andere Manipulationen.* Zurich: Museum für Gestaltung, 1995.

Rose, Frank. *Into the Heart of the Mind.* New York: Harper and Row, 1984.

Rosenberg, Charles. *No Other Gods*. Baltimore: Johns Hopkins University Press, 1961/78.

Rosenblueth, Arturo, Norbert Wiener, and Julian Bigelow. "Behavior, Purpose and Teleology." *Philosophy of Science* 10 (1943): 18–24.

Rosenheim, Shawn James. *The Cryptographic Imagination: Secret Writings from Edgar Poe to the Internet*. Baltimore: Johns Hopkins University Press, 1997.

Rottman, Fritz, and Marshall Nirenberg. "RNA Codons and Protein Synthesis, XI. Template Activity of Modified RNA Codons." *Journal of Molecular Biology* 21 (1966): 555–70.

Rouse, Joseph. "Foucault and the Natural Sciences." In John Caputo and Mark Yount, eds., *Foucault and the Critique of Institutions*, pp. 137–64. University Park: Pennsylvania State University Press, 1993.

Rowe, Mona S., ed. *The First Forty Years, 1947–1987*. Upton, N.Y.: Brookhaven National Laboratory, 1987.

Rudy, Stephen. "Roman Jakobson: A Brief Chronology." In Howard Gardner, ed., *The Mind's New Science: A History of the Cognitive Revolution*, pp. 196–205. New York: Basic Books, 1987.

Rychlik, I., and F. Šorm. "Replacements of Amino Acids in Proteins and Ribonucleic Acid Coding." *Collection of Czechoslovakian Chemical Communications* 27 (1962): 2686–91.

Sambrook, J. F., D. P. Fan, and S. Brenner. "A Strong Suppressor Specific for UGA." *Nature* 214 (1967): 452–53.

Sanger, Frederick, and E. O. P. Thompson. "The Amino-Acid Sequence in the Glycyl Chain of Insulin." Pts. I and II. *Biochemical Journal* 53 (1953): 353–66; 366–74.

Sapolsky, Harvey M. *The Polaris System Development: Bureaucratic and Programmatic Success in Government*. Cambridge: Harvard University Press, 1972.

———. *Science and the Navy: The History of the Office of Naval Research*. Princeton: Princeton University Press, 1990.

Sapp, Jan. *Beyond the Gene: Cytoplasmic Inheritance and the Struggle for Authority in Genetics*. New York: Oxford University Press, 1987.

———. *Where the Truth Lies: Franz Moewus and the Origins of Molecular Biology*. New York: Cambridge University Press, 1990.

Sarabhai, A. S., A. O. W. Stretton, and S. Brenner. "Co-Linearity of the Gene with the Polypeptide Chain." *Nature* 201 (1964): 13–17.

Sarkar, Sahotra. "Biological Information: A Skeptical Look at Some Central Dogmas of Molecular Biology." *The Philosophy and History of Molecular Biology: New Perspectives, Boston Studies in the Philosophy of Science* 183 (1996): 187–233.

———. "The Boundless Ocean of Unlimited Possibilities: Logic in Carnap's Logical Syntax of Language." *Synthese* 93 (1992): 191–237.

———. "Reductionism and Molecular Biology: A Reappraisal." Ph.D. diss., Department of Philosophy, University of Chicago, 1989.

Saumjan, Sebastian Konstantinovic. "La Cybernétique et la Langue." In *Collection Diogène*, pp. 137–52. Paris: Gallimard, 1965.

Schaffner, Kenneth. "Logic of Discovery and Justification in Regulatory Genetics." *Studies in History and Philosophy of Science* 4 (1974): 349–85.

Scheffler, I. "Thoughts on Teleology." *British Journal of Philosophy of Science* 9 (1959): 265–84.

Schmeck, Harold M., Jr. "Mutation Agent Held Clue to Life." *New York Times*, 26 January 1960, p. 30.

Schmitt, Francis O., ed. *Macromolecular Specificity and Biological Memory*. Cambridge: MIT Press, 1962.

Schönberger, Martin. *The I Ching and the Genetic Code: The Hidden Key to Life*. Santa Fe, N.M.: Aurora Press, 1992.

Schrödinger, Erwin. *What Is Life?* Cambridge: Cambridge University Press, 1944.

Schultz, Jack. "Aspects of the Relation between Genes and Development in *Drosophila*." *American Naturalist* 69 (1935): 30–31.

Schuster, Von Heinz, and Gerhard Schramm. "Stimmung der biologisch wirksamen Einheit in der Ribosenucleinsaure des Tabakmosaikvirus auf chemischen Wege." *Zeitschrigt für Naturforschung* 13b (1958): 697–704.

Schwartz, Drew. "Coding Problem in Proteins." *Nature* 181 (1958): 769.

———. "Speculations on Gene Action and Protein Specificity." *Proceedings of the National Academy of Sciences* 41 (1955): 300–307.

"Scientists in the News." *Science* 119 (1954): 540.

See, Richard. "Mechanical Translation and Related Research." *Science* 144 (1964): 621–32.

Semon, Richard. *Die Mneme als erhaltendes Prinzip*. Leipzig, Ger.: Engemann, 1904.

Serafini, Anthony. *Linus Pauling: The Man and His Science*. New York: Paragon House, 1989.

Sereno, Martin I. "DNA and Language: The Nature of the Symbolic-Representation System in Cellular Protein Synthesis and Human Language Comprehension." Ph.D. diss., Department of Philosophy, University of Chicago, 1984.

Shannon, Claude E. "Communication Theory of Secrecy Systems." *Bell System Technical Journal* 28, no. 4 (1949): 656–715.

———. "The Mathematical Theory of Communication." *Bell System Technical Journal* 27, nos. 3 and 4 (1948): 379–423; 623–56.

———. "Prediction and Entropy of Printed English." *Bell System Technical Journal* 30 (1951): 50–64.

———. "A Symbolic Analysis of Relay and Switching Circuits." *Transactions of the American Institute of Electrical Engineers* 57 (1938): 713–23.

Shannon, Claude E., and J. McCarthy, eds. *Automata Studies*. Princeton: Princeton University Press, 1956.

Shannon, Claude, and Warren Weaver. *The Mathematical Theory of Communication*. Urbana: University of Illinois Press, 1949.

Shannon, James A. "The Advancement of Medical Research: A Twenty-Year View of the Role of the National Institutes of Health." *Journal of Medical Education* 42 (1967): 97–108.

Sherrington, Sir Charles. *Man and His Nature: Gifford Lectures, 1937–38.* Cambridge: Cambridge University Press, 1940.
Sherry, Michael S. *In the Shadow of War: The United States Since the 1930s.* New Haven: Yale University Press, 1995.
———. *Planning for the Next War: American Plans for Postwar Defense.* New Haven: Yale University Press, 1977.
Sigurdsson, Skuli. "Physics, Life, and Contingency: Born, Schrödinger, and Weyl in Exile." In Mitchell G. Ash and Alfons Sollner, eds., *Forced Migration and Scientific Change: Emigre German-Speaking Scientists and Scholars after 1933,* pp. 48–71. Cambridge: Cambridge University Press, 1996.
Silverstein, Arthur M. *A History of Immunology.* San Diego, Cal.: Academic Press, 1989.
———. "History of Immunology." In W. E. Paul, ed., *Fundamentals of Immunology,* pp. 23–40. New York: Raven Press, 1984.
Simpson, Christopher. *Science of Coercion: Communication Research and Psychological Warfare, 1945–1960.* New York: Oxford University Press, 1994.
Singer, Maxine F. "1968 Nobel Laureate in Medicine or Physiology." *Science* 162 (1968): 433–36.
Singer, M. F., L. Heppel, and R. J. Hilmoe. "Oligonucleotides as Primers for Polynucleotide Phosphorylase." *Biochemica et Biophysica Acta* 26 (1957): 447–48.
———. "Oligonucleotides as Primers for Polynucleotide Phosphorylase." *Journal of Biological Chemistry* 235 (1960): 738–50.
Sinnott, E. W., and L. C. Dunn. *Principles of Genetics.* 3d ed. London: McGraw-Hill, 1939.
Sinsheimer, Robert L. "The Action of Pancreatic Desoxyribosnuclease." Pt. I. *Journal of Biological Chemistry* 208 (1954): 445–59.
———. "The Action of Pancreatic Desoxyribosnuclease." Pt. II. *Journal of Biological Chemistry* 215 (1955): 579–83.
———. *The Book of Life.* Reading, Mass.: Addison-Wesley, 1967.
———. "Is the Nucleic Acid Message in a Two-Symbol Code?" *Journal of Molecular Biology* 1 (1959): 218–20.
———. "The Prospect of Designed Genetic Change." *Engineering and Science* 32 (1969): 8–13.
———. *The Strands of Life: The Science of DNA and the Art of Education.* Berkeley and Los Angeles: University of California Press, 1994.
Smart, Barry. In David Couzens Hoy, ed., *Foucault: A Critical Reader,* pp. 157–71. Cambridge, Mass.: Basil Blackwell, 1986.
Smith, Merritt Roe, ed. *Military Enterprise and Technological Change: Perspectives on the American Experience.* Cambridge: MIT Press, 1985.
Sonneborn, Tracy. "Nucleotide Sequence of a Gene: First Complete Specification." *Science* 148 (1965): 1410.
———, ed. *The Control of Human Heredity and Evolution.* New York: Macmillan Press, 1965.
Spencer, Herbert. "The Social Organism." In *Essays. Scientific, Political, and Speculative,* Vol. I, pp. 265–307. New York: D. Appleton, 1892.

Speyer, Joseph F., Peter Lengyel, Carlos Basilio, and Severo Ochoa. "Synthetic Polynucleotides and the Amino Acid Code, II." *Proceedings of the National Academy of Sciences* 48 (1962): 63–68.
———. "Synthetic Polynucleotdies and the Amino Acid Code, IV." *Proceedings of the National Academy of Sciences* 48 (1962): 441–48.
Spiegelman, Sol. "On the Nature of the Enzyme-Forming System." In Oliver H. Gaebler, ed., *Enzymes: Units of Biological Structure and Function*, pp. 67–89. New York: Academic Press, 1956.
———. "The Relation of Informational RNA to DNA." *Cold Spring Harbor Symposia on Quantitative Biology* 26 (1961): 75–90.
Spiegelman, Sol, and O. E. Landman. "Genetics in Microorganism." *Annual Review of Microbiology* 8 (1954): 181–236.
Standskot, H. H. "Physiological Aspects of Human Genetics. Five Human Blood Characteristics." *Physiological Review* 24 (1944): 445–66.
Stanley, Wendell M. "Isolation of Crystalline Protein Possessing the Properties of Tobacco Mosaic Virus." *Science* 81 (1935): 644–45.
———. "The Regulation and Transfer of Biological Information." *Proceedings of the Robert A. Welch Foundation Conferences on Chemical Research* December (1961): 131–57.
Star, Susan L., and James R. Griesemer. "Institutional Ecology, 'Translations,' and Boundary Objects: Amateurs and Professionals in Berkeley's Museum of Vertebrate Zoology." *Social Studies of Science* 19 (1989): 387–420.
Steelman, John R. "A Report to the President." *The Nation's Medical Research, Vol. 5, Science and Policy*. Washington, D.C.: U.S. Government Printing Office, 1947.
Steigerwald, Joan. "The Cultural Enframing of Nature: Environmental Histories During the German Romantic Period." In Elinor G. K. Melville and Richard G. Hoffmann, eds., *Human and Ecosystems before Global Development*, 1999.
Stent, Gunther. *The Coming of the Golden Age: A View of the End of Progress*. New York: Natural History Press, 1969.
———. "Induction and Repression of Enzyme Synthesis." In G. C. Quarton, T. Melnechuk, and F. O. Schmitt, eds., *Neurosciences*, pp. 152–61. New York: Rockefeller University Press, 1967.
———. "Prematurity and Uniqueness in Scientific Discovery." *Scientific American* 227 (1972): 84–93.
———. "That Was the Molecular Biology That Was." *Science* 160 (1968): 390–95.
Stent, Gunther S., and Richard Calendar, *Molecular Genetics: An Introductory Narrative*. San Francisco: W. H. Freeman, 1978.
Stepan, Nancy Leys. "Race and Gender: The Role of Analogy in Science" *Isis* 77 (1986): 261–77.
Stern, Kurt G. "Nucleoproteins and Gene Structure." *Yale Journal of Biology and Medicine* 19 (1947): 937–49.
Stetten, DeWitt, Jr., ed. *NIH: An Account of Research in Its Laboratories and Clinics*. Orlando, Fla.: Academic Press, 1984.
Stewart, Irwin. *Organizing Scientific Research for the War: The Administrative*

History of the Office of Scientific Research and Development. Boston: Little, Brown, 1948.

Stock, Brian. *The Implications of Literacy: Written Language and Models of Interpretation in the Eleventh and Twelfth Century*. Princeton: Princeton University Press, 1983.

Strickland, Stephen. *Politics, Science, and Dread Disease*. Cambridge: Harvard University Press, 1972.

Sturtevant, A. H. "Can Specific Mutations Be Induced by Serological Methods?" *Proceedings of the National Academy of Sciences* 30 (1944): 176–78.

Sueoka, Noboru. "Correlation Between Base Composition of Deoxyribonucleic Acid and Amino Acid Composition of Protein." *Proceedings of the National Academy of Sciences* 47 (1961): 1141–49.

Symonds, Neville. "What Is Life? Schrödinger's Influence on Biology." *Quarterly Review of Biology* 61, no. 2 (1986): 221–26.

Szilard, Leo. "Über die Entropieverminderung in einem thermodynamischen System bei Eingriffen intelligenter Wesen." *Zeitschrift für Physik* 53 (1929): 840–56.

Tauber, Alfred I., ed. *Organism and the Origins of Self*. Boston: Kluwer Academic Publishers, 1991.

Tauber, Alfred I., and Leon Chernyak. *Metchnikoff and the Origins of Immunology*. New York: Oxford University Press, 1991.

Teich, Mikulas. "A Single Path to the Double Helix?" *History of Science* XIII (1975): 264–83.

Thieffry, Denis. "Contributions of the 'Rouge-Clôitre Group' to the Notion of 'Messenger RNA.'" *History and Philosophy of Life Sciences* 19 (1997): 89–113.

———. "Escherichia coli as a Model System with Which to Study Differentiation." *History and Philosophy of Life Science* 18 (1996): 163–93.

Thieffry, Denis, and Richard Burian. "Jean Brachet's Scheme for Protein Synthesis." *Trends in Biochemical Sciences* 21 (1996): 114–17.

Thieffry, Denis, and Sahotra Sarkar. "Forty Years under the Central Dogma." *Trends in Biochemical Sciences* 23 (1998): 312–16.

Timofeff-Ressovsky, N. W., K. Zimmer, and M. Delbrück. "Über die Natur der Genmutation and Genstruktur." *Göttingen Nachrichten, Mathematische-Physikalische Klasse, Fachgruppe* 6 (1935): 189–245.

Tissières, A., D. Schlesinger, and Françoise Gros. "Amino Acid Incorporation into Proteins by Escherichia Coli Ribosomes." *Proceedings of the National Academy of Sciences, U.S.A.* 46 (1960): 1450–63.

Trifonov, Edward N., and Volker Brendel. *Gnomic: A Dictionary of Genetic Codes*. Philadelphia and Rehovot: Balaban, 1986.

Trupin, Joel S., Fritz M. Rottman, and Richard L. C. Brimacombe. "RNA Codewords and Protein Synthesis, VI. On the Nucleotide Sequences of Degenerate Codeword Sets for Isoleucine, Tyrosine, Asparagine, and Lysine." *Proceedings of the National Academy of Sciences* 53 (1965): 807–11.

Tsugita, A., and Heinz Fraenkel-Conrat. "The Amino Acid Composition and C-Terminal Sequence of a Chemically Evoked Mutant of TMV." *Proceedings of the National Academy of Sciences* 46 (1960): 636–42.

Tsugita, A., Heinz Fraenkel-Conrat, and M. W. Nirenberg. "Demonstration of the Messenger Role of Viral RNA." *Proceedings of the National Academy of Sciences, U.S.A.* 48 (1962): 846–53.

Tsugita, A., D. T. Gish, J. Young, H. Fraenkel-Conrat, C. A. Knight, and W. M. Stanley. "The Complete Amino Acid Sequence of the Protein of Tobacco Mosaic Virus." *Proceedings of the National Academy of Sciences, U.S.A.* 46 (1960): 1463–69.

Tweedell, Kenyon S. "Identical Twinning and the Information Content of Zygotes." In *Symposium on Information Theory in Biology*, pp. 215–50. New York: Pergamon Press, 1956.

Umbarger, H. Edwin Umbarger. "Evidence for a Negative-Feedback Mechanism in Biosynthesis of Isoleucine." *Science* 123 (1955): 848.

van Helvoort, Ton. "History of Virus Research in the 20th Century: The Problem of Conceptual Continuity." *History of Science* 32 (1994): 185–235.

———. "What Is a Virus? The Case of Tobacco Mosaic Disease." *Studies in History and Philosophy of Science* 22 (1991): 557–88.

Varela, Francisco, Evan Thompson, and Eleanor Rosch. *The Embodied Mind: Cognitive Science and Human Experience*. Cambridge: MIT Press, 1993.

Vogel, Henry J., Vernon Bryson, and J. Oliver Lampen, eds. Preface to *Informational Macromolecules*. New York: Academic Press, 1963.

Volkin, Elliot, and Lazarus Astrachan. "Intracellular Distribution of Labeled Ribonucleic Acid After Phage Infection of Escherichia coli." *Virology* 2 (1956): 433–37.

———. "RNA Metabolism in T2–Infected Escherichia Coli." In William D. McElroy and Bentley Glass, eds., *The Chemical Basis of Heredity*, pp. 686–95. Baltimore: Johns Hopkins University Press, 1957.

Volkin, Elliot, Lazarus Astrachan, and Joan L. Countryman. "Metabolism of RNA Phosphorus in Escherichia coli Infected with Bacteriophage T7." *Virology* 6 (1958): 545–55.

von Foerster, Heinz. "Epistemology of Communication." In Kathleen Woodward, ed., *The Myths of Information: Technology and Postindustrial Culture*, pp. 18–27. Madison: Coda Press, 1980.

———. Interview with author. Pescadero, Cal., 26 June 1994.

von Franz, Marie-Louise. "Symboll des Unus Mundus." In W. Bitter, ed., *Dialog uber den Menschen*, pp. 231ff. and 249ff. Stuttgart: Klett Verlag, 1968.

von Goethe, Johann Wolfgang. *Faust*. Trans. Walter Kaufmann. New York: Doubleday, 1961.

von Humboldt, Wilhelm. *On Language: The Diversity of Human Language Structure and Its Influence on the Mental Development of the Mind*. Cambridge: Cambridge University Press, 1988.

Vonnegut, Kurt. *Player Piano*. New York: Charles Scribner, 1952.

von Neumann, John. "The General and Logical Theory of Automata." In Lloyd Jeffress, ed., *Cerebral Mechanisms in Behavior*, pp. 1–31. New York: Hafner, 1951.

Waddington, C. H. "Form and Information." In C. H. Waddington, ed., *Toward a Theoretical Biology, 14*. Edinburgh: Edinburgh University Press, 1972.

Wahba, Albert J., Carlos Basilio, Joseph Speyer, Peter Lengyel, Robert S. Miller, and Severo Ochoa. "Synthetic Polynucleotides and the Amino Acid Code, VI." *Proceedings of the National Academy of Sciences, U.S.A.* 48 (1962): 1683–86.

Wahba, Albert J., Robert S. Gardner, Carlos Basilio, Robert S. Miller, Joseph Speyer, and Peter Lengyel. "Synthetic Polynucleotides and the Amino Acid Code, VIII." *Proceedings of the National Academy of Sciences, U.S.A.* 49 (1963): 116–22.

Wall, Robert. "Overlapping Genetic Code." *Nature* 193 (1962): 1268–70.

Walter, Katya. *DNA & the I Ching: The Code of the Universe.* Rockport, Mass.: Element, 1996.

Wang, Jessica. "American Science in an Age of Anxiety: Scientists, Civil Liberties, and the Cold War, 1945–1950." Ph.D. diss., Program in Science, Technology, and Society, MIT, 1995.

———. *American Science in an Age of Anxiety: Scientists, Anti-Communism, and the Cold War.* Chapel Hill: University of North Carolina Press, 1999.

Wang, Zuoyue. "The Politics of Big Science in the Cold War: PSAC and the Funding of SLAC." *Historical Studies in the Physical and Biological Sciences* 25 (1995): 329–57.

Watson, James. "The Biological Properties of X-ray Inactivated Bacteriophage." *Journal of Bacteriology* 60 (1950): 697–718.

———. *The Double Helix.* New York: W. W. Norton, 1980.

———. "The Involvement of RNA in the Synthesis of Proteins." In David Baltimore, ed., *Nobel Lectures in Molecular Biology, 1933–1975*, pp. 181–93. New York: Elsevier North-Holland, 1977.

———. "Values from Chicago Upbringing." In Donald A. Chambers, ed., *DNA: The Double Helix: Perspective and Prospective at Forty Years*, p. 197. New York: New York Academy of Science, 1995.

Watson, James D., and Francis H. C. Crick. "Genetical Implications of the Structure of Deoxyribonucleic Acid." *Nature* 171 (1953): 964–67.

———. "Molecular Structure of Nucleic Acids. A Structure for Deoxyribose Nucleic Acid." *Nature* 171 (1953): 737–38.

Weart, Spencer. *Nuclear Fear: A History of Images.* Cambridge: Harvard University Press, 1988.

Weaver, Warren. "A Quarter Century in the Natural Sciences." *Rockefeller Foundation Annual Report* (1958): 28–34.

Weber, Max. "Politics as a Vocation." In H. Gerth and C. W. Mills, eds., *Max Weber.* London: Routledge & Kegan Paul, 1970.

———. "Science as a Vocation." In H. Gerth and C. W. Mills, eds., *Max Weber.* London: Routledge & Kegan Paul, 1970.

Weigert, Martin G., and Alan Garen. "Base Composition of Nonsense Codons in E. coli." *Nature* 206 (1965): 992–98.

Weiner, Charles. "Anticipating the Consequences of Genetic Engineering: Past, Present, and Future." In Carl F. Cranor, ed., *The Social Consequences of the New Genetics*, pp. 2–31. New Brunswick, N.J.: Rutgers University Press, 1994.

Weingart, Peter, J. Kroll, and K. Bayertz. *Rasse, Blut und Gene: Geschichte der Eugenik und Rassenhygiene in Deutschland*. Frankfurt am Mein: Suhrkamp Verlag, 1988.
Weiss, Paul. *Principles of Development*. New York: Henry Holt, 1939.
———. "Principles of Development." *Yale Journal of Experimental Medicine* 19 (1947): 235–78.
Welchman, Gordon. *The Hut Six Story: Breaking the Enigma Codes*. New York: McGraw-Hill, 1982.
West, J. C. "Forty Years in Control." *Institution of Electrical Engineers Proceedings* 132, no. 1 (1985): 1–8.
Wexler, Immanuel. *The Marshall Plan Revisited*. Westport, Conn.: Greenwood Press, 1983.
Whitfield, Stephen J. *The Culture of the Cold War*. Baltimore: Johns Hopkins University Press, 1991.
Wiener, Norbert. "Cybernetics." *Scientific American* 179 (1948): 14–19.
———. *Cybernetics: Or Control and Communication in the Animal and the Machine*. 2d ed. Cambridge: MIT Press, 1961; and New York: John Wiley and Sons, 1961.
———. *Extrapolation, Interpolation and Smoothing of Stationary Time Series*. Cambridge: MIT Press, 1949.
———. *The Human Use of Human Beings: Cybernetics and Society*. Boston: Houghton Mifflin, 1950.
———. "A Scientist Rebels." *Atlantic Monthly* 170 (1947): 46.
Wimsatt, William C. "Some Problems with the Concept of 'Feedback.'" *Boston Studies in the Philosophy of Science* VIII (1971): 241–56.
Winner, Langdon. *The Whale and the Reactor: A Search for Limits in the Age of High Technology*. Chicago: University of Chicago Press, 1986.
Witkowski, J. A. "Schrödinger's *What Is Life?* Entropy, Order, and Hereditary Code-Script." *Trends in Biochemical Sciences* 11 (1986): 266–68.
Wittman, H. G. "Comparison of the Tryptic Peptides of Chemically Induced and Spontaneous Mutants of Tobacco Mosaic Virus." *Virology* 12 (1960): 609–12.
Wittner, Lawrence S. *Cold War America: From Hiroshima to Watergate*. New York: Praeger Publishers, 1974.
Woese, Carl R. "Coding Ratios for the Ribonucleic Acid Viruses." *Nature* 190 (1961): 697–98.
———. "Composition of Various Ribonucleic Acid Fractions from Microorganisms of Different Deoxynucleic Acid Composition." *Nature* 189 (1961): 920–21.
———. *The Genetic Code: The Molecular Basis for Genetic Expression*. New York: Harper and Row, 1967.
———. "Nature of the Biological Code." *Nature* 194 (1962): 1114–15.
———. "A Nucleotide Triplet Code for Amino Acids." *Biochemical and Biophysical Research Communications* 5 (1961): 88–93.
Wollman, Elie L. "Bacterial Conjugation." In John Cairns, Gunther S. Stent, and James D. Watson, eds., *Phage and the Origins of Molecular Biology*, pp. 216–

25. Cold Spring Harbor: Cold Spring Harbor Laboratory of Quantitative Biology, 1966.
Wollman, Elie, and François Jacob. "Sexuality in Bacteria." *Scientific American* 195, no. 1 (1956): 109–18.
Wolstenholme, Gordon, ed. *Man and His Future*. Boston: Little, Brown, 1963.
Wyatt, H. V. "When Does Information Become Knowledge?" *Nature* 239 (1972): 234.
Yanofsky, C., B. C. Carlton, J. R. Guest, and D. R. Helinski. "On the Colinearity of Gene Structure and Protein Structure." *Proceedings of the National Academy of Science* 51 (1964): 266–72.
Yates, Richard A., and Arthur B. Pardee. "Control of Pyrimidine Biosynthesis in Escherichia Coli by Feed-Back Mechanism." *Journal of Biological Chemistry* 221 (1956): 757–70.
———. "Pyrimidine Biosynthesis in Escherichia Coli." *Journal of Biological Chemistry* 221 (1956): 743–56.
Yčas, Martynas. *The Biological Code*. New York: Elsevier North-Holland, 1969.
———. "Biological Coding and Information Theory." In H. L. Lucas, ed., *The Cullowhee Conference on Training in Biomathematics*, pp. 245–58. Raleigh: North Carolina State College, 1961.
———. "Correlation of Viral Ribonucleic Acid and Protein Composition." *Nature* 188 (1960): 209–12.
———. Interview with author. Syracuse, N.Y., 6 October 1993.
———. "The Protein Text." In Hubert P. Yockey, ed., *Symposium on Information Theory in Biology*, pp. 70–100. New York: Pergamon Press, 1956.
———. "Replacement of Amino Acids in Proteins." *Journal of Theoretical Biology* 2 (1961): 244–57.
Yčas, Martynas, and Walter S. Vincent. "A Ribonucleic Acid Fraction from Yeast Related in Composition to Deoxyribonucleic Acid." *Proceedings of the National Academy of Sciences (U.S.A)* 46 (1960): 804–11.
Yockey, Hubert P. "Some Introductory Ideas Concerning the Application of Information Theory in Biology." In Hubert P. Yockey, ed., *Symposium on Information Theory in Biology*, pp. 50–60. New York: Pergamon Press, 1956.
———, ed. *Symposium on Information Theory in Biology*. New York: Pergamon Press, 1956.
Yoxen, Edward J. "Giving Life a New Meaning: The Rise of the Molecular Biology Establishment." In N. Elias, H. Martins, and R. Whitly, eds., *Scientific Establishments and Hierarchies: Sociology of the Sciences, IV*, pp. 123–43. Dordrecht: D. Reidel, 1982.
———. "Life as a Productive Force: Capitalizing the Science and Technology of Molecular Biology." In Robert M. Young and Les Levidow, eds., *Studies in the Labor Process*, Vol. 1, pp. 66–112. London: CSE Books, 1981.
———. "The Social Impact of Molecular Biology." Ph.D. diss., King's College, University of Cambridge, Eng., 1977.
———. "Where Does Schrödinger's *What Is Life?* Belong in the History of Molecular Biology?" *History of Science* 17 (1979): 17–52.

Zallen, Doris. "Louis Rapkine and the Restoration of French Science after the Second World War." *French Historical Studies* 17 (1991): 6–37.

Zamecnik, Paul C. "A Historical Account of Protein Synthesis, with Current Overtones—A Personalized View." *Cold Spring Harbor Symposia on Quantitative Biology* 34 (1969): 1–16.

Zimmerman, Joan. Telephone conversation with author. New York, 2 June 1994.

Zimmerman, Joanna. Telephone conversation with author. Pittsburgh, Penn., 6 and 14 June 1994.

Zubay, Geoffrey. "A Possible Mechanism for the Initial Transfer of the Genetic Code from Deoxyribonucleic Acid to Ribonucleic Acid." *Nature* 182 (1958): 112–13.

INDEX

In this index an "f" after a number indicates a separate reference on the next page, and an "ff" indicates separate references on the next two pages. A continuous discussion over two or more pages is indicated by a span of page numbers, e.g., "57–59." *Passim* is used for a cluster of references in close but not consecutive sequence.

AAF, *see* Army Air Force
Abelson, Philip H., 106
Aberdeen Proving Grounds, 102
Abir-Am, Pnina, 173
Academy of Science (Moscow), 307
Acheson, Dean, 9
Adaptation, 201–2
Adaptor hypothesis, 165, 176, 243
Adler, Julius, 282
AEC, *see* Atomic Energy Commission
Ageno, Mario, 270
Aiken, Henry, 80
Aiken, Howard, 102
Air Force Scientific Advisory Board, 135
Alan of Lille, 32
AMA, *see* American Medical Association
American Black Chambers, 134
American Medical Association (AMA), 235
Ames, Bruce, 223, 236, 238, 240
Amino Acid Code, 276, 282
Amino acids, 8, 12, 141, 160, 283; cryptanalysis of, 125–26; and genetic code, 145–46, 153; and protein synthesis, 147, 168–69, 286–87; and triplet code, 266–67
AMP, *see* Applied Mathematics Panel
Antibodies, 43, 51
Anticodon, 287
Anti-communism, *see* McCarthyism
Aperiodic crystals, 61, 186
Applied Mathematics Panel (AMP), 78–79
Aquinas, St. Thomas, 31, 295
Arbib, Michael, 26
Argonne National Laboratory, 10, 76, 116, 123–24
Aristotle, 38, 41
Arms race, 104, 166, 169
Army Air Force (AAF), 76. *See also* Military; U.S. Air Force
Army Office of Operations Research, 135
Army Security Agency, 134
Arrhenius, Svante, 43
Ashby, Ross W., 90, 119
Aspray, Bill, 20, 77
Astrachan, Lazarus, 227
Atlan, Henri, 26
Atomic bombs, 75, 132, 275–76
Atomic Energy Commission (AEC), 10, 76–77, 102, 124; and life sciences, 132, 235–36
ATP synthesis, 281
Audureau, Alice, 198
Augenstine, Leroy, 305
Augustine, St., 31, 295
Automata, 101–2; biological, 106–7, 109–12, 321
Autopoiesis, 35
Autosynthesis, 69
Avery, Oswald T., 39, 42, 55–56

Bacteria, 195f, 199; enzyme adaptation in, 53, 198, 201; conjugation of, 208–10, 211–12. *See also* B. cereus; E. coli
Bacteriophage, 9. *See also* Phages
Ballistics Research Laboratories, 102
Bar-Hillel, Yehoshua, 21, 99–100, 115, 352n99
Barnett, Leslie, 266
Bastide, Françoise, 318
Bates, Marston, 116

Baudrillard, Jean, 23, 34, 114, 322
B. cereus, 241–42, 243, 247
Beadle, George W., 39, 41, 51–52, 53, 106, 109, 132, 170, 206, 290–91, 294, 305; *The Language of Life*, 17, 291, 311
Beatty, John, 10
Behavior, behaviorism, 49, 80
"Behavior, Purpose, and Teleology" (Wiener, Bigelow, and Rosenblueth), 81, 83
Bell Laboratories, 20, 75, 112; and military contracts, 91–92; and Shannon, 92–102
Belozersky, A. N., 175
Benzer, Seymour, 168, 177, 224, 266f, 275, 312
Bergmann, Max, 50, 182
Betatron, 116, 118
Biagioli, Mario, 11
Bigelow, Julian, 79, 81
Biochemical and Biophysical Research Communication, 248
Biochemistry, 194, 257, 269, 277; semiotics of, 27–28; at NIH, 236, 237–38; and genetic code, 255–56; as information theory, 273, 274–75
Biological engineering, 279–80, 292
Biological specificity, *see* Specificity
Biology, 61; as concept, 40–41; information theory in, 78, 118–27; and cybernetics, 90, 105–7, 114–15; and linguistics, 308–10; models in, 314–15
Biomedical research, 108, 293
Biopower, 19, 48
Biosemiotics, 175, 195–96
Biosynthesis, 198–99, 211, 212–13
Biotechnology, 200, 322
Biowriting, 297
Birkoff, Garrett, 79
Birkoff, George D., 80
Black, Max, 22
Bletchley Park, 134
Bohr, Niels, 63, 106
Bohr's complementarity principle, 302
Bonner, James, 33, 109
Bono, James, 29
Book of Changes, *see* I Ching
Book of Life, 1–5 *passim*, 14, 39, 326; metaphor of, 31–32; signification of, 35, 294–97; authorship of, 36, 310; as chimera, 318–19
Book of Life, The (Sinsheimer), 17, 291
Book of Nature, 2, 31, 32–34
Boring, E. G., 80
Born, Max, 63
Borsook, Henry, 109
Boulding, Kenneth E., 90
Boyd, Richard, 24
Brachet, Jean, 55, 139, 226
Branson, Howard, 120
Brendel, Volker, 36, 324
Brenner, Sydney, 142f, 155, 168, 173, 176f, 193f, 224, 229, 252, 266, 269, 272, 285, 329, 365n1; code developed by, 147–48, 149; disproof of overlapping codes, 161–62; and messenger RNA, 196, 227–28, 230, 232
Bretcher, M. S., 271, 275
Bridgman, P. W., 80
Brillouin, Leon, 17, 65, 80, 119, 220
British Department of Scientific and Industrial Budget, 278
Bronowski, Jacob, 306f
Brookhaven National Laboratory, 10, 76, 124–25, 176–77
Brown, Amos, 43
Bulletin of Atomic Scientists, 91
Bureau of Ordnance, 105, 135
Burian, Richard, 194
Burnet, F. Macfarlane, 25, 51, 115
Bush, Vannevar, 74–75, 78, 92

Cairns, John, 289
California Institute of Technology (Caltech), 36, 75, 109, 196. *See also* Jet Propulsion Laboratory
Calvin, M., 141
Cambridge University, 58, 173
Canguilhem, Georges, 26f, 38, 40, 221
Cannon, Walter, 47, 63, 79, 233
Carnap, Rudolph, 99, 105
Carnegie Institution of Washington, D.C., 282
Caskey, Thomas, C., 290
Caspersson, Torbjorn O., 55
"Cellular Regulatory Mechanisms" seminar, 231–33
Center for Communication Sciences (MIT), 300–301, 303
"Central Dogma," 30, 36, 174–75, 179, 217, 273, 275
Central Intelligence Agency, 301
Centre National de la Recherche Scientifique (CNRS), 200, 203, 278
"Cerebral Mechanisms in Behavior" symposium, 109
Chadarevian, Soraya de, 173
Champollion, Jean-François, 190

Chance and Necessity (Monod), 17, 197
Chantrenne, Hubert, 139
"Characteristics and Stabilization of DNAase-Sensitive Protein Synthesis in E. coli Extracts" (Matthaei and Nirenberg), 253–54
Chargaff, Erwin, 27, 39, 42, 115, 139, 141, 176, 180, 256; on nucleic acids, 56–57; and information discourse, 58, 276
Chargaff's rule, 57f
Chemical Basis of Heredity, The (McElroy and Glass), 240
"Chemically Synthesized Deoxynucleotides as Templates for Ribonucleic Acid Polymerase" (Khorana, Falaschi, and Adler), 282
"Chemical Specificity of Nucleic Acids and Mechanism of Their Enzymatic Degradation" (Chargaff), 57
Chemostat, 199
Cherry, Colin, 21, 100f, 115, 119, 127, 301
Chimera, 318–19
China, 75
Chomskian linguistics, 319, 324
Chomsky, Noam, 1, 301, 304, 323
Chromosomes, 61, 118–19, 186
Churchill, Winston, 75
CIBA Foundation symposia, 279
Cipher(s), 134, 139, 151–52, 178, 225
Cistrons, 312
CNRS, *see* Centre National de la Recherche Scientifique
Code-script, 61–62, 65, 336n7
Coding, codes, 9, 14, 39, 42, 113, 151, 178, 224f, 270; concept of, 66–67, 71–72; Stern's development of, 67–69; Dounce's concepts of, 69–71; secrecy, 92–93; binary, 96–97, 175–76, 231; Gamow and colleagues' work on, 128–29, 144–50, 329; computers and, 146–47; combination, 155–56, 157; disproof of overlapping, 161–62; commaless, 163, 165–66, 170, 249–50, 255, 264, 267; DNA variation and, 176–77; approaches to, 193–94; and linguistics, 283, 304–5; information theory and, 302–3
Codons, 272, 283–84, 286, 287–88, 312
Cohen, George, 204, 206–7, 278
Cohn, Melvin, 197–202 *passim*, 213, 216f, 275
Cold Spring Harbor Symposia in Quantitative Biology, 16, 36, 56, 282; in 1961, 193, 231–33; in 1963, 282–83; in 1966, 13, 288–90
Cold war, 6–9 *passim*, 30, 72, 228; and science, 75–76, 132–33; and computers, 102–3
Collado-Vides, Julio, 324
Colloquium on Information in Biology, 275
Columbia University, 75, 79, 205
Combination code, 155–56
Commas, 165–66. *See also* Punctuation; *and under* Coding, codes: commaless
Communication sciences, communication theory, 25, 78, 94–95, 220, 303–4; and molecular biology, 5, 28; and information, 20–21; Shannon on, 92–98, 134–35, 351n78; Weaver on, 98–99
Communication systems, 114, 185–86, 275, 321
"Communication Theory of Secrecy Systems" (Shannon), 92–93
Communism, 75f
Complementarity, 43, 50–51, 302
Computers, computing machines, 10, 78; cold war and, 102–3, 133; military and, 103–4; von Neumann and, 104–5; and genetic coding, 140–41, 146–47, 148–49, 158
Condillac, Etienne de, 296
Control, 85
Control Systems Laboratory, 116, 119, 121
Convair, 135
Conway, J. H., 321
Cooperation, 49
Cori, Carl F., 106, 257
Cori, Gery, 257
Cornell University, 246
Courant, Richard, 79
Creager, Angela, 180
Crick, Francis C.H., 8, 16, 36, 42, 128f, 131–32, 142f, 147, 150, 155, 185, 213, 254, 266, 277, 279, 284, 287f, 311, 329; on protein synthesis, 29–30; on DNA structure, 58–59, 156–57, 176–77; and Gamow, 138, 141, 159; and genetic code, 148, 152, 161, 164, 193, 271–72; and commaless code, 165–66, 264; on protein synthesis, 173–74; "Central Dogma" of, 174–75, 179, 217, 273; and Pasteur Institute, 223–24; and Zamecnik, 243f; triplet model, 267–68
Cryptanalysis, 6, 9, 131, 301; and protein synthesis, 125–26; and genetic code, 132, 149, 152–53, 193; development of, 133–34
Cryptics, 206

Cryptography, 133, 151–52, 295–96
Cryptology, 93, 134
Cuénot, Lucien, 201–2
Culture: writing and, 32–33; printing and, 33–34
"Cybernetical Aspect of Genetics, A" (Kalmus), 87
Cybernetics, cyborg, 5–6, 15, 27, 58, 72, 118, 196, 337n23; origins of, 78ff; Wiener's book on, 84–85, 86–87; information transfer and, 88–89; adoption of, 89–90, 114–15; discourse on, 90–91; military use of, 105–6; and biology, 106–7; enzyme formation and, 114, 212–13; linguistics of, 297–98, 300
Cybernetics (Wiener), 84–85, 86–88
Cystic fibrosis, 326

Dancoff, Sydney M., 118–19, 125
Darwinism, 202, 319–22; molecular, 322f
Davis, Bernard, 232f
De Chardin, Teilhard, 201
De Gaulle, Charles, 205, 222
Delbrück, Max, 16, 30, 36, 38f, 50, 61, 63, 106f, 109, 115, 129, 140, 166, 170, 175, 185, 198, 207, 254, 269, 275, 294; and Gamow, 131, 159; and RNA Tie Club, 141f; on transposability, 171–72
Délégation Général à La Recherche Scientifique et Technique (DGRST), 222
DeMars, Robert, 238, 240
Demerec, Milislav, 106, 289
Deoxyribonucleic acid (DNA), 55, 197, 232f, 239, 281, 294, 317; nucleotide-assembling enzymes and, 12–13, 57; language and linguistics of, 17, 37, 178–79, 319f, 324–25, 326, 330–31; as informational code, 42, 52, 58–59, 137–38, 327–28; Gamow's mathematical view of, 138–39; directionality of, 156–57; ordered structure of, 156–57; transposability of, 171–72; variability in base, 175, 176–77
Department of Defense, *see* U.S. Department of Defense
"Dependence of Cell-Free Protein Synthesis in E. coli upon Naturally Occurring or Synthetic Polyribonucleotides, The" (Nirenberg and Matthaei), 253–54
"Dependence of Cell-Free Protein Synthesis in E. coli upon RNA Prepared from Ribosomes, The" (Nirenberg and Matthaei), 248–49
Derrida, Jacques, 23, 26, 30, 32f, 297, 304
Descartes, René, 33
Deutsch, Karl, 80, 90
DGRST, *see* Délégation Général à La Recherche Scientifique et Technique
Diamond code, 136f, 140, 144
Dictionary, 170–73 *passim*, 290
Differential equations, 78
Dipeptide sequences, 162
Discourse: signification and, 18–19
Discursive practices, 15–17, 26
DNA, *see* Deoxyribonucleic acid
DNAase, 13, 245
Dobzhanksy, Theodosius, 56, 291
DOD, *see* U.S. Department of Defense
Donnan, Frederick G., 63
Doty, Paul, 141, 255
Doublet code, 270, 285–86
Dounce, Alexander, 27, 42, 65, 69, 141, 160, 269; publications by, 70–71
Drosophila genetics, 51
Dubois-Reymond, Emil, 63
Dulbecco, Renato, 168
Dulles, John Foster, 9, 132
Dunham Lectures: Monod's, 216–17
"Duplicating Mechanisms for Peptide Chain and Nucleic Acid Synthesis" (Dounce), 70

Eck, Richard V., 191, 269, 286
Eckert, Presper J., 103
E. coli, 6, 9, 12, 53, 193, 211, 214, 230f, 245, 282, 289; and DNA sequencing, 13, 245; enzymatic feedback inhibition, 169, 199, 201; mutant, 206–7; and polynucleotide synthesis, 247–54, 260–61
Edge, David, 28
Edsall, John T., 106, 109
EDVAC, *see* Electronic Discrete Variable Arithmetic Computer
Edwards, Paul, 10, 104
Ehrlich, Paul, 43
Eigen, Manfred, 112, 156, 305, 330–31; on origins of life, 319–21; and molecular evolution, 321–23; and linguistics, 323–24
Eisenhower, Dwight, 9, 76, 132, 205, 235
Eisenstein, Elizabeth, 33
Electronic Discrete Variable Arithmetic Computer (EDVAC), 103
Electronic Numerical Integrator and Calculator (ENIAC), 102f
EMBO, *See* European Institute for Molecular Biology
Embryology, 219–20, 240
Emerson, Sterling, 52, 54

Encoding, 302–3
Endocrinology, 25
Energy: and Maxwell's demon, 64–65
England, *see* Great Britain
ENIAC, *see* Electronic Numerical Integrator and Calculator
Enlightenment, 33, 296
Entropy, 64–65, 94, 122f, 153–54
Enzyme adaptation, 53, 198, 201–2, 344n58
Enzymes, 6, 48–53 *passim*, 114, 207, 233, 242; discourse on, 195, 197–98; biosynthesis of, 198–99, 329–30; induction of, 210, 212–13, 216–18; information transfer in, 218–20; as activators, 239–40
Enzymology, 48, 53, 238, 239–40, 366n13
Ephrussi, Boris, 53, 58, 203
Episteme, 35f
Ethics, 292–93
Eugenics, 48, 279, 291–92
Europe, 278–79. *See also individual countries by name*
European Institute for Molecular Biology (EMBO), 222, 278
"Evidence for a Negative-Feedback Mechanism in the Biosynthesis of Isoleucine" (Umbarger), 211
Evolution: and specificity, 43–44; human, 279–80; in molecular biology, 321–22
"Extrapolation, Interpolation, and Smoothing of Stationary Time Series with Engineering Applications, The" (Wiener), 81

Falaschi, Arturo, 282
Faust, 295
FCP, *see* French Communist Party
Feedback, 80, 83, 85–86, 199, 242. *See also* Negative feedback
Feller, Will, 87
Fermi, Giulio, 158
Feynman, Richard, 140f, 144, 254
Figlio, Karl, 40
Fischer, Emil, 140
Florkin, Marcel, 27–28, 34
Forman, Paul, 8, 10, 76
Foucault, Michel, 19, 40, 84
Fox, Maurice, 166
Fraenkel-Conrat, Heinz, 179, 181, 191, 250–51, 266; TMV research, 27, 182–85, 187, 193, 250; virus research, 191–92
France, 9, 15, 201–6 *passim*, 222, 278, 307–8. *See also* Pasteur Institute
Franck, James, 106
Frank, Phillip, 80
Frazer, Dean, 181
French Communist Party (FCP), 204
Friedman, Wolfe, 134
Frimmel, Franz, 62
Fruton, Joseph, 28, 274
Fulbright, J. William, 10–11
Functional interdependency, 122
Fussell, Paul, 30

Gabor, Denis, 119
Galileo Galilei, 33
Galison, Peter, 24, 146
Game theory, 320
Gamow, George, 8, 16, 42, 53, 106, 121, 130, 135, 152, 162, 167, 169, 255, 283, 314, 320, 329, 356n4; and RNA Tie Club, 6, 57, 72, 128, 142f, 176; and genetic code, 128–29, 131–32, 136, 138–39, 140–41, 144–51, 155–56, 159, 329; collaboration with Yčas, 125–26, 140, 157–58, 159, 160–61, 165–66, 180; on randomness, 154–55; and TMV, 184, 186, 189
Garen, Alan, 275, 285
Gatlin, Lila, 26
Gaudillière, Jean-Paul, 194, 199, 222
"General and Logical Theory of Automata, The" (von Neumann), 110
General Electric Research Laboratory, 75, 191
General Motors, 75
"General Nature of the Genetic Code for Proteins" (Crick et al.), 266
Generative grammar, 1, 304, 324
Gene therapy, 326–27
Genetic code(s), 3, 11, 30, 137, 248, 262, 288–89, 290; Schrödinger and, 4–5; information theory and, 6, 321; Gamow and, 128–29, 136, 138–39, 140–41, 144–50; information and, 150–51, 327–28; as language, 151–52, 164–65, 192, 306–7, 310–13; randomness and, 154–55, 158–59; punctuation in, 163–64, 170–71; transposability of, 171–72; dictionary of, 172–73; as binary, 175–76; black box approaches to, 193–94; biochemical approach to, 231, 255–56; competition in, 256–57; theoretical approaches to, 263–64; triplet model in, 266–68, 270, 283–84; protein synthesis and, 272–73; universality of, 276–77, 330–31; social implications of, 279–80; and linguistics, 282–83, 284–85, 294–

95; Cold Spring Harbor symposium and, 288–91; as revealed knowledge, 296–97; linguistic approach to, 304–5; and I Ching, 315–18; as chimera, 318–19
Genetic Code, The (Woese), 16
"Genetic Control and Cytoplasmic Expression of 'Inducibility' in the Synthesis of β-Galactosidase by E. Coli" (PaJaMa), 221–222
Genetic engineering, 3, 293
Genetics, 10, 44–45, 48, 50, 54, 112–13, 266
Genetic Society of America, 10
Genomics, 1, 294f, 326
Germany, 9. *See also* Max Planck Institute
Gierer, Alfred, 187
Gilbert, Scott, 43
Gilbert, Walter, 1, 196, 254, 327
Glass, H. Bentley, 240, 292
Glycogen phosphorylase, 239f
Gnomic, 37, 324
Gnomic: A Dictionary of Genetic Codes (Trifonov and Brendel), 36–37, 324
Goal-directed systems, 196. *See also* Cybernetics; Feedback; Teleology; Teleonomy
Goethe, Johann Wolfgang von, 33, 295
Goldstine, Herman, 103
Golomb, Solomon W., 170–73 *passim*
Gordon, Basil, 170f
Gordon, H., 141
Gramophone records, 66–67
Great Britain, 9, 15, 133–34, 157, 278
"Great Society," 277
Great War and Modern Memory, The (Fussell), 30
Griffith, John S., 161, 164f
Gros, François, 196, 226, 275; and messenger RNA, 228–29, 230, 232, 244
Gros, Françoise, 229, 244, 255
Grunberg-Manago, Marianne, 242, 259, 271, 278

Hacking, Ian, 11, 35, 77
Haldane, John B. S., 50, 67, 80, 86–87, 115
Halle, Morris, 301–2
Haraway, Donna, 10
Hartley, R. V., 95f
Harvard University, 75, 88, 194
Harvey Lecture, 214
Heidegger, Martin, 35
Heims, Steve, 104
Hemoglobin, 43–44, 179
Henderson, L. J., 80
Heppel, Leon, 236f, 247, 259, 262, 285
Heredity, 55, 86, 308; code-script of, 62, 336n7; information transfer in, 128–29; as communication system, 185–86; human, 279–80; and information theory, 313–14
Herriott, Roger, 245
Hersh, R. T., 269–70
Hershey, Alfred, 227
Hesse, Mary, 22
HEW, *see* U.S. Department of Health, Education, and Welfare
Hiatt, H., 196
Hierarchy, 46–47
Hieroglyphs, 33, 189–90, 311
Hill, Lester S., 134
Hinegardner, Ralph T., 290
Hinshelwood, Cyril, 42, 65, 69, 72
History of Sexuality, The (Foucault), 19
Hixon symposium: on biological automata, 109–11
Hoagland, Mahlon, 194
Holley, Robert W., 14, 244, 286–87, 293, 330
Holton, Gerald, 80
Hood, Leroy, 327
Hotelling, Harold, 79
House Committee on Un-American Activities (HUAC), 280
Hugh of St. Victor, 32
Human Genome Project, 1, 17, 37, 297, 327, 330
Human Use of Human Beings, The (Wiener), 88–89
Humphrey, Hubert, 222
Huntington Memorial Hospital, Collis P., 194, 243
Huxley, Thomas H., 48
Hypercycles, 322–23, 324

IAS, *see* Institute for Advanced Study
IBM, 105
I Ching: and genetic code, 315–18
Immunochemistry, 53–54
Immunology, 25, 42–43
"Induction of Two Similar Enzymes by One Inducer: A Test Case for Shared Genetic Information, The" (Nirenberg), 238
Industrial-military-academic complex, 11, 72, 77–78; Bell Labs and, 91–92
Information, 39, 41, 85, 298, 318; as metaphor, 2–3, 20, 21–22, 58–59, 194–95, 238–39; discourse of, 16–20, 275–76,

329–30, 370n80; and communication, 20–21; discursive politics of, 23–24, 58; and specificity, 52–53; military-industrial complex and, 77–78; and cybernetics, 88–89, 337n23; and communication theory, 95–96, 98–99; and semantics, 99–100; and genetic coding, 150–51; and enzyme induction, 218–19; and RNA, 230–31; origin of life as, 319–21, 331
"Information Content and Error Rate of Living Things, The" (Quastler and Dancoff), 118–19
Information theory, 5–6, 9, 14–15, 20–21, 58, 154, 162, 273, 321; military and, 23–24; molecular biology and, 25–27, 39, 115, 118–27, 129, 273–74; binary code and, 96–97; Bar-Hillel on, 99–100; and biology, 115, 118–27; Monod on, 220–21; biochemistry and, 274–75; elements of, 302–3; and heredity, 313–14
"Information Theory in Biology" symposium, 119–20
Information transfer, 154, 196, 244; in heredity, 128–29; in phage induction, 214–15, 224–25; and enzymes, 216–17, 218–20, 233; messenger RNA and, 226–27, 230
"Information Transfer in the Living Cell" (Gamow), 154
Ingram, Vernon, 173, 179
Institute for Advanced Study (IAS), 63, 82
Instructional theory, instructive theory, 51
Insulin, 8, 138ff
Inter-bacterial information, 58
International Congress of Biochemistry, 254, 283
Irwin, M. R., 121–22
Iskandar, Kai Kā'us ibn, 157
Isoleucine synthesis, 213

Jackson, David, 297
Jacob, François, 6, 16, 18, 42, 59, 122, 194ff, 202, 212, 224, 232f, 278, 312, 324, 329, 369n58; *The Logic of Life*, 17, 40–41, 197, 313; collaboration with Monod, 207–8, 222–23; collaboration with Wollman, 208–9; and repressor hypothesis, 213–14; and phage induction, 214–16; on operons, 222–23, 225–26; and messenger RNA, 227–28, 244–45, 255; work on the genetic code, 229, 231; on genetics and language, 294, 307–9, 318; on biological models, 314–15
Jacobson, Homer, 112
Jakobson, Roman, 1, 7, 80, 88, 294, 323f; and cybernetics, 297–98; on linguistics, 299–300; at MIT, 300–301; phoneme analysis, 301–2; and genetic code, 304–5; and molecular biology, 306–7, 308; on genetics and language, 309, 310–14, 318–19
Jakoby, William, 138
Jeener, Raymond, 226
Jeffress, Lloyd, 109
Jessup Lectures, 205
Jet Propulsion Laboratory (JPL), 169–70
Jewett, Frank B., 92
John, St., 31, 295
Johns Hopkins Univeristy, 245
Johnson, Lyndon B., 277
Johnson, Mark, 22
Joint Chiefs of Staff, 76
Jordan, Pascual, 63
Journal of Molecular Biology, 221–22
JPL, *see* Jet Propulsion Laboratory

Kahn, David, 178
Kalckar, Herman, 237
Kalmus, J., 87, 115, 177–78
Kant, Immanuel, 33
Keller, Evelyn Fox, 129, 216
Kemeny, John, 111, 113
Kendrew, John, 173, 221–22
Kennedy, John F., 277
Kepes, Gyorgy, 80, 90
Khorana, Har Gobind, 13f, 277, 283, 287, 293, 330; nucleic acid research by, 280–81; polymerase synthesis by, 281–82
Kipling, R. J., 89
Kittler, Friedrich, 295
Knight, Arthur C., 180, 183
Knowledge, 133, 279, 338n41; signification of, 18–19, 34–35; of nature, 32–33; revealed, 296–97
Kober, Alice B., 178
Korean war, 9
Kornberg, Arthur, 194, 236, 257, 282
Kurland, C. G., 196

Laboratory of Molecular Biology, 173
Lac– mutations, 206–7, 211–12. *See also* Operons
Lakoff, George, 22, 338n41
Lamborg, Marvin R., 244f
Landsteiner, Karl, 43, 50
Langmuir, Irving, 108

Language, 297; DNA, 1–2, 17, 330; genetic code as, 151–53, 164–65, 192, 307–8; and entropy, 153–54; genomic, 295–96; and communication theory, 303–4; de Saussure on, 298–99; Jakobson on, 310–13; and molecular biology, 323–24
Language of Life, The (Beadle), 17, 291, 311
Lashley, K. S., 80
LeCorbeiller, Philippe, 80
Leder, Philip, 236, 282–85 *passim*
Lederberg, Esther, 198–99
Lederberg, Joshua, 52–53, 115, 121, 206, 279; on reproduction, 112–13; and biological specificity, 113–14; and enzyme biosynthesis, 198–99, 201; and human genome, 276–77
Ledley, Robert, 141, 148–49
Lehninger, Albert L., 305
Leibnitz, Gottfried W., 33
Lengyel, Peter, 259, 260–61, 275, 374n58
Leontief, Wassily, 80
Leopold, Urs, 58
Leslie, Stuart, 10, 76
Lettvin, Jerome, 301
Levene, Phoebus A. T., 55
Levinthal, Cyrus, 177, 224
Lévi-Strauss, Claude, 34, 307, 309–10
L'Héretier, Philippe, 34, 308, 310
Lillie, Frank R., 43
Linderstrom-Lang, K. V., 182, 184
Linear B, 178
Linguistics, 7, 18, 23, 134, 296, 356n4; and DNA, 1–2, 178–79, 320, 324–25; of molecular biology, 26–27; of biochemistry, 27–28; of nature, 32–34; Chomskian, 36–37; and genetic code, 152–53, 282–83, 294–95, 310–14; and cybernetics, 297–98; de Saussure on, 298–99, 303; Jakobson on, 299–300; and biology, 308–10, 323–24
"Linguistique et Génétique" (Bastide), 318
Linschitz, Henry, 122f
Lipmann, Fritz, 142, 246
Loeb, Jacques, 43f
Logic (Condillac), 296
"Logical Calculus of the Ideas Immanent in Nervous Activity, A" (McCulloch and Pitts), 105
Logic of Life, The (Jacob), 17, 40–41, 197, 313
Los Alamos National Laboratory, 9, 135, 141
Loyalty oath, 76, 132, 280
Lucretius, 32
Luria, Salvador, 109, 118, 198, 279
Lwoff, Andre, 17, 195, 203, 207f, 278
Lyotard, Jean-Françoise, 23
Lysenkoism, 201f
Lysogeny, 207–9

Maas, Werner, 212
McCarthy, John, 301
McCarthyism, 10, 76, 132, 204, 235
McCarty, Maclyn, 55
McClintock, Barbara, 118
McCormick, Thomas J., 9
McCulloch, Warren S., 87, 101–2, 105f, 108, 118, 125–26
McElheny, Victor, 278
McElroy, William D., 240
Mach, Ernst, 80
Machines: maze-solving, 101f. *See also* Automata
MacKay, Donald M., 302
MacKenzie, Donald, 24
MacLeod, Colin M., 55
McLuhan, Marshall, 114
McMillan, Brockway, 171
Macy, Josiah Jr. Foundation, 82, 99, 102, 123
Main Trends of Research in the Social and Human Studies, 310–11
"Man and His Future" symposium, 279
Mandelbrot, Benoit, 87, 119
MANIAC, 9, 141; in genetic code production, 146–47, 148–49, 158f
Mao Zedong, 75
Mark I, 102–3
Marshall Plan, 75
Martin, Robert G., 16, 236ff, 240, 257, 262, 282, 293
Mason, Max, 45, 304
Massachusetts Institute of Technology (MIT), 75, 95, 231, 300–301, 303f, 352n99
Mathematical Theory of Communication, 96–98
Mathematical Theory of Communication, The (Shannon and Weaver), 98–99, 153, 300
Mathematics, 87, 122, 338n41; and communication theory, 96–98, 100, 351n78; of genetic coding, 138–39, 171
Matthaei, Heinrich, 7f, 12, 192, 234, 271; and Nirenberg, 245–55, 262, 330, 373n41; and triplet model, 268–69
Mauchly, John, 103
Max Planck Institute, 184, 186, 271, 278

Maxwell, James Clerk, 64–65, 86
Maxwell's demon, 64–65, 86, 201, 207
Mayan Codex, 291
Medawar, Peter, 26–27, 220
Medical Research Council (MRC), 173, 278
Medical sciences, 45
Medvedev, Zhores Alexandrovich, 270
Mendelism, 308
Meselson, Matthew, 196, 227–28, 254
Messages, 23, 85
Messenger RNA (mRNA), 12, 196–97, 212, 225, 233–34, 330; and protein synthesis, 223–31, 249f; information transfer and, 226–27; defining, 227–31, 232, 244–45
Metaphor(s), 11, 23; information as, 2–3, 19–20, 21–22, 226–27; and social control, 28–29; of writing, 31–32; specificity as, 45–46
Metropolis, Nicholas, 141, 146–47, 156, 158
Militarization, 9, 73–74
Military, 301; and science, 10–11, 73, 75–77, 329; and information theory, 23–24; and Bell Labs, 91–102; and computer development, 102–4; cybernetics and, 105–6
Military-industrial-scientific complex, 7, 11, 72, 77–78, 80–81; Bell Labs and, 91–92
Milton, John, 121, 153
Minsky, Marvin, 301
Mirsky, Alfred E., 49f
Missiles, 169
Mr. Tompkins Learns the Facts of Life (Gamow), 131, 135
MIT, *see* Massachusetts Institute of Technology
Mneme principle, 62
"Modele Linguistique en Biologie, Le" (Jacob), 313
Molecular biology, 5, 45, 61, 173, 237, 240, 269, 339n48; information theory and, 15, 24–27, 39, 118–27, 194–95, 222; discursive practices of, 15–17; dialectics of, 35–36; Rockefeller Foundation and, 48, 49–50; as information science, 129, 273; funding for, 277–78; Jakobson and, 306–7; and linguistics, 310–11; evolution in, 321–22
Molecular ecology, 54
Monod, Jacques, 6, 8, 16, 36, 39, 41, 53, 169, 194ff, 215, 232f, 237, 278, 329, 369n58, 371n7; publications by, 17, 197, 199; and biological specificity, 54–55; collaboration with Szilard, 200–201; at Pasteur Institute, 202–4, 205–6; and U.S. security issues, 204–5; and George Cohen, 206–7; and François Jacob, 207–8, 222–23; PaJaMa experiments and, 212f; enzyme induction and, 216–20; and information theory, 220–21; on operons, 222–23, 225–26, 312
Monte Carlo simulations, 9, 146
Moore, E. F., 112
Moore, Walter, 62
Morgan, Thomas Hunt, 44–45
Morison, Robert S., 83, 88
Morse, Samuel F. B., 133
Morse code, 66, 152, 313
Moulin, Anne-Marie, 318
MRC, *see* Medical Research Council
Muller, Hermann J., 106, 109, 279
Mundry, Karl-Wolfgang, 187
Mutagens, 177, 187
Mutations, 87, 177, 285; of TMV, 186, 264; Lac−, 206–7, 211–12; FCO, 267–68

NACA, *see* National Advisory Committee on Aeronautics
Narita, K., 184
NASA, *see* National Aeronautics and Space Administration
National Academy of Sciences (U.S.A.), 138, 279–80
National Advisory Committee on Aeronautics (NACA), 169
National Aeronautics and Space Administration (NASA), 10, 169f
National Bureau of Standards, 301
National Defense Act (1947), 76
National Defense Research Council (NDRC), 78
National Defense Security Act (1957), 10, 76, 169–70
National Foundation, 183, 279
National Institute of Hygiene, 278
National Institutes of Health (NIH), 6, 10, 13, 191f, 203, 235, 257, 282, 330; and biochemical research, 222f, 236–38; funding for, 235f, 277; Nirenberg at, 239–41, 242–43, 330; Matthaei at, 246–48
National Science Foundation (NSF), 10, 167, 203, 235, 301
National security (U.S.), 9, 204–5
National Security Agency (NSA), 134
NATO, *see* North Atlantic Treaty Organization

Naturalism, 32
Natural sciences, 45
Nature: knowledge of, 32–34
"Nature of Intermolecular Forces Operating in Biological Process, The" (Delbrück and Pauling), 50
Naturphilosophie, 33–34
Navy Bureau of Ordnance, 105
NDRC, *see* National Defense Research Council
Negative entropy, negentropy, 64f
Negative feedback, 81, 196, 199ff, 211, 330
Neighbor distribution plots, 149
Neo-Darwinian evolution, 320
Neo-Lamarckianism, 201
Neuroscience, 79–80
Neuroscience Research Program (NRP), 304, 319
Neurospora, 52f, 206
New Landscape in Art and Science, The (Kepes), 90
New York Times, 279, 290; information discourse of, 275–76; on codons, 284–85
New York University, 90, 194, 258–59
Neyman, Jerzy, 79
Niemann, Carl, 50
NIH, *see* National Institutes of Health
Nirenberg, Marshall, 6–7, 8, 12, 14, 27, 30, 192, 212, 243, 257, 265–66, 271f, 282, 305, 330, 371n11; collaboration with Matthaei, 234, 245–51, 252–55, 330, 373n41; at National Institutes of Health, 235f, 238–41; and penicillinase synthesis, 241–42; competition with Ochoa, 259–60, 261–62; and genetic code, 263–65, 275, 280, 283–84; and triplet model, 268–69; doublet vs. triplet coding, 285–86; 1966 Cold Spring Harbor symposium, 288, 290; on biological engineering, 292f
Nirenberg-Ochoa U-rich code, 276
Niu, C.-I., 184
Nobel Prize, 278; Nirenberg, Khorana, and Holley, 13–14, 239, 330; Stanley, 179; Fraenkel-Conrat, 189; Lwoff, Monod, and Jacob, 195, 278; Crick, Watson, and Wilkins, 271; Eigen, Norrish, and Porter, 319
Noble, David, 24, 89
NORC, 105
Norrish, Ronald, 319
North Atlantic Treaty Organization (NATO), 9, 246
Northrop, F. C. S., 106
Novick, Aaron, 118, 166, 169, 199f, 205, 212
NRP, *see* Neuroscience Research Program
NSA, *see* National Security Agency
NSF, *see* National Science Foundation
Nucleic acids, 42; synthesis of, 12, 240, 259; and biological specificity, 55–56; Avery's and Chargaff's studies on, 56–57; Stern's studies of, 67–69; Dounce's studies of, 70–71. *See also* Deoxyribonucleic acid; Ribonucleic acid
"Nucleic Acid Template Hypothesis" (Dounce), 70–71
Nucleotides: and DNA, 57, 249–59; and genetic code, 262, 264; triplet model, 263, 266–67
Nucleotide triplets, 262, 271. *See also* Triplet code; Triplet model
Nyquist, Harry, 95

Oak Ridge National Laboratory, 10, 76, 227; information theory symposium at, 124–26, 160–61, 162–63
Ochoa, Severo, 194, 237, 242, 272, 374n58; work on genetic code, 257–59; competition with Nirenberg, 259–62, 265, 330; on triplet codes, 270–71; on redundancy, 305f
O'Connor, Basil, 279
Office of Naval Research (ONR), 119, 121, 186
Office of Ordnance Research, 125
Office of Science Research and Development (OSRD), 74–75, 78
Olby, Robert, 43
"One gene–one enzyme" hypothesis, 52
ONR, *see* Office of Naval Research
Operons, 55, 212–23, 225–26, 237, 312
Oppenheimer, J. Robert, 132, 144–45, 204
Order of life, 46–47
Order of Things, The (Foucault), 40
Organism as a Whole, The (Loeb), 44
Organization, 46–48, 49, 123
Orgel, Leslie, 141f, 144, 161, 164f
Origin of life: Eigen on, 319–22, 331
OSRD, *see* Office of Science Research and Development

PaJaMa experiments, 195f, 207, 212–14, 215, 221–22, 226, 368n41
Paradise Lost (Milton), 121, 153
Pardee, Arthur, 195, 200, 213, 226; and Pasteur Institute, 209–12
Parsons, Talcott, 80, 90

Pasteur Institute, 192, 194ff, 199ff, 222, 231, 278, 329; Monod at, 54, 202–4, 205–6; Jacob at, 207–9; Pardee at, 210–12; Crick and, 223–24
Patents, 200
Patronage, 10–11
Paul, St., 31
Pauling, Linus, 39, 41, 49–51, 109, 138, 291–92
Pax Americana, 9
Pencillinase, 241–42
Penrose, Lionel S., 112, 115
Phages: genomes in, 12, 39; mating of, 193, 224; study of, 201, 207; induction of, 214–15
"Phenomenon of Enzymatic Adaptation and Its Bearings on Problems of Genetics and Cellular Differentiation, The" (Monod), 54, 199
Phonemes, 298–99, 301–2, 304–5
Phonemic theory, 301
PHS, *see* U.S. Public Health Service
Physics: and molecular biology, 64–65, 129
"Physics of Living Matter" conference, 131
Pike, Sumner T., 141
Pitts, Walter, 105, 108
Platt, John R., 273
Player Piano (Vonnegut), 90–91
Pneumococcus bacteria, 55
Poisson distribution, 149–50, 152, 283
Pollack, Robert, 2
Pollock, Martin R., 197, 241
Poly-A, 249–50, 251, 374n58
Polymerase, 12f, 281, 282
Polymers, 251, 259
Polynucleotide phosphorylase, 12–13
Polynucleotides, 12–13, 237, 248, 251–52
Polypeptides, 285
Poly-U effect, 251–52, 254f, 257, 260, 262, 271
Porter, George, 319
Porter, R. R., 182, 184
"Possible Mathematical Relation Between Deoxynucleic Acid and Proteins" (Gamow), 138
Poster, Mark, 32
Primum Mobile, 38
Princeton, *see* Institute for Advanced Study
"Problem of Information Transfer from the Nucleic Acids to Proteins" (Gamow, Rich, and Yčas), 150, 152
Prophage induction, 208
Proteins, protein paradigm, 42, 118, 120f, 139; and specificity, 43–44, 50–51, 52; and Rockefeller Foundation research, 48–49; X-ray crystallography of, 108–9; decoding, 146–47; in TMV, 184–85
Protein synthesis, 6, 8, 12f, 70, 125, 194, 224, 232, 275, 324; cryptanalysis and, 125–26, 193; Szilard on, 168–69; Crick on, 173–75; and the genetic code, 177–78, 272–73; phage, 227–28; cell-free systems, 241–42, 243; and messenger RNA, 249f; and tRNA, 286–87
"Protein Text, The" (Yčas), 264
Protocybernetics, 82
Protoplasmic theory of life, 48
Public Health Service, *see* U.S. Public Health Service
Puck, Theodore, 167
Pugwash committee, 213
Punctuation: in genetic code, 163–64, 165–66
Pyrimidine synthesis, 213, 367n37

Quartermaster Corps, 143
Quastler, Gertrud, 117, 123
Quastler, Henry, 6, 50, 78, 114, 116f, 126, 129, 161, 174, 221, 263f, 314, 321; and information-based biology, 25–26, 115, 118–25, 127
Quine, W. V., 80

Radiobiology, 116
Radioisotopes, 12
Rand Corporation, 91, 135
Randomness, 154–55, 158–59
Recognition process, 286
Reddy, Michael J., 23, 304
Redundancy, 134–35, 305–6, 351n81
Reichert, Edward, 43
Replication, 108–9, 172, 232. *See also* Deoxyribonucleic acid
Repressor hypothesis, 213–14
Reproduction, 112–13. *See also* Genetics; Heredity
Research: and military, 74–77
Research and development, 76–77
Rheinberger, Hans-Jörg, 13, 36, 238, 243
Ribonucleic acid (RNA), 13, 139, 172, 232, 243, 281, 328; and viruses, 8, 183, 184, 186, 189, 250; messenger, 12, 195; and DNA templates, 70–71; composition of, 159, 160; as binary code, 175–76; TMV, 183, 186–87, 254; informational, 230–31; synthetic polymers of, 259, 260–61.

See also Messenger RNA; Soluble RNA; Transfer RNA
"Ribonucleotide Composition of the Genetic Code" (Nirenberg et al.), 262
Rich, Alexander, 53, 121, 140ff, 148f, 155, 185, 236, 255, 314; on information transfer, 150ff
Rickover, Hyman, 11
Riley, Monica, 226
Risebrough, R. W., 196
RNA, *see* Ribonucleic acid
RNAase, 13
"RNA Codewords and Protein Synthesis, VII. The General Nature of the RNA Code," 285
RNA Tie Club, 6, 27, 71f, 128, 141, 147, 156, 161f, 165, 201, 240–41, 256, 263, 283, 329; members of, 57, 141–42; operations of, 142–44, 193; on heredity, 185–86
Roberts, Richard B., 245, 270, 305
Rockefeller Foundation, 5, 10, 20, 45, 78, 88, 116, 183, 222, 300, 328; and molecular life sciences, 39, 41–42, 49–50, 55; protein paradigm of, 48–49; grants from, 82f, 180; and Pasteur Institute, 203f
Rose, Frank, 105
Rosenberg, Charles, 28–29
Rosenberg trial, 205
Rosenbleuth, Arturo, 79, 81f, 83–84, 108
Rosenblith, Walter, 301
Rosetta stone, 13, 17, 189–90, 291, 294
Rossignol, Antoine, 133
Rothschild family, 203
Rouge-Clôitre group, 139, 195, 226
Russia, *see* Soviet Union
Russian-bath code, 145–46
Rutgers University (Institute of Microbiology), 274
Rychlik, I., 270

Salk, Jonas, 206
Salk Institute for Biological Studies, 306
Sanger, Frederick, 8, 138, 182, 184
Santillana, Gorgio, 80
Sapp, Jan, 44, 201
Saussure, Ferdinand de, 298–99, 303, 318
Schmitt, Francis O., 106, 304, 319
Schönberger, Martin, 317–18
School for War Training, 92
Schramm, Gerhard, 185, 187, 193, 305–6, 363n142
Schrödinger, Erwin, 3, 34, 71–72, 85, 199, 269; *What Is Life?*, 4–5, 42, 59–61, 63–64, 336n7; and code-script, 61–62, 65
Schultz, Jack, 52
Schumpeter, Joseph, 80
Schüster, Heinz, 187
Schwartz, Drew, 160
Science, 279; and cold war, 7–14, 132–33; and military, 10–11, 73, 75–77, 91, 329; and World War II, 74–75, 329; in United States, 204–5; and language, 310–11
Science: The Endless Frontier (Bush), 75
"Scientist Rebels, A" (Wiener), 73
Secrecy systems, 92–93, 296
Security clearance, 132
Seitz, Frederick, 279–80
Selection: Darwinian, 321–22
Self-regulation, 47
Semantics: information content and, 99–100
Semon, Richard, 62
Serratia bacteria, 231
Servomechanisms, 81–82
Shannon, Claude, 6, 64, 72, 78, 85, 123, 153, 171, 220, 301, 351n78; and information theory, 20–21, 120f, 126–27, 303; publications by, 92–94, 98–99, 153, 300; on mathematical theory of communication, 96–97, 134–35, 351n81; on machine communication and, 97–98; on automata, 101–2, 112
Sherrington, Charles, 63
Side-chain theory, 43
Signal Intelligence Service (SIS), 134
Signals, 23, 127
Signification, 57, 85–86, 296
Sign theory, 298f
Silverstein, Arthur, 42
Simon, Herbert, 25, 160
Simons, N., 141
Simulacra, 114, 322. *See also* Automata
Singer, Maxine, 236, 247, 262
Sinsheimer, Robert, 17, 34, 115, 172, 291f, 294; and binary code, 175–76
SIS, *see* Signal Intelligence Service
Skinner, B. F., 80
Smythe, Henry D., 124
Social control, 28–29, 45
Social order, 45, 47
Society for Experimental Biology, 173–74
Sonneborn, Tracy, 118
Soluble RNA (sRNA), 243–44. *See also* Transfer RNA

Šorm, F., 270
Soviet Union (USSR), 10, 15, 75, 228, 236
Space program, 236. *See also* Jet Propulsion Laboratory; National Aeronautic and Space Administration
Specialization model, 46–47
Speciation, 321
Specificity, 41, 54; biological, 40, 113–15, 120, 122, 341n11; and immunology, 42–43; and proteins, 43–44, 48, 50–51; genetic, 44–45, 51–52, 328; as metaphor, 45–46; and information, 52–53, 122; of nucleic acids, 55–56
Specificity of Serological Reactions, The (Landsteiner), 50
Speyer, Joseph, 257, 259, 260–61, 288
Spiegelman, Sol, 27, 106, 107–8, 109, 114f, 118, 194, 197, 227, 229, 275, 370n80; on messenger RNA, 230–31; at Cold Spring Harbor symposia, 232, 288
Spirin, A. S., 175
Sputnik, 10, 236
sRNA, *see* Soluble RNA
Stanier, Roger Y., 197, 216
Stanley, Wendell M., 106, 185, 191, 250, 363n142; and TMV, 48, 179–80; virus research by, 180–81, 182–83, 186, 188f
Statistical analysis, 9, 149–50, 190–91
"Statistical Correlation of Protein and Ribonucleic Acid Composition" (Gamow and Yčas), 159
Stent, Gunther, 39, 59, 141, 147, 181, 185, 210, 277, 315
Stepan, Nancy Leys, 29
Stereocomplementarity, 43, 50
Stern, Kurt, 42, 65, 67–69
Stetten, DeWitt, Jr., 238
Stewart, Frederick C., 246
Stock, Brian, 32
Stratton, Julius, 301
Streisinger, George, 168
"Structure of Native, Denatured, and Coagulated Proteins, On the" (Pauling and Mirsky), 50
Sturtevant, Alfred H., 51, 109
Sturtevant, Julian M., 87
Sueoka, Noburo, 270
Sumner, John B., 70
Sweden, 278
Sweet, William H., 305f
Symbolic logic, 148
Symonds, Neville, 61
"Symposium on Informational Macromolecules," 274, 276
Symposium on Information Theory in Biology, 264
Synthetic polynucleotides, 259, 260–61
"Synthetic Polynucleotides and the Amino Acid Code" (Lengyel, Speyer, and Ochoa), 260–61
System of representations, 19
Szent-Gyorgyi, Albert, 109, 142, 147, 273
Szilard, Leo, 8, 16f, 106, 114, 166–67, 199, 200, 212f, 220, 232, 278, 368n41; on Maxwell's demon, 64f; on protein synthesis, 168–69; collaboration with Monod, 200–201; and national security issues, 204f
Szilard, Trude, 166–67

Tatum, Edward C., 52, 206
Techne, 35f
Technology, 35
Telecommunications, 100
Telegraphic communication, 133
Teleology, 202, 221
Teleonomy, 196, 221, 369n58
Teller, Edward, 106, 141, 144–46, 204
"Terminology of Enzyme Formation," 197
Tetranucleotide hypothesis, 55
Theism, 295
"Theory of Structure and Process of Formation of Antibodies, The" (Pauling), 51
Thermodynamics, 64–65
Thomas, Lewis, 236
Tiselius, Arne, 275–76
Tissières, Alfred, 229, 244, 247, 254f
Tisza, Laszlo, 80
Tobacco Mosaic Virus (TMV), 6, 27, 48, 149, 179, 193, 251, 254, 266, 270; coat sequencing in, 8, 12; Fraenkel-Conrat's work on, 182–85; RNA in, 186, 254; amino acid sequencing of, 187–89; as decoding apparatus, 190–91, 250; redundancy of, 305–6
Todd, Alexander, 280
Tomkins, Gordon, 234, 237, 242f, 247, 252, 255, 262
Torriani, Annamaria, 199
Transcription, 232
"Transfer and Expression of Genetic Information in Escherichia Coli K12" (Jacob), 215

Transfer RNA (tRNA), 243–44, 283–84, 286–87, 290
Translation, 232
Treguer, Michel, 308
Triangular code, 144
Trifonov, Edward, 36–37, 324
Triplet code, 70, 144, 261–62, 271, 287–88; and doublet conversion, 270, 285–86; tRNA and, 283–84
Triplet model, 262, 329; development of, 266–69
Troubetskoy, Nikolay, 299
Truman, Harry, 75, 134
Truman Doctrine, 75
Tsugita, A., 187–88, 266
Turing, Alan M., 105, 134
Turing machine, 110
Turnip yellow virus (TYV), 149, 250
Tweedell, Kenyon, 122
Twinning, 122
Tyler, Albert, 51, 139
TYV, *see* Turnip yellow virus

Uhlenbeck, George, 80
Ulam, Stanislaw, 156, 159, 321
Umbarger, Edwin H., 200, 211, 212–13
Uncertainty principle, 99
United States, 76, 200; foreign policy of, 9–10; science in, 15, 75, 157; anti-communism in, 204–5; biomedical research in, 277–78; on biological engineering, 279–80. *See also individual agencies, institutions, and universities by name*
U.S. Air Force, 277, 301
U.S. Army, 143, 277, 301
U.S. Department of Defense (DOD), 9–10, 76, 132–33, 235–36
U.S. Department of Health, Education, and Welfare (HEW), 235
U.S. Navy, 105, 277
U.S. Public Health Service (PHS), 183, 235, 277. *See also* National Institutes of Health; Virus Laboratory
Unit for Molecular Biology, 58, 148
Unit for the Study of Molecular Structures of Biological Systems (Cambridge), 58, 173
Unity of Science Movement, 80
Universality, 276–77, 330–31. *See also under* Genetic code
University of California, Berkeley, 180
University of Chicago, 166
University of Illinois, 116, 194, 227
"Unstable Intermediate Carrying Information from Genes to Ribosomes for Protein Synthesis, An" (Brenner et al.), 229–30
"Unstable Ribonucleic Acid Revealed by Pulse Labeling of Escherichia Coli" (Gros et al.), 229–30

Van Harreveld, Anthony, 109
Van Niel, Cornelius, 166
Vanuxem Lectures, 111
Veblen, Oswald, 79
Ventris, Michael, 178
Verlaine, Paul, 209
Vienna circle, in exile, 80
Vietnam War, 280
Virchow, Rudolph, 63
Viruses, 149, 179; RNA and amino acids in, 8, 183, 184; research on, 180–81, 191–92; chemical modification of, 186–87; amino acid sequencing in, 187–89. *See also* Bacteriophage; Tobacco Mosaic Virus
Viruses and the Nature of Life, 192
Virus Laboratory, 179ff, 186, 251; Fraenkel-Conrat at, 182–84, 250; Pardee and, 210–11
Vischer, E., 56
"Vivre et Parler," 307–9
Volkin, Eliott, 227
Voluntary action, 80
Von Foerster, Heinz, 21, 23–24, 26, 116, 118, 123, 125, 127
Vonnegut, Kurt, Jr.: *Player Piano*, 90–91
Von Neumann, John, 6, 27, 50, 72, 78f, 82, 91, 113, 129, 135, 320; computer development and, 102–5 *passim*; biological automata and, 106f, 109–12, 321f; and Spiegelman, 207–8; X-ray crystallography and, 108–9; and Gamow, 158f

Waddington, C. H., 275
Wall, Robert, 269
Warfare: automated, 103–4
War games, 101–2
Washington Conferences on Theoretical Physics, 106
Washington University (St. Louis), 194, 201
Watson, James D., 8, 37, 39, 42, 131–32, 141f, 147, 194, 228, 288, 297; and DNA structure, 58–59, 271
Watson-Crick model, 125
Watts-Tobin, R. J., 266
Weaver, Warren, 14, 23, 45, 78f, 83, 94, 116, 304; and Rockefeller Foundation, 48–49,

88; and communication theory, 98–99, 153, 300
Weigle, Jean J., 58, 228
Weiss, Paul, 46, 54, 63, 233
Welch, Lloyd R., 170ff
Western Regional Research Laboratory, 182
Westmoreland, William, 104
What Is Life? (Schrödinger), 4–5, 42, 63–64, 65, 336n7; concepts in, 59–61
Wiener, Norbert, 6, 17, 21, 50, 64f, 72f, 78, 81, 108, 129, 174, 220, 320; and neurology, 79–80; cybernetics, 84–85, 86–89, 90f; and Shannon, 93–94; and communication theory, 94–95, 297
Wiener-Shannon theory of communication, 95, 99, 115, 118f, 123, 126, 302, 339n48
Wiersma, Cornelius A. G., 109
Wiesner, Jerome, 301
Wilkins, Maurice, 271
William of Conches, 32
Williams, R., 141
Wimsatt, William, 83
Winkler, Ruthilde, 323
Wittmann, H. G., 266, 275
Wobble hypothesis, 287
Woese, Carl R., 16, 27, 129, 191, 255f, 270, 275, 286, 289
Wollman, Elie, 208–9, 215
Wollman, Elizabeth, 208
Wollman, Eugène, 208
Woods Hole, 147
World War I, 30, 133
World War II, 20, 23–24, 30, 74–75, 76, 134, 329
Wrench, Dorothy, 109
Writing: and the Book of Life/Nature, 32–34; genomic, 34–36, 295–96

X-ray crystallography, 108–9

Yanofsky, Charles, 193, 285, 311
Yardley, Herbert Osborne, 134
Yčas, Martynas, 53, 115, 121, 139–40, 141, 147f, 264, 267, 283; collaboration with Gamow by, 125–26, 149–52 *passim*, 157–58, 159, 160–61, 180, 184, 186; and RNA Tie Club, 141, 142–43; and information theory symposia, 162–63; and combination code, 165–66; and TMV, 190–91
Yockey, Hubert L., 125, 161

Zamecnik, Paul, 194, 243ff, 255, 286
Zog, King (Ahmed Bey Zogu), 116
Zygotic technology, 209–10, 212

WRITING SCIENCE

Lily E. Kay, *Who Wrote the Book of Life? A History of the Genetic Code*

Jean Petitot, Francisco J. Varela, Bernard Pachoud, and Jean-Michel Roy, eds., *Naturalizing Phenomenology: Issues in Contemporary Phenomenology and Cognitive Science*

Francisco J. Varela, *Ethical Know-How: Action, Wisdom, and Cognition*

Bernhard Siegert, *Relays: Literature as an Epoch of the Postal System*

Friedrich A. Kittler, *Gramophone, Film, Typewriter*

Dirk Baecker, ed., *Problems of Form*

Felicia McCarren, *Dance Pathologies: Performance, Poetics, Medicine*

Timothy Lenoir, ed., *Inscribing Science: Scientific Texts and the Materiality of Communication*

Niklas Luhmann, *Observations on Modernity*

Dianne F. Sadoff, *Sciences of the Flesh: Representing Body and Subject in Psychoanalysis*

Flora Süssekind, *Cinematograph of Words: Literature, Technique, and Modernization in Brazil*

Timothy Lenoir, *Instituting Science: The Cultural Production of Scientific Disciplines*

Klaus Hentschel, *The Einstein Tower: An Intertexture of Dynamic Construction, Relativity Theory, and Astronomy*

Richard Doyle, *On Beyond Living: Rhetorical Transformations of the Life Sciences*

Hans-Jörg Rheinberger, *Toward a History of Epistemic Things: Synthesizing Proteins in the Test Tube*

Nicolas Rasmussen, *Picture Control: The Electron Microscope and the Transformation of Biology in America, 1940–1960*

Helmut Müller-Sievers, *Self-Generation: Biology, Philosophy, and Literature Around 1800*

Karen Newman, *Fetal Positions: Individualism, Science, Visuality*

Peter Galison and David J. Stump, eds., *The Disunity of Science: Boundaries, Contexts, and Power*

Niklas Luhmann, *Social Systems*

Hans Ulrich Gumbrecht and K. Ludwig Pfeiffer, eds., *Materialities of Communication*

Library of Congress Cataloging-in-Publication Data

Kay, Lily E.
Who wrote the book of life? : a history of the genetic code / Lily E. Kay.
p. cm. — (Writing science)
Includes bibliographical references and index.
ISBN 0-8047-3384-8 (cloth : alk. paper). — ISBN 0-8047-3417-8 (pbk. : alk. paper)
1. Genetic code—Research—History. I. Title. II. Series.
QH450.2.K39 2000
572.8′633—dc21 99-39446

∞ This book is printed on acid-free, recycled paper.

Original printing 2000
Last figure below indicates year of this printing:
09 08 07 06 05 04 03 02 01